The Most Successful Way of JEE Preparation

DPP

Daily Practice Problems

JEE Main & Advanced

Functions, Matrices & Determinants

with Relations and Inverse Trigonometric Functions

MATHEMATICS Vol-6

Amt M Agarwal

arihant
ARIHANT PRAKASHAN (SERIES), MEERUT

DPP
Daily Practice Problems

arihant
ARIHANT PRAKASHAN (SERIES), MEERUT

卐 Administrative & Production Offices

Corporate Office 'Ramchhaya' 4577/15, Agarwal Road, Darya Ganj, New Delhi -110002
Tele: 011- 47630600, 43518550; Fax: 011- 23280316

Head Office Kalindi, TP Nagar, Meerut (UP) - 250002
Tel: 0121-2401479, 2512970, 4004199; Fax: 0121-2401648

卐 Sales & Support Offices

Agra, Ahmedabad, Bengaluru, Bareilly, Chennai, Delhi, Guwahati, Hyderabad, Jaipur, Jhansi, Kolkata, Lucknow, Meerut, Nagpur & Pune

卐 ISBN : 978-93-13193-54-8

Published by Arihant Publications (I) Ltd.

Production Team

Publishing Managers Mahendra Singh Rawat & Keshav Mohan
Project Head Sorabh Chaudhary
Cover Designer Shanu Mahad
Inner Designer Mazher Chaudhary
Page Layouting Sunil Sharma, Hemant
Proof Reading Himanshu

For further information about the products from Arihant, log on to www.arihantbooks.com or email to info@arihantbooks.com

/arihantpub /@arihantpub Arihant Publications /arihantpub

ABOUT DPP

Daily Practice Problems

Indian Institutes of Technology better known as IITs have always been the dream destination of each science group students who want to pursue their career as Engineers. These are the institutes which have got very high value not even in India but in the whole world. To get the admission in such institutes one need to face entrance Examinations; JEE Main & JEE Advanced, out of which JEE Main is considered as the preliminary exam & JEE Advanced is the final exam to get into the IITs. Approximately per year 12 lakh students enroll themselves for JEE Main and only the top 2.25 lakh in JEE Main merit can attempt for JEE Advanced, the fact clearly express the degree of toughness of JEE Main & JEE Advanced.

To have success in such an exam one needs to be very planned & systematic in its preparation. A very popular wise saying 'Rome was not built in a day' suggests that to achieve a big target like IITs we need to work continuously. For continuous work, one need to have proper planning that how much work he/she has to do on regular basis, i.e., on daily basis. Each student preparing for JEE has many books but unfortunately none of them gives the students a plan for daily studies, which is the most effective success key for JEE.

Keeping the daily practice needs of the students in mind we have come up with DPP Series (Daily Practice Problems) having approximately 26 books. The primary focus of this series is on achieving success through practice & proper planning. For that we have divided the whole syllabus of JEE Main & JEE Advanced; Physics, Chemistry & Mathematics, into different volumes (around 8 volumes of Mathematics & 9 volumes each of Physics & Chemistry), each volume has Daily Practice Problems Sheets having specific questions on various Topics of the individual chapter of the volume ensuring the complete Practice & Assessment of the topic.

DPP

Daily Practice Problems

In this way we have tried to give you an effective daily planning for JEE keeping in mind the daily hours that a student can devote to his practice, you are just required to practice one sheet of each subject on daily basis. This habit of daily practice will keep you away from exam stress and thoughts like; how much I have prepared/am I preparing properly or not/ do I need to study more on daily basis or what I am doing is ok, etc., which regularly tease your mind and can give you panic at any stage of your preparation.

Some special features that we have tried to incorporate in our Daily Practice Problems (DPP) series are discussed below

1. Micro Level coverage of all the topics of a Chapter/ Unit given in JEE Main & Advanced Syllabus.
2. Each DPP has various types of questions; Subjective Questions, Single/ Multiple Correct Options, Statements type, Comprehension based, Integer Value Answer type, Matching, etc., making you practice & ready for all the question formats of JEE.
3. New & innovative questions in each DPP challenging & improving your problem solving skills.
4. Along with Topical Coverage, Revisal DPPs for JEE Main & JEE Advanced with each Chapter, covering multidisciplinary questions proving comprehensive practice.
5. JEE Main & JEE Advanced Archive (collection of last 13 years' exams questions) with each chapter.
6. Complete Solutions for each DPP.

We hope that DPP series will prove to be very beneficial for the students preparing for JEE, by giving them daily practice in a systematic & planned manner. The suggestions/criticism from the students, teachers & parents about DPP series will be welcomed whole heartedly and we promise to update this series accordingly from time to time.

DPP
Daily Practice Problems

DAILY TASKS...

1 INVERSE TRIGONOMETRIC FUNCTIONS

2 RELATIONS AND FUNCTIONS

DPP
Daily Practice Problems

3 MATRICES AND DETERMINANTS

1. Inverse Trigonometric Functions

DPP
DAILY PRACTICE PROBLEMS

DPP-1 Domain and Range of ITF, Trig. $(\text{Trig}^{-1}(x)) = x$, $\text{Trig}^{-1}(\text{Trig}(x)) = x$ and Interconversion between ITF's

Subjective Questions

Working Space

Direction *(Q. Nos. 1-7) These questions are subjective in nature, need to be solved completely on notebook.*

1 Draw the graph for the following functions and discuss their domain and range.

i. $y = \sin^{-1} x$

ii. $y = \cos^{-1} x$

iii. $y = \tan^{-1} x$

iv. $y = \operatorname{cosec}^{-1} x$

v. $y = \sec^{-1} x$

vi. $y = \cot^{-1} x$

2 Find the value of

i. $\sin^{-1}\left[\sin\left(-\frac{2\pi}{3}\right)\right]$

ii. $\sin\left[\frac{\pi}{3} - \sin^{-1} - \frac{1}{2}\right]$

iii. $\tan^{-1}\left(\tan\frac{3\pi}{4}\right)$

iv. $\cot[\sec^{-1}(-\sqrt{2}) + \cos^{-1}(-1)]$

v. $\tan\left[\cos^{-1}\left(\frac{1}{2}\right) + \tan^{-1}\left(-\frac{1}{\sqrt{3}}\right)\right]$

3 Solve for x :

i. $\sin^{-1} x > \frac{\pi}{6}$

ii. $-\frac{\pi}{4} < \cos^{-1} x < \frac{2\pi}{3}$

iii. $\log_{\frac{\pi}{4}} \tan^{-1}(x-1) \geq 1$

iv. $(\sin^{-1} x)^2 - 3(\sin^{-1} x) + 2 \leq 0$

4 Draw graph for

i. $\sin(\sin^{-1} x)$

ii. $\cos(\cos^{-1} x)$

iii. $\tan(\tan^{-1} x)$

iv. $\operatorname{cosec}(\operatorname{cosec}^{-1} x)$

v. $\sec(\sec^{-1} x)$

vi. $\cot(\cot^{-1} x)$

5 Draw graph for

i. $\sin^{-1}(\sin x)$

ii. $\cos^{-1}(\cos x)$

iii. $\tan^{-1}(\tan x)$

iv. $\operatorname{cosec}^{-1}(\operatorname{cosec} x)$

v. $\sec^{-1}(\sec x)$

vi. $\cot^{-1}(\cot x)$

Working Space

6 Find the value of

i. $\sin^{-1}\left(\sin\frac{\pi}{3}\right)$ ii. $\cos^{-1}\left(\cos\frac{2\pi}{3}\right)$

iii. $\tan^{-1}\left(\tan\frac{\pi}{4}\right)$ iv. $\sin^{-1}\left(\sin\frac{2\pi}{3}\right)$

v. $\cos^{-1}\left(\cos\frac{7\pi}{6}\right)$ vi. $\tan^{-1}\left(\tan\frac{3\pi}{4}\right)$

vii. $\cos\left\{\cos^{-1}-\frac{\sqrt{3}}{2}+\pi/6\right\}$

7 Evaluate the following:

i. $\cos\left(\sin^{-1}\frac{3}{5}\right)$ ii. $\sin\left[\cos^{-1}\left(\frac{4}{5}\right)\right]$

iii. $\cos\left[\sin^{-1}\left(-\frac{3}{5}\right)\right]$ iv. $\tan\left(\cos^{-1}\frac{8}{17}\right)$

v. $\text{cosec}\left[\cos^{-1}\left(-\frac{12}{13}\right)\right]$

Only One Option Correct Type

***Direction** (Q. Nos. 8-20) This section contains 13 multiple choice questions. Each question has four choices (a), (b), (c) and (d), out of which ONLY ONE option is correct.*

8 $\cot\left[\cos^{-1}\left(\frac{7}{25}\right)\right]$ is equal to

a. $\frac{25}{24}$ b. $\frac{25}{7}$

c. $\frac{24}{25}$ d. None of these

9 $\sin(\cot^{-1}x)$ is equal to

a. $\sqrt{1+x^2}$ b. x

c. $(1+x^2)^{-3/2}$ d. $(1+x^2)^{-1/2}$

10 The value of $1+\cot^2(\sin^{-1}x)$ is

a. $\frac{1}{2x}$ b. x^2

c. $\frac{1}{x^2}$ d. $\frac{2}{x}$

11 The value of $\cos(\tan^{-1}x)$ is

a. $\sqrt{1+x^2}$ b. $\frac{1}{\sqrt{1+x^2}}$

c. $1+x^2$ d. None of these

12 The value of $\tan(\sec^{-1}\sqrt{1+x^2})$ is

a. $\frac{1}{x}$ b. x

c. $\frac{1}{\sqrt{1+x^2}}$ d. $\frac{x}{\sqrt{1+x^2}}$

Working Space

13 The value of $\sin[\cot^{-1}\{\tan(\cos^{-1} x)\}]$ is

a. x
b. $\pi/2$
c. 1
d. None of these

14 The value of $\sin^{-1}\left(\frac{\sqrt{x}}{\sqrt{x+a}}\right)$ is

a. $\cos^{-1}\left(\sqrt{\frac{x}{a}}\right)$
b. $\text{cosec}^{-1}\left(\sqrt{\frac{x}{a}}\right)$
c. $\tan^{-1}\left(\sqrt{\frac{x}{a}}\right)$
d. None of these

15 If $\sin^{-1} x = \theta + \beta$ and $\sin^{-1} y = \theta - \beta$, then $(1 + xy)$ is equal to

a. $\sin^2\theta + \sin^2\beta$
b. $\sin^2\theta + \cos^2\beta$
c. $\cos^2\theta + \cos^2\beta$
d. $\cos^2\theta + \sin^2\beta$

16 $\sec^2(\tan^{-1} 2) + \text{cosec}^2(\cot^{-1} 3)$ is equal to

a. 5
b. 13
c. 15
d. 16

17 If $\pi \le x \le 2\pi$, then $\cos^{-1}(\cos x)$ is

a. x
b. $-x$
c. $2\pi + x$
d. $2\pi - x$

18 The value of x which satisfies the equation $\tan^{-1} x = \sin^{-1}\left(\frac{3}{\sqrt{10}}\right)$ is

a. 3
b. -3
c. $\frac{1}{3}$
d. $\frac{-1}{3}$

19 If $\cos^{-1} x + \cos^{-1} y + \cos^{-1} z = 3\pi$, then $xy + yz + zx$, is equal to

a. 0
b. 1
c. 3
d. -3

20 If $\cos^{-1} x + \cos^{-1} y + \cos^{-1} z + \cos^{-1} t = 4\pi$, then find the value of $x^2 + y^2 + z^2 + t^2$, is

a. $xy + yz + zx$
b. $1 - 2xyt$
c. 4
d. 6

DPP-2 Identities on Addition and Subtraction

Subjective Questions

Direction *(Q. Nos. 1and 2) These questions are subjective in nature, need to be solved completely on notebook.*

Working Space

1 Prove that

i. $\tan^{-1}\left(\frac{1}{7}\right)+\tan^{-1}\left(\frac{1}{13}\right)=\tan^{-1}\left(\frac{2}{9}\right)$

ii. $\tan^{-1}(2)+\tan^{-1}(3)=\frac{3\pi}{4}$

iii. $\tan^{-1}\left(\frac{1}{2}\right)+\tan^{-1}\left(\frac{1}{5}\right)+\tan^{-1}\left(\frac{1}{8}\right)=\frac{\pi}{4}$

iv. $\sin^{-1}\left(\frac{3}{5}\right)+\sin^{-1}\left(\frac{8}{17}\right)=\sin^{-1}\left(\frac{77}{85}\right)$

v. $\sin^{-1}\left(\frac{4}{5}\right)+\sin^{-1}\left(\frac{5}{13}\right)+\sin^{-1}\left(\frac{16}{65}\right)=\frac{\pi}{2}$

2 Solve the following equations for x.

i. $\tan^{-1}(2x)+\tan^{-1}(3x)=n\pi+\frac{3\pi}{4}$

ii. $\tan^{-1}(x+1)+\tan^{-1}(x-1)=\tan^{-1}\left(\frac{8}{31}\right)$

iii. $\sin^{-1}x+\sin^{-1}2x=\frac{\pi}{3}$

iv. $\cos^{-1}x+\sin^{-1}\frac{x}{2}=\frac{\pi}{6}$

Only One Option Correct Type

Direction *(Q. Nos. 3-22) This section contains 20 multiple choice questions. Each question has four choices (a), (b), (c) and (d), out of which ONLY ONE option is correct.*

3 If $\sin\left(\sin^{-1}\frac{1}{5}+\cos^{-1}x\right)=1$, then x is equal to

a. 1 b. 0 c. $\frac{4}{5}$ d. $\frac{1}{5}$

4 The value of $\sin\left(\sin^{-1}\frac{1}{3}+\sec^{-1}3\right)+\cos\left(\tan^{-1}\frac{1}{2}+\tan^{-1}2\right)$ is

a. 1
b. 2
c. 3
d. 4

5 The value of $\cos\left[\cos^{-1}\left(-\frac{1}{7}\right)+\sin^{-1}\left(-\frac{1}{7}\right)\right]$ is

a. $\frac{-1}{3}$
b. 0
c. $\frac{1}{3}$
d. $\frac{4}{9}$

6 If $4\sin^{-1}x+\cos^{-1}x=\frac{\pi}{2}$, then x is equal to

a. 0
b. $\frac{1}{2}$
c. $\frac{-\sqrt{3}}{2}$
d. $\frac{1}{\sqrt{2}}$

Working Space

7 If $\cot^{-1} x + \tan^{-1} 3 = \frac{\pi}{2}$, then x is

a. $\frac{1}{3}$ b. $\frac{1}{4}$ c. 3 d. 4

8 The value of $\left(\cot^{-1} 9 + \operatorname{cosec}^{-1} \frac{\sqrt{41}}{4}\right)$ is

a. 0
b. $\tan^{-1} 2$
c. $\frac{\pi}{4}$
d. $\frac{\pi}{2}$

9 If $\sin^{-1}\left(\frac{1}{3}\right) + \sin^{-1}\left(\frac{2}{3}\right) = \sin^{-1} x$, then x is

a. 0
b. $\frac{\sqrt{5} - 4\sqrt{2}}{9}$
c. $\frac{\sqrt{5} + 4\sqrt{2}}{9}$
d. $\frac{\pi}{2}$

10 If $\tan^{-1} x + 2\cot^{-1} x = \frac{2\pi}{3}$, then x is

a. $\sqrt{2}$
b. 3
c. $\sqrt{3}$
d. $\frac{\sqrt{3} - 1}{\sqrt{3} + 1}$

11 The value of $\cos^{-1}\left(\frac{15}{17}\right) + 2\tan^{-1}\left(\frac{1}{5}\right)$ is

a. $\frac{\pi}{2}$
b. $\cos^{-1}\left(\frac{171}{221}\right)$
c. $\frac{\pi}{4}$
d. None of these

12 The value of $\sin^{-1}\left(\frac{3}{5}\right) + \tan^{-1}\left(\frac{1}{7}\right)$ is

a. $\frac{\pi}{4}$ b. $\frac{\pi}{2}$ c. $\cos^{-1}\left(\frac{4}{5}\right)$ d. π

13 The value of $\tan^{-1}\left(\frac{a - b}{1 + ab}\right) + \tan^{-1}\left(\frac{b - c}{1 + bc}\right)$ is

a. $\tan^{-1} a - \tan^{-1} b$
b. $\tan^{-1} a - \tan^{-1} c$
c. $\tan^{-1} b - \tan^{-1} c$
d. $\tan^{-1} c - \tan^{-1} a$

14 If $\tan^{-1} x + \tan^{-1} y + \tan^{-1} z = \frac{\pi}{2}$, then

a. $x + y + z - xyz = 0$
b. $x + y + z + xyz = 0$
c. $xy + yz + zx + 1 = 0$
d. $xy + yz + zx - 1 = 0$

15 If $\sin^{-1} x + \sin^{-1} y + \sin^{-1} z = \frac{\pi}{2}$, then the value of $x^2 + y^2 + z^2 + 2xyz$ is

a. 0
b. 1
c. 2
d. 3

16 The trigonometric equation $\sin^{-1} x = 2\sin^{-1} a$ has a solution for **[AIEEE 2013]**

a. $|a| \le \frac{1}{\sqrt{2}}$
b. $\frac{1}{2} < |a| < \frac{1}{\sqrt{2}}$
c. $a \in R$
d. $|a| < \frac{1}{2}$

Working Space

17 If $\cos^{-1} x - \cos^{-1} \frac{y}{2} = \alpha$, then the value of $4x^2 - 4xy \cos \alpha + y^2$ is **[AIEEE 2005]**

a. $2 \sin 2\alpha$
b. 4
c. $4 \sin^2 \alpha$
d. $-4 \sin^2 \alpha$

18 If $\sin^{-1}\left(\frac{x}{5}\right) + \text{cosec}^{-1}\left(\frac{5}{4}\right) = \frac{\pi}{2}$, then the value of x is **[AIEEE 2007]**

a. 1
b. 3
c. 4
d. 5

19 The value of $\cot\left(\text{cosec}^{-1} \frac{5}{3} + \tan^{-1} \frac{2}{3}\right)$ is **[AIEEE 2008]**

a. 6/17
b. 3/17
c. 4/17
d. 5/17

20 The number of real solutions of $\tan^{-1} \sqrt{x(x+1)} + \sin^{-1} \sqrt{x^2 + x + 1} = \frac{\pi}{2}$ is **[IIT JEE 1994]**

a. 0
b. 1
c. 2
d. ∞

21 If $\sin^{-1}\left(x - \frac{x^2}{2} + \frac{x^3}{4} \cdots\right) + \cos^{-1}\left(x^2 - \frac{x^4}{2} + \frac{x^6}{4} \cdots\right) = \frac{\pi}{2}$, for $0 < |x| < \sqrt{2}$, then x equals **[IIT JEE 2001]**

a. $\frac{1}{2}$
b. 1
c. $\frac{-1}{2}$
d. -1

22 If x, y, z are in an AP and $\tan^{-1} x$, $\tan^{-1} y$ and $\tan^{-1} z$ are also in an AP, then **[JEE Main 2013]**

a. $x = y = z$
b. $2x = 3y = 6z$
c. $6x = 3y = 2z$
d. $6x = 4y = 3z$

One or More than One Options Correct Type

Direction (Q. Nos. 23-26) *This section contains 4 multiple choice questions. Each question has four choices (a), (b), (c) and (d), out of which ONE or MORE THAN ONE are correct.*

23 If $\tan^{-1} x + \sin^{-1} x = \tan^{-1} 2x$, then x cannot be less than

a. $-1/2$
b. 0
c. 1/2
d. 1

24 If $\sin^{-1}(x^2 - 4x + 5) + \cos^{-1}(y^2 - 2y + 2) = \pi/2$, then

a. $x + y = 3$
b. $xy = 2$
c. $x - y = 1$
d. $x - y = 0$

25 If x and y are positive integers such that $\tan^{-1} x + \cot^{-1} y = \tan^{-1} 3$, then

a. $x > 2, y < 2$
b. $x < 3, y \geq 2$
c. $x \leq 2, y \leq 7$
d. $x > 2, y > 7$

26 If $\cos^{-1}\left(\frac{7}{|x|}\right) + \cos^{-1}\left(\frac{4\sqrt{15}}{|x|}\right) = \frac{\pi}{2}$, then

a. $x = \pm 16$
b. $x = \pm 17$
c. $\tan^{-1}\left(\sqrt{\frac{|x|}{|x| - 12}}\right) = \frac{\pi}{3}$
d. $\tan^{-1}\left(\frac{10|x| - 1}{(|x| - 4)^2}\right) = \frac{\pi}{4}$

Working Space

Comprehension Type

Direction (Q. Nos. 27 and 28) *This section contains a paragraph, describing theory, experiments, data, etc. Two questions related to the paragraph have been given. Each question has only one correct answer among the four given options (a), (b), (c) and (d).*

Passage

Consider the system of equations

$$\cos^{-1} x + (\sin^{-1} y)^2 = \frac{p\pi^2}{4} \text{ and } (\cos^{-1} x)(\sin^{-1} y)^2 = \frac{\pi^4}{16}, p \in Z.$$

27 The value of p for which system of equations has a solution

a. 1
b. 2
c. 0
d. -1

28 The value of x which satisfies the system of equations, is

a. $\cos^2 \frac{\pi}{8}$
b. $\sin \frac{\pi^2}{4}$
c. $\cos \frac{\pi^2}{2}$
d. $\cos \frac{\pi^2}{4}$

One Integer Value Correct Type

Direction (Q. Nos. 29-35) *This section contains 7 questions. When worked out will result in an integer from 0 to 9 (both inclusive).*

29 Sum of the series to infinity

$\tan^{-1}\left(\frac{1}{1+1+1^2}\right) + \tan^{-1}\left(\frac{1}{1+2+2^2}\right) + \tan^{-1}\left(\frac{1}{1+3+3^2}\right) + \cdots \infty = \frac{\pi}{k}$, then the value of k is

30 Sum of the series to infinity

$\tan^{-1}\left(\frac{1}{2\cdot 1^2}\right) + \tan^{-1}\left(\frac{1}{2\cdot 2^2}\right) + \tan^{-1}\left(\frac{1}{2\cdot 3^2}\right) + \cdots \infty = \frac{\pi}{k}$, then the value of k is

31 If $\sum_{m=1}^{n} \tan^{-1}\left(\frac{2m}{m^4 + m^2 + 2}\right) = \tan^{-1}\left(\frac{n^2 + n}{an^2 + bn + c}\right)$, then $(a + b + c)$ is

32 The sum of the series to infinity

$\cot^{-1}\left(2^2 + \frac{1}{2}\right) + \cot^{-1}\left(2^3 + \frac{1}{2^2}\right) + \cot^{-1}\left(2^4 + \frac{1}{2^3}\right) + \cdots \infty = \cot^{-1}(k)$, then the value of k is

33 The sum of the series,

$\cot^{-1}\left(1 + \frac{3}{4}\right) + \cot^{-1}\left(2^2 + \frac{3}{4}\right) + \cot^{-1}\left(3^2 + \frac{3}{4}\right) + \cdots \infty = \tan^{-1}(k)$, then the value of k is

34 The sum of $\sin^{-1}\left(\frac{1}{\sqrt{2}}\right) + \sin^{-1}\left(\frac{\sqrt{2}-1}{\sqrt{6}}\right) + \cdots + \sin^{-1}\left(\frac{\sqrt{n} - \sqrt{n-1}}{\sqrt{n(n+1)}}\right) + \cdots \infty = \frac{\pi}{k}$, then the value of k is

35 If $x = \operatorname{cosec}[\tan^{-1}\{\cos(\cot^{-1}(\sec(\sin^{-1} a)))\}]$ and

$y = \sec[\cot^{-1}\{\sin(\tan^{-1}(\operatorname{cosec}(\cos^{-1} a)))\}]$, where $a \in [0, 1]$, then $(x^2 - y^2)$ is

DPP-3 Special Formulae

Only One Option Correct Type

Working Space

Direction *(Q. Nos. 1-17) This section contains 17 multiple choice questions. Each question has four choices (a), (b), (c) and (d), out of which ONLY ONE option is correct.*

1 The graph for $y = \sin^{-1}(2x\sqrt{1-x^2})$ is

a. 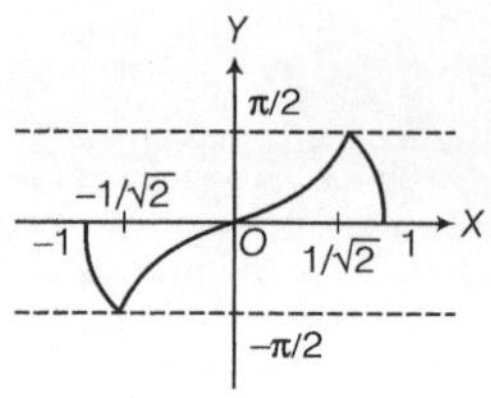

b. 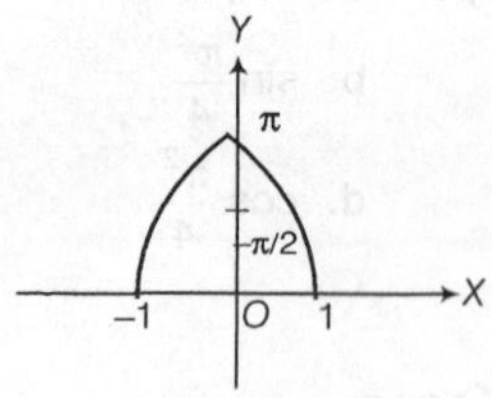

c. 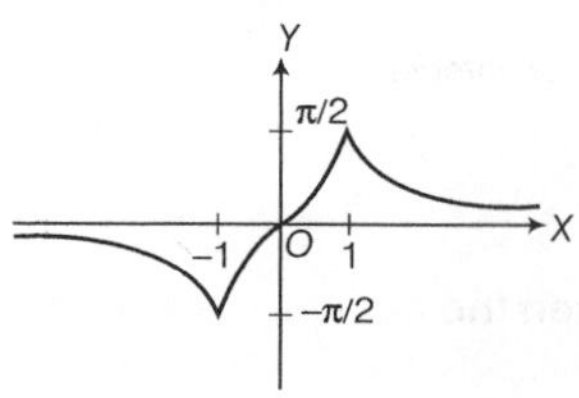

d. 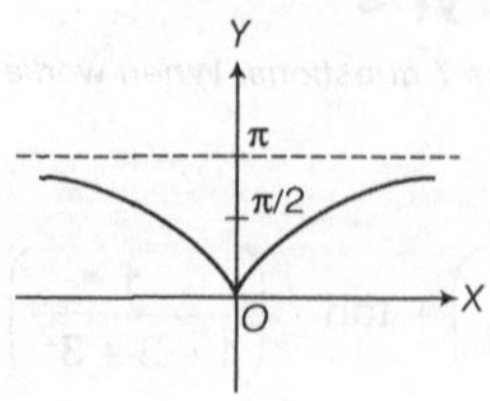

2 The graph for $y = \cos^{-1}(2x^2 - 1)$ is

a. 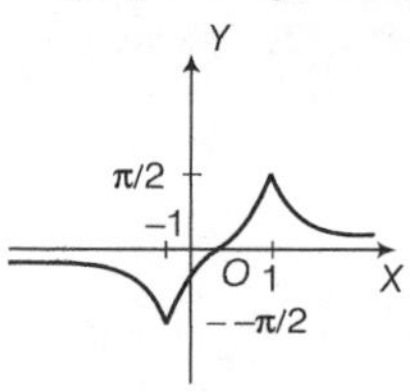

b. 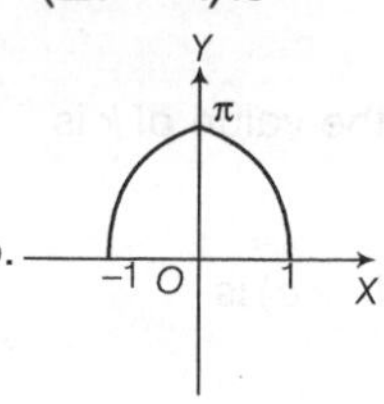

c. 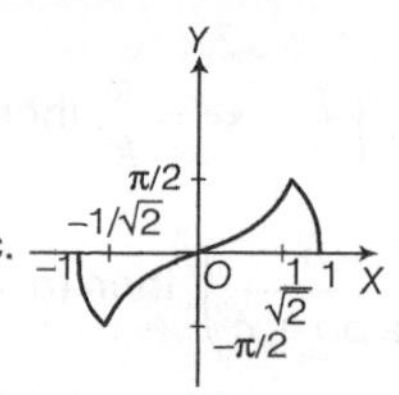

d. 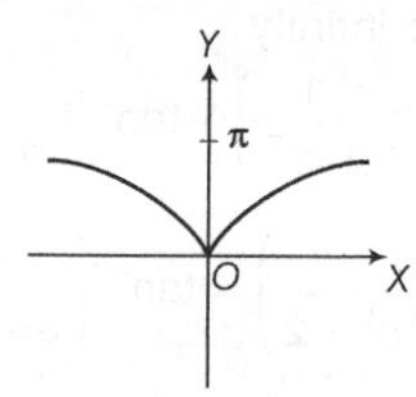

3 The graph for $y = \sin^{-1}\left(\dfrac{2x}{1+x^2}\right)$ is

a. 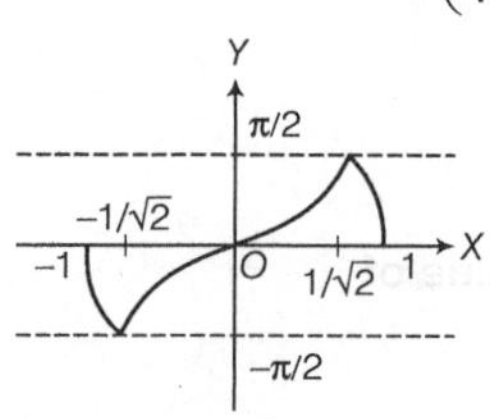

b. 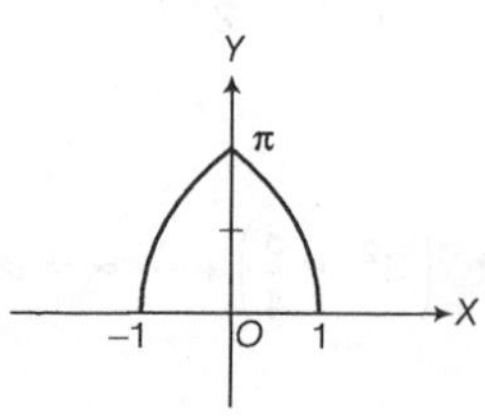

c. 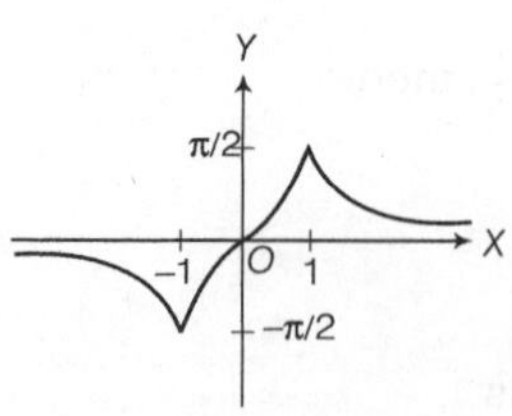

d. 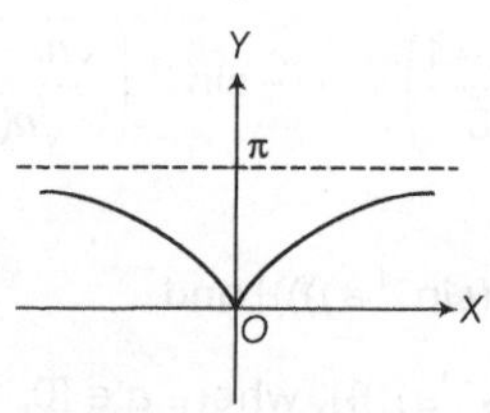

Working Space

4 The graph for $y = \cos^{-1}\left(\dfrac{1-x^2}{1+x^2}\right)$ is

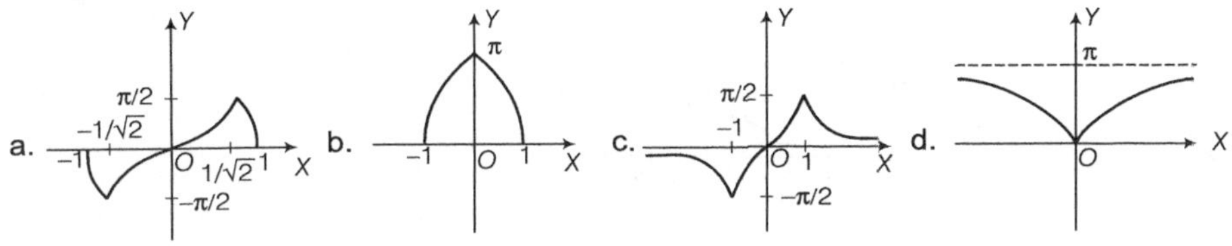

5 The graph for $y = \tan^{-1}\left(\dfrac{2x}{1-x^2}\right)$, is

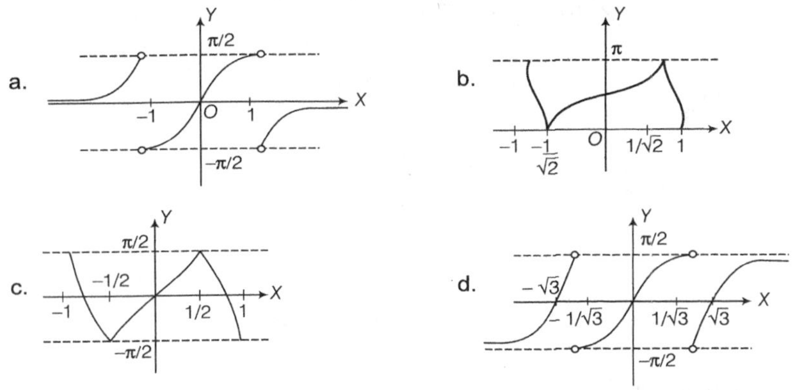

6 The graph for $y = \cos^{-1}(4x^3 - 3x)$ is

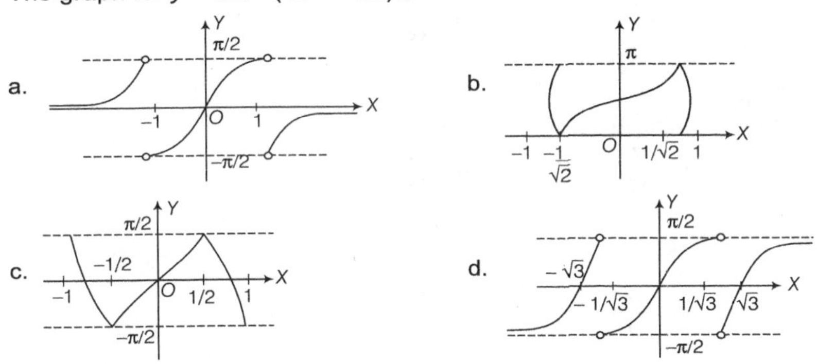

7 The graph for $y = \sin^{-1}(3x - 4x^3)$ is

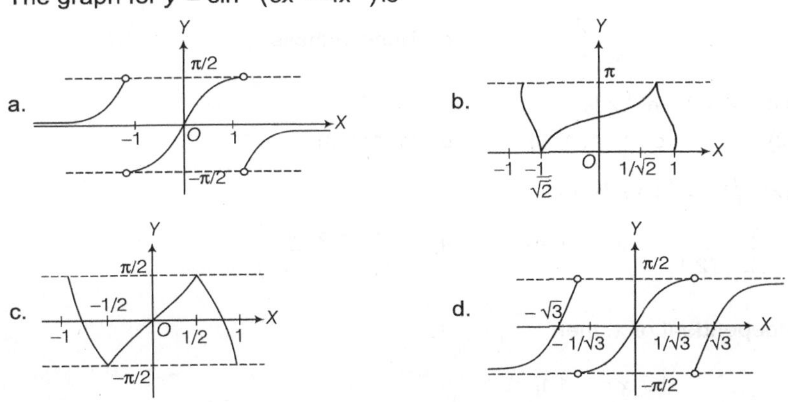

Working Space

8 The graph for $y = \tan^{-1}\left(\frac{3x - x^3}{1 - 3x^2}\right)$ is

a. 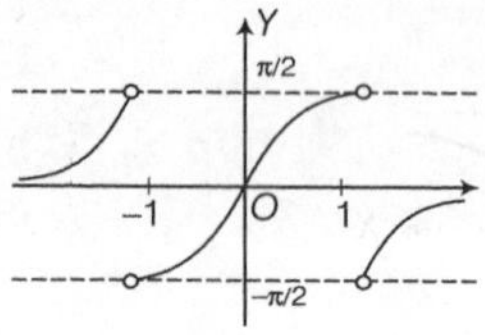

b. 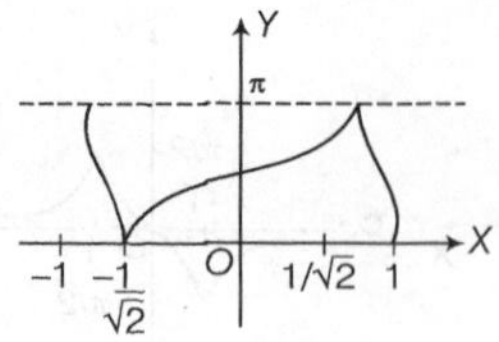

c. 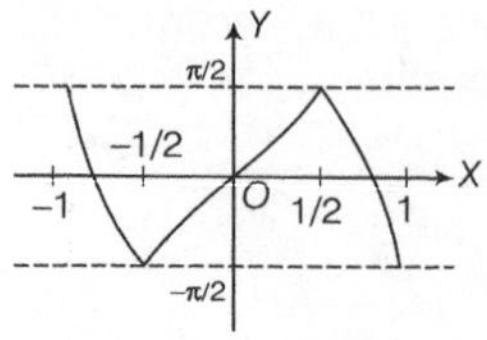

d. 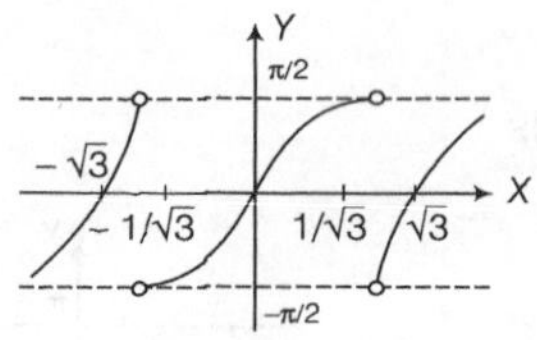

9 The value of $\cos\left(2\cos^{-1}\frac{1}{5} + \sin^{-1}\frac{1}{5}\right)$ is

a. $\frac{2\sqrt{6}}{5}$ b. $-\frac{2\sqrt{6}}{5}$ c. $\frac{1}{5}$ d. $\frac{-1}{5}$

10 The value of $\sin\left(4\tan^{-1}\frac{1}{3}\right)$ is

a. $\frac{12}{25}$ b. $\frac{24}{25}$ c. $\frac{1}{5}$ d. None of these

11 The value of $\sin\left[2\tan^{-1}\left(\frac{1}{3}\right)\right] + \cos(\tan^{-1} 2\sqrt{2})$ is

a. $\frac{16}{15}$ b. $\frac{14}{15}$ c. $\frac{12}{15}$ d. $\frac{11}{15}$

12 The value of $\tan\left[2\tan^{-1}\left(\frac{1}{5}\right) - \frac{\pi}{4}\right]$ is

a. $\frac{17}{7}$ b. $\frac{-17}{7}$ c. $\frac{7}{17}$ d. $\frac{-7}{17}$

13 The value of $\tan\left(2\cos^{-1}\frac{3}{5}\right)$ is

a. $\frac{7}{25}$ b. $\frac{24}{25}$ c. $\frac{-24}{7}$ d. $\frac{8}{3}$

14 If $\cos(2\sin^{-1} x) = \frac{1}{9}$, then x is

a. $\frac{2}{3}$ b. $\frac{-2}{3}$ c. $\frac{2}{3}, \frac{-2}{3}$ d. None of these

15 The solution set of equation $\sin^{-1} x = 2\tan^{-1} x$ is

a. {1, 2} b. {– 1, 2} c. {– 1, 1, 0} d. {1, 1/2, 0}

16 The formula $2\sin^{-1} x = \sin^{-1}(2x\sqrt{1 - x^2})$ holds for

a. $x \in [0, 1]$ b. $x \in \left[-\frac{1}{\sqrt{2}}, \frac{1}{\sqrt{2}}\right]$ c. $x \in (-1, 0)$ d. $x \in \left(-\frac{\sqrt{3}}{2}, \frac{\sqrt{3}}{2}\right)$

17 If $2\tan^{-1} x + \sin^{-1}\left(\frac{2x}{1 + x^2}\right)$ is independent of x, then

a. $x \in [0, \infty)$ b. $x \in [-1, 1]$
c. $x \in (-\infty, -1] \cup [1, \infty)$ d. None of these

Working Space

One or More than One Options Correct Type

Direction (Q. Nos. 18-20) *This section contains 3 multiple choice questions. Each question has four choices (a), (b), (c) and (d), out of which ONE or MORE THAN ONE are correct.*

18 For the equation $2x = \tan(2\tan^{-1} a) + 2\tan(\tan^{-1} a + \tan^{-1} a^3)$, which of the following is invalid?

a. $a^2x + 2a = x$ b. $a^2 + 2ax + 1 = 0$

c. $a \neq 0$ d. $a \neq 1, -1$

19 $2\tan[\tan^{-1}(x) + \tan^{-1}(x^3)]$, where $x \in R - \{-1, 1\}$ is equal to

a. $\dfrac{2x}{1-x^2}$ b. $\tan(2\tan^{-1} x)$

c. $\tan[\cot^{-1}(-x) - \cot^{-1}(x)]$ d. $\tan(2\cot^{-1} x)$

20 The value of $\cos\left[\frac{1}{2}\cos^{-1}\left\{\cos\left(-\frac{14}{5}\pi\right)\right\}\right]$ is

a. $\cos\left(\frac{-7\pi}{5}\right)$ b. $\sin\left(\frac{\pi}{10}\right)$ c. $\cos\left(\frac{2\pi}{5}\right)$ d. $-\cos\left(\frac{3\pi}{5}\right)$

Matching List Type

Direction (Q. Nos. 21 and 22) *Choices for the correct combination of elements from Column I and Column II are given as options (a), (b), (c) and (d), out of which one is correct.*

21 Match the principal values of $\cos^{-1}(8x^4 - 8x^2 + 1)$ given in column I with the corresponding intervals of x given in column II, for which it holds.

Match the column I with column II and mark the correct option from the codes given below.

	Column I		Column II
i.	$4\cos^{-1} x$	p.	$0 \le x \le \frac{1}{\sqrt{2}}$
ii.	$4\cos^{-1} x - 2\pi$	q.	$\frac{1}{\sqrt{2}} \le x \le 1$
iii.	$2\pi - 4\cos^{-1} x$	r.	$-1 \le x \le -\frac{1}{\sqrt{2}}$
iv.	$4\pi - 4\cos^{-1} x$	s.	$-\frac{1}{\sqrt{2}} \le x \le 0$

Codes

	i	ii	iii	iv			i	ii	iii	iv
a.	p	q	s	r		b.	q	s	p	r
c.	r,s	q	p	s		d.	s	p	q	r

22 Match the column I with column II and mark the correct option from the codes given below.

	Column I		Column II
i.	$\cos^{-1}(4x^3 - 3x) = 3\cos^{-1} x$, then x	p.	$\in [\sqrt{3}/2, 1]$
ii.	$\sin^{-1}(3x - 4x^3) = 3\sin^{-1} x$, then x	q.	$\in [-1/2, 0]$
iii.	$\cos^{-1}(4x^3 - 3x) = 3\sin^{-1} x$, then x	r.	$\in [0, \sqrt{3}/2]$
iv.	$\sin^{-1}(3x - 4x^3) = 3\cos^{-1} x$, then x	s.	$\in [0, 1/2]$

Codes

	i	ii	iii	iv			i	ii	iii	iv
a.	p	q,s	r	p,q		b.	r	p,q	p,s	r,s
c.	p	q,s	r,s	p		d.	p	q	r,s	q,s

Comprehension Type

Working Space

Direction (Q. Nos. 23-27) *This section contains a paragraph, describing theory, experiments, data, etc. Five questions related to the paragraph have been given. Each question has only one correct answer among the four given options (a), (b), (c) and (d).*

Passage

While defining a inverse trigonometric function, a new system is followed where codomains have been redefined as follows :

Function	Domain	Codomain
$\sin^{-1}(x)$	$[-1,1]$	$[\pi/2, 3\pi/2]$
$\cos^{-1}(x)$	$[-1, 1]$	$[\pi, 2\pi]$
$\tan^{-1}(x)$	R	$(\pi/2, 3\pi/2)$

23 $\sin^{-1}(-x)$ is equal to

a. $-\sin^{-1}x$
b. $\pi+\sin^{-1}x$
c. $2\pi-\sin^{-1}x$
d. None of these

24 If $f(x)=3\sin^{-1}x-2\cos^{-1}x$, then $f(x)$ is

a. even function
b. odd function
c. neither even nor odd
d. even as well as odd function

25 The minimum value of $(\sin^{-1}x)^3-(\cos^{-1}x)^3$ is

a. $\frac{-63\pi^3}{8}$
b. $\frac{63}{8}\pi^3$
c. $\frac{125}{32}\pi^3$
d. $-\frac{125}{32}\pi^3$

26 The value of $\sin^{-1}x+\cos^{-1}x$ is equal to

a. $\pi/2$
b. $3\pi/2$
c. $5\pi/2$
d. $7\pi/2$

27 If in another such system, codomain of $\sin^{-1}x$ is $\left[-\frac{5\pi}{2}, -\frac{3\pi}{2}\right]$ and that of $\sin^{-1}x+\cos^{-1}x=\frac{5\pi}{2}$, then in the same system, codomain of $\cos^{-1}x$ will be

a. $[4\pi, 5\pi]$
b. $[3\pi, 4\pi]$
c. $[6\pi, 7\pi]$
d. $[5\pi, 6\pi]$

One Integer Value Correct Type

Direction (Q. Nos. 28-30) *This section contains 3 questions. When worked out will result in an integer from 0 to 9 (both inclusive).*

28 If $\cos^{-1}(4x^3-3x)=a+b\cos^{-1}x$ for $-1<x<-\frac{1}{2}$, then $|[a+b+2]|$ is equal to (where $[\cdot]$ is greatest integer function).

29 The minimum integral value of α, for which the quadratic equation $(\cot^{-1}\alpha)x^2-(\tan^{-1}\alpha)^{3/2}\cdot x+2(\cot^{-1}\alpha)^2=0$, has both positive roots, is

30 If $k=\tan^{-1}\left(\frac{1}{2}\tan 2A\right)+\tan^{-1}(\cot A)+\tan^{-1}(\cot^3 A)$ and $\pi/4<A<\pi/2$, then $[k]$ is (where $[\cdot]$ denotes the greatest integer function).

SET - 1

Revisal Problems
for JEE Main

Only One Option Correct Type

Working Space

Direction (Q. Nos. 1-32) *This section contains 32 multiple choice questions. Each question has four choices (a), (b), (c) and (d), out of which ONLY ONE option is correct.*

1 The value of $1+\cot^2(\sin^{-1}x)$ is

a. $1/2x$ b. x^2
c. $1/x^2$ d. $2/x$

2 $\sec^2(\tan^{-1}2)+\text{cosec}^2(\cot^{-1}3)$ is equal to

a. 1 b. 5
c. 10 d. 15

3 The value of $\tan^2(\sec^{-1}2)+\cot^2(\text{cosec}^{-1}3)$ is

a. 13 b. 15
c. 11 d. None of these

4 $\sin\left(\frac{1}{2}\cos^{-1}\frac{4}{5}\right)$ is equal to

a. $\frac{1}{\sqrt{10}}$ b. $\frac{-1}{\sqrt{10}}$
c. $\frac{1}{10}$ d. $-\frac{1}{10}$

5 If $\sin^{-1}x+\sin^{-1}y=2\pi/3$, then $\cos^{-1}x+\cos^{-1}y$ is equal to

a. $2\pi/3$ b. $\pi/3$
c. $\pi/6$ d. π

6 If $2\tan^{-1}(\cos x)=\tan^{-1}(\text{cosec}^2 x)$, then x is equal to

a. $\pi/2$ b. π
c. $\pi/6$ d. $\pi/3$

7 The number of solutions of equation $\tan^{-1}(x-1)+\tan^{-1}(x)+\tan^{-1}(x+1)=\tan^{-1}(3x)$, if $-\sqrt{2}<x<\sqrt{2}$, is

a. 1 b. 2
c. 3 d. 4

8 If $\sin^{-1}x>\cos^{-1}x$, then

a. all values of x b. $x\in\left(0,\frac{1}{\sqrt{2}}\right)$
c. $x\in\left[\frac{1}{\sqrt{2}},1\right]$ d. None of these

9 If $3\cos^{-1}\left(x^2-7x+\frac{25}{2}\right)=\pi$, then x is equal to

a. Only 3 b. Only 4
c. 3 or 4 d. None of these

10 If $\sin^{-1}x+\cot^{-1}\frac{1}{2}=\frac{\pi}{2}$, then x is equal to

a. 0 b. $\frac{1}{\sqrt{5}}$
c. $\frac{2}{\sqrt{5}}$ d. $\frac{\sqrt{3}}{2}$

Working Space

11 If $\sin^{-1} x + \sin^{-1} y + \sin^{-1} z = 3\pi/2$, then the value of $x^{100} + y^{100} + z^{100} - \dfrac{9}{x^{100} + y^{100} + z^{100}}$ is

a. 0
b. 1
c. 2
d. 3

12 If $\tan^{-1} x + \cos^{-1} \dfrac{y}{\sqrt{1+y^2}} = \sin^{-1} \dfrac{3}{\sqrt{10}}$ and both x and y are positive integers, then x and y are

a. (1, 2) and (2, 7)
b. (1, 2) and (1, 7)
c. (1, 7) and (2, 7)
d. (1, 7) and (2, 1)

13 The value of $\cos^{-1}\left(\cos \dfrac{5\pi}{3}\right) + \sin^{-1}\left(\sin \dfrac{5\pi}{3}\right)$ is

a. 0
b. $\pi/2$
c. $2\pi/3$
d. $10\pi/3$

14 If $\cot^{-1}(\sqrt{\cos \alpha}) - \tan^{-1}(\sqrt{\cos \alpha}) = x$, then $\sin x$ is **[AIEEE 2013]**

a. $\tan^2(\alpha/2)$
b. $\cot^2(\alpha/2)$
c. $\tan \alpha$
d. None of these

15 The solution of the equation $\sin^{-1}\left(\tan \dfrac{\pi}{4}\right) - \sin^{-1}(\sqrt{3/x}) = \pi/6$, is

a. $x = 2$
b. $x = -4$
c. $x = 4$
d. None of these

16 The value of $\sin^{-1}[\cos\{\cos^{-1}(\cos x) + \sin^{-1}(\sin x)\}]$, where $x \in (\pi/2, \pi)$ is

a. $\pi/2$
b. $\pi/4$
c. $-\pi/4$
d. $-\pi/2$

17 The value of $\cos[\tan^{-1}(\tan 2)]$ is

a. $1/\sqrt{5}$
b. $-1/\sqrt{5}$
c. $\cos 2$
d. $-\cos 2$

18 The value of $\sin^{-1}(\sin 10)$ is

a. 10
b. $10 - 3\pi$
c. $3\pi - 10$
d. None of these

19 The value of $\cos^{-1}(\cos 12) - \sin^{-1}(\sin 14)$ is

a. -2
b. $8\pi - 26$
c. $4\pi + 2$
d. None of these

20 The value of $\cos^{-1}\left(\sqrt{\dfrac{2}{3}}\right) - \cos^{-1}\left(\dfrac{\sqrt{6}+1}{2\sqrt{3}}\right)$ is

a. $\pi/3$
b. $\pi/4$
c. $\pi/2$
d. $\pi/6$

21 The value of $\cos^{-1}[\cos\{2\cot^{-1}(\sqrt{2}-1)\}]$ is

a. $\sqrt{2}-1$
b. $\pi/4$
c. $3\pi/4$
d. None of these

22 If $x_1 = 2\tan^{-1}\left(\dfrac{1+x}{1-x}\right)$, $x_2 = \sin^{-1}\left(\dfrac{1-x^2}{1+x^2}\right)$, where $x \in (0, 1)$, then $x_1 + x_2$ is equal to

a. 0
b. 2π
c. π
d. None of these

23 If $x \in (-\pi/2, \pi/2)$, then the value of $\tan^{-1}\left(\dfrac{\tan x}{4}\right) + \tan^{-1}\left(\dfrac{3\sin 2x}{5 + 3\cos 2x}\right)$ is

a. $x/2$
b. $2x$
c. $3x$
d. x

Working Space

24 If $\tan^{-1} x + \tan^{-1} y + \tan^{-1} z = \pi$, then $x + y + z$ is

a. $\frac{1}{x} + \frac{1}{y} + \frac{1}{z}$
b. xyz
c. $xy + yz + zx$
d. None of these

25 If $3\sin^{-1} x = \pi + \sin^{-1}(3x - 4x^3)$, then

a. $\frac{1}{2} \le x \le 1$
b. $|x| \le 1$
c. $|x| \le 1/2$
d. None of these

26 The complete solution set of $\sin^{-1}(\sin 5) > x^2 - 4x$ is

a. $|x-2| < \sqrt{9-2\pi}$
b. $|x-2| > \sqrt{9-2\pi}$
c. $|x| < \sqrt{9-2\pi}$
d. $|x| > \sqrt{9-2\pi}$

27 The value of $\tan^{-1}\sqrt{\frac{a\lambda}{bc}} + \tan^{-1}\sqrt{\frac{b\lambda}{ca}} + \tan^{-1}\sqrt{\frac{c\lambda}{ab}}$, where $a, b, c, \in R^+$ and $\lambda = a + b + c$ is equal to

a. $\pi/4$
b. π
c. $\pi/2$
d. None of these

28 Complete solution set of $\tan^2 (\sin^{-1} x) > 1$, is

a. $\left[-1, -\frac{1}{\sqrt{2}}\right) \cup \left(\frac{1}{\sqrt{2}}, 1\right]$
b. $\left(-\frac{1}{\sqrt{2}}, \frac{1}{\sqrt{2}}\right) - \{0\}$
c. $[-1, 1) - \{0\}$
d. None of these

29 If $x \in \left(\frac{3\pi}{2}, 2\pi\right)$, then $\sin^{-1} [\cos \{\cos^{-1} (\cos x) + \sin^{-1} (\sin x)\}]$ is equal to

a. $-\pi/2$
b. $\pi/2$
c. 0
d. π

30 If α, β, γ are the roots of the equation $x^3 + mx^2 + 3x + m = 0$, then the general value of $\tan^{-1} \alpha + \tan^{-1} \beta + \tan^{-1} \gamma$ is

a. $(2n + 1) \pi/2$
b. $n\pi$
c. $n\pi/2$
d. None of these

31 If $\alpha = \sin^{-1}[\cos (\sin^{-1} x)]$ and $\beta = \cos^{-1} [\sin(\cos^{-1} x)]$, then

a. $\tan \alpha = \tan \beta$
b. $\tan \alpha = -\cot \beta$
c. $\tan \alpha = \cot \beta$
d. $\tan \alpha = -\tan \beta$

32 Which of the following statement is meaningless?

a. $\cos^{-1}\left[\log\left(\frac{2e+4}{3}\right)\right]$
b. $\text{cosec}^{-1} (\pi/3)$
c. $\cot^{-1}(\pi/2)$
d. $\sec^{-1}(\pi)$

Statement Type

Direction *(Q. Nos. 33-40)* ***This section is based on Statement I and Statement II. Select the correct answer from the codes given below.***

a. Both Statement I and Statement II are correct and Statement II is the correct explanation of Statement I
b. Both Statement I and Statement II are correct but Statement II is not the correct explanation of Statement I
c. Statement I is correct but Statement II is incorrect
d. Statement II is correct but Statement I is incorrect

33 **Statement I** The value of $\cot^{-1}(-1)$ is $\frac{3\pi}{4}$.

Statement II $\cot^{-1} (-x) = \pi - \cot^{-1} x, x \in R$.

34 **Statement I** If $x + \frac{1}{x} = 2$, then the principal value of $\sin^{-1} x$ is $\frac{\pi}{2}$.

Statement II $\sin^{-1} (\sin x) = x, \forall x \in R$.

Working Space

35 **Statement I** Number of ordered pairs (*x*, *y*) satisfying $\sin^{-1} x = \pi - \sin^{-1} y$ and $\cos^{-1} x + \cos^{-1} y = 0$ simultaneously is 1.

Statement II Ordered pairs (*x*, *y*) satisfying $\sin^{-1} x = \pi - \sin^{-1} y$ and $\cos^{-1} x + \cos^{-1} y = 0$ will lie one $x^2 + y^2 = 2$.

36 **Statement I** If α, β are roots of $6x^2 + 11x + 3 = 0$, then $\cos^{-1}\alpha$ exists but not $\cos^{-1}\beta$, $(\alpha > \beta)$.

Statement II Domain of $\cos^{-1} x$ is $[-1, 1]$.

37 **Statement I** $\tan^2(\sec^{-1} 2) + \cot^2(\operatorname{cosec}^{-1} 3) = 11$.

Statement II $\tan^2\theta + \sec^2\theta = 1 = \cot^2\theta + \operatorname{cosec}^2\theta$.

38 **Statement I** If $a > 0, b > 0, \tan^{-1}\left(\frac{a}{x}\right) + \tan^{-1}\left(\frac{b}{x}\right) = \frac{\pi}{2} \Rightarrow x = \sqrt{ab}$.

Statement II If $m, n \in N, n \geq m$, then $\tan^{-1}\left(\frac{m}{n}\right) + \tan^{-1}\left(\frac{n-m}{n+m}\right) = \frac{\pi}{4}$.

39 **Statement I** The number of solutions of the equation $\sin x + \cos x = \sin^{-1} x + \cos^{-1} x$ is zero.

Statement II $\sin^{-1} x + \cos^{-1} x = \frac{\pi}{2}, \forall x \in [-1, 1]$ and maximum value of $\sin x + \cos x$ is $\sqrt{2}$ for $x \in R$.

40 **Statement I** If $\tan^{-1} x + \tan^{-1} y = \frac{\pi}{4} - \tan^{-1} z$ and $x + y + z = 1$, then arithmetic mean of odd powers of *x*, *y*, *z* is equal to (1/3).

Statement II For any *x*, *y*, *z*, we have

$$xyz - xy - yz - zx + x + y + z = 1 + (x-1)(y-1)(z-1)$$

SET - 2

Revisal Problems
for JEE Advanced

Only One Option Correct Type

Working Space

Direction (Q. Nos. 1-28) *This section contains 28 multiple choice questions. Each question has four choices (a), (b), (c) and (d), out of which ONLY ONE option is correct.*

1 The number of solutions of equation $|\tan^{-1}|x|| = \sqrt{(x^2+1)^2 - 4x^2}$, is

a. 2 b. 3
c. 4 d. None of these

2 If $\sin^{-1}(\cos^{-1}x) < 1$ and $\cos^{-1}(\cos^{-1}x) < 1$, then x belongs to

a. $(\sin 1, \sin(\sin 1))$ b. $(\cos(\cos 1), \cos(\sin 1))$
c. $(\cos(\sin 1), \cos(\cos 1))$ d. ϕ

3 Given, $0 \le x \le 1/2$, then the value of $\tan\left[\sin^{-1}\left(\frac{x}{\sqrt{2}} + \sqrt{\frac{1-x^2}{2}}\right) - \sin^{-1}x\right]$, is

a. -1 b. 1
c. $\frac{1}{\sqrt{3}}$ d. $\sqrt{3}$

4 If α is the only real root of $x^3 + bx^2 + cx + 1 = 0\ (b < c)$, then $\tan^{-1}\alpha + \tan^{-1}\left(\frac{1}{\alpha}\right)$, is

a. $-\pi/2$ b. $\pi/2$
c. 0 d. Can'not be determined

5 If $x > 1$ and $\cos^{-1}\left(\frac{x^2-1}{x^2+1}\right) + \tan^{-1}\left(\frac{2x}{x^2-1}\right) = \frac{2\pi}{3}$, then x is equal to

a. 2 b. $\sqrt{3}$
c. 4 d. 3

6 The number of solutions to equation $\sqrt{|\sin^{-1}|\cos x|| + |\cos^{-1}|\sin x||}$ $= \sin^{-1}|\cos x| - \cos^{-1}|\sin x|$, if $x \in [0, 4\pi]$, is

a. 1 b. 2
c. 4 d. 8

7 The number of solutions of the equation $\sin^{-1}\left(\frac{1+x^2}{2x}\right) = \frac{\pi}{2}\sec(x-1)$, is

a. 1 b. 2
c. 3 d. infinite

8 If $u = \cot^{-1}\sqrt{\tan\alpha} - \tan^{-1}\sqrt{\tan\alpha}$, then $\tan(\pi/4 - u/2)$ is

a. $\sqrt{\tan\alpha}$ b. $\sqrt{\cot\alpha}$
c. $\tan\alpha$ d. $\cot\alpha$

9 The number of solution(s) of the equation, $\sin^{-1}x + \cos^{-1}(1-x) = \sin^{-1}(-x)$, is/are

a. 0 b. 1
c. 2 d. more than 2

10 The number of real solutions of (x, y) where, $|y| = \sin x$ and $y = \cos^{-1}(\cos x)$, $-2\pi \le x \le 2\pi$, is

a. 2 b. 1
c. 3 d. 4

Working Space

11 All possible values of p and q for which $\cos^{-1}\sqrt{p}+\cos^{-1}\sqrt{1-p}+\cos^{-1}\sqrt{1-q}=\frac{3\pi}{4}$, holds is

a. $p=1, q=1/2$
b. $q>1, p=1/2$
c. $0\le p\le 1, q=1/2$
d. None of these

12 The set of values of a for which $x^2+ax+\sin^{-1}(x^2-4x+5)+\cos^{-1}(x^2-4x+5)=0$, has atleast one solution is

a. $(-\infty,-\sqrt{2\pi}]\cup[\sqrt{2\pi},\infty)$
b. $(-\infty,-\sqrt{2\pi})\cup(\sqrt{2\pi},\infty)$
c. R
d. None of these

13 The set of values of k, for which $x^2-kx+\sin^{-1}(\sin 4)>0$ for all real x, is

a. $\{0\}$
b. $(-2,2)$
c. R
d. None of these

14 $\cos^{-1}\left\{\frac{1}{2}x^2+\sqrt{1-x^2}\cdot\sqrt{1-\frac{x^2}{4}}\right\}=\cos^{-1}\frac{x}{2}-\cos^{-1}x$, holds for

a. $|x|\le 1$
b. $x\in R$
c. $0\le x\le 1$
d. $-1\le x\le 0$

15 If $[\cot^{-1}x]+[\cos^{-1}x]=0$, where $[\cdot]$ denotes the greatest integer function, then complete set of values of x is

a. $(\cos 1, 1]$
b. $(\cot 1, \cos 1)$
c. $(\cot 1, 1]$
d. None of these

16 The complete solution set of the inequality $[\cot^{-1}x]^2-6[\cot^{-1}x]+9\le 0$, where $[\cdot]$ denotes greatest integer function, is

a. $(-\infty, \cot 3]$
b. $[\cot 3, \cot 2]$
c. $[\cot 3, \infty)$
d. None of these

17 The number of real values of x satisfying the inequation $\frac{(\sin^{-1}x)^3+(\cos^{-1}x)^3}{(\tan^{-1}x+\cot^{-1}x)^3}=7$, is

a. 1
b. 2
c. 3
d. 4

18 The solution set of inequality $(\cot^{-1}x)(\tan^{-1}x)+\left(2-\frac{\pi}{2}\right)\cot^{-1}x-3\tan^{-1}x-3\left(2-\frac{\pi}{2}\right)>0$, is

a. $x\in(\tan 2, \tan 3)$
b. $x\in(\cot 3, \cot 2)$
c. $x\in(-\infty, \tan 2)\cup(\tan 3, \infty)$
d. None of these

19 If $2\cos^{-1}x=\cot^{-1}\left(\frac{2x^2-1}{2x\sqrt{1-x^2}}\right)$, then x belongs to

a. $(0,1)$
b. $(-1,1)-\{0\}$
c. $(-1,0)$
d. $[-1,1]$

20 The range of values of p for which equation $\sin[\cos^{-1}\{\cos(\tan^{-1}x)\}]=p$ has a solution is

a. $\left[-\frac{1}{\sqrt{2}},\frac{1}{\sqrt{2}}\right]$
b. $[0,1)$
c. $\left[\frac{1}{\sqrt{2}},1\right)$
d. $(-1,1)$

21 The number of values of x satisfying equation $\sin^{-1}\left(\frac{5}{x}\right)+\sin^{-1}\left(\frac{12}{x}\right)=\frac{\pi}{2}$, is

a. 0
b. 1
c. 2
d. more than 2

22 Which of the following is the solution set of equation $\sin^{-1}x=\cos^{-1}x+\sin^{-1}(3x-2)$?

a. $\left\{\frac{1}{2},1\right\}$
b. $\left[\frac{-1}{2},1\right]$
c. $\left[\frac{1}{3},1\right]$
d. $\left\{\frac{-1}{3},1\right\}$

Working Space

23 The solution set of equation $\sin^{-1}\sqrt{1-x^2}+\cos^{-1}x=\cot^{-1}\left(\frac{\sqrt{1-x^2}}{x}\right)-\sin^{-1}x$, is

a. $[-1,1]-\{0\}$
b. $[0,1]\cup\{-1\}$
c. $[-1,0)\cup\{1\}$
d. $[-1,1]$

24 The range of function, $f(x)=(1+\sec^{-1}x)(1+\cos^{-1}x)$ is

a. $(-\infty,\infty)$
b. $(-\infty,0]\cup[4,\infty)$
c. $\{1,(1+\pi)^2\}$
d. None of these

25 Let $\cos^{-1}(x)+\cos^{-1}(2x)+\cos^{-1}(3x)=\pi$. If x satisfies the cubic equation $ax^3+bx^2+cx-1=0$, then $(a+b+c)$ has the value equal to

a. 24
b. 25
c. 26
d. 27

26 The product of all real values of x, satisfying the equation

$$\sin^{-1}\left[\cos\left(\frac{2x^2+10|x|+4}{x^2+5|x|+3}\right)\right]=\cot\left[\cot^{-1}\left(\frac{2-8|x|}{9|x|}\right)\right]+\frac{\pi}{2},\text{ is}$$

a. 9
b. -9
c. -3
d. -1

27 Let $f(x)=\sin x+\cos x+\text{arc}\sin x+\text{arc}\cos x+\text{arc}\tan x$. If M and m are maximum and minimum values of $f(x)$, then their arithmetic mean is equal to

a. $\frac{\pi}{2}+\cos 1$
b. $\frac{\pi}{2}+\sin 1$
c. $\frac{\pi}{4}+\tan 1+\cot 1$
d. None of these

28 There exists a positive real number x, satisfying $\cos(\tan^{-1}x)=x$. Then, the value of $\cos^{-1}\left(\frac{x^2}{2}\right)$ is

a. $\pi/10$
b. $\pi/5$
c. $2\pi/5$
d. $4\pi/5$

One or More than One Options Correct Type

Direction (Q. Nos. 29-32) *This section contains 4 multiple choice questions. Each question has four choices (a), (b), (c) and (d), out of which ONE or MORE THAN ONE are correct.*

29 If $\cos^{-1}x=\tan^{-1}x$, then

a. $x^2=\frac{\sqrt{5}-1}{2}$
b. $x^2=\frac{\sqrt{5}+1}{2}$
c. $\sin(\cos^{-1}x)=\frac{\sqrt{5}-1}{2}$
d. $\tan(\cos^{-1}x)=\frac{\sqrt{5}-1}{2}$

30 If $f(x)=\cos^{-1}x+\cos^{-1}\left\{\frac{x}{2}+\frac{1}{2}\sqrt{3-3x^2}\right\}$, then

a. $f\left(\frac{2}{3}\right)=\frac{\pi}{3}$
b. $f\left(\frac{2}{3}\right)=2\cos^{-1}\left(\frac{2}{3}\right)-\frac{\pi}{3}$
c. $f\left(\frac{1}{3}\right)=\frac{\pi}{3}$
d. $f\left(\frac{1}{3}\right)=2\cos^{-1}\left(\frac{1}{3}\right)-\frac{\pi}{3}$

31 Which of the following is/are correct?

a. $\cos(\cos(\cos^{-1}1))<\sin(\sin^{-1}(\sin(\pi-1)))<\sin(\cos^{-1}(\cos(2\pi-2)))$
b. $\sum_{t=1}^{5000}\cos^{-1}(\cos(2t\pi-1))=\sum_{t=1}^{2500}\cot^{-1}(\cot(t\pi+2)),t\in I$
c. $\cot^{-1}(\cot\,\text{cosec}^{-1}\text{cosec}\,\sec^{-1}\tan\tan^{-1}\cos\cos^{-1}\sin^{-1}\sin 4)=4-\pi$
d. None of the above

Working Space

32 Let x_1, x_2, x_3 and x_4 be four non-zero numbers satisfying the equation

$$\tan^{-1}\left(\frac{a}{x}\right) + \tan^{-1}\left(\frac{b}{x}\right) + \tan^{-1}\left(\frac{c}{x}\right) + \tan^{-1}\left(\frac{d}{x}\right) = \frac{\pi}{2}.$$

Then, which of the following relation(s) hold good?

a. $\sum_{i=1}^{4} x_i = a + b + c + d$

b. $\sum_{i=1}^{4} \frac{1}{x_i} = 0$

c. $\prod_{i=1}^{4} x_i = abcd$

d. $(x_1 + x_2 + x_3)(x_2 + x_3 + x_4)(x_3 + x_4 + x_1)(x_4 + x_1 + x_2) = abcd$

Matching List Type

Direction (Q. Nos. 33-36) *Choices for the correct combination of elements from Column I and Column II are given as options (a), (b), (c) and (d), out of which one is correct.*

33 Let (x, y) be such that $\sin^{-1}(ax) + \cos^{-1}(y) + \cos^{-1}(bxy) = \frac{\pi}{2}$. Match the statements in column I with statements in column II and indicate your answer by marking the appropriate option from given codes. **[IIT JEE 2007]**

	Column I		Column II
i.	If $a = 1$ and $b = 0$, then (x, y)	p.	lies on the circle $x^2 + y^2 = 1$
ii.	If $a = 1$ and $b = 1$, then (x, y)	q.	lies on $(x^2 - 1)(y^2 - 1) = 0$
iii.	If $a = 1$ and $b = 2$, then (x, y)	r.	lies on $y = x$
iv.	If $a = 2$ and $b = 2$, then (x, y)	s.	lies on $(4x^2 - 1)(y^2 - 1) = 0$

Codes

	i	ii	iii	iv
a.	p	q	p	s
b.	p	q	r	s
c.	q	p	s	r
d.	r	p	q	s

34 Match the column I with column II and mark the correct option from the given codes.

	Column I		Column II
i.	$\left[\frac{1}{y^2}\left(\frac{\cos(\tan^{-1} y) + y\sin(\tan^{-1} y)}{\cot(\sin^{-1} y) - \tan(\sin^{-1} y)}\right)^2 + y^4\right]^{1/2}$	p.	$\frac{1}{2}\sqrt{\frac{5}{3}}$
ii.	If $\cos x + \cos y + \cos z = 0 = \sin x + \sin y + \sin z$, then possible value of $\cos\left(\frac{x-y}{2}\right)$ is	q.	$\sqrt{2}$
iii.	If $\cos\left(\frac{\pi}{4} - x\right)\cos 2x + \sin x \sin 2x \sec x = \cos x \sin 2x \cdot \sec x + \cos\left(\frac{\pi}{4} - x\right)\cdot \cos 2x$, then possible value of $\sec x$ is	r.	$\frac{1}{2}$
iv.	If $\cot(\sin^{-1}\sqrt{1-x^2}) = \sin[\tan^{-1}(x\sqrt{6})]$, $x \neq 0$, then possible value of x is	s.	1

Codes

	i	ii	iii	iv
a.	p	q	r	s
b.	s	r	q	p
c.	q	r	p	s
d.	r	p	q	s

Working Space

35 Let $t_1 = (\sin^{-1} x)^{\sin^{-1} x}, t_2 = (\sin^{-1} x)^{\cos^{-1} x}, t_3 = (\cos^{-1} x)^{\sin^{-1} x}$ and $t_4 = (\cos^{-1} x)^{\cos^{-1} x}$

Match the column I with column II and mark the correct option from the given codes.

	Column I		Column II
i.	$x \in (0, \cos 1)$	p.	$t_1 > t_2 > t_4 > t_3$
ii.	$x \in \left(\cos 1, \frac{1}{\sqrt{2}}\right)$	q.	$t_4 > t_3 > t_1 > t_2$
iii.	$x \in \left(\frac{1}{\sqrt{2}}, \sin 1\right)$	r.	$t_1 > t_2 > t_3 > t_4$
iv.	$x \in (\sin 1, 1)$	s.	$t_3 > t_4 > t_1 > t_2$

Codes

	i	ii	iii	iv
a.	p	q	s	r
b.	q	p	p	s
c.	p	p	s	q
d.	q	s	p	p

36 Let the range of $f(x)$ be $[a, b]$. Match the column I with column II and mark the correct option from the given codes.

	Column I $f(x)$		Column II Range $[a, b]$, then $(b - a)$ is
i.	$\frac{1}{\pi^2}[(\cos^{-1} x)^2 - (\sin^{-1} x)^2]$	p.	27/32
ii.	$\frac{1}{\pi^2}[(\cos^{-1} x)^2 + (\sin^{-1} x)^2]$	q.	1
iii.	$\frac{1}{\pi^3}[(\cos^{-1} x)^3 - (\sin^{-1} x)^3]$	r.	9/8
iv.	$\frac{1}{\pi^3}[(\cos^{-1} x)^3 + (\sin^{-1} x)^3]$	s.	5/4

Codes

	i	ii	iii	iv		i	ii	iii	iv
a.	p	q	s	r	b.	r	s	p	q
c.	q	r	s	p	d.	q	r	s	p

Comprehension Type

Direction (Q. Nos. 37-45) This section contains 3 paragraphs, each describing theory, experiments, data, etc. Nine questions related to the paragraphs have been given. Each question has only one correct answer among the four given options (a), (b), (c) and (d).

Passage I

Suppose $f(x) = 3x^3 - 13x^2 + 14x - 2$. It is assumed that $f(x) = 0$ will have 3 roots say α, β and γ, where $\alpha < \beta < \gamma$.

37 $[\alpha]\,[\beta]\,[\gamma]$ (where $[\cdot]$ denotes the GIF) will be in

a. AP
b. GP
c. HP
d. None of these

38 $\lim_{n \to \infty} (\alpha)^{n!} (\beta)^{1/n}$ will be equal to

a. 1
b. e
c. 0
d. None of these

39 The value of $\tan^{-1} \alpha + \tan^{-1} \beta + \tan^{-1} \gamma$ is

a. $\pi/2$
b. $3\pi/2$
c. $\pi/4$
d. $3\pi/4$

Working Space

Passage II

$$\tan^{-1}(\tan\theta)=\begin{cases}\pi+\theta, & -3\pi/2<\theta<-\pi/2,\\ \theta, & -\pi/2<\theta<\pi/2,\\ -\pi+\theta, & \pi/2<\theta<3\pi/2\end{cases}\quad \sin^{-1}(\sin\theta)=\begin{cases}-\pi-\theta, & -3\pi/2\le\theta<-\pi/2\\ \theta, & -\pi/2\le\theta\le\pi/2\\ \pi-\theta, & \pi/2<\theta\le 3\pi/2\end{cases}$$

and $\cos^{-1}(\cos\theta)=\begin{cases}-\theta, & -\pi\le\theta<0\\ \theta, & 0\le\theta\le\pi\\ 2\pi-\theta, & \pi<\theta\le 2\pi\end{cases}$

40 $\cos^{-1}x$ is equal to

a. $\sin^{-1}\sqrt{1-x^2}$, if $-1<x<1$

b. $-\sin^{-1}\sqrt{1-x^2}$, if $-1<x<0$

c. $\sin^{-1}\sqrt{1-x^2}$, if $-1<x<0$

d. $\sin^{-1}\sqrt{1-x^2}$, if $0<x<1$

41 $\sin^{-1}x$ is equal to

a. $\cos^{-1}\sqrt{1-x^2}$, if $-1<x<0$

b. $\cos^{-1}\sqrt{1-x^2}$, if $-1<x<1$

c. $\cos^{-1}\sqrt{1-x^2}$, if $0<x<1$

d. $-\cos^{-1}\sqrt{1-x^2}$, if $0<x<1$

42 $\cos^{-1}x$ is equal to

a. $-\tan^{-1}\left(\frac{\sqrt{1-x^2}}{x}\right)$, if $-1<x<0$

b. $\tan^{-1}\left(\frac{\sqrt{1-x^2}}{x}\right)$, if $-1<x<0$

c. $-\tan^{-1}\left(\frac{\sqrt{1-x^2}}{x}\right)$, if $0<x<1$

d. $\pi+\tan^{-1}\left(\frac{\sqrt{1-x^2}}{x}\right)$, if $-1<x<0$

Passage III

Let $a=\cos^{-1}(\cos 20)$, $b=\cos^{-1}(\cos 30)$ and $c=\sin^{-1}[\sin(a+b)]$, then

43 The maximum value of $\sin[2(a+b+c)x]+\cos^2[(a+b+c)x]$ is

a. $\frac{\sqrt{5}+1}{2}$

b. $\frac{\sqrt{5}-1}{2}$

c. $\frac{\sqrt{3}-1}{2}$

d. None of these

44 The largest integer x for which $\sin^{-1}(\sin x)\ge|x-(a+b+c)|$, is

a. 1 b. 2 c. 3 d. 4

45 If $5\sec^{-1}x+10\sin^{-1}y=10(a+b+c)$, then the value of $\tan^{-1}x+\cos^{-1}(y-1)$ is

a. $\pi/2$ b. $\pi/4$ c. π d. 0

One Integer Value Correct Type

***Direction** (Q. Nos. 46-50) This section contains 5 questions. When worked out will result in an integer from 0 to 9 (both inclusive).*

46 The least value of n for which $(n-2)x^2+8x+(n+4)>\sin^{-1}(\sin 12)+\cos^{-1}(\cos 12)$, $\forall\, x\in R$, where $n\in N$, is

47 For $x, y, z, t\in R$, if $\sin^{-1}x+\cos^{-1}y+\sec^{-1}z\ge t^2-\sqrt{2\pi}\cdot t+3\pi$, then the value of $\tan^{-1}x+\tan^{-1}y+\tan^{-1}z+\tan^{-1}\left(\sqrt{\frac{2}{\pi}}t\right)$ is

48 Let $f(x)=\frac{1}{\pi}(\sin^{-1}x+\cos^{-1}x+\tan^{-1}x)+\frac{(x+1)}{x^2+2x+10}$, if the absolute maximum value $f(x)=M$, then the integral part of $\frac{1}{M}$ is

49 The number of rational numbers in the domain of function $y=\sin^{-1}x+\cos^{-1}x+\tan^{-1}x+\text{cosec}^{-1}x+\sec^{-1}x+\cot^{-1}x$, is

50 If $\sin^{-1}x+\sin^{-1}y+\sin^{-1}z=\frac{3\pi}{2}$, then $A=x^2+y^2+z^2$. If $(\sin^{-1}x)^2+(\sin^{-1}y)^2+(\sin^{-1}z)^2=\frac{3\pi^2}{4}$, then $B=|(x+y+z)_{min}|$. Then, the value of $(A+B)$ is

JEE Main & AIEEE Archive

(Compilation of Last 13 Years' Questions)

1 If x, y and z are in an AP and $\tan^{-1} x$, $\tan^{-1} y$ and $\tan^{-1} z$ are also in an AP, then **(JEE Main 2013)**

a. $x = y = z$
b. $2x = 3y = 6z$
c. $6x = 3y = 2z$
d. $6x = 4y = 3z$

2 The value of $\cot\left\{\sum_{n=1}^{23} \cot^{-1}\left(1 + \sum_{k=1}^{n} 2k\right)\right\}$ is **(JEE Main 2013)**

a. $\frac{23}{25}$
b. $\frac{25}{23}$
c. $\frac{23}{24}$
d. $\frac{24}{23}$

3 The value of $\cot\left(\operatorname{cosec}^{-1}\frac{5}{3} + \tan^{-1}\frac{2}{3}\right)$ is **(AIEEE 2008)**

a. $\frac{5}{17}$
b. $\frac{6}{17}$
c. $\frac{3}{17}$
d. $\frac{4}{17}$

4 If $\sin^{-1}\left(\frac{x}{5}\right) + \operatorname{cosec}^{-1}\left(\frac{5}{4}\right) = \frac{\pi}{2}$, then the value of x is **(AIEEE 2007)**

a. 1
b. 3
c. 4
d. 5

5 If $\cos^{-1} x - \cos^{-1}\frac{y}{2} = \alpha$, then $4x^2 - 4xy\cos\alpha + y^2$ is equal to **(AIEEE 2005)**

a. $-4\sin^2\alpha$
b. $4\sin^2\alpha$
c. 4
d. $2\sin 2\alpha$

6 The trigonometric equation $\sin^{-1} x = 2\sin^{-1} a$ has a solution for **(AIEEE 2003)**

a. $\frac{1}{2} < |a| < \frac{1}{\sqrt{2}}$
b. all real values of a
c. $|a| \le \frac{1}{\sqrt{2}}$
d. $|a| \ge \frac{1}{\sqrt{2}}$

7 $\cot^{-1}(\sqrt{\cos\alpha}) - \tan^{-1}(\sqrt{\cos\alpha}) = x$, then $\sin x$ is equal to **(AIEEE 2002)**

a. $\tan^2\left(\frac{\alpha}{2}\right)$
b. $\cot^2\left(\frac{\alpha}{2}\right)$
c. $\tan\alpha$
d. $\cot\left(\frac{\alpha}{2}\right)$

8 $\tan^{-1}\left(\frac{1}{4}\right) + \tan^{-1}\left(\frac{2}{9}\right)$ is equal to **(AIEEE 2002)**

a. $\frac{1}{2}\cos^{-1}\left(\frac{3}{5}\right)$
b. $\frac{1}{2}\sin^{-1}\left(\frac{3}{5}\right)$
c. $\frac{1}{2}\tan^{-1}\left(\frac{3}{5}\right)$
d. $\tan^{-1}\left(\frac{1}{2}\right)$

JEE Advanced & IIT JEE Archive

(Compilation of Last 10 Years' Questions)

1 If $\alpha = 3\sin^{-1}\left(\frac{6}{11}\right)$ and $\beta = 3\cos^{-1}\left(\frac{4}{9}\right)$, where the inverse trigonometric functions take only the principal values, then the correct option(s) is(are)

a. $\cos\beta > 0$ b. $\sin\beta < 0$
c. $\cos(\alpha + \beta) > 0$ d. $\cos\alpha < 0$

(2015 Adv., One or More than One Options Correct Type)

2 Let $f: [0, 4\pi] \to [0, \pi]$ be defined by $f(x) = \cos^{-1}(\cos x)$. The number of points $x \in [0, 4\pi]$ satisfying the equation $f(x) = \frac{10 - x}{10}$ is

(2015 Adv., Integer Type)

3 Match column I with column II and select the correct answer using the codes given below the list. **(2013 Adv. Matching Type)**

	Column I		Column II
i.	$\left[\frac{1}{y^2}\left\{\frac{\cos(\tan^{-1}y) + y\sin(\tan^{-1}y)}{\cot(\sin^{-1}y) + \tan(\sin^{-1}y)}\right\}^2 + y^4\right]^{1/2}$ takes value	p.	$\frac{1}{2}\sqrt{\frac{5}{3}}$
ii.	If $\cos x + \cos y + \cos z = 0 = \sin x + \sin y + \sin z$, then possible value of $\cos\frac{x-y}{2}$ is	q.	$\sqrt{2}$
iii.	If $\cos\left(\frac{\pi}{4} - x\right)\cos 2x + \sin x \sin 2x \sec x$ $= \cos x \sin 2x \sec x + \cos\left(\frac{\pi}{4} + x\right)\cos 2x$, then possible value of $\sec x$ is	r.	$\frac{1}{2}$
iv.	If $\cot\left(\sin^{-1}\sqrt{1 - x^2}\right) = \sin[\tan^{-1}(x\sqrt{6})]$, $x \neq 0$. Then, possible value of x is	s.	1

Codes

	i	ii	iii	iv
a.	s	r	p	q
b.	s	r	q,s	p
c.	r	s	q	p
d.	r	s	p	q

4 If $0 < x < 1$, then $\sqrt{1 + x^2}\,[\{x\cos(\cot^{-1}x) + \sin(\cot^{-1}x)\}^2 - 1]^{1/2}$ is equal to **(2008, Only One Option Correct Type)**

a. $\frac{x}{\sqrt{1 + x^2}}$ b. x
c. $x\sqrt{1 + x^2}$ d. $\sqrt{1 + x^2}$

5 Let (x, y) be such that $\sin^{-1}(ax) + \cos^{-1}(y) + \cos^{-1}(bxy) = \frac{\pi}{2}$.

(2007, Matching Type)

	Column I		Column II
i.	If $a = 1$ and $b = 0$, then (x, y)	p.	lies on the circle $x^2 + y^2 = 1$
ii.	If $a = 1$ and $b = 1$, then (x, y)	q.	lies on $(x^2 - 1)(y^2 - 1) = 0$
iii.	If $a = 1$ and $b = 2$, then (x, y)	r.	lies on $y = x$
iv.	If $a = 2$ and $b = 2$, then (x, y)	s.	lies on $(4x^2 - 1)(y^2 - 1) = 0$

Codes

	i	ii	iii	iv
a.	r	p	q	q
b.	p	q	p	s
c.	p,s	q	p	r,s
d.	q	r	p,q	r

2. Relations and Functions

DPP
DAILY PRACTICE PROBLEMS

DPP-1 Relations

Only One Option Correct Type

Working Space

Direction (*Q. Nos. 1-35*) *This section contains 10 multiple choice questions. Each question has four choices (a), (b), (c) and (d), out of which ONLY ONE is correct.*

1 If number of elements in sets A and B are m and n respectively, then the number of relations from A to B is

a. 2^{m+n} b. 2^{mn}
c. $m+n$ d. mn

2 Let R be a reflexive relation in a finite set having n elements and let there be m ordered pairs in R. Then,

a. $m \geq n$ b. $m \leq n$
c. $m = n$ d. None of these

3 If R is a relation from a set P to set Q, then

a. $R \subseteq P \times Q$ b. $R \subseteq Q \times P$
c. $R = P \times Q$ d. $R = P \cup Q$

4 If A is a finite set having n elements, then the number of relations which can be defined in A is

a. 2^n b. n^2
c. 2^{n^2} d. n^n

5 For real numbers x and y, we define $x\,R\,y$ iff $x - y + \sqrt{5}$ is an irrational number. The relation R is

a. reflexive b. symmetric
c. transitive d. None of these

6 Let A be the set of first ten natural numbers and let R be a relation in A defined by if and only if $x + 2y = 10$. Which of the following is false?

a. $R = \{(2, 4), (4, 3), (6, 2), (8, 1)\}$
b. Domain of $R = \{2, 4, 6, 8\}$
c. Range of $R = \{1, 2, 3, 4\}$
d. At least one is false

7 Let $A = \{1, 2, 3, 4\}$ and let $R = \{(2, 2)\ (3, 3), (4, 4), (1, 2)\}$ be a relation in A. Then, R is

a. reflexive b. symmetric
c. transitive d. None of these

8 A relation R is defined in the set Z of integers as follows: $x, y \in R$ iff $x + y = 9$. Which of the following is false?

a. $R = \{(0, 3), (0, -3), (3, 0), (-3, 0)\}$
b. Domain of $R = \{-3, 0, 3\}$
c. Range of $R = \{-3, 0, 3\}$
d. At least one is false

Working Space

9 The relation R defined in $A = \{1, 2, 3\}$ by aRb, if $|a^2 - b^2| \le 5$. Which of the following is false?
a. $R = \{(1, 1), (2, 2), (3, 3), (2, 1), (1, 2), (2, 3), (3, 2)\}$
b. $R^{-1} = R$
c. Domain of $R = \{1, 2, 3\}$
d. Range of $R = \{5\}$

10 Let R be a relation in N defined by $R = \{(1 + x, 1 + x^2) : x \le 5, x \in N\}$. Which of the following is false?
a. $R = \{(2, 2), (3, 5), (4, 10), (5, 17), (6, 25)\}$
b. Domain of $R = \{2, 3, 4, 5, 6\}$
c. Range of $R = \{2, 5, 10, 17, 26\}$
d. At least one is false

11 Let $A = \{1, 2, 3\}$, $B = \{1, 3, 5\}$. If relation R from A to B is given by $\{(1, 3), (2, 5), (3, 3)\}$, then R^{-1} is
a. $\{(3, 3), (3, 1), (5, 3)\}$
b. $\{(1, 3), (2, 5), (3, 3)\}$
c. $\{(1, 3), (5, 2)\}$
d. None of these

12 The relation R defined on the set $A = \{1, 2, 3, 4, 5\}$ by $R = \{(x, y) : |x^2 - y^2| < 16\}$ is given by
a. $\{(1, 1), (2, 1), (3, 1), (4, 1), (2, 3)\}$
b. $\{(2, 2), (3, 2), (4, 2), (2, 4)\}$
c. $\{(3, 3), (4, 3), (5, 4), (3, 4)\}$
d. None of these

13 Let R be a relation in N defined by $R = \{(x, y) : x + 2y = 8\}$. The range of R is
a. $\{2, 4, 6\}$
b. $\{1, 2, 3\}$
c. $\{1, 2, 3, 4, 6\}$
d. None of these

14 If $R = \{(x, y): x, y \in Z, x^2 + y^2 \le 4\}$ is a relation in Z, then domain of R is
a. $\{0, 1, 2\}$
b. $\{-2, -1, 0\}$
c. $\{-2, -1, 0, 1, 2\}$
d. None of these

15 If $A = \{1, 2, 3\}$, then the relation $R = \{(2, 3)\}$ in A is
a. symmetric and transitive
b. symmetric
c. transitive
d. not transitive

16 Let X be a family of sets and R be a relation in X, defined by 'A is disjoint from B'. Then, R is
a. reflexive
b. symmetric
c. anti-symmetric
d. transitive

17 A relation R from C to R is defined by $x\ R\ y$ iff $|x| = y$. Which of the following is correct?
a. $(2 + 3i)\ R 13$
b. $3\ R\ (-3)$
c. $(1 + i)\ R 2$
d. $i\ R\ 1$

18 The void relation in a set A is
a. reflexive
b. reflexive and symmetric
c. reflexive and transitive
d. trivially symmetric and transitive

19 If R is a relation from a set A to the set B and S is a relation from B to C, then the relation $S o R$
a. is from C to A
b. is from A to C
c. does not exist
c. None of these

Working Space

20 Let R and S be two equivalence relations in a set A. Then,

a. $R \cup S$ is an equivalence relation in A
b. $R \cap S$ is an equivalence relation in A
c. $R - S$ is an equivalence relation in A
d. None of the above

21 Let a relation R be defined by $R = \{(4, 5), (1, 4), (4, 6), (7, 6), (3, 7)\}$. The relation $R^{-1}oR$ is given by

a. $\{(1, 1), (4, 4), (7, 4), (4, 7), (7, 7)\}$
b. $\{(1, 1), (4, 4), (4, 7), (7, 4), (7, 7), (3, 3)\}$
c. $\{(1, 5), (1, 6), (3, 6)\}$
d. None of the above

22 Which of the following is not an equivalence relation in Z?

a. $aRb \Leftrightarrow a + b$ is an even integer
b. $aRb \Leftrightarrow a - b$ is an even integer
c. $aRb \Leftrightarrow a < b$
d. $aRb \Leftrightarrow a = b$

23 Let $A = \{p, q, r\}$. Which of the following is an equivalence relation in A?

a. $R_1 = \{(p, q), (q, r), (p, r), (p, p)\}$
b. $R_2 = \{(r, q), (r, p), (r, r), (q, q)\}$
c. $R_3 = \{(p, p), (q, q), (r, r), (p, q)\}$
d. None of the above

24 Let R and S be two non-void relations in a set A. Which of the following statements is false ?

a. R and S transitive $\Rightarrow R \cup S$ is transitive
b. R and S transitive $\Rightarrow R \cap S$ is transitive
c. R and S symmetric $\Rightarrow R \cup S$ is symmetric
d. R and S reflexive $\Rightarrow R \cap S$ is reflexive

25 Let N denotes the set of natural numbers and R is a relation in $N \times N$. Which of the following is not an equivalence relation in $N \times N$?

a. $(a, b) R (c, d) \Leftrightarrow a + d = b + c$
b. $(a, b) R (c, d) \Leftrightarrow ad = bc$
c. $(a, b) R (c, d) \Leftrightarrow ad(b + c) = bc(a + d)$
d. $(a, b) R (c, d) \Leftrightarrow bc(b + c) = ad(a + d)$

26 If $R \subseteq A \times B$ and $A \subseteq B \times C$ be two relations, then $(SoR)^{-1}$ is equal to

a. $S^{-1}oR^{-1}$ b. RoS
c. $R^{-1}oS^{-1}$ d. None of these

27 Let A be the non-empty set of children in a family. The relation 'x is a brother of y' in A is

a. reflexive b. symmetric
c. transitive d. an equivalence relation

28 R is a relation defined in Z by aRb if and only if , then R is

a. reflexive b. symmetric
c. transitive d. equivalence

29 Let $A = Z \cup \{\sqrt{2}\}$. Define a relation R in A by aRb if and only if $a + b \in z$. The relation R is

a. reflexive b. symmetric and transitive
c. only transitive d. None of these

Working Space

30 The relation "congruence modulo m" is

a. reflexive only
b. transitive only
c. symmetric only
d. an equivalence relation

31 R is a relation over the set of integers and it is given by $(x, y) \in R \Leftrightarrow |x - y| \le 1$. Then, R is

a. reflexive and transitive
b. reflexive and symmetric
c. symmetric and transitive
d. an equivalence relation

32 Let r be a relation over the set $N \times N$ and it is defined by $(a, b)\, r\, (c, d) \Rightarrow a + d = b + c$. Then, r is

a. reflexive
b. symmetric
c. transitive
d. In equivalence relation

33 Let R be the relation over the set of straight lines of a plane, such that $l_1 R l_2 \Leftrightarrow l_1 \perp l_2$. Then, R is

a. symmetric
b. reflexive
c. transitive
d. an equivalence relation

34 Let R be the relation over the set of integers, such that $m R n$ if and only if m is a multiple of n. Then, R is

a. reflexive and transitive
b. symmetric
c. transitive
d. an equivalence relation

35 Let $A = \{1, 2, 3, 4\}$ and R be a relation in A given by $R = \{(1, 1), (2, 2), (3, 3), (4, 4), (1, 2), (2, 1), (3, 1), (1, 3)\}$. Then, R is

a. reflexive
b. symmetric
c. transitive
d. an equivalence relation

DPP-2 Definition of Functions

Subjective Questions

Working Space

Direction (Q. Nos. 1-4) *These questions are subjective in nature, need to be solved completely on notebook.*

1 Let $A = \{a, b\}$ and $B = \{\alpha, \beta, \gamma\}$. Draw pictorial diagrams of all possible functions from set A to set B.

2 Which one of the following graphs are graphs of function.

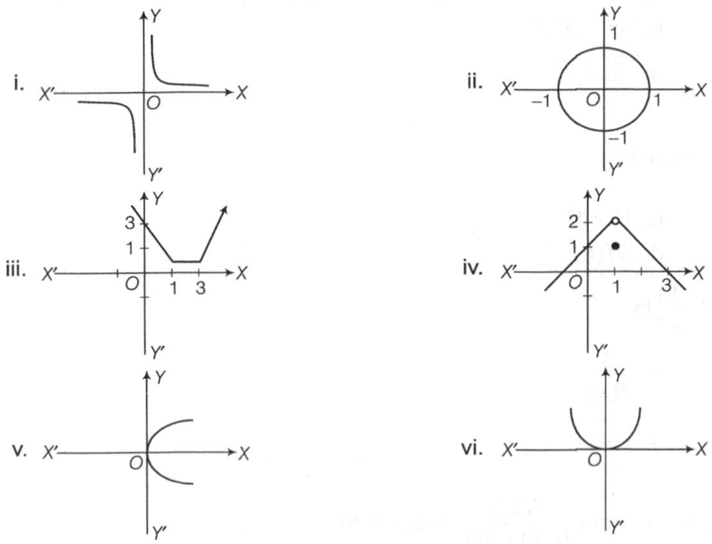

3 If $f(x) = \dfrac{4^x}{4^x + 2}$, then show that $f(x) + f(1 - x) = 1$.

4 If S is set of all triangles and $f : S \to R^+$, $f(\Delta) =$ area of triangle, then f is a function or not.

Only One Option Correct Type

Direction (Q. Nos. 5-12) *This section contains 8 multiple choice questions. Each question has four choices (a), (b), (c) and (d), out of which ONLY ONE is correct.*

5 If $f(1) = 1$, $f(n + 1) = 2f(n) + 1$ and $n \geq 1$, then $f(n)$ is equal to

a. $2^n + 1$ b. 2^n

c. $2^n - 1$ d. $2^{n-1} - 1$

Working Space

6 If $y = f(x)$ satisfy the condition $f\left(x+\frac{1}{x}\right) = x^2 + \frac{1}{x^2}$ $(x \neq 0)$, then $f(x)$ is

a. $-x^2 + 2$
b. $-x^2 - 2$
c. $x^2 - 2,\ x \in R - \{0\}$
d. $x^2 - 2,\ |x| \in [2, \infty)$

7 If $f(x) = \cos(\log_e x)$, then the value of $f(x), f(y) - \frac{1}{2}\left[f\left(\frac{x}{y}\right) + f(xy)\right]$ is

a. -1
b. $\frac{1}{2}$
c. -2
d. None of these

8 If $f(x) = \frac{ax - c}{cx - a} = y$, then $f(y)$ equals

a. x
b. $\frac{1}{x}$
c. 1
d. 0

9 If $f(x + ay, x - ay) = axy$, then $f(x, y)$ is equal to

a. xy
b. $x^2 - a^2y^2$
c. $\frac{x^2 - y^2}{4}$
d. $\frac{x^2 - y^2}{a^2}$

10 If $f(x) = \log\left(\frac{1+x}{1-x}\right)$, then $f\left(\frac{2x}{1+x^2}\right)$ is equal to

a. $[f(x)]^2$
b. $[f(x)]^3$
c. $2f(x)$
d. $3f(x)$

11 If $f(x) = 4x^3 + 3x^2 + 3x + 4$, then $x^3 f\left(\frac{1}{x}\right)$ is

a. $f(-x)$
b. $\frac{1}{f(x)}$
c. $\left[f\left(\frac{1}{x}\right)\right]^2$
d. $f(x)$

12 If function $f : N \to N$ is defined by $f(x) = \sqrt{x}$, then $\frac{f(25)}{f(16) + f(1)}$ is equal to

a. $\frac{5}{6}$
b. $\frac{5}{7}$
c. $\frac{5}{3}$
d. 1

Comprehension Type

***Direction** (Q. Nos. 13-15) This section contains a paragraph, each describing theory, experiments, data, etc. There questions related to the paragraph have been given. Each question has only one correct answer among the four given options (a), (b), (c) and (d).*

Passage

Let $f : N \times N \to N$ be a function such that $f(1, 1) = 2$, $f(\alpha + 1, \beta) = f(\alpha, \beta) + \alpha$ and $f(\alpha, \beta + 1) = f(\alpha, \beta) - \beta$, $\forall\ \alpha, \beta \in N$ and $f(a, b) = 2001$.

13 The number of ordered pairs (a, b) is

a. 1
b. 2
c. 3
d. infinite

Working Space

14 The maximum value of $(a+b)$ is

a. 4000
b. 3999
c. 2999
d. 2000

15 The minimum value of $(a+b)$ is

a. 2999
b. 1999
c. 2000
d. 1000

One Integer Value Correct Type

Direction (Q. No. 16-20) *This section contains 5 questions. When worked out will result in an integer from 0 to 9 (both inclusive).*

16 If $f(x+y)=f(x)\cdot f(y)$, $\forall x, y \in N$ and $f(1)=2$, then $|\sum_{n=1}^{10} f(n) - 2050|$ is

17 Let $f(x)$ be a polynomial function satisfying the relation $f(x)\cdot f\left(\frac{1}{x}\right)=f(x)+f\left(\frac{1}{x}\right)$, $\forall x \in R-\{0\}$ and $f(3)=-26$, then $|f(2)|$ is

18 Let f be a real valued function such that $f(x)+f\left(\frac{2002}{x}\right)=3x$, $\forall x>0$. The value of $\left(\frac{f(2)}{2000}\right)$ is

19 Let f be a function satisfying $f(xy)=\frac{f(x)}{y}$ for all positive real numbers x and y. If $f(30)=20$, then the value of $|f(40)-10|$ is

20 If $f(x, y)=\{\max(x, y)\}^{\min(x, y)}$ and $g(x, y)=\max(x, y)-\min(x, y)$, then

$$[f\{g(-1, -3/2), g(-4, -1\cdot 75)\}] \text{ equals,}$$

(where [·] denotes greatest integer function)

DPP-3 Domain

Subjective Questions

Working Space

Direction *(Q. Nos. 1 and 2) These questions are subjective in nature, need to be solved completely on notebook.*

1 Find the domain of the following functions.

i. $f(x)=\dfrac{x^3-5x+3}{x^2-1}$ ii. $f(x)=\dfrac{\sin^{-1}x}{x}$

iii. $f(x)=\dfrac{1}{\sqrt{x+|x|}}$ iv. $f(x)=e^{x+\sin x}$

v. $f(x)=\dfrac{1}{\log_{10}(1-x)}+\sqrt{x+2}$ vi. $f(x)=\sqrt{1-2x}+3\sin^{-1}\left(\dfrac{3x-1}{2}\right)$

vii. $f(x)=2^{\sin^{-1}x}+\dfrac{1}{\sqrt{x-2}}$ viii. $f(x)=\log_x\log_2\left(\dfrac{1}{x-\dfrac{1}{2}}\right)$

2 Find the domain of definitions of the following functions.

i. $f(x)=\sqrt{3-2^x-2^{1-x}}$ ii. $f(x)=\sqrt{1-\sqrt{1-x^2}}$

iii. $f(x)=(x^2+x+1)^{-3/2}$ iv. $f(x)=\sqrt{\dfrac{x-2}{x+2}}+\sqrt{\dfrac{1-x}{1+x}}$

v. $f(x)=\sqrt{\tan x-\tan^2 x}$ vi. $f(x)=\dfrac{1}{\sqrt{1-\cos x}}$

vii. $f(x)=\sqrt{\log_{1/4}\left(\dfrac{5x-x^2}{4}\right)}$ viii. $f(x)=\log_{10}[1-\log_{10}(x^2-5x+16)]$

Only One Option Correct Type

Direction *(Q. Nos. 3-18) This section contains 16 multiple choice questions. Each question has four choices (a), (b), (c) and (d), out of which ONLY ONE is correct.*

3 The domain of real valued function $f(x)=\sqrt{5-4x-x^2}+x^2\log(x+4)$ is

a. $-5\le x\le 1$ b. $-5\le x$ and $x\ge 1$

c. $-4<x\le 1$ d. ϕ

4 The domain of $f(x)=\sin^{-1}5x$ is

a. $\left(-\dfrac{1}{5},\dfrac{1}{5}\right)$ b. $\left[-\dfrac{1}{5},\dfrac{1}{5}\right]$

c. R d. $\left(0,\dfrac{1}{5}\right)$

5 The domain of $f(x)=\dfrac{\sin^{-1}(3-x)}{\log(|x|-2)}$ is

a. $[2,4]$ b. $(2,\ 3)\cup(3,\ 4]$

c. $[2,\ \infty)$ d. None of these

Working Space

6 The domain of $\sin^{-1}\left[\log_3\left(\frac{x}{3}\right)\right]$ is **[AIEEE 2003]**

a. [1, 9] b. [−1, 9] c. [−9, 1] d. [−9, 1]

7 The domain of $f(x) = \frac{\log_2(x+3)}{x^2 + 3x + 2}$ is **[IIT 2001]**

a. $R - \{-1, -2\}$
b. $(-2, \infty)$
c. $R - \{-1, -2, -3\}$
d. $(-3, \infty) - \{-1, -2\}$

8 The domain of $f(x) = \sqrt{\log \frac{1}{|\sin x|}}$ is

a. $R - \{2n\pi, n \in I\}$ b. $R - \{n\pi, n \in I\}$ c. $R - \{-\pi, \pi\}$ d. $(-\infty, \infty)$

9 The domain of $f(x) = \log(\sqrt{x-4} + \sqrt{6-x})$ is

a. $[4, \infty)$ b. $(-\infty, 6]$ c. $[4, 6]$ d. None of these

10 The domain of $f(x) = \log_{x+3}(x^2 - 1)$ is

a. $(-3, -1) \cup (1, \infty)$
b. $[-3, -1) \cup [1, \infty)$
c. $(-3, -2) \cup (-2, -1) \cup (1, \infty)$
d. None of these

11 The domain of $f(x) = \sqrt{x - x^2} + \sqrt{4 + x} + \sqrt{4 - x}$ is

a. $[-4, \infty)$ b. $[-4, 4]$ c. $[0, 4]$ d. $[0, 1]$

12 The domain of $f(x) = \sqrt{\log(x^2 - 6x + 6)}$ is

a. $(-\infty, \infty)$
b. $[3 - \sqrt{3}, 3 + \sqrt{3}]$
c. $(-\infty, 1] \cup [5, \infty)$
d. $[0, \infty)$

13 The domain of $f(x) = \sin^{-1}(|x - 1| - 2)$ is

a. $[-2, 0] \cup [2, 4]$ b. $(-2, 0) \cup (2, 4)$ c. $[-2, 0] \cup [1, 3]$ d. None of these

14 The domain of $f(x) = \frac{\sqrt{-\log_{0.3}(x-1)}}{\sqrt{x^2 + 2x + 8}}$ is

a. (1, 4) b. (−2, 4) c. (2, 4) d. $[2, \infty)$

15 The function $f(x) = \cot^{-1}\sqrt{(x+3)x} + \cos^{-1}\sqrt{x^2 + 3x + 1}$ is defined on S, where S is equal to

a. {0, 3} b. (0, 3) c. {−3, 0} d. [−3, 0]

16 The domain of definition of function $f(x) = \sqrt{\sin^{-1}(2x) + \frac{\pi}{6}}$ for real values of x, is **[IIT JEE 2003]**

a. $\left[-\frac{1}{4}, \frac{1}{2}\right]$ b. $\left[-\frac{1}{2}, \frac{1}{2}\right]$ c. $\left(-\frac{1}{2}, \frac{1}{4}\right)$ d. $\left[-\frac{1}{4}, \frac{1}{4}\right]$

17 The domain of definition of function, $y(x)$ given by the equation $2^x + 2^y = 2$, is **[IIT 2000]**

a. $0 < x \le 1$ b. $0 \le x \le 1$ c. $0 < x < 1$ d. $-\infty < x < 1$

18 Let $f : (-1, 1) \to IR$ be such that $f(\cos 4\theta) = \frac{2}{2 - \sec^2\theta}$ for $\theta \in \left(0, \frac{\pi}{4}\right) \cup \left(\frac{\pi}{4}, \frac{\pi}{2}\right)$, then the value(s) of $f\left(\frac{1}{3}\right)$ is (are) **[IIT 2012]**

a. $2 + \sqrt{\frac{3}{2}}$ b. $1 + \sqrt{\frac{3}{2}}$ c. $1 - \sqrt{\frac{2}{3}}$ d. $1 + \sqrt{\frac{2}{3}}$

Working Space

One or More than One Options Correct Type

Direction (Q. Nos. 19 and 20) This section contains 2 multiple choice questions. Each question has four choices (a), (b), (c) and (d), out of which ONE or MORE THAN ONE are correct.

19 Which of the following is/are transcendental?

a. $f(x) = 5\sin\sqrt{x}$

b. $f(x) = \dfrac{2\sin 3x}{x^2 + 2x - 1}$

c. $f(x) = \sqrt{x^2 + 2x + 1}$

d. $f(x) = (x^2 + 2)\cdot 2^x$

20 The domain of definition of function,

$$f(x) = [2\tan \pi x]\cdot^{\log_{[2\tan\pi x]}\left(\frac{x^2+2x-3}{4x^2-4x-3}\right)}$$

where [·] denotes the greatest integer function is given by $\left[n + \dfrac{1}{4}, n + \dfrac{1}{2}\right)$; where $n \in I$, then n can be equal to

a. −5 b. −1 c. 0 d. 1

Comprehension Type

Direction (Q. Nos. 21-23) This section contains a paragraph, describing theory, experiments, data, etc. Three questions related to the paragraph have been given. Each question has only one correct answer among the four given options (a), (b), (c) and (d).

Passage

If $\{f(x)\}^2 \times f\left(\dfrac{1-x}{1+x}\right) = 64x,\ x \neq 0, 1$, then

21 $f(x)$ is equal to

a. $4x^{2/3}\left(\dfrac{1+x}{1-x}\right)^{\frac{1}{3}}$

b. $x^{1/3}\left(\dfrac{1-x}{1+x}\right)^{\frac{1}{3}}$

c. $x^{2/3}\left(\dfrac{1-x}{1+x}\right)^{\frac{2}{3}}$

d. None of these

22 The domain of $f(x)$ is

a. $[0, \infty)$

b. $R - \{-1, 1\}$

c. $(-\infty, \infty)$

d. None of these

23 The value of $f\left(\dfrac{9}{7}\right)$ is

a. $8\left(\dfrac{7}{4}\right)^{2/3}$ b. $4\left(\dfrac{9}{7}\right)^{1/3}$ c. $-8\left(\dfrac{9}{7}\right)^{2/3}$ d. None of these

One Integer Value Correct Type

Direction (Q. Nos. 24 and 25) This section contains 2 questions. When worked out will result in an integer from 0 to 9 (both inclusive).

24 Let a and b be constants and $f(x) = a\sin x + bx\cos x + 2x^2$. If $f(2) = 15$, then the value of $f(-2)$ is

25 If $f(x) = [x] + \sum_{r=1}^{2015} \dfrac{\{x + r\}}{2015}$, then the value of $f(3)$ is

DPP-4 Range of Functions

Subjective Questions

Direction (Q. Nos. 1 and 2) *These questions are subjective in nature, need to be solved completely on notebook.*

Working Space

1 Find the range of the following functions.

i. $f(x) = |x - 3|$

ii. $f(x) = \dfrac{x}{1 + x^2}$

iii. $f(x) = \sqrt{16 - x^2}$

iv. $f(x) = \dfrac{|x - 4|}{x - 4}$

v. $f(x) = 5 + \sin x + \cos x$

vi. $f(x) = \dfrac{1}{1 + \sqrt{x}}$

2 Find the domain and range of each of the following functions.

i. $f(x) = \dfrac{1}{\sqrt{4 + 3\sin x}}$

ii. $f(x) = x!$

iii. $f(x) = \dfrac{x^2 - 9}{x - 3}$

iv. $f(x) = \sin^2(x^3) + \cos^2(x^3)$

Only One Option Correct Type

Direction (Q. Nos. 3-20) *This section contains 18 multiple choice questions. Each question has four choices (a), (b), (c) and (d), out of which ONLY ONE option is correct.*

3 The range of $f(x) = \sec\left(\dfrac{\pi}{4}\cos^2 x\right)$, $x \in R$ is

a. $[1, \sqrt{2}]$
b. $[1, \infty)$
c. $[-\sqrt{2}, -1] \cup [1, \sqrt{2}]$
d. R

4 The range of $f(x) = \dfrac{x^2 + x + 2}{x^2 + x + 1}$, $x \in R$ is **[IIT JEE 2003]**

a. $(1, \infty)$
b. $\left(1, \dfrac{11}{7}\right)$
c. $\left(1, \dfrac{7}{3}\right]$
d. $\left(1, \dfrac{7}{5}\right]$

5 If $f(x) = a\cos(bx + c) + d$, then the range of $f(x)$ is

a. $[d + a, d + 2a]$
b. $[a - d, a + d]$
c. $[d + a, a - d]$
d. $[d - a, d + a]$

6 The range of $f(x) = [x] - x$ is

a. $[0, 1]$
b. $(-1, 0]$
c. R
d. $(-1, 1)$

7 The range of $f(x) = \cos\left(\dfrac{x}{3}\right)$ is

a. $\left(-\dfrac{1}{3}, \dfrac{1}{3}\right)$
b. $[-1, 1]$
c. R
d. $(-3, 3)$

Working Space

8 The range of $f(x)=\dfrac{x+2}{|x+2|}$ is

a. $\{0, 1\}$
b. $\{-1, 1\}$
c. R
d. $R-\{-2\}$

9 The range of $f(x)=\cos 2x-\sin 2x$ is

a. $[2, 4]$ b. $[-\sqrt{2}, \sqrt{2}]$ c. $[-2, 2]$ d. $[-4, 4]$

10 The range of $f(x)=\cos x-\sin x$ is

a. $(-1, 1)$ b. $[-1, 1)$ c. $\left[-\dfrac{\pi}{2}, \dfrac{\pi}{2}\right]$ d. $[-\sqrt{2}, \sqrt{2}]$

11 The range of $f(x)=\dfrac{1}{2-\sin 3x}$ is

a. $[1, 3]$ b. $\left[\dfrac{1}{3}, 1\right]$ c. $(1, 3)$ d. $\left(\dfrac{1}{3}, 1\right)$

12 If $x\in R$, then the value of expression $\dfrac{x^2+14x+9}{x^2+2x+3}$ lies between

a. 5 and 4
b. -5 and 4
c. $(-\infty, -4)\cup(5, \infty)$
d. None of these

13 For $\theta>\dfrac{\pi}{3}$, then value of $f(\theta)=\sec^2\theta+\cos^2\theta$ always lies in the interval

a. $[0, 2)$ b. $[0,1]$ c. $(1, 2)$ d. $[2, \infty)$

14 The range of $f(x)=\log_e\sqrt{4-x^2}$ is

a. $(0, \infty)$ b. $(-\infty, \infty)$ c. $(-\infty, \log_e 2]$ d. $(\log_e 2, \infty)$

15 The range of $f(x)=\tan\sqrt{\dfrac{\pi^2}{9}-x^2}$ is

a. $[0, 3]$ b. $[0, \sqrt{3}]$ c. $[3, 5]$ d. $[\sqrt{3}, 3]$

16 The range of $f(x)=\log_e(3x^2-4x+5)$ is

a. $\left[\log_e\dfrac{11}{3}, \infty\right)$
b. $[\log_e 10, \infty)$
c. $\left[\log_e\dfrac{11}{6}, \infty\right)$
d. None of these

17 The range of $f(x)=4^x+2^x+1$ is

a. $(0, \infty)$ b. $(1, \infty)$ c. $(2, \infty)$ d. $(3, \infty)$

18 The range of $f(x)=\log_{\sqrt{5}}\{\sqrt{2}(\sin x-\cos x)+3\}$ is

a. $[0, 1]$
b. $[0, 2]$
c. $\left[0, \dfrac{3}{2}\right]$
d. None of these

19 The range of $f(x)={}^{7-x}P_{x-3}$ is **[AIEEE 2004]**

a. $\{1, 2, 3, 4\}$
b. $\{1, 2, 3, 4, 5, 6\}$
c. $\{1, 2, 3\}$
d. $\{1, 2, 3, 4\}$

20 The range of $f(x)=\dfrac{x}{1+|x|}$, $x\in R$ is **[AIEEE 2012]**

a. $[-1, 1]$ b. R c. $R-\{0\}$ d. $(-1, 1)$

Working Space

One or More than One Options Correct Type

Direction (Q. Nos. 21 and 22) *This section contains 2 multiple choice questions. Each question has four choices (a), (b), (c) and (d), out of which ONE or MORE THAN ONE are correct.*

21 Let $f(x) = |x-1| + |x-2| + |x-3| + |x-4|$, then

a. the least value of $f(x)$ is 4
b. the least value is not attained at a unique point
c. the number of integral solution of $f(x) = 4$ is 2
d. the least value of $f(x)$ is 2

22 Let $f(x) = \sec^{-1}[1+\cos^2 x]$, where [·] denotes the greatest integer function, then

a. the domain of f is R
b. the domain of f is [1, 2]
c. the range of f is [1, 2]
d. the range of f is $\{\sec^{-1}1, \sec^{-1}2\}$

Comprehension Type

Direction (Q. Nos. 23-25) *This section contains a paragraph which describing theory, experiments, data, etc. Three questions related to the paragraph have been given. Each question has only one correct answer among the four given options (a), (b), (c) and (d).*

Passage

A tower has the shape as : A truncated right circular cone (one with radii $2R$ (the lower base) and R (the upper base) and the height R bears a right circular cylinder whose radius is R, the height being $2R$. Finally, a semisphere of radius R is mounted on the cylinder, suppose that the cross-sectional area S of the tower is given by $f(x)$, where x is the distance of the cross-section from the lower base of the cone.

23 The domain of $f(x)$ is

a. $(-\infty, \infty)$
b. $[0, 4R]$
c. $[0, R]$
d. $[R, 4R]$

24 For $0 \le x \le R$, the function $f(x)$ is given by

a. $\pi(2R-x)^2$
b. $\pi(R-x)^2$
c. πR^2
d. $4\pi x^2$

25 The range of $f(x)$ is

a. $[0, 4\pi R^2]$
b. $[\pi R^2, 4\pi R^2]$
c. $[0, \pi R^2]$
d. None of these

One Integer Value Correct Type

Direction (Q. Nos. 26-30) *This section contains 5 questions. When worked out will result in an integer from 0 to 9 (both inclusive).*

26 The range of $f(x) = \log_e(\sin^{-1} x)$ is $\left(-\infty, \log_e \frac{\pi}{k}\right)$, then the value of k is

27 The range of $f(x) = 2 - 3x - 5x^2$ is $(-\infty, \lambda)$. Then, the greatest natural number satisfying λ is

28 The range of $f(x) = 3|\sin x| - 4|\cos x|$ is $[a, b]$, then the value of $|a+b|$ is

29 The range of $f(x) = \frac{\sin x}{\sqrt{1+\tan^2 x}} + \frac{\cos x}{\sqrt{1+\cot^2 x}}$ is $[a, b]$, then the value of $|a+b|$ is

30 The range of $f(x) = \log_{\sqrt{2}}(2 - \log_2(16\sin^2 x + 1))$ is $(-\infty, a]$, then the value of a is

DPP-5 Identical, Even and Odd Functions

Only One Option Correct Type

Direction *(Q. Nos. 1-4) This section contains 4 multiple choice questions. Each question has four choices (a), (b), (c) and (d), out of which ONLY ONE is correct.*

Working Space

1 Which of the following pairs of functions are identical?

a. $f(x) = \sin^{-1} x + \cos^{-1} x$ and $g(x) = \dfrac{\pi}{2}$

b. $f(x) = \tan^{-1} x + \cot^{-1} x$ and $g(x) = \dfrac{\pi}{2}$

c. $f(x) = \sec^{-1} x + \operatorname{cosec}^{-1} x$ and $g(x) = \dfrac{\pi}{2}$

d. All of the above

2 Which of the following pairs of functions are identical?

a. $f(x) = e^{\log_e (\sec^{-1} x)}$ and $g(x) = \sec^{-1} x$

b. $f(x) = \tan(\tan^{-1} x)$ and $g(x) = \cot(\cot^{-1} x)$

c. $f(x) = \operatorname{sgn}(x)$ and $g(x) = \operatorname{sgn}(\operatorname{sgn}(x))$

d. $f(x) = \cot^2 x \cdot \cos^2 x$ and $g(x) = \cot^2 x - \cos^2 x$

3 Which of the following pairs of functions are identical?

a. $f(x) = \sqrt{x^2}$ and $g(x) = (\sqrt{x})^2$

b. $f(x) = \sec(\sec^{-1} x)$ and $g(x) = \operatorname{cosec}(\operatorname{cosec}^{-1} x)$

c. $f(x) = \sqrt{\dfrac{1 + \cos 2x}{2}}$ and $g(x) = \cos x$

d. $f(x) = x$ and $g(x) = e^{\log_e x}$

4 Which of the following pairs of functions are identical?

a. $f(x) = \sqrt{1 + \sin x}$ and $g(x) = \sin \dfrac{x}{2} + \cos \dfrac{x}{2}$

b. $f(x) = \sin^{-1}\left(\dfrac{2x}{1 + x^2}\right)$ and $g(x) = 2\tan^{-1} x$

c. $f(x) = \sqrt{x^2}$ and $g(x) = (\sqrt{x})^2$

d. $f(x) = \log_e x^3 + \log_e x^5$ and $g(x) = 8\log_e x$

Subjective Questions

Direction (Q. No. 5) *This question is subjective in nature, need to be solved completely on notebook.*

5 Determine whether the following functions are even or odd or neither even nor odd.

i. $\tan x$

ii. $\cos x$

iii. $\sin(x^2 + 1)$

iv. $x + x^2$

v. $x - x^3$

vi. $f(x) = x \cdot \left(\frac{a^x - 1}{a^x + 1}\right)$

vii. $f(x) = \log(x + \sqrt{x^2 + 1})$

viii. $f(x) = \sin x + \cos x$

ix. $f(x) = (x^2 - 1)|x|$

x. $f(x) = \frac{(1 + 2^x)^7}{2^x}$

xi. $f(x) = \frac{\sec x + x^2 - 9}{x \sin x}$

xii. $f(x) = \sqrt{1 + x + x^2} - \sqrt{1 - x + x^2}$

xiii. $f(x) = \begin{cases} x\,|x|, & x \le -1 \\ [1 + x] + [1 - x], & -1 < x < 1 \\ -x\,|x|, & x \ge 1 \end{cases}$

xiv. $f(x) = \frac{2x(\sin x + \tan x)}{2\left[\frac{x + 2\pi}{\pi}\right] - 3}$

DPP-6 Periodicity of Functions

Only One Option Correct Type

Working Space

Direction (*Q. Nos. 1-15*) *This section contains 15 multiple choice questions. Each question has four choices (a), (b), (c) and (d), out of which ONLY ONE is correct.*

1 The period of $f(x) = x - [x]$, where $[\cdot]$ denotes greatest integer function is

a. π b. 2
c. 1 d. None of these

2 The period of $f(x) = \sin^2 x$ is

a. $\frac{\pi}{2}$ b. π
c. 2π d. None of these

3 The period of $f(x) = \sin^{-1}(\sin x)$ is

a. π b. 2π
c. $\frac{\pi}{2}$ d. None of these

4 The period of $f(x) = |\cos x|$ is

a. $\frac{\pi}{2}$ b. π
c. $\frac{3\pi}{2}$ d. 2π

5 The period of $f(x) = |\cos 2x|$ is

a. $\frac{\pi}{2}$ b. π
c. 2π d. None of these

6 The period of $f(x) = \tan 4x$ is

a. $\frac{\pi}{4}$ b. $\frac{\pi}{2}$
c. π d. $\frac{3\pi}{2}$

7 The period of $f(x) = \tan(3x + 2)$ is

a. π b. $\frac{2\pi}{3}$
c. $\frac{\pi}{3}$ d. None of these

8 The period of $f(x) = \sin\left(\frac{x}{3}\right) + \cos\left(\frac{x}{2}\right)$ is

a. 2π b. 3π
c. 6π d. 12π

Working Space

9 The period of $f(x) = |\sin x| + |\cos x|$ is

a. $\frac{\pi}{2}$ b. π
c. 2π d. None of these

10 The period of $f(x) = \sin^4 x + \cos^4 x$ is

a. π b. $\frac{\pi}{2}$
c. 2π d. None of these

11 The period of $f(x) = \sec(\sin x)$ is

a. $\frac{\pi}{2}$ b. 2π
c. π d. None of these

12 If $f(x) = \sin\sqrt{[a]}x$, where [·] denotes the greatest integer function, has π as its period, then

a. $a = 1$ b. $a = 9$
c. $a \in [1, 2)$ d. $a \in [4, 5)$

13 The fundamental period of $f(x) = x + a - [x + b] + \sin \pi x + \cos 2\pi x + \sin 3\pi x + \cos 4\pi x + \cdots + \sin(2n - 1)\pi x + \cos 2n\pi x$ for every $a, b \in R$ (where, [·] denotes greatest integer function), is

a. 2 b. 4
c. 1 d. 3

14 The period of $f(x) = \sin\frac{\pi}{4}[x] + \cos\frac{\pi x}{2} + \cos\frac{\pi}{3}[x]$, where, [·] denotes greatest integer function, is

a. 8 b. 12
c. 24 d. None of these

15 The period of function $f(x) = [x] + \left[x + \frac{1}{3}\right] + \left[x + \frac{2}{3}\right] - 3x + 15$ is

a. $\frac{1}{3}$ b. $\frac{2}{3}$
c. 1 d. non-periodic

One or More than One Option Correct Type

***Direction** (Q. Nos. 16-20) This section contains 5 multiple choice questions. Each question has four choices (a), (b), (c) and (d), out of which ONE or MORE THAN ONE are correct.*

16 The period of $f(x) = \sin^4 3x + \cos^4 3x$ is

a. $\pi/6$ b. $\pi/3$
c. $\pi/2$ d. $\pi/12$

17 The functions which are aperiodic, are

a. $y = [x + 1]$ b. $y = \sin(x^2)$
c. $y = \sin^2 x$ d. $y = \sin^{-1} x$

18 If $f : R \to [-1, 1]$, where $f(x) = \sin\frac{\pi}{2}[x]$ (where, [·] denotes greatest integer function), then

a. $f(x)$ is even b. $f(x)$ is odd
c. $f(x)$ is periodic d. None of these

Working Space

19 Identify the statement(s) which is/are incorrect?

a. The function $f(x)=\cos(\cos^{-1}x)$ is neither odd nor even

b. The fundamental period of $f(x)=\cos(\sin x)+\cos(\cos x)$ is π

c. The range of the function $f(x)=\cos(3\sin x)$ is $[-1, 1]$

d. None of the above

20 If $f(x)=\dfrac{\sin\pi[x]}{\{x\}}$, then $f(x)$ is

a. periodic with fundamental period 1
b. even
c. range is singleton
d. None of these

Matching List Type

Direction *(Q. No. 21) Choices for the correct combination of elements from Column I and Column II are given as options (a), (b), (c) and (d), out of which one is correct.*

21 Match the column I and column II and mark the correct option from the options given below.

	Column I		Column II
i.	The period of function, $f(x)=\sin\left(2\pi t+\frac{\pi}{3}\right)+2\sin\left(3\pi t+\frac{\pi}{4}\right)+3\sin 5\pi t$	p.	2
ii.	The fundamental period of the function, $f(x)=\frac{1}{2}\left(\frac{\lvert\sin(\pi/4)x\rvert}{\cos(\pi/4)x}+\frac{\sin(\pi/4)x}{\lvert\cos(\pi/4)x\rvert}\right)$	q.	π
iii.	The period of $f(x)=\tan\frac{\pi}{2}[x]$ is	r.	3
iv.	The period of $f(x)=\frac{\sin x+\sin 3x}{\cos x+\cos 3x}$ is	s.	8

Codes

	i	ii	iii	iv
a.	p	s	p	q
b.	q	q,r	p	s
c.	q	s	r,s	p
d.	s	p	q	r

Comprehension Type

Direction *(Q. Nos. 22-24) This section contains a paragraph describing theory, experiments, data, etc. Three questions related to the paragraph have been given. Each question has only one correct answer among the four given options (a), (b), (c) and (d).*

Passage

Given a function $f_1(x)=e^{\tan\left\{\frac{x}{4}\right\}}+\cos\pi\left(\frac{(1-2[x])}{2}\right)+\sin\left(\frac{\pi[x]}{2}\right)$, whose fundamental period is p,

(where [·] and {·} represents greatest integer function and fractional part functions respectively)

and $y=\sqrt{2p+\frac{p}{2}[x]-[x]^2}$. The domain of y is $[q, r)$ one another function $f_2(x)=\begin{cases}2+x, & x\geq 0\\ 2-x, & x<0\end{cases}$.

Working Space

22 The period p of $f_1(x)$ is

a. an irrational number
b. a prime number
c. a composite number
d. neither a prime nor a composite number

23 The value of $(r - q - 1)$ is

a. 6
b. 7
c. 8
d. 9

24 The range of $f_2[f_2(x)]$ in terms of p, q, r, is

a. $[p, \infty)$
b. $[q, r)$
c. $[r, \infty)$
d. $(-\infty, -r) \cup (p, \infty)$

One Integer Value Correct Type

Direction (*Q. Nos.* 25-29) *This section contains 5 questions. When worked out will result in an integer from 0 to 9 (both inclusive).*

25 The period of $f(x) = \sin\left(\frac{\pi}{2}[x]\right)$ (where, [·] denotes greatest integer function), is

26 The fundamental period of $f(x) = e^{\cos^4 \pi x + x - [x] + \cos \pi x}$

(where, [·] denotes greatest integer function), is

27 The period of $f(x)$ satisfying $f(x-1) + f(x+3) = f(x+1) + f(x+5)$, is

28 The period of $f(x)$ satisfying

$f(x+p) = 1 + \{1 - 3f(x) + 3f^2(x) - f^3(x)\}^{1/3}$ is (λp). Then, the value of λ is

29 The period of $f(x)$ satisfying $f(x+a) = \frac{1}{2} + \sqrt{f(x) - (f(x))^2}$ is (λa). Then, the value of λ is

DPP-7 Composition of Functions

Subjective Questions

Direction (Q. Nos. 1-3) *These questions are subjective in nature, need to be solved completely on notebook.*

Working Space

1 Let $f(x) = x^2 + x + 1$ and $g(x) = \sin x$. Show that $fog \neq gof$.

2 Let $f(x) = x^2$, $g(x) = \sin x$ and $h(x) = \sqrt{x}$, then verify that $[fo(goh)](x)$ and $[(fog)oh]\ (x)$ are equal.

3 Find fog and gof, if

i. $f(x) = e^x$ and $g(x) = \log x$

ii. $f(x) = |x|$ and $g(x) = \sin x$

iii. $f(x) = \sin^{-1} x$ and $g(x) = x^2$

iv. $f(x) = x^2 + 2$ and $g(x) = 1 - \dfrac{1}{1-x}$, $x \neq 1$

Only One Option Correct Type

Direction (Q. Nos. 4-18) *This section contains 15 multiple choice questions. Each question has four choices (a), (b), (c) and (d), out of which ONLY ONE is correct.*

4 If $f(x) = e^{2x}$ and $g(x) = \log(\sqrt{x})$, $x > 0$, then $(fog)\ (x)$ is equal to

a. e^{2x} b. $\log\sqrt{x}$ c. $e^{2x}\log\sqrt{x}$ d. x

5 If $f : [-6, 6] \to R$ is defined by $f(x) = x^2 - 3$ for $x \in R$, then $(fofof)\ (-1) + (fofof)\ (0) + (fofof)(1)$ is equal to

a. $f(4\sqrt{2})$ b. $f(3\sqrt{2})$ c. $f(2\sqrt{2})$ d. $f(\sqrt{2})$

6 If $f(x) = \dfrac{x}{\sqrt{1+x^2}}$, then $(fofof)\ (x)$ is equal to

a. $\dfrac{3x}{\sqrt{1+x^2}}$ b. $\dfrac{x}{\sqrt{1+3x^2}}$ c. $\dfrac{3x}{1-x^2}$ d. None of these

7 If $g(x) = x^2 + x - 2$ and $\dfrac{1}{2}(gof)\ (x) = 2x^2 - 5x + 2$, then $f(x)$ is equal to

a. $2x - 3$ b. $2x + 3$ c. $2x^2 + 3x + 1$ d. $2x^2 - 3x - 1$

8 If $f(x) = e^x$ and $g(x) = \log_e x$, then which of the following is true?

a. $f\{g(x)\} \neq g\{f(x)\}$ b. $f\{g(x)\} = g\{f(x)\}$

c. $f\{g(x)\} + g\{f(x)\} = 0$ d. $f\{g(x)\} - g\{f(x)\} = 1$

9 Let f and g be the functions defined by $f(x) = \dfrac{x}{x+1}$, $g(x) = \dfrac{x}{1-x}$, then $(fog)\ (x)$ is equal to

a. $\dfrac{1}{x}$ b. $\dfrac{1}{x-1}$ c. $x - 1$ d. x

Working Space

10 If from $R \to R$, $f(x) = (x+1)^2$, $g(x) = x^2 + 1$, then $(fog)(-3)$ is equal to

a. 121
b. 112
c. 211
d. None of these

11 Suppose $g(x) = 1 + \sqrt{x}$ and $f\{g(x)\} = 3 + 2\sqrt{x} + x$, then $f(x)$ is equal to

a. $1 + 2x^2$
b. $2 + x^2$
c. $1 + x$
d. $2 + x$

12 Let $g(x) = 1 + x - [x]$ and $f(x) = \begin{cases} -1, & x < 0 \\ 0, & x = 0 \\ 1, & x > 0 \end{cases}$, then for all x, $f\{g(x)\}$ is equal to

a. x
b. 1
c. $f(x)$
d. $g(x)$

13 If $f(x) = \dfrac{\alpha x}{x+1}$, $x \neq -1$, then for what value of α is $f\{f(x)\} = x$? **[IIT JEE 2001]**

a. $\sqrt{2}$
b. $-\sqrt{2}$
c. 1
d. -1

14 If $f(x) = \dfrac{2x+1}{3x-2}$, then $(fof)(2)$ is equal to

a. 1
b. 2
c. 3
d. 4

15 If $f(x) = (a - x^n)^{1/n}$, where $a > 0$ and n is a positive integer, then $f\{f(x)\}$ is

a. x^3
b. x^2
c. x
d. None of these

16 Let $f(x)$ be a function whose domain is $[-5, 7]$ and $g(x) = |2x + 5|$. Then, domain of $(fog)(x)$ is

a. $[-4, 1]$
b. $[-5, 1]$
c. $[-6, 1]$
d. None of these

17 If $f(x) = \dfrac{ax+b}{cx+d}$, then $(fof)(x) = x$, provided that

a. $d + a = 0$
b. $d - a = 0$
c. $a = b = c = d = 1$
d. $a = b = 1$

18 If $f(x) = \cot^{-1} x : R^+ \to \left(0, \dfrac{\pi}{2}\right)$ and $g(x) = 2x - x^2 : R \to R$. Then, the range of function $f\{g(x)\}$ wherever define is

a. $\left(0, \dfrac{\pi}{2}\right)$
b. $\left(0, \dfrac{\pi}{4}\right)$
c. $\left[\dfrac{\pi}{4}, \dfrac{\pi}{2}\right)$
d. $\left\{\dfrac{\pi}{4}\right\}$

One Integer Value Correct Type

Directions *(Q. Nos. 19 and 20) This section contains 2 questions. When worked out will result in an integer from 0 to 9 (both inclusive).*

19 If $f(x) = \begin{cases} 1 + x^2, & x \leq 1 \\ x + 1, & 1 < x \leq 2 \end{cases}$ and $g(x) = 1 - x$, $-2 \leq x \leq 1$, thus

$(fog)(x) = \begin{cases} 2 - 2x + x^2, & a \leq x \leq b \\ 2 - x, & c \leq x < a \end{cases}$. Then, the value $(a + b + c)$ is

20 Let $f(x) = \begin{cases} 1 + x, & 0 \leq x \leq 2 \\ 3 - x, & 2 < x \leq 3 \end{cases}$. Thus, $(fof)(x) = \begin{cases} 2 + x, & a \leq x \leq b \\ 2 - x, & b < x \leq c \\ 4 - x, & c < x \leq d \end{cases}$, then the value of $(a + b + c + d)$ is

DPP-8 Types of Functions

Only One Option Correct Type

Working Space

Directions (Q. Nos. 1-17) *This section contains 17 multiple choice questions. Each question has four choices (a), (b), (c) and (d), out of which ONLY ONE is correct.*

1 The function $f : R \to R$ is given by $f(x) = x^3 - 1$, is

a. one-one function b. an onto function
c. a bijective d. neither one-one nor onto

2 If for two functions g and f, gof is both injective and surjective, then which of the following is true?

a. g and f should be injective and surjective
b. g should be injective and surjective
c. f should be injective and surjective
d. None of the above

3 Mapping $f : R \to R$ which is defined as $f(x) = \cos x, x \in R$, is

a. one-one but not onto b. onto but not one-one
c. both one-one and onto d. neither one-one nor onto

4 The function $f : R \to R$ defined as $f(x) = (x - 1)(x - 2)(x - 3)$, is

a. one-one but not onto b. onto but not one-one
c. both one-one and onto d. neither one-one nor onto

5 Let R and C denotes the set of real numbers and complex numbers respectively. The function $f : C \to R$ defined by $f(z) = |z|$ is

a. one-one b. onto
c. bijective d. neither one-one nor onto

6 For real x, let $f(x) = x^3 + 5x + 1$, then

a. f is one-one but not onto R b. f is onto R but not one-one
c. f is one-one and onto R d. f is neither one-one nor onto R

7 Set A has 3 elements and set B has 4 elements. The number of injective that can be defined from set A to B, is

a. 144 b. 12 c. 24 d. 64

8 Let $f : R \to R$ be a function defined by $f(x) = \dfrac{x - m}{x - n}$, where $m \neq n$. Then,

a. f is one-one and onto b. f is one-one and into
c. f is many-one and onto d. f is many-one and into

9 Which one of the following is a bijection function on the set of real numbers?

a. $2x - 5$ b. $|x|$ c. x^2 d. $x^2 + 1$

10 Let $f(x) = \dfrac{x^2 - 4}{x^2 + 4}$ for $|x| > 2$, then the function $f : (-\infty, -2] \cup [2, \infty) \to (-1, 1)$ is

a. one-one into b. one-one onto
c. many-one into d. many-one onto

Working Space

11 Let the function $f: R \to R$ be defined by $f(x) = 2x + \sin x$, $x \in R$. Then, f is

a. one-one and onto
b. one-one but not onto
c. onto but not one-one
d. neither one-one nor onto

12 A function f from the set of natural numbers to integers defined by

$$f(n) = \begin{cases} \dfrac{n-1}{2}, & \text{if } n \text{ is odd} \\ -\dfrac{n}{2}, & \text{if } n \text{ is even} \end{cases}, \text{ is}$$

a. one-one but not onto
b. onto but not one-one
c. one-one and onto
d. neither one-one nor onto

13 If $f: [0, \infty) \to [0, \infty)$ and $f(x) = \dfrac{x}{1+x}$, then f is

a. one-one and onto
b. one-one but not onto
c. onto but not one-one
d. neither one-one nor onto

14 If R denotes the set of all real numbers, then the function $f: R \to R$ defined as $f(x) = [x]$ will be

a. only one-one
b. only onto
c. both one-one and onto
d. neither one-one nor onto

15 Let $f(x) = x + \sqrt{x^2}$ is a function from $R \to R$, then $f(x)$ is

a. injective
b. surjective
c. bijective
d. None of these

16 A mapping from N to N is defined as follows :
$f: N \to N, f(n) = (n+5)^2, n \in N$. Then, f is

a. one-one
b. onto
c. one-one and onto
d. one-one but not onto

17 Let $f: (-1, 1) \to B$ be a function defined by $f(x) = \tan^{-1}\left(\dfrac{2x}{1-x^2}\right)$, then f is both one-one and onto, when B in the interval

a. $\left[-\dfrac{\pi}{2}, \dfrac{\pi}{2}\right]$
b. $\left(-\dfrac{\pi}{2}, \dfrac{\pi}{2}\right)$
c. $\left(0, \dfrac{\pi}{2}\right)$
d. $\left[0, \dfrac{\pi}{2}\right)$

Comprehension Type

Directions *(Q. Nos. 18-24) This section contains 2 paragraphs, each describing theory, experiments, data, etc. Seven questions related to the paragraphs have been given. Each question has only one correct answer among the four given options (a), (b), (c) and (d).*

Passage I

Let $f(x) = \dfrac{x^3}{3} + \dfrac{x^2}{2} + ax + b, \forall x \in R$.

18 The least value of a for which $f(x)$ is injective function, is

a. $\dfrac{1}{4}$
b. 1
c. $\dfrac{1}{2}$
d. $\dfrac{1}{8}$

Working Space

19 If $a = -1$, then $f(x)$ is

a. bijective
b. many-one and onto
c. one-one and into
d. many-one and into

20 $f(x)$ is invertible iff

a. $a \in \left[\frac{1}{4}, \infty\right), b \in R$
b. $a \in \left[\frac{1}{8}, \infty\right), b \in R$
c. $a \in \left[-\infty, \frac{1}{4}\right), b \in R$
d. $a \in \left[-\infty, \frac{1}{4}\right), b \in R$

Passage II

Let $f(x) = x^2 - 2x - 1, \forall x \in R$ and $f : (-\infty, a] \to [b, \infty)$, where a is the largest real number for which $f(x)$ is bijective.

21 The value of $(a + b)$ is

a. -2 b. -1 c. 0 d. 1

22 Let $f : R \to R$, $g(x) = f(x) + 3x - 1$, then the least value of function $y = g(|x|)$ is

a. $-9/4$ b. $-5/4$ c. -2 d. -1

23 Let $f : [a, \infty) \to [b, \infty)$, then the $f^{-1}(x)$ is given by

a. $1 + \sqrt{x+2}$ b. $1 - \sqrt{x+3}$ c. $1 - \sqrt{x+2}$ d. $1 + \sqrt{x+3}$

24 Let $f : R \to R$, then the range of values of k for which equation $f(|x|) = k$ has 4 distinct real roots, is

a. $(-2, -1)$ b. $(-2, 0)$ c. $(-1, 0)$ d. $(0, 1)$

One Integer Value Correct Type

Direction (Q. Nos. 25-30) This section contains 6 questions. When worked out will result in an integer from 0 to 9 (both inclusive).

25 The function $f : [2, \infty) \to Y$ defined by $f(x) = = x^2 - 4x + 5$ is both one-one and onto, if $Y \in [a, \infty)$, then the value of a is

26 If $f(x) = \left(\frac{4a-7}{3}\right)x^3 + (a-3)x^2 + x + 5$ is one-one function, where $a \in [\lambda, \mu]$, then the value of $|\lambda - \mu|$ is

27 Let f be a one-one function with domain $\{x, y, z\}$ and range $\{1, 2, 3\}$. It is given that exactly one of the following statements is true and remaining two are false, $f(x) = 1, f(y) \neq -1, f(z) \neq 2$, then $f(y)$ is

28 If $n(A) = 4$, $n(B) = 5$, number of functions from A to B such that range contains exactly 3 elements is k, then the value of $\frac{k}{60}$ is

29 If $A = \{1, 2, 3\}$, $B = \{1, 3, 5, 7, 9\}$, then $\frac{\text{Number of one-one functions}}{\text{Number of strictly monotonic functions}}$ is equal to

30 If $f(x)$ be a polynomial of degree 4 with leading coefficient 1 satisfying $f(1) = 10$, $f(2) = 20$, $f(3) = 30$, where $f(12) + f(-8) = (3968)\lambda$, then the value of λ is

DPP-9 Inverse of Functions

Only One Option Correct Type

Direction (Q. Nos. 1-10) *This section contains 10 multiple choice questions. Each question has four choices (a), (b), (c) and (d), out of which ONLY ONE is correct.*

Working Space

1 The inverse of function $f(x) = \dfrac{e^x - e^{-x}}{e^x + e^{-x}} + 2$ is

a. $\log_e\left(\dfrac{x-2}{x-1}\right)^{1/2}$ b. $\log_e\left(\dfrac{x-1}{3-x}\right)^{1/2}$ c. $\log_e\left(\dfrac{x}{2-x}\right)^{1/2}$ d. $\log_e\left(\dfrac{x-1}{x+1}\right)^{-1/2}$

2 If the function $f:[1,\infty) \to [1,\infty)$ is defined by $f(x) = 2^{x(x-1)}$, then $f^{-1}(x)$ is

a. $\left(\dfrac{1}{2}\right)^{x(x-1)}$ b. $\dfrac{1}{2}(1+\sqrt{1+4\log_2 x})$

c. $\dfrac{1}{2}(1-\sqrt{1+4\log_2 x})$ d. not defined

3 Let $f: N \to Y$ be a function defined as $f(x) = 4x + 3$, where $Y = \{y \in N : y = 4x + 3 \text{ for some } x \in N\}$, then its inverse is

a. $4 + \dfrac{x+3}{4}$ b. $\dfrac{x+3}{4}$

c. $\dfrac{x-3}{4}$ d. $\dfrac{3x+4}{3}$

4 If $f: R \to R$ is defined by $f(x) = 3x - 4$, then $f^{-1}(x)$ is

a. $4 - 3x$ b. $\dfrac{x+4}{3}$ c. $\dfrac{1}{3x-4}$ d. $\dfrac{3}{x+4}$

5 Which of the following functions have inverse function?

a. $f(x) = \dfrac{1}{x-1}, f: R - \{1\} \to R - \{0\}$ b. $f(x) = x^2, f: R \to R$

c. $f(x) = x^2, f: R^+ \to R$ d. $f(x) = x^2, f: R^- \to R$

6 The inverse of the function $f(x) = \dfrac{10^x - 10^{-x}}{10^x + 10^{-x}}$ is

a. $\dfrac{1}{2}\log_{10}\left(\dfrac{1+x}{1-x}\right)$ b. $\dfrac{1}{2}\log_{10}\left(\dfrac{1-x}{1+x}\right)$

c. $\dfrac{1}{4}\log_{10}\left(\dfrac{2x}{2-x}\right)$ d. None of these

7 Let $f:(2,4) \to (1,3)$ be a function defined by $f(x) = x - \left[\dfrac{x}{2}\right]$ (where, [.] denotes greatest integer function), then $f^{-1}(x)$ is equal to

a. $2x$ b. $x + \left[\dfrac{x}{2}\right]$

c. $x + 1$ d. $x - 1$

Working Space

8 Let $f:[-\sqrt{2}+1, \sqrt{2}+1] \to \left[\frac{-\sqrt{2}+1}{2}, \frac{\sqrt{2}+1}{2}\right]$ be a function defined by $f(x)=\frac{1-x}{1+x^2}$, then $f^{-1}(x)$ is

a. $\sqrt{4x-4x^2+1}, x \neq 0$

b. $\begin{cases} \frac{-1+\sqrt{4x+1-4x^2}}{2x}, & x \neq 0 \\ 1, & x = 0 \end{cases}$

c. does not exist

d. None of these

9 Let $f: N \to N$, where $f(x) = x + (-1)^{x-1}$, then $f^{-1}(x)$ is

a. x b. $x-1$ c. $x-(-1)^{x-1}$ d. $x+(-1)^{x-1}$

10 Let $a > 1$ be a real number and $f(x) = \log_a x^2$, for $x > 0$. If $f^{-1}(x)$ is the inverse function of f and b, c are real numbers, then $f^{-1}(b+c)$ is equal to

a. $f^{-1}(b) \cdot f^{-1}(c)$ b. $f^{-1}(b)+f^{-1}(c)$ c. $\frac{1}{f(b+c)}$ d. None of these

Comprehension Type

Direction (Q. Nos. 11-13) *This section contains a paragraph, describing theory, experiments, data, etc. Three questions related to the paragraph have been given. Each question has only one correct answer among the four given options (a), (b), (c) and (d).*

Passage

If $f(x)=\frac{x+a}{bx^2+cx+2}$, where $a, b, c \in R$, $f(-1)=0$ and $y=1$ is an asymptote of $y=f(x)$, $y=f^{-1}(x)$ is inverse function of $f(x)$.

11 $f(x)$ is equal to

a. $\frac{x+1}{x+2}$ b. $\frac{x+1}{x^2+x+2}$ c. $\frac{x+1}{2-x}$ d. $\frac{x-1}{2x+1}$

12 If $f^{-1}: R-\{1\} \to R$, then $f^{-1}(x)$ is

a. one-one and onto

b. one-one and into

c. many-one and into

d. many-one and onto

13 Area bounded between asymptotes of curve $f(x)$ and $f^{-1}(x)$ is

a. 4 b. 9 c. 16 d. 25

One Integer Value Correct Type

Directions (Q. Nos. 14 and 15) *This section contains 2 questions. When worked out will result in an integer from 0 to 9 (both inclusive).*

14 Let $f:\left[\frac{\pi}{3}, \frac{\pi}{6}\right] \to B$. Defined by $f(x) = 2\cos^2 x + \sqrt{3}\sin 2x + 1$, where $B \in [a, b]$ such that $f^{-1}(x)$ exists, then $(a+b)$ is

15 Let R be the region in the first quadrant bounded by the X and Y-axes and the graph of $f(x)=\frac{9}{25}x+b$ and $y=f^{-1}(x)$. If the area of R is 49, where the value of b is $\frac{\lambda}{\mu}$, then $|\lambda - 5\mu|$ is

SET - 1

Revisal Problems
for JEE Main

Only One Option Correct Type

Working Space

Direction (Q. Nos. 1-53) *This section contains 53 multiple choice questions. Each question has four choices (a), (b), (c) and (d), out of which ONLY ONE option is correct.*

1 The domain of function $f(x)=\sqrt{x-\sqrt{1-x^2}}$ is

a. $\left[-1,-\frac{1}{\sqrt{2}}\right]\cup\left[\frac{1}{\sqrt{2}},1\right]$ b. $[-1,1]$

c. $\left(-\infty,\frac{-1}{2}\right)\cup\left[\frac{1}{\sqrt{2}},\infty\right)$ d. $\left[\frac{1}{\sqrt{2}},1\right]$

2 Let $f(x)$ be a function whose domain is $[-5,7]$ and $g(x)=|2x+5|$. Then, the domain of $(fog)(x)$ is

a. $[-5,1]$ b. $[-4,0]$ c. $[-6,1]$ d. None of these

3 The domain of definition of $f(x)=\sin^{-1}[2-4x^2]$ (where, $[\cdot]$ denotes the greatest integer function), is

a. $\left[\frac{-\sqrt{3}}{2},\frac{\sqrt{3}}{2}\right]$ b. $\left[\frac{-\sqrt{3}}{2},0\right]$

c. $\left[\frac{-\sqrt{3}}{2},0\right)\cup\left(0,\frac{\sqrt{3}}{2}\right]$ d. $\left[\frac{-\sqrt{3}}{2},\infty\right)$

4 The domain of the function $f(x)=\log_2[\log_3\{\log_4(x^2-3x+6)\}]$ is

a. $(1,2)$ b. $[1,2]$

c. $(-\infty,1)\cup(2,\infty)$ d. $(-\infty,1]\cup[2,\infty)$

5 The domain of definition of $f(x)=\log_3|\log_e x|$ is

a. $(1,\infty)$ b. $(0,\infty)$ c. (e,∞) d. None of these

6 The domain of $f(x)=x^{\frac{1}{\log_{10}x}}$ is

a. $(0,1)\cup(1,\infty)$ b. $(0,\infty)$ c. $[0,\infty)$ d. $[0,1)\cup(1,\infty)$

7 The domain of $f(x)=\sqrt{e^{\cos^{-1}(\log_4 x^2)}}$ is

a. $\left[\frac{1}{2},2\right]$ b. $\left[-2,\frac{-1}{2}\right]\cup\left[\frac{1}{2},2\right]$

c. $\left[-2,\frac{-1}{2}\right]$ d. None of these

8 The domain of $f(x)=\cos^{-1}\left\{\frac{x}{\sqrt{x^2-[x]^2}}\right\}$ is

a. $R-\{\pm\sqrt{n}:n\in N\}$ b. $R-\{\pm\sqrt{n};n\geq 0,n\in Z\}$

c. R d. $R-\{0\}$

9 The domain of definition of the function $f(x)=\sqrt{3-2^x-2^{1-x}}+\sqrt{\sin^{-1}x}$ is

a. $[0,1]$ b. $(0,1]$ c. $[0,1)$ d. None of these

10 If $f(x)$ is defined on $(0,1)$, then the domain of the definition of $f(\sin x)$, is

a. $(2n\pi,(2n+1)\pi),n\in Z$ b. $\left((2n+1)\frac{\pi}{2},(2n+3)\frac{\pi}{2}\right),n\in Z$

c. $((n-1)\pi,(n+1)\pi),n\in Z$ d. None of these

Working Space

11 Let $f(x) = \cos^{-1}\left(\dfrac{x^2}{x^2+1}\right)$, then the range of f is

a. $\left(0, \dfrac{\pi}{2}\right]$ b. $\left[-\dfrac{\pi}{2}, \dfrac{\pi}{2}\right]$ c. $\left[-\dfrac{\pi}{2}, 0\right]$ d. None of these

12 The range of function $f(x) = x^2 + \dfrac{1}{x^2+1}$ is

a. $[1, \infty)$ b. $[2, \infty)$ c. $\left[\dfrac{3}{2}, \infty\right)$ d. None of these

13 The range of function $f(x) = \log_5(25 - x^2)$ is

a. $[0, 5]$ b. $[0, 2)$ c. $(0, 2)$ d. None of these

14 The range of the function $f : [0, 1] \to R$, given by $f(x) = x^3 - x^2 + 4x + 2\sin^{-1} x$, is

a. $[-\pi - 2, 0]$ b. $[2, 3]$ c. $[0, 4 + \pi]$ d. $[0, 2 + \pi]$

15 The range of $f(x) = \dfrac{x^2 + x + 2}{x^2 + x + 1}$, $x \in R$ is

a. $(1, \infty)$ b. $\left(1, \dfrac{11}{7}\right]$ c. $\left[1, \dfrac{7}{3}\right]$ d. $\left(1, \dfrac{7}{5}\right]$

16 Let $f(x) = [9^x - 3^x + 1]$ for all $x \in (-\infty, 1)$, then the range of $f(x)$ (where, [·] denotes the greatest integer function), is

a. {0, 1, 2, 3, 4, 5, 6, 7}
b. {0, 1, 2, 3, 4, 5, 6}
c. {1, 2, 3, 4, 5, 6, 7}
d. {1, 2, 3, 4, 5, 6}

17 The domain of $f(x) = \sin^{-1}\left(\dfrac{2-|x|}{4}\right) + \cos^{-1}\left(\dfrac{2-|x|}{4}\right) + \tan^{-1}\left(\dfrac{2-|x|}{4}\right)$ is

a. [0 3] b. $[-6, 6]$ c. $[-1, 1]$ d. $[-3, 3]$

18 The domain of $f(x) = \sqrt[4]{\log_3\left\{\dfrac{1}{|\cos x|}\right\}}$ is

a. R
b. $R - \{n\pi : n \in Z\}$
c. $R - \left\{(2n+1)\dfrac{\pi}{2} : n \in Z\right\}$
d. None of these

19 The domain of $f(x) = \cos^{-1}(x + [x])$ is

a. $[0, 1)$ b. $R - Z$ c. $(0, \infty)$ d. None of these

20 The domain of $f(x) = \underbrace{\log_{10}\log_{10}\log_{10}\cdots\log_{10}}_{n \text{ times}} x$, is

a. $(10^n, \infty)$ b. $(10^{n-1}, \infty)$ c. $(10^{n-2}, \infty)$ d. None of these

21 The domain of $f(x) = \dfrac{1}{\sqrt{[x]^2 - [x] - 6}}$ is

a. $(-\infty, -2) \cup [4, \infty)$
b. $(-\infty, -2] \cup [4, \infty)$
c. $(-\infty, -2) \cup (4, \infty)$
d. None of these

22 The domain of $f(x) = \dfrac{\log_2(x+3)}{x^2 + 3x + 2}$ is

a. $R - \{-1, -2\}$ b. $(-2, \infty)$ c. $(-3, \infty) - \{-1, -2\}$ d. None of these

23 Let a, b, c be rational numbers and $f : Z \to Z$ be a function given by $f(x) = ax^2 + bx + c$. Then, $(a + b)$ is

a. a negative integer
b. an integer
c. a non-integral rational number
d. None of these

24 If $f : R \to R$ given by $f(x) = 2x + |\cos x|$ is

a. one-one and into
b. one-one and onto
c. many-one and into
d. many-one and onto

Working Space

25 The set of values of parameter a for which the function $f: R \to R$ defined by $f(x) = ax + \sin x$ is bijective, is

a. $[-1, 1]$ b. R c. $R - [-1, 1]$ d. $(-1, 1)$

26 Let $A = \left\{x : 0 \le x < \frac{\pi}{2}\right\}$ and $f: R \to A$ be an onto function given by $f(x) = \tan^{-1}(x^2 + x + \lambda)$, where λ is a constant, then

a. $\lambda \le \frac{1}{4}$ b. $\lambda \ge \frac{1}{4}$ c. $\lambda < \frac{1}{4}$ d. $0 \le \lambda \le 1$

27 The function $f: R \to R$ defined by $f(x) = 2^x + 2^{|x|}$ is

a. one-one and onto b. many-one and into
c. one-one and into d. many-one and onto

28 If $f: R \to R$ is given by $f(x) = 3x - 5$, then $f^{-1}(x)$

a. is given by $\frac{1}{3x - 5}$ b. is given by $\frac{x + 5}{3}$
c. does not exist, because f is not one-one d. does not exists because f is not onto

29 Let $A = \{x \in R : x \le 1\}$ and $f: A \to A$ be defined as $f(x) = x(2 - x)$. Then, $f^{-1}(x)$ is

a. $1 + \sqrt{1 - x}$ b. $1 - \sqrt{1 - x}$
c. $\sqrt{1 - x}$ d. $1 \pm \sqrt{1 - x}$

30 If the function $f: R \to R$ be such that $f(x) = x - [x]$, where $[x]$ denotes the greatest integer less than or equal to x, then $f^{-1}(x)$ is

a. $\frac{1}{x - [x]}$ b. $[x] - x$
c. not defined d. None of these

31 Let $A = \left\{x \in R : x \ge \frac{1}{2}\right\}$ and $B = \left\{x \in R : x \ge \frac{3}{4}\right\}$. If $f: A \to B$ is defined as $f(x) = x^2 - x + 1$, then the solution set of equation $f(x) = f^{-1}(x)$, is

a. $\{1\}$ b. $\{2\}$ c. $\{1, 2\}$ d. None of these

32 Let $E = \{1, 2, 3, 4\}$ and $F = \{1, 2\}$. Then, the number of onto functions from E to F is

a. 14 b. 16 c. 12 d. 8

33 Let $g(x) = 1 + x - [x]$ and $f(x) = \begin{cases} -1, & x < 0 \\ 0, & x = 0 \\ 1, & x > 0 \end{cases}$, where $[x]$ denotes the greatest integer less than or equal to x. Then, for all x, $f[g(x)]$ is equal to

a. x b. 1 c. $f(x)$ d. $g(x)$

34 Let the function $f: R \to R$ be defined by $f(x) = 2x + \sin x$. Then, f is

a. one-one and onto b. one-one and into
c. many-one and onto d. many-one and into

35 The function $f: R \to R$ given by $f(x) = 5 - 3\sin x$ is

a. one-one b. onto c. one-one and onto d. None of these

36 The function $f: (-\infty, -1] \to (0, e^5]$ defined by $f(x) = e^{x^3 - 3x + 2}$ is

a. one-one and onto b. one-one and into
c. many one and into d. many one and onto

37 The total number of onto functions from the set $\{1, 2, 3, 4\}$ to the set $\{3, 4, 7\}$, is

a. 18 b. 36 c. 64 d. None of these

38 The value of parameter α, for which the function $f: R \to R$ given by $f(x) = 1 + \alpha x$, $\alpha \ne 0$ is the inverse of itself, is

a. -2 b. -1 c. 1 d. 2

39 Let $f: [2, \infty) \to x$ be defined by $f(x)$, as $f(x) = 4x - x^2$, then f is invertible, if x is equal to

a. $[2, \infty)$ b. $(-\infty, 2]$ c. $(-\infty, 4]$ d. $[4, \infty)$

Working Space

40 If $f(x)=\begin{cases}|x|, & x\le 1\\ 2-x, & x>1\end{cases}$, then $(fof)(x)$ is equal to

a. $\begin{cases}2-|x|, & x<-1\\ |x|, & -1\le x\le 1\\ |2-x|, & x>1\end{cases}$

b. $\begin{cases}|x|, & x<-1\\ 2-|x|, & -1\le x\le 1\\ |2-x|, & x>1\end{cases}$

c. $\begin{cases}|2-x|, & x<-1\\ |x|, & -1\le x\le 1\\ 2-|x|, & x>1\end{cases}$

d. None of these

41 The function $f: A\to B$ defined by $f(x)=x^2+6x-8$ is a bijection, if

a. $A=[-\infty, 3)$ and $B=(-\infty, 1]$
b. $A=[-3, \infty)$ and $B\in[-17, \infty)$
c. $A=[3, \infty)$ and $B=[1, \infty)$
d. None of these

42 If $f(x)$ is real valued function satisfying $f(x+y)=f(x)+f(y)-xy-1, \forall x, y\in R$ such that $f(1)=1$, then the number of solution, of $f(n)=n, n\in N$, is

a. 1 b. 2 c. 3 d. 4

43 If $f(x)=\sin(\sqrt{[\lambda]}\,x)$ is a periodic function with period π, where $[\lambda]$ denotes the greatest integer less than or equal to λ, is π. Then,

a. $\lambda\in[4, 5)$
b. $\lambda\in[4, 5]$
c. $\lambda=4, 5$
d. None of these

44 The function $f(x)=2\cos 5x+3\sin\sqrt{5}x$ is

a. a periodic function with period 2π
b. a periodic function with period $\frac{2\pi}{5}$
c. a periodic function with period $\frac{2\pi}{\sqrt{5}}$
d. not a periodic function

45 The function $f(x)=\min\{x-[x], -x-[-x]\}$ is a periodic function with period

a. 1 b. $\frac{1}{2}$ c. 2 d. Non-periodic

46 If $f(x)$ is odd periodic function with periodic 2, then $f(4)$ is equal to

a. -4 b. 4 c. 2 d. 0

47 If a real valued function $f(x)$ satisfies the equation $f(x+y)=f(x)+f(y)\forall x, y\in R$, then $f(x)$ is

a. a periodic function
b. an even function
c. an odd function
d. None of these

48 Period of $f(x)=\frac{1}{3}\{\sin 3x+|\sin 3x|+[\sin 3x]\}$ (where, $[x]$ denotes greatest integer function) is

a. $\frac{\pi}{3}$
b. $\frac{2\pi}{3}$
c. $\frac{4\pi}{3}$
d. π

49 Let $f(x)=\frac{\sin 2nx}{1+\cos^2 nx}$, $n\in N$ has $\frac{\pi}{6}$ as its period, then n is equal to

a. 2
b. 4
c. 6
d. None of these

50 If the graph of the function $y=f(x)$ is symmetrical about the line $x=2$, then

a. $f(x)=-f(-x)$
b. $f(2+x)=f(2-x)$
c. $f(x)=f(-x)$
d. $f(x+2)=f(x-2)$

51 If $[\cdot]$ and $\{\cdot\}$ denotes greatest integer and fractional part of x respectively, then period of $f(x)=e^{\sin 3\pi\{x\}+\tan\pi[x]}$ is

a. 2/3
b. 1
c. 3
d. None of these

52 Let $[\cdot]$ denotes greatest integer function. If $f(x) = \tan(\sqrt{[n]}x)$ has period $\frac{\pi}{3}$, then

a. $n \in (1, 3)$
b. $n \in (9, 10)$
c. $n \in [9, 10)$
d. $n \in [9, \infty)$

53 If $[\cdot]$ denotes the greatest integer function and $n \in N$, then $f(x) = nx + n - [nx + n] + \tan\frac{\pi x}{2}$, is periodic with period

a. 1
b. 2
c. 4
d. 3

Statement Type

Direction **(Q. Nos. 54-60)** ***This section is based on Statement I and Statement II. Select the correct answer from the codes given below.***

a. Both Statement I and Statement II are correct and Statement II is the correct explanation of Statement I
b. Both Statement I and Statement II are correct but Statement II is not the correct explanation of Statement I
c. Statement I is correct but Statement II is incorrect
d. Statement II is correct but Statement I is incorrect

54 **Statement I** If $f(x)$ and $g(x)$ both are one-one, then $f[g(x)]$ is also one-one.

Statement II If $f_1(x) = f_2(x) \Leftrightarrow x_1 = x_2$, then $f(x)$ is one-one.

55 **Statement I** If $y = f(x)$ is increasing in $[\alpha, \beta]$, then its range is $[f(\alpha), f(\beta)]$.

Statement II Every increasing function need not to be continuous.

56 **Statement I** Let $f : [0, 3] \to [1, 13]$ is defined by $f(x) = x^2 + x + 1$, then inverse is $f^{-1}(x) = \frac{-1 + \sqrt{4x - 3}}{2}$.

Statement II Many one function is not invertible.

57 **Statement I** All points of intersection of $y = f(x)$ and $y = f^{-1}(x)$ lie on $y = x$ only.

Statement II If a point $P(\alpha, \beta)$ lies on $y = f(x)$, then $Q(\beta, \alpha)$ lie on $y = f^{-1}(x)$.

58 **Statement I** If $a > b > c$ and $a^3 + b^3 + c^3 = 3abc$, then the quadratic equation $ax^2 + bx + c = 0$ has roots of opposite sign.

Statement II If roots of a quadratic equation $ax^2 + bx + c = 0$ are of opposite sign, then product of roots < 0 and $|\text{sum of roots}| \geq 0$.

59 Consider the function $f(x) = ({}^{x+1}C_{2x-8})({}^{2x-8}C_{x+1})$.

Statement I Domain of $f(x)$ is singleton.

Statement II Range of $f(x)$ is singleton.

60 Consider a cubic function $f(x) = ax^3 + bx + c$, where $a, b, c \in R$.

Statement I $f(x)$ cannot have 3 non-negative real roots.

Statement II Sum of roots is equal to zero.

Working Space

SET - 2

Revisal Problems
for JEE Advanced

Only One Option Correct Type

Direction *(Q. Nos. 1-29) This section contains 29 multiple choice questions. Each question has four choices (a), (b), (c) and (d), out of which ONLY ONE is correct.*

Working Space

1 Let f be a real valued function such that $f(x)+2f\left(\frac{2002}{x}\right)=3x$, for all $x>0$. The value of $f(2)$, is

a. 1000 b. 2000
c. 3000 d. 4000

2 If $a^2+b^2+c^2=1$, then $ab+bc+ca$, where lies in the interval, where ($a, b, c \in$ are sides of Δ).

a. $\left[\frac{1}{2}, 2\right]$ b. $[-1, 2]$
c. $\left(\frac{1}{2}, 1\right]$ d. $\left[-1, \frac{1}{2}\right]$

3 Given $f(x)=\sqrt{\frac{8}{1-x}+\frac{8}{1+x}}$ and $g(x)=\frac{4}{f(\sin x)}+\frac{4}{f(\cos x)}$. Then, $g(x)$ is

a. periodic with period 2 b. periodic with period $\frac{\pi}{2}$
c. periodic with period 2π d. Non periodic

4 Suppose $f(x, n)=\sum_{k=1}^{n}\log_x\left(\frac{k}{x}\right)$, then the value of x satisfying the equation $f(x, 10)=f(x, 11)$, is

a. 9 b. 10
c. 11 d. None of these

5 Let f be a function satisfying $f(xy)=\frac{f(x)}{y}$ for all positive real numbers x and y. If $f(30)=20$, then value of $f(40)$ is

a. 15 b. 20
c. 40 d. 60

6 Let $f(x)=\sin^2 x+\cos^4 x+2$ and $g(x)=\cos(\cos x)+\cos(\sin x)$. Also, let period of $f(x)$ and $g(x)$ be T_1 and T_2 respectively, then

a. $T_1=2T_2$ b. $2T_1=T_2$
c. $T_1=T_2$ d. $T_1=4T_2$

7 The minimum value of expression $|x-p|+|x-15|+|x-p-15|$ for x in the range $p\le x\le 15$, where $0<p<15$, is

a. 10 b. 15
c. 30 d. 0

8 If $f(x, y)=[\max(x, y)]^{\min(x,y)}$ and $g(x, y)=\max(x, y)-\min(x, y)$, then $f[g(-1, -3/2), g(-4, -1.75)]$ equals

a. -0.5 b. 0.5
c. 1 d. 1.5

9 If the solution set of $f(x)<3$ is $(0, \infty)$ and the solution set of $f(x)>-2$ is $(-\infty, 5)$, then the solution set for $(f(x))^2\ge f(x)+6$ is

a. R b. $(-\infty, 0]$
c. $[0, 5]$ d. $(-\infty, 0]\cup[5, \infty)$

Working Space

10 Consider the functions $f : X \to Y$ and $g : Y \to Z$, then which of the following is/are incorrect?

a. If f and g both are injective, then $gof : X \to Z$ is injective
b. If f and g both are surjective, then $gof : X \to Z$ is surjective
c. If $gof: X \to Z$ is bijective, then f is injective and g is surjective
d. None of the above

11 The graph of function $y = g(x)$ is shown, The number of solutions of equation $||g(x)| - 1| = \frac{1}{2}$ is

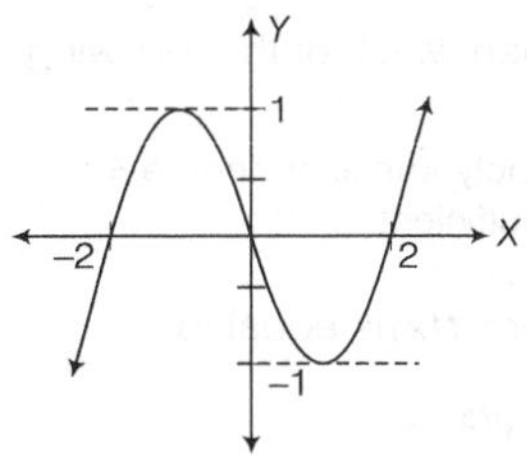

a. 4
b. 5
c. 6
d. 8

12 Which of the following statements are incorrect?

I. If $f(x)$ and $g(x)$ are one-one, then $f(x) + g(x)$ is also one-one.
II. If $f(x)$ and $g(x)$ are one-one, then $f(x) \cdot g(x)$ is also one-one.
III. If $f(x)$ is odd, then it is necessarily one-one.

a. I and II
b. II and III
c. III and I
d. I, II and III

13 The function f is one to one and the sum of all the intercepts of the graph is 5. Then, sum of all the intercept of the graph of $y = f^{-1}(x)$ is

a. 5
b. 1/5
c. 2/5
d. –5

14 If $f(x) = x^2 + bx + c$ and $f(2 + t) = f(2 - t)$ for all real numbers t, then which of the following is true?

a. $f(1) < f(2) < f(4)$
b. $f(2) < f(1) < f(4)$
c. $f(2) < f(4) < f(1)$
d. $f(4) < f(2) < f(1)$

15 The solution of set of $[x]\{x\} = 1$, where $[x]$ and $\{x\}$ are greatest integer and fractional part of x, is

a. $R^+ - (0, 1)$
b. $R^+ - \{1\}$
c. $\left\{m + \frac{1}{m} : m \in I\right\}$
d. $\left\{m + \frac{1}{m} : m \in N - \{1\}\right\}$

16 The range of the function $f(x) = \frac{\{x\}}{1 + \{x\}}$, where $\{x\}$ denotes the fractional part of x, is

a. $[0, 1)$
b. $\left[0, \frac{1}{2}\right]$
c. $\left[0, \frac{1}{2}\right)$
d. $\left(0, \frac{1}{2}\right)$

17 Let $f(x) = \frac{x}{1 + x}$ and $g(x) = \frac{rx}{1 - x}$. Let S be the set of all real numbers r such that $f[g(x)] = g[f(x)]$ for infinitely many real number x. Then, the number of elements in set S is

a. 1
b. 2
c. 3
d. 5

Working Space

18 Let $f(k) = \dfrac{k}{2009}$ and $g(k) = \dfrac{f^4(k)}{[1 - f(k)]^4 + [f(k)]^4}$, then the sum $\sum_{k=0}^{2009} g(k)$ is

a. 2009
b. 2008
c. 1005
d. 1004

19 A polynomial of 6th degree $f(x)$ satisfies $f(x) = f(2 - x)$, $\forall x \in R$, if $f(x) = 0$ has 4 distinct and 2 equal roots, then the sum of roots of $f(x) = 0$, is

a. 4
b. 5
c. 6
d. 7

20 Let $f(x)$ be a polynomial satisfying $f(9x) = f'(3x) \cdot f''(3x)$, then which of the following is correct?

a. $f(x)$ is a polynomial of degree 3
b. $f(x)$ is a polynomial of degree 4
c. $f(x)$ is a polynomial of degree 5
d. Data insufficient

21 Let $f : R - \{-1, 0, 1\} \to R$ satisfies $f^2(x) \cdot f\left(\dfrac{1-x}{1+x}\right) = 64x$, then $f(x)$ is equal to

a. $4\left(\dfrac{x(1+x)}{(1-x)}\right)^{1/3}$
b. $\left(\dfrac{4(1-x)}{1+x}\right)^{1/3}$
c. $4\left(\dfrac{x^2(1+x)}{1-x}\right)^{1/3}$
d. None of these

22 Let $f(x)$ be a polynomial of degree n such that $1 + f(x) = \dfrac{f(x-1) + f(x+1)}{2}$, $\forall x \in R$, then n satisfies

a. $n < 2$
b. $n > 2$
c. $n = 2$
d. None of these

23 Let $f : R \to R$ be a many one function and there exists a function $g : R \times R \to R$ such that $f(x + y) = g[f(x), y] \; \forall \; x, y \in R$, then

a. f is non-periodic
b. g is non-periodic
c. $f \cdot g$ may or may not be periodic
d. None of these

24 The difference between maximum and minimum value of x that satisfy $2[x]^2 + 3[x] = 4x$ (where $[\cdot]$ denotes greatest integer function), is

a. 3/2
b. 3/4
c. 5/2
d. 5/4

25 A function f is defined for all positive integers and satisfies $f(1) = 2005$ and $f(1) + f(2) + f(3) + \cdots + f(n) = n^2 \cdot f(n)$, $\forall \; n > 1$, then, the value of $f(2004)$ is

a. $\dfrac{1}{1002}$
b. $\dfrac{1}{2004}$
c. $\dfrac{2004}{2005}$
d. 2004

26 Let $f(x)$ and $g(x)$ are two non-periodic functions, then which of the following may be periodic functions?

a. $f(x) \cdot g(x)$
b. $f(x) + g(x)$
c. $f(x) - g(x)$
d. $f[g(x)]$

27 Let $f : R - \{0, 1\} \to R$ satisfying $f(x) + f\left(\dfrac{1}{1-x}\right) = \dfrac{2(1-2x)}{x(1-x)}$, then

a. $y = f(x)$ crosses X-axis at $x = 1$
b. $f(2) + f(3) = 5$
c. $f(2) + f(3) = 4$
d. $f(2) + f(3) = 6$

28 If $f(x) = 0$ be a polynomial whose coefficients are all ± 1 and whose roots are all real, then degree of $f(x)$ can be

a. 1
b. 2
c. 3
d. 4

29 The function g defined by $g(x) = \sin\alpha + \cos\alpha - 1$, $\alpha = \sin^{-1}\sqrt{\{x\}}$, $\{\cdot\}$ denotes fractional part function is

a. an even function
b. a periodic function
c. an odd function
d. neither even nor odd

Working Space

Statement Type

Direction *(Q. Nos. 30 and 31) This section is based on Statement I and Statement II. Select the correct answer from the codes given below.*

a. Both Statement I and Statement II are correct and Statement II is the correct explanation of Statement I
b. Both Statement I and Statement II are correct but Statement II is not the correct explanation of Statement I
c. Statement I is correct but Statement II is incorrect
d. Statement II is correct but Statement I is incorrect

30 **Statement I** Function $f(x) = \sin(x + 3\sin x)$ is periodic.

StatemenII $f[g(x)]$ is periodic, if $g(x)$ is periodic.

31 **Statement I** Let $f : r - \{3\}$ be a function given by $f(x + 10) = \dfrac{f(x) - 5}{f(x) - 3}$, then $f(10) = f(50)$.

Statement II $f(x)$ is a periodic function.

Comprehension Type

Direction *(Q. Nos. 32-40) This section contains 3 paragraphs, each describing theory, experiments, data, etc. Nine questions related to the paragraphs have been given. Each question has only one correct answer among the four given options (a), (b), (c) and (d).*

Passage I

Let $f : R \to R$, $f(x) = x^2 - 2x + a$ and $g : R \to R$, $g(x) = |x - 1|$.

32 Let $f(x) = g(x)$ has four distinct real roots, then a belongs to

a. $\left(-\infty, \dfrac{5}{4}\right)$ b. $\left(0, \dfrac{5}{4}\right)$
c. $\left(1, \dfrac{5}{4}\right)$ d. $\left(-1, \dfrac{5}{4}\right)$

33 If $f(x) = g(x)$ has three real roots, then a is equal to

a. 1 b. $\dfrac{5}{4}$
c. -1 d. None of these

34 Let $g(x)$ is tangent to the curve $f(x)$, then a is equal to

a. 1 b. $\dfrac{5}{4}$
c. -1 d. None of these

Passage II

Let $f(x) = \log_{\{x\}}[x]$, $g(x) = \log_{[x]}\{x\}$, $h(x) = \log_{\{x\}}\{x\}$, (where $[\cdot]$, $\{\cdot\}$ denotes greatest integer function and fractional part).

35 For $x \in (1, 5)$, $f(x)$ is not defined at how many points?

a. 5 b. 4 c. 3 d. 2

36 If $A = \{x : x \in \text{domain of } f(x)\}$ and $B = \{x : x \in \text{domain of } g(x)\}$, then $(A - B)$ will be

a. (2,3) b. (1, 3)
c. (1, 2) d. None of these

37 The domain of $h(x)$ is

a. R b. I c. $R - I$ d. $R^+ - I$

Passage III

For $x \neq 0, 1$, $f_1(x) = x$, $f_2(x) = \dfrac{1}{x}$, $f_3(x) = 1 - x$, $f_4(x) = \dfrac{1}{1 - x}$, $f_5(x) = \dfrac{x - 1}{x}$, $f_6(x) = \dfrac{x}{x - 1}$, this family of function is closed under compositions i.e. composition of any two of these functions is again one of these.

Working Space

38 Let F be a function such that $f_1 oF = f_4$, then F is equal to

a. f_1
b. f_2
c. f_3
d. f_4

39 Let G be a function such that $G of_3 = f_6$, then G is

a. f_5
b. f_4
c. f_3
d. f_2

40 Let J be a function such that $f_3 \, o \, J of_2 = f_4$, then J is

a. f_6
b. f_5
c. f_4
d. f_3

One Integer Value Correct Type

Direction *(Q. Nos. 41-44) This section contains 4 questions. When worked out will result in an integer from 0 to 9 (both inclusive).*

41 The number of pairs (x, y) of real numbers with $0 < x < \frac{\pi}{2}$ such that

$$\frac{(\sin x)^{2y}}{(\cos x)^{y^2/2}} + \frac{(\cos x)^{2y}}{(\sin x)^{y^2/2}} = \sin 2x, \text{ is}$$

42 Let $f(x)$ be a polynomial of degree 4 such that $f([i]) = \frac{1}{1+[i]}$ for $0 \le i \le 4$ (where $[x]$ denotes greatest integer function), then $f(6)$ is

43 Let $f(r)$ denotes the number of zeroes in $r\,!$. If $f(m) - f(n) = 3$, then the value of

$$\frac{(m-n)_{\max} - (m-n)_{\min}}{2} \text{ is}$$

44 If p is a positive integer and f be a function defined for positive numbers and attains only positive values such that $f\,[xf(y)] = x^p y^4$, then p is

JEE Main & AIEEE Archive

(Compilation of Last 13 Years' Questions)

1 If $f_k(x) = \frac{1}{k}(\sin^k x + \cos^k x)$, where $x \in R$ and $k \geq 1$, then $f_4(x) - f_6(x)$ is equal to **(JEE Main 2014)**

a. 1/6 b. 1/3
c. 1/4 d. 1/12

2 If g is the inverse of a function f and $f'(x) = \frac{1}{1+x^5}$, then $g'(x)$ is equal to **(JEE Main 2014)**

a. $1+x^5$ b. $5x^4$
c. $\frac{1}{1+\{g(x)\}^5}$ d. $1+\{g(x)\}^5$

3 Let R be the set of real numbers.
Statement I $A = \{(x, y) \in R \times R : y - x$ is an integer$\}$ is an equivalence relation on R.
Statement II $B = \{(x, y) \in R \times R : x = \alpha y$ for some rational number $\alpha\}$ is an equivalence relation on R. **(AIEEE 2011)**

a. Both Statement I and Statement II are correct and Statement II is the correct explanation of Statement I
b. Both Statement I and Statement II are correct but Statement II is not the correct explanation of Statement I
c. Statement I is correct but Statement II is incorrect
d. Statement II is correct but Statement I is incorrect

Condition for equivalence relation A relation which is symmetric, reflexive and transitive is equivalence relation.

4 The domain of the function $f(x) = \frac{1}{\sqrt{|x| - x}}$ is **(AIEEE 2011)**

a. $(0, \infty)$
b. $(-\infty, 0)$
c. $(-\infty, \infty) - \{0\}$
d. $(-\infty, \infty)$

5 Let for $a \neq a_1 \neq 0$, $f(x) = ax^2 + bx + c$, $g(x) = a_1x^2 + b_1 x + c_1$ and $p(x) = f(x) - g(x)$. If $p(x) = 0$ only for $x = -1$ and $p(-2) = 2$, then the value of $p(2)$ is **(AIEEE 2011)**

a. 18 b. 3
c. 9 d. 6

6 Consider the following relation R on the set of real square matrices of order 3.
$R = \{(A, B) : A = P^{-1}BP$ for some invertible matrix $P\}$.
Statement I R is an equivalence relation.
Statement II For any two invertible 3×3 matrices M and N, $(MN)^{-1} = N^{-1}M^{-1}$.

a. Both Statement I and Statement II are correct and Statement II is the correct explanation of Statement I
b. Both Statement I and Statement II are correct but Statement II is not the correct explanation of Statement I
c. Statement I is correct but Statement II is incorrect
d. Statement II is correct but Statement I is incorrect

Condition for equivalence relation A relation which is symmetric, reflexive and transitive is equivalence relation.

7 Let f be a function defined by $f(x) = (x-1)^2 + 1$, $(x \geq 1)$.
Statement I The set $\{x : f(x) = f^{-1}(x)\} = \{1, 2\}$
Statement II f is bijection and $f^{-1}(x) = 1 + \sqrt{x-1}, x \geq 1$ **(AIEEE 2011)**

a. Both Statement I and Statement II are correct and Statement II is the correct explanation of Statement I
b. Both Statement I and Statement II are correct but Statement II is not the correct explanation of Statement I
c. Statement I is correct but Statement II is incorrect
d. Statement II is correct but Statement I is incorrect

8 Consider the following relations:
$R = \{(x, y) \mid x, y$ are real numbers and $x = wy$ for some rational number $w\}$;
$S = \left\{\left(\frac{m}{n}, \frac{p}{q}\right) \middle| m, n, p \text{ and } q \text{ are integers such that } n, q \neq 0 \text{ and } qm = pn\right\}$. Then, **(AIEEE 2010)**

a. R is an equivalence relation but S is not an equivalence relation
b. Neither R nor S is an equivalence relation
c. S is an equivalence relation but R is not an equivalence relation
d. R and S both are equivalence relations

JEE Archive

9 For real x, let $f(x) = x^3 + 5x + 1$, then

a. f is one-one but not onto R **(AIEEE 2009)**

b. f is onto R but not one-one

c. f is one-one and onto R

d. f is neither one-one nor onto R

10 Let $f(x) = (x+1)^2 - 1, x \geq -1$

Statement I The set $\{x : f(x) = f^{-1}(x)\}$

$= \{0, -1\}$

Statement II f is a bijection. **(AIEEE 2009)**

a. Both Statement I and Statement II are correct and Statement II is the correct explanation of Statement I

b. Both Statement I and Statement II are correct but Statement II is not the correct explanation of Statement I

c. Statement I is correct but Statement II is incorrect

d. Statement II is correct but Statement I is incorrect

11 Let R be the real line. Consider the following subsets of the plane $R \times R$

$S = \{(x, y) : y = x + 1 \text{ and } 0 < x < 2\}$

$T = \{(x, y) : x - y \text{ is an integer}\}$ **(AIEEE 2008)**

Which one of the following is true?

a. T is an equivalence relation on R but S is not

b. Neither S nor T is an equivalence relation on R

c. Both S and T are equivalence relations on R

d. S is an equivalence relation on R but T is not

12 Let $f : N \to Y$ be a function defined as $f(x) = 4x + 3$, where $Y = \{y \in N : y = 4x + 3 \text{ for some } x \in N\}$. **(AIEEE 2008)**

Show that f is invertible and its inverse is

a. $f^{-1}(y) = \dfrac{y-3}{4}$ b. $f^{-1}(y) = \dfrac{3y+4}{3}$

c. $f^{-1}(y) = 4 + \dfrac{y+3}{4}$ d. $f^{-1}(y) = \dfrac{y+3}{4}$

13 The largest interval lying in $\left(-\dfrac{\pi}{2}, \dfrac{\pi}{2}\right)$ for which the function

$f(x) = 4^{-x^2} + \cos^{-1}\left(\dfrac{x}{2} - 1\right) + \log(\cos x)$ is defined,

is **(AIEEE 2007)**

a. $[0, \pi]$ b. $\left(-\dfrac{\pi}{2}, \dfrac{\pi}{2}\right)$ c. $\left[-\dfrac{\pi}{4}, \dfrac{\pi}{2}\right)$ d. $\left[0, \dfrac{\pi}{2}\right)$

14 Let W denotes the words in the English dictionary define the relation R by $R = \{(x, y) \in W \times W :$ the words x and y have atleast one letter in common$\}$. Then, R is **(AIEEE 2006)**

a. reflexive, symmetric and not transitive

b. reflexive, symmetric and transitive

c. reflexive, not symmetric and transitive

d. not reflexive, symmetric and transitive

15 Let $R = \{(3, 3), (6, 6), (9, 9), (12, 12), (6, 12), (3, 9), (3, 12), (3, 6)\}$ be a relation on the set $A = \{3, 6, 9, 12\}$. The relation is **(AIEEE 2005)**

a. reflexive and symmetric only

b. an equivalence relation

c. reflexive only

d. reflexive and transitive only

16 Let $f : (-1, 1) \to B$ be a function defined by $f(x) = \tan^{-1}\left(\dfrac{2x}{1-x^2}\right)$, then f is both one-one and onto, when B is in the interval **(AIEEE 2005)**

a. $\left(-\dfrac{\pi}{2}, \dfrac{\pi}{2}\right)$ b. $\left[-\dfrac{\pi}{2}, \dfrac{\pi}{2}\right]$

c. $\left[0, \dfrac{\pi}{2}\right)$ d. $\left(0, \dfrac{\pi}{2}\right)$

17 A real valued function $f(x)$ satisfies the functional equation

$f(x - y) = f(x)f(y) - f(a - x)f(a + y)$,

where a is a given constant and $f(0) = 1$, then $f(2a - x)$ is equal to **(AIEEE 2005)**

a. $f(-x)$ b. $f(a) + f(a - x)$

c. $f(x)$ d. $-f(x)$

18 The graph of the function $y = f(x)$ is symmetrical about the line $x = 2$, then **(AIEEE 2004)**

a. $f(x+2) = f(x-2)$

b. $f(2+x) = f(2-x)$

c. $f(x) = f(-x)$

d. $f(x) = -f(-x)$

19 The domain of the function, $f(x) = \dfrac{\sin^{-1}(x-3)}{\sqrt{9-x^2}}$ is **(AIEEE 2004)**

a. $[2, 3]$ b. $[2, 3)$

c. $[1, 2]$ d. $[1, 2)$

20 A function f from the set of natural numbers to integers defined by

$$f(n) = \begin{cases} \dfrac{n-1}{2}, & \text{when } n \text{ is odd} \\ -\dfrac{n}{2}, & \text{when } n \text{ is even} \end{cases}, \text{ is}$$

a. one-one but not onto **(AIEEE 2003)**

b. onto but not one-one

c. one-one and onto both

d. neither one-one nor onto

21 If $f : R \to R$ satisfies $f(x + y) = f(x) + f(y)$, for all $x, y \in R$ and $f(1) = 7$, then $\sum_{r=1}^{n} f(r)$ is **(AIEEE 2003)**

a. $\dfrac{7n}{2}$ b. $\dfrac{7(n+1)}{2}$

c. $7n(n+1)$ d. $\dfrac{7n(n+1)}{2}$

22 The domain of definition of the function $f(x) = \frac{3}{4 - x^2} + \log_{10}(x^3 - x)$, is **(AIEEE 2003)**

a. (1, 2)
b. $(-1, 0) \cup (1, 2)$
c. $(1, 2) \cup (2, \infty)$
d. $(-1, 0) \cup (1, 2) \cup (2, \infty)$

23 The function $f(x) = \log(x + \sqrt{x^2 + 1})$, is **(AIEEE 2003)**

a. an even function
b. an odd function
c. a periodic function
d. neither an even nor an odd function

24 The period of $\sin^2 \theta$ is **(AIEEE 2002)**

a. π^2 b. π
c. 2π d. $\frac{\pi}{2}$

25 The domain of $\sin^{-1}\left[\log_3\left(\frac{x}{3}\right)\right]$ is **(AIEEE 2002)**

a. [1, 9] b. (–1, 9)
c. (–9, 1) d. (–9, –1)

26 The period of function $f(x) = \sin^4 x + \cos^4 x$ is **(AIEEE 2002)**

a. π
b. $\frac{\pi}{2}$
c. 2π
d. None of the above

27 The domain of definition of the function $f(x) = \sqrt{\log_{10}\left(\frac{5x - x^2}{4}\right)}$ is **(AIEEE 2002)**

a. [1, 4] b. (1, 0)
c. (0, 5) d. (5, 0)

JEE Advanced & IIT JEE Archive

(Compilation of Last 10 Years' Questions)

1 Let $f(x)=\sin\left(\frac{\pi}{6}\sin\left(\frac{\pi}{2}\sin x\right)\right)$ for all $x\in R$ and $g(x)=\frac{\pi}{2}\sin x$ for all $x\in R$. Let $(fog)(x)$ denotes $f[g(x)]$ and $(gof)(x)$ denotes $g[f(x)]$.

Then, which of the following is (are) true?

(2015 Adv., One or More than One Options Correct Type)

a. Range of f is $\left[-\frac{1}{2},\frac{1}{2}\right]$

b. Range of fog is $\left[-\frac{1}{2},\frac{1}{2}\right]$

c. $\lim_{x\to 0}\frac{f(x)}{g(x)}=\frac{\pi}{6}$

d. There is $x\in R$ such that $(gof)(x)=1$

2 Let $f:\left(-\frac{\pi}{2},\frac{\pi}{2}\right)\to R$ be given by $f(x)=[\log(\sec x+\tan x)]^3$. Then,

(2014 Adv., One or More than One Options Correct Type)

a. $f(x)$ is an odd function

b. $f(x)$ is a one-one function

c. $f(x)$ is an onto function

d. $f(x)$ is an even function

3 For every pair of continuous function $f,\ g:[0,1]\to R$ such that

$\max\{f(x):x\in[0,1]\}=\max\{g(x):x\in[0,1]\}$.

The correct statement(s) is (are)

(2014 Adv., One or More than One Options Correct Type)

a. $[f(c)]^2+3f(c)=[g(c)]^2+3g(c)$ for some $c\in[0,1]$

b. $[f(c)]^2+f(c)=[g(c)]^2+3g(c)$ for some $c\in[0,1]$

c. $[f(c)]^2+3f(c)=[g(c)]^2+g(c)$ for some $c\in[0,1]$

d. $[f(c)]^2=[g(c)]^2$ for some $c\in[0,1]$

4 Let $f_1:R\to R, f_2:[0,\infty]\to R, f_3:R\to R$ and $f_4:R\to[0,\infty)$ be defined by **(2014 Adv., Matching Type)**

$f_1(x)=\begin{cases}|x|, & \text{if } x<0\\ e^x, & \text{if } x\ge 0\end{cases}$; $f_2(x)=x^2$; $f_3(x)=\begin{cases}\sin x, & \text{if } x<0\\ x, & \text{if } x\ge 0\end{cases}$

and $f_4(x)=\begin{cases}f_2[f_1(x)], & \text{if } x<0\\ f_2[f_1(x)]-1, & \text{if } x\ge 0\end{cases}$

	Column I		Column II
i.	f_4 is	p.	onto but not one-one
ii.	f_3 is	q.	neither continuous nor one-one
iii.	f_2of_1 is	r.	differentiable but not one-one
iv.	f_2 is	s.	continuous and one-one

Codes

	i	ii	iii	iv
a.	r	p	s	q
b.	p	r	s	q
c.	r	p	q	s
d.	p	r	q	s

5 For every integer n, let a_n and b_n be real numbers. Let function $f:R\to R$ be given by $f(x)=\begin{cases}a_n+\sin\pi x, & \text{for } x\in[2n,2n+1]\\ b_n+\cos\pi x, & \text{for } x\in(2n-1,2n)\end{cases}$, for all integers n.

If f is continuous, then which of the following hold(s) for all n?

(2012, One or More than One Options Correct Type)

a. $a_{n-1}-b_{n-1}=0$

b. $a_n-b_n=1$

c. $a_n-b_{n+1}=1$

d. $a_{n-1}-b_n=-1$

6 The function $f:[0,3]\to[1,29]$, defined by $f(x)=2x^3-15x^2+36x+1$, is

(2012, Only One Option Correct Type)

a. one-one and onto

b. onto but not one-one

c. one-one but not onto

d. neither one-one nor onto

7 Let $f:(0,1)\to R$ be defined by $f(x)=\frac{b-x}{1-bx}$, where b is a constant such that $0<b<1$. Then,

(2011, One or More than One Options Correct Type)

a. f is not invertible on (0, 1)

b. $f\ne f^{-1}$ on (0, 1) and $f'(b)=\frac{1}{f'(0)}$

c. $f=f^{-1}$ on (0, 1) and $f'(b)=\frac{1}{f'(0)}$

d. f^{-1} is differentiable on (0, 1)

8 Let $f(x)=x^2$ and $g(x)=\sin x$ for all $x\in R$. Then, the set of all x satisfying

$(fogogof)(x)=(gogof)(x)$, where $(fog)(x)=f(g(x))$, is **(2011, Only One Option Correct Type)**

a. $\pm\sqrt{n\pi}, n\in\{0,1,2,\ldots\}$

b. $\pm\sqrt{n\pi}, n\in\{1,2,\ldots\}$

c. $\pi/2+2n\pi, n\in\{\ldots,-2,-1,0,1,2,\ldots\}$

d. $2n\pi, n\in\{\ldots,-2,-1,0,1,2,\ldots\}$

9 Let $S=\{1,2,3,4\}$. The total number of unordered pairs of disjoint subsets of S is equal to **(2010, Only One Option Correct Type)**

a. 25

b. 34

c. 42

d. 41

10 Let $f(x) = \dfrac{x^2 - 6x + 5}{x^2 - 5x + 6}$.

Match the Column I with Column II and mark the correct option from the codes given below.

	Column I		Column II
i.	If $-1 < x < 1$, then $f(x)$ satisfies	p.	$0 < f(x) < 1$
ii.	If $1 < x < 2$, then $f(x)$ satisfies	q.	$f(x) < 0$
iii.	If $3 < x < 5$, then $f(x)$ satisfies	r.	$f(x) > 0$
iv.	If $x > 5$, then $f(x)$ satisfies	s.	$f(x) < 1$

(2007, Matching Type)

Codes

	i	ii	iii	iv
a.	q	p	r	r
b.	p	q	q	p
c.	s	p	q	r
d.	p	r	q,r	s

Direction *(Q. Nos. 11 -13) These questions are based on Statement I and Statement II. Select the correct answer from the codes given below.*

11 Let $F(x)$ be an indefinite integral of $\sin^2 x$.

Statement I The function $F(x)$ satisfies $F(x + \pi) = F(x)$ for all real x.

Because

Statement II $\sin^2(x + \pi) = \sin^2 x$ for all real x.

(2007, Statement Type)

12 Let $f(x) = 2 + \cos x$ for all real x.

Statement I For each real t, there exists a point c in $[t, t + \pi]$ such that $f'(c) = 0$.

Because

Statement II $f(t) = f(t + 2\pi)$ for each real t.

(2007, Statement Type)

13 **Statement I** The curve $y = -\dfrac{x^2}{2} + x + 1$ is symmetric with respect to the line $x = 1$.

Statement II A parabola is symmetric about its axis.

(2007, Statement Type)

14 Let $f(x) = \begin{cases} x, & \text{if } x \text{ is rational} \\ 0, & \text{if } x \text{ is irrational} \end{cases}$

$g(x) = \begin{cases} 0, & \text{if } x \text{ is rational} \\ x, & \text{if } x \text{ is irrational} \end{cases}$

Then, $f - g$ is **(2005, Only One Option Correct Type)**

a. one-one and into

b. neither one-one nor onto

c. many one and onto

d. one-one and onto

15 If X and Y are two non-empty sets, where $f : X \to Y$, is function is defined such that

$f(C) = \{f(x) : x \in C\}$ for $C \subseteq X$

and $f^{-1}(D) = \{x : f(x) \in D\}$ for $D \subseteq Y$,

for any $A \subseteq Y$ and $B \subseteq Y$, then

(2005, Only One Option Correct Type)

a. $f^{-1}\{f(A)\} = A$

b. $f^{-1}\{f(A)\} = A$ only if $f(X) = Y$

c. $f\{f^{-1}(B)\} = B$ only if $B \subseteq f(x)$

d. $f\{f^{-1}(B)\} = B$

16 Find the range of values of t for which

$$2 \sin t = \frac{1 - 2x + 5x^2}{3x^2 - 2x - 1},$$

$$t \in \left[-\frac{\pi}{2}, \frac{\pi}{2}\right].$$

(2005, Descriptive Type)

3. Matrices and Determinants

DPP

DAILY PRACTICE PROBLEMS

DPP-1 Types of Matrices, Addition and Subtraction of Matrices

Only One Option Correct Type

Working Space

Direction (Q. Nos. 1-15) *This section contains 15 multiple choice questions. Each question has four choices (a), (b), (c) and (d), out of which ONLY ONE is correct.*

1 The number of matrices having 12 elements is

a. 3 b. 1
c. 6 d. None of these

2 As 3×2 matrix whose elements are given by $a_{ij} = 2i - j$, is

a. $\begin{bmatrix} 1 & 1 \\ 3 & 2 \\ 5 & 4 \end{bmatrix}$ b. $\begin{bmatrix} 1 & 0 \\ 3 & 2 \\ 5 & 4 \end{bmatrix}$

c. $\begin{bmatrix} 1 & 1 \\ -3 & 2 \\ -5 & 4 \end{bmatrix}$ d. None of these

3 If $a_{ij} = \frac{1}{2}(3i - 2j)$ and $A = \{a_{ij}\}_{2 \times 2}$, then A is

a. $\begin{bmatrix} 1/2 & -1/2 \\ 2 & 1 \end{bmatrix}$ b. $\begin{bmatrix} -1/2 & 2 \\ 1/2 & 1 \end{bmatrix}$

c. $\begin{bmatrix} 2 & 2 \\ 1/2 & -1/2 \end{bmatrix}$ d. None of these

4 If $A = \begin{bmatrix} 1 & 0 & 0 \\ 2 & 3 & 0 \\ 4 & 5 & 6 \end{bmatrix}$, then A is

a. an upper triangular matrix b. a null matrix
c. a lower triangular matrix d. None of these

5 If A and B are square matrix of same order, then

a. $A + B = B + A$ b. $A + B = A - B$
c. $A - B = B - A$ d. $AB = BA$

6 Choose the correct answer.

a. Every identity matrix is a scalar matrix
b. Every scalar matrix is an identity matrix
c. Every diagonal matrix is an identity matrix
d. A square matrix whose each element is 1 is an identity matrix

Working Space

7 The minimum number of zero's in an upper triangular matrix of order $n \times n$, is

a. $\frac{n(n-1)}{2}$ b. $\frac{n(n+1)}{2}$

c. $\frac{n(n^2-1)}{2}$ d. None of these

8 If $A = \begin{bmatrix} 1 & -5 & 7 \\ 0 & 7 & 9 \\ 11 & 8 & 9 \end{bmatrix}$, then the trace of A is

a. 17 b. 25

c. 3 d. 12

9 If every row of matrix A contains p elements and its column contains q elements, then the order of A is

a. $p \times p$ b. $q \times q$

c. $p \times q$ d. $q \times p$

10 If $\begin{bmatrix} 2 & -3 \\ 4 & 0 \end{bmatrix} - \begin{bmatrix} a & c \\ b & d \end{bmatrix} = \begin{bmatrix} 1 & 4 \\ 2 & -5 \end{bmatrix}$, then (a, b, c, d) is

a. (1, 6, 2, 5) b. (1, 2, 7, 5)

c. (1, 2, – 7, 5) d. None of these

11 If $2A + 3B = \begin{bmatrix} 2 & -1 & 4 \\ 3 & 2 & 5 \end{bmatrix}$ and $A + 2B = \begin{bmatrix} 5 & 0 & 3 \\ 1 & 6 & 2 \end{bmatrix}$, then B is

a. $\begin{bmatrix} 8 & -1 & 2 \\ -1 & 10 & -1 \end{bmatrix}$ b. $\begin{bmatrix} 8 & 1 & 2 \\ -1 & 10 & -1 \end{bmatrix}$

c. $\begin{bmatrix} 8 & 1 & -2 \\ -1 & 10 & -1 \end{bmatrix}$ d. $\begin{bmatrix} 8 & 1 & 2 \\ 1 & 10 & 1 \end{bmatrix}$

12 If $A = \begin{bmatrix} 1 & -2 \\ 3 & 0 \end{bmatrix}$, $B = \begin{bmatrix} -1 & 4 \\ 2 & 3 \end{bmatrix}$, $C = \begin{bmatrix} 0 & -1 \\ 1 & 0 \end{bmatrix}$, then $5A - 3B - 2C$, is

a. $\begin{bmatrix} 8 & 20 \\ 7 & 9 \end{bmatrix}$ b. $\begin{bmatrix} 8 & -20 \\ 7 & -9 \end{bmatrix}$

c. $\begin{bmatrix} -8 & 20 \\ -7 & 9 \end{bmatrix}$ d. None of these

13 If $\begin{bmatrix} x & 0 \\ 1 & y \end{bmatrix} + \begin{bmatrix} -2 & 1 \\ 3 & 4 \end{bmatrix} = \begin{bmatrix} 3 & 5 \\ 6 & 3 \end{bmatrix} - \begin{bmatrix} 2 & 4 \\ 2 & 1 \end{bmatrix}$, then

a. $x = -3, y = -2$ b. $x = 3, y = -2$

c. $x = 3, y = 2$ d. $x = -3, y = 2$

14 If $\begin{bmatrix} x+y & 2x+z \\ x-y & 2z+w \end{bmatrix} = \begin{bmatrix} 4 & 7 \\ 0 & 10 \end{bmatrix}$, then values of x, y, z, w are

a. 2, 2, 3, 4 b. 2, 3, 1, 2

c. 3, 3, 0, 1 d. None of these

15 Each element of a 25×25 matrix is either 1 or –1. Let a_i be the product of all elements of the ith row and b_j be the product of all elements in the jth column, then $a_1 \cdot a_2 \cdot a_3 \cdot a_4 \ldots a_{25}$ is

a. $a_1 + a_2 + a_3 + \ldots + a_{25}$ b. $b_1 + b_2 + b_3 + \ldots + b_{25}$

c. $b_1 b_2 b_3 \ldots b_{25}$ d. 1

DPP-2 Multiplication and Transpose of a Matrix

Only One Option Correct Type

Working Space

Direction (Q. Nos. 1-25) This section contains 25 multiple choice questions. Each question has four choices (a), (b), (c) and (d), out of which ONLY ONE is correct.

1 If $A=\begin{bmatrix}1\\2\\3\end{bmatrix}$ and $B=\begin{bmatrix}-5&4&0\\0&2&-1\\1&-3&2\end{bmatrix}$, then

a. $AB=\begin{bmatrix}-5&8&0\\0&4&-2\\3&-9&6\end{bmatrix}$

b. $AB=[-2\ \ -1\ \ 4]$

c. $AB=\begin{bmatrix}-1\\1\\1\end{bmatrix}$

d. AB does not exists

2 If A and B are square matrices of order 2, then $(A+B)^2$ is

a. $A^2+2AB+B^2$

b. $A^2+AB+BA+B^2$

c. $A^2+2BA+B^2$

d. None of these

3 If $A=\begin{bmatrix}3&-4\\1&-1\end{bmatrix}$, then A^3 is

a. $\begin{bmatrix}7&-12\\3&-5\end{bmatrix}$

b. $\begin{bmatrix}-7&12\\-3&5\end{bmatrix}$

c. $\begin{bmatrix}7&12\\3&5\end{bmatrix}$

d. None of these

4 If $A=\begin{bmatrix}0&-\tan\alpha/2\\\tan\alpha/2&0\end{bmatrix}$, then $(I-A)\begin{bmatrix}\cos\alpha&-\sin\alpha\\\sin\alpha&\cos\alpha\end{bmatrix}$ is

a. $I+A$

b. $2I+2A$

c. $I-A$

d. $I+2A$

5 If $A=\text{diag}\{2,-1,3\}$ and $B=\text{diag}\{-1,3,2\}$, then A^2B is

a. diag {5, 4, 11}

b. diag {−4, 3, 18}

c. diag {3, 1, 8}

d. B

6 If $A=\begin{bmatrix}2&-1\\-7&4\end{bmatrix}$ and $B=\begin{bmatrix}4&1\\7&2\end{bmatrix}$, then B^TA^T is

a. null matrix

b. an identity matrix

c. scalar but not an identity matrix

d. such that $T_r\ (B^TA^T)=4$

Working Space

7 If the matrix AB is zero matrix, then

a. $A = 0$ or $B = 0$
b. $A = 0$ and $B = 0$
c. It is not necessary that either $A = 0$ or $B = 0$
d. All the above statements are wrong

8 Which relation is true for $A = \begin{bmatrix} 2 & -1 \\ -1 & 2 \end{bmatrix}, B = \begin{bmatrix} 1 & 4 \\ -1 & 1 \end{bmatrix}$?

a. $(A + B)^2 = A^2 + 2AB + B^2$
b. $(A - B)^2 = A^2 - 2AB + B^2$
c. $AB = BA$
d. None of the above

9 If $AB = 0$ for the matrices $A = \begin{bmatrix} \cos^2\theta & \cos\theta\sin\theta \\ \cos\theta\sin\theta & \sin^2\theta \end{bmatrix}$

and $B = \begin{bmatrix} \cos^2\phi & \cos\phi\sin\phi \\ \cos\phi\sin\phi & \sin^2\phi \end{bmatrix}$, then $(\theta - \phi)$ is

a. an odd multiple of $\pi/2$
b. an odd multiple of π
c. an even multiple of $\pi/2$
d. 0

10 If $X = \begin{bmatrix} 3 & -4 \\ 1 & -1 \end{bmatrix}$, then X^n is

a. $\begin{bmatrix} 3n & -4n \\ n & n \end{bmatrix}$
b. $\begin{bmatrix} 2+n & 5-n \\ n & -n \end{bmatrix}$
c. $\begin{bmatrix} 3^n & (-4)^n \\ 1 & (-1)^n \end{bmatrix}$
d. None of these

11 If $AB = A$ and $BA = B$, then B^2 is

a. B
b. A
c. 1
d. 0

12 If A and B are symmetric matrices, then ABA is

a. symmetric matrix
b. skew-symmetric
c. diagonal matrix
d. None of these

13 If A is skew-symmetric matrix and n is an even positive integer, then A^n is

a. a symmetric matrix
b. a skew-symmetric matrix
c. a diagonal matrix
d. None of these

14 If A is a non-singular matrix and A^T denotes the transpose of A, then

a. $|A| \neq |A^T|$
b. $|AA^T| \neq |A|^2$
c. $|A^T \cdot A| \neq |A^T|^2$
d. $|A| + |A^T| \neq 0$

15 Which of the following is incorrect?

a. $A^2 - B^2 = (A + B)(A - B)$
b. $(A^T)^T = A$
c. $(AB)^n = A^n B^n$, where AB commute
d. $(A - I)(I + A) = 0 \Leftrightarrow A^2 = I$

16 If $A = \begin{bmatrix} a & b \\ c & d \end{bmatrix}$ satisfies the equation $x^2 - (a + d)x + k = 0$, then

a. $k = bc$
b. $k = ad$
c. $k = a^2 + b^2 + c^2 + d^2$
d. $k = ad - bc$

Working Space

17 Which of the following is nilpotent matrix?

a. $\begin{bmatrix} 1 & 0 \\ 0 & 1 \end{bmatrix}$

b. $\begin{bmatrix} \cos\theta & -\sin\theta \\ \sin\theta & \cos\theta \end{bmatrix}$

c. $\begin{bmatrix} 0 & 0 \\ 1 & 0 \end{bmatrix}$

d. $\begin{bmatrix} 1 & 1 \\ 1 & 1 \end{bmatrix}$

18 If $P = \begin{bmatrix} \sqrt{3}/2 & 1/2 \\ -1/2 & \sqrt{3}/2 \end{bmatrix}$, $A = \begin{bmatrix} 1 & 1 \\ 0 & 1 \end{bmatrix}$ and $Q = PAP^T$ and $X = P^T\ Q^{2005}P$, then X is **[IIT JEE 2005]**

a. $\begin{bmatrix} 1 & 2005 \\ 0 & 1 \end{bmatrix}$

b. $\begin{bmatrix} 4+2005\sqrt{3} & 6015 \\ 2005 & 4-2005\sqrt{3} \end{bmatrix}$

c. $\begin{bmatrix} 1 & 0 \\ 2005 & 1 \end{bmatrix}$

d. $\begin{bmatrix} 1 & 0 \\ 0 & 1 \end{bmatrix}$

19 If A and B are square matrices of equal order, then which one is correct? **[IIT JEE 1995]**

a. $A+B=B+A$

b. $A+B=A-B$

c. $A-B=B-A$

d. $AB=BA$

20 If $A = \begin{bmatrix} \alpha & 0 \\ 1 & 1 \end{bmatrix}$ and $B = \begin{bmatrix} 1 & 0 \\ 5 & 1 \end{bmatrix}$. Then value of α for which $A^2 = B$, is **[IIT JEE 2003]**

a. 1

b. −1

c. 4

d. no real values

21 If A is a square matrix for which $a_{ij} = i^2 - j^2$, then A is

a. zero matrix

b. unit matrix

c. symmetric matrix

d. skew-symmetric matrix

22 The matrix $A = \begin{bmatrix} 1/\sqrt{2} & 1/\sqrt{2} \\ -1/\sqrt{2} & -1/\sqrt{2} \end{bmatrix}$, is

a. unitary

b. orthogonal

c. nilpotent

d. involuntary

23 Let $A = \begin{bmatrix} 0 & 0 & -1 \\ 0 & -1 & 0 \\ -1 & 0 & 0 \end{bmatrix}$. The only correct statements about the matrix A is

a. $A^2 = I$

b. $A = -I$

c. $A^2 = A$

d. $A^2 = 0$

24 For the matrix $A = \begin{bmatrix} 1 & 1 & 0 \\ 1 & 2 & 1 \\ 2 & 1 & 0 \end{bmatrix}$, which of the following is correct?

a. $A^3 + 3A^2 - I = 0$

b. $A^3 - 3A^2 - I = 0$

c. $A^3 + 2A^2 - I = 0$

d. $A^3 - A^2 + I = 0$

25 An orthogonal matrix is

a. $\begin{bmatrix} \cos\alpha & 2\sin\alpha \\ -2\sin\alpha & \cos\alpha \end{bmatrix}$

b. $\begin{bmatrix} \cos\alpha & \sin\alpha \\ -\sin\alpha & \cos\alpha \end{bmatrix}$

c. $\begin{bmatrix} \cos\alpha & \sin\alpha \\ \sin\alpha & \cos\alpha \end{bmatrix}$

d. $\begin{bmatrix} 1 & 1 \\ 1 & 1 \end{bmatrix}$

One or More than One Options Correct Type

Working Space

Direction *(Q. Nos. 26-30) This section contains 5 multiple choice questions. Each question has four choices (a), (b), (c) and (d), out of which ONE or MORE THAN ONE are correct.*

26 If A and B are two matrices such that $AB = BA$, then for every $n \in N$

a. $(AB)^n = A^n B^n$

b. $A^n B = BA^n$

c. $A^{2n} - B^{2n} = (A^n - B^n)(A^n + B^n)$

d. $(A+B)^n = A^n + {}^nC_1 A^{n-1}B + {}^nC_2 A^{n-2}B^2 + \cdots + B^n$

27 Which one of the following is wrong?

a. The elements on the main diagonal of a symmetric matrix are all zero

b. The elements on the main diagonal of a skew-symmetric matrix are all zero

c. For any square matrix A, $\frac{1}{2}(A + A^T)$ is symmetric

d. For any square matrix A, $\frac{1}{2}(A - A^T)$ is skew-symmetric

28 If B is an idempotent matrix and $A = I - B$, then

a. $A^2 = A$ b. $A^2 = I$ c. $AB = 0$ d. $BA = 0$

29 Which of the following is true for matrix $A = \begin{bmatrix} 1 & -1 \\ 2 & 3 \end{bmatrix}$?

a. $A + 4I$ is a symmetric matrix

b. $A^2 - 4A + 5I_2 = 0$

c. $A - B$ is a diagonal matrix for any value of $\alpha, \beta = \begin{bmatrix} \alpha & -1 \\ 2 & 5 \end{bmatrix}$

d. $A - 4I$ is a skew-symmetric matrix

30 If the product of n matrices

$\begin{bmatrix} 1 & 1 \\ 0 & 1 \end{bmatrix}\begin{bmatrix} 1 & 2 \\ 0 & 1 \end{bmatrix}\begin{bmatrix} 1 & 3 \\ 0 & 1 \end{bmatrix}\cdots\begin{bmatrix} 1 & n \\ 0 & 1 \end{bmatrix} = \begin{bmatrix} 1 & 378 \\ 0 & 1 \end{bmatrix}$, then the value of n is

a. 26 b. 27 c. 377 d. 378

One Integer Value Correct Type

Direction *(Q. Nos. 31-35) This section contains 5 questions. When worked out will result in an integer from 0 to 9 (both inclusive).*

31 If $A = \begin{bmatrix} a & b & c \\ b & c & a \\ c & a & b \end{bmatrix}$, where a, b, c are real positive numbers, $abc = 1$ and $A^T A = I$, then find the value of $a^3 + b^3 + c^3$ is, **[IIT JEE 2003]**

32 If $\begin{bmatrix} 1 & 2 \\ 2 & 3 \end{bmatrix}$ and $M^2 - \lambda M - I_2 = 0$, then the value of λ is

33 If $A = \begin{bmatrix} 1 & 1 \\ 1 & 1 \end{bmatrix}$ and $A^{100} = \lambda^{99} A$, then the value of λ is

34 If $A = \begin{bmatrix} 0 & -1 \\ 1 & 0 \end{bmatrix}$ and $A^{16} = \lambda I$, then the value of λ is

35 If A is 2×2 matrix such that $A\begin{bmatrix} 1 \\ -1 \end{bmatrix} = \begin{bmatrix} -1 \\ 2 \end{bmatrix}$ and $A^2\begin{bmatrix} 1 \\ -1 \end{bmatrix} = \begin{bmatrix} 1 \\ 0 \end{bmatrix}$, the sum of the elements of A is

DPP-3 Properties of Determinant

Only One Option Correct Type

Working Space

Direction *(Q. Nos. 1-27) This section contains 27 multiple choice questions. Each question has four choices (a), (b), (c) and (d), out of which ONLY ONE is correct.*

1 If $\begin{vmatrix} 4 & 2 \\ 1 & 1 \end{vmatrix}^2 = \begin{vmatrix} 3 & 2 \\ 1 & x \end{vmatrix} - \begin{vmatrix} x & 3 \\ -2 & 1 \end{vmatrix}$, then x is

a. -14
b. 2
c. 6
d. 7

2 The sum of the products of the elements of any row of a determinant A with the cofactor of same row is always equal to

a. 1
b. 0
c. $|A|$
d. $\frac{1}{2}|A|$

3 If A and B are square matrices of the same order, then

a. $(AB)^T = A^T B^T$
b. $(AB)^T = B^T A^T$
c. $AB = 0$; if $|A| = 0$ or $|B| = 0$
d. $AB = 0$; if $A = I$ or $B = I$

4 The value of $\begin{vmatrix} a-b & b-c & c-a \\ x-y & y-z & z-x \\ p-q & q-r & r-p \end{vmatrix}$ is equal to

a. $a(x+y+z) + b(p+q+r) + c$
b. 0
c. $abc + xyz + pqr$
d. None of these

5 The roots of $\begin{vmatrix} 1 & 4 & 20 \\ 1 & -2 & 5 \\ 1 & 2x & 5x^2 \end{vmatrix} = 0$, are

a. $-1, -2$
b. $-1, 2$
c. $1, -2$
d. $1, 2$

6 If $\begin{vmatrix} x+1 & 3 & 5 \\ 2 & x+2 & 5 \\ 2 & 3 & x+4 \end{vmatrix} = 0$, then x is equal to

a. 1, 9
b. $-1, 9$
c. $-1, -9$
d. $1, -9$

7 The value of $\begin{vmatrix} a+b & a+2b & a+3b \\ a+2b & a+3b & a+4b \\ a+4b & a+5b & a+6b \end{vmatrix}$, is

a. $a^2 + b^2 + c^2 - 3abc$
b. $3ab$
c. $3a + 5b$
d. 0

Working Space

8 The value of $\begin{vmatrix} 1 & 1 & 1 \\ a & b & c \\ a^3 & b^3 & c^3 \end{vmatrix}$ is equal to

a. $a^3 + b^3 + c^3 - 3abc$
b. $a^3 + b^3 + c^3 + 3abc$
c. $(a + b + c)(a - b)(b - c)(c - a)$
d. None of these

9 If $f(x) = \begin{vmatrix} x - 3 & 2x^2 - 18 & 3x^3 - 81 \\ x - 5 & 2x^2 - 50 & 4x^3 - 500 \\ 1 & 2 & 3 \end{vmatrix}$, then $f(1) \cdot f(3) + f(3) \cdot f(5) \cdot f(5) \cdot f(1)$ is equal to

a. $f(1)$
b. $f(3)$
c. $f(1) + f(3)$
d. $f(1) + f(5)$

10 The value of $\begin{vmatrix} a & b & c \\ b & c & a \\ c & a & b \end{vmatrix}$ is equal to

a. $3abc + a^3 + b^3 + c^3$
b. $3abc - a^3 - b^3 - c^3$
c. $abc - a^3 + b^3 + c^3$
d. $abc + a^3 - b^3 - c^3$

11 The determinant $\begin{vmatrix} a & b & a\alpha + b \\ b & c & b\alpha + c \\ a\alpha + b & b\alpha + c & 0 \end{vmatrix} = 0$, if a, b, c are in

[IIT JEE 1996]

a. AP
b. GP
c. HP
d. None of these

12 The value of $\begin{vmatrix} 1+x & 1 & 1 \\ 1 & 1+y & 1 \\ 1 & 1 & 1+z \end{vmatrix}$, is equal to

a. $xyz\left(1+\frac{1}{x}+\frac{1}{y}+\frac{1}{z}\right)$
b. xyz
c. $1+\frac{1}{x}+\frac{1}{y}+\frac{1}{z}$
d. $\frac{1}{x}+\frac{1}{y}+\frac{1}{z}$

13 If a, b, c are unequal, then what is the condition that the value of the given determinant is zero, $\Delta = \begin{vmatrix} a & a^2 & a^3 + 1 \\ b & b^2 & b^3 + 1 \\ c & c^2 & c^3 + 1 \end{vmatrix}$?

a. $1 + abc = 0$
b. $a + b + c + 1 = 0$
c. $(a - b)(b - c)(c - a) = 0$
d. None of these

14 If $\begin{vmatrix} 1 + \sin^2\theta & \sin^2\theta & \sin^2\theta \\ \cos^2\theta & 1 + \cos^2\theta & \cos^2\theta \\ 4\sin 4\theta & 4\sin 4\theta & 1 + 4\sin 4\theta \end{vmatrix} = 0$, then $\sin 4\theta$ is equal to

a. $1/2$
b. 1
c. $-1/2$
d. -1

15 If $\begin{vmatrix} x + 1 & x + 2 & x + 3 \\ x + 2 & x + 3 & x + 4 \\ x + a & x + b & x + c \end{vmatrix} = 0$, then a, b, c are in

a. AP
b. GP
c. HP
d. None of these

Working Space

16 $\begin{vmatrix} 1 & 1 & 1 \\ \cos(nx) & \cos(n+1)x & \cos(n+2)x \\ \sin(nx) & \sin(n+1)x & \sin(n+2)x \end{vmatrix}$ is not dependent

a. on x
b. on n
c. Both on x and n
d. None of these

17 The value of $\begin{vmatrix} b^2-ab & b-c & bc-ac \\ ab-a^2 & a-b & b^2-ab \\ bc-ac & c-a & ab-a^2 \end{vmatrix}$ is

a. $abc\,(a+b+c)$
b. $3a^2b^2c^2$
c. 0
d. None of these

18 The value of $\begin{vmatrix} b^2+c^2 & a^2 & a^2 \\ b^2 & c^2+a^2 & b^2 \\ c^2 & c^2 & a^2+b^2 \end{vmatrix}$ is

a. abc
b. $4abc$
c. $4a^2b^2c^2$
d. $a^2b^2c^2$

19 If $\begin{vmatrix} a & b & a+b \\ b & c & b+c \\ a+b & b+c & 0 \end{vmatrix} = 0$, then a, b, c are in

a. AP
b. GP
c. HP
d. None of these

20 The value of $\begin{vmatrix} (a^x+a^{-x})^2 & (a^x-a^{-x})^2 & 1 \\ (b^x+b^{-x})^2 & (b^x-b^{-x})^2 & 1 \\ (c^x+c^{-x})^2 & (c^x-c^{-x})^2 & 1 \end{vmatrix}$ is

a. 0
b. $2abc$
c. $a^2b^2c^2$
d. None of these

21 If n is an integer and $\begin{vmatrix} x^n & x^{n+2} & x^{n+3} \\ y^n & y^{n+2} & y^{n+3} \\ z^n & z^{n+2} & z^{n+3} \end{vmatrix} = (x-y)(y-z)(z-x)\cdot\left(\frac{1}{x}+\frac{1}{y}+\frac{1}{z}\right)$.

Then, n is equal to

a. 1
b. -1
c. 2
d. None of these

22 If a, b, c are respectively the pth, qth, rth terms of an AP, then $\begin{vmatrix} a & p & 1 \\ b & q & 1 \\ c & r & 1 \end{vmatrix}$ is

a. 1
b. -1
c. 0
d. pqr

23 If x is a positive integer, then $\Delta = \begin{vmatrix} x! & (x+1)! & (x+2)! \\ (x+1)! & (x+2)! & (x+3)! \\ (x+2)! & (x+3)! & (x+4)! \end{vmatrix}$, is

a. $2(x)!(x+1)!$
b. $2(x!)(x+1)!(x+2)!$
c. $2(x!)(x+3)!$
d. None of these

24 The value of $\begin{vmatrix} 1 & 1 & 1 \\ 1 & \omega^2 & \omega \\ 1 & \omega & \omega^2 \end{vmatrix}$ is (where ω is cube root of unity)

a. $3\sqrt{3}i$
b. $-3\sqrt{3}i$
c. $i\sqrt{3}$
d. 3

Working Space

25 The value of $\sum_{n=1}^{N} U_n$, if $U_n = \begin{vmatrix} n & 1 & 5 \\ n^2 & 2N+1 & 2N+1 \\ n^3 & 3N^2 & 3N \end{vmatrix}$ is

a. 0
b. 1
c. –1
d. None of these

26 If $D_r = \begin{vmatrix} 2^{r-1} & 2\cdot 3^{r-1} & 4\cdot 5^{r-1} \\ x & y & z \\ 2^n - 1 & 3^n - 1 & 4^n - 1 \end{vmatrix}$, then the value of $\sum_{r=1}^{n} D_r$ is

a. 1
b. –1
c. 0
d. None of these

27 If A and B are square matrices of order 3 such that $|A| = -1$, $|B| = 3$, then $|3AB|$ is equal to

a. –9
b. –81
c. –27
d. 81

One or More than One Options Correct Type

Direction (Q. Nos. 28-32) *This section contains 5 multiple choice questions. Each question has four choices (a), (b), (c) and (d), out of which ONE or MORE THAN ONE are correct.*

28 If $\Delta = \begin{vmatrix} x & 2y - z & -z \\ y & 2x - z & -z \\ z & 2y - z & 2x - 2y - z \end{vmatrix}$, then

a. $(x - y)$ is a factor of Δ
b. $(x - y)^2$ is a factor of Δ
c. $(x - y)^3$ is a factor of Δ
d. Δ is dependent on z

29 Let $a, b > 0$ and $\Delta = \begin{vmatrix} -x & a & b \\ b & -x & a \\ a & b & -x \end{vmatrix}$, then

a. $(a + b - x)$ is a factor of Δ
b. $x^2 + (a + b)x + a^2 + b^2 - ab$ is a factor of Δ
c. $\Delta = 0$ has three real roots, if $a = b$
d. None of the above

30 If $\Delta = \begin{vmatrix} b & c & b\alpha + c \\ c & d & c\alpha + d \\ b\alpha + c & c\alpha + d & a\alpha^3 - c\alpha \end{vmatrix} = 0$, if **[IIT JEE 1997]**

a. b, c, d are in an AP
b. b, c, d are in GP
c. b, c, d are in HP
d. α is root of $ax^3 - bx^2 - 3cx - d = 0$

31 Let $\Delta = \begin{vmatrix} a & a^2 & 0 \\ 1 & 2a + b & (a + b)^2 \\ 0 & 1 & 2a + 3b \end{vmatrix}$, then

a. $(a + b)$ is a factor of Δ
b. $(a + 2b)$ is a factor of Δ
c. $(2a + 3b)$ is a factor of Δ
d. a^2 is a factor of Δ

32 If for the matrix $A = \begin{bmatrix} \alpha & 2 \\ 2 & \alpha \end{bmatrix}$; $|A^3| = 125$, then the value of α is **[IIT JEE 2004]**

a. 1
b. –1
c. 3
d. –3

Working Space

Comprehension Type

Direction (Q. Nos. 33 and 34) *This sections contain a paragraph, describing theory, experiments, data, etc. Two questions related to the paragraph have been given. Each question has only one correct answer among the four given options (a), (b), (c) and (d).*

Passage

Let A is a matrix of order 3×3 and a_{ij} is its element of ith row and jth column. T_r is the arithmetic mean of elements of rth row and $a_{ij} + a_{jk} + a_{ki} = 0$ holds for all $1 \le i, j, k \le 3$.

33 $\sum_{1 \le i, j \le 3} a_{ij}$ is not equal to

a. $t_1 + t_2 + t_3$
b. 0
c. $|A|^2$
d. $t_1 t_2 t_3$

34 A is

a. non-singular
b. symmetric
c. skew-symmetric
d. neither symmetric nor skew-symmetric

One Integer Value Correct Type

Direction (Q. Nos. 35-40) *This section contains 6 questions. When worked out will result in an integer from 0 to 9 (both inclusive).*

35 If $A = \begin{bmatrix} 1 & 1 \\ 1 & 1 \end{bmatrix}$ and $\det(A^n - I) = 1 - \lambda^n, n \in N$. Then, the value of λ, is

36 Let the matrix A and B be defined as $A = \begin{bmatrix} 3 & 2 \\ 2 & 1 \end{bmatrix}$ and $B = \begin{bmatrix} 3 & 1 \\ 7 & 3 \end{bmatrix}$. Then, the value of $|\det(2A^9 B^{-1})|$ is

37 If $A = \begin{bmatrix} 1 & \tan x \\ -\tan x & 1 \end{bmatrix}$, then let us define a function $f(x) = \det(A^T A^{-1})$, then which of the following cannot be the value of $f(f(f(f \cdots f(x))))$ is $(n \ge 2)$?

38 Let $f(x) = \begin{vmatrix} 1+\sin^2 x & \cos^2 x & 4\sin 2x \\ \sin^2 x & 1+\cos^2 x & 4\sin 2x \\ \sin^2 x & \cos^2 x & 1+4\sin 2x \end{vmatrix}$, then the maximum value of $f(x)$, is

39 If $a^2 + b^2 + c^2 = -2$ and $f(x) = \begin{vmatrix} 1+a^2 x & (1+b^2)x & (1+c)^2 x \\ (1+a^2)x & 1+b^2 x & (1+c^2)x \\ (1+a^2)x & (1+b^2)x & 1+c^2 x \end{vmatrix}$, then $f(x)$ is a polynomial of degree

40 If A, B and C are $n \times n$ matrices and $\det(A) = 2$, $\det(B) = 3$ and $\det(C) = 5$, then the value of $\frac{5}{3}\det(A^2 BC^{-1})$ is equal to

DPP-4 Adjoint and Inverse of a Matrix

Only One Option Correct Type

Working Space

Direction *(Q. Nos. 1-25) This section contains 25 multiple choice questions. Each question has four choices (a), (b), (c) and (d), out of which ONLY ONE is correct.*

1 $(\text{adj } A^T) - (\text{adj } A)^T$ is equal to

a. $2|A|$ b. $2|A|I$
c. 0 d. I

2 If A and B are square matrices of same order, then adj (AB) is

a. $(\text{adj}A)\cdot(\text{adj}B)$ b. $(\text{adj}B)\cdot(\text{adj}A)$
c. $\text{adj } A + \text{adj } B$ d. $(\text{adj } A) - (\text{adj } B)$

3 If $A = \begin{bmatrix} 4 & 2 \\ 3 & 3 \end{bmatrix}$, then adj (adj$A$) is equal to

a. $\begin{bmatrix} 3 & -2 \\ -3 & 4 \end{bmatrix}$ b. $\begin{bmatrix} 4 & 2 \\ 3 & 3 \end{bmatrix}$
c. $6\begin{bmatrix} 4 & 2 \\ 3 & 3 \end{bmatrix}$ d. None of these

4 If A is a square matrix of order n, then the value of $|\text{adj } A|$, is

a. $|A|^{n-1}$ b. $|A|^n$
c. $|A|^{n+1}$ d. $|A|^{n+2}$

5 The adjoint of symmetric matrix, is

a. symmetric matrix
b. skew-symmetric matrix
c. diagonal matrix
d. None of the above

6 The adjoint of the matrix $\begin{bmatrix} 1 & -2 & 3 \\ 0 & 2 & -1 \\ -4 & 5 & 2 \end{bmatrix}$, is

a. $\begin{bmatrix} 9 & 19 & -4 \\ 4 & 14 & 1 \\ 8 & 3 & 2 \end{bmatrix}$ b. $\begin{bmatrix} 9 & 4 & 8 \\ 19 & 14 & 3 \\ -4 & 1 & 2 \end{bmatrix}$
c. $\begin{bmatrix} 9 & -19 & -4 \\ -4 & 14 & -1 \\ 8 & -3 & 2 \end{bmatrix}$ d. None of these

7 For any square matrix A, which statement is wrong?

a. $(\text{adj } A)^{-1} = \text{adj}(A^{-1})$ b. $(A^T)^{-1} = (A^{-1})^T$
c. $(A^3)^{-1} = (A^{-1})^3$ d. None of these

Working Space

8 The inverse of the matrix $\begin{bmatrix} 4 & 7 \\ 1 & 2 \end{bmatrix}$ is

a. $\begin{bmatrix} 2 & -7 \\ -1 & 4 \end{bmatrix}$ b. $\begin{bmatrix} 2 & -1 \\ -7 & 4 \end{bmatrix}$

c. $\begin{bmatrix} -2 & 7 \\ 1 & -4 \end{bmatrix}$ d. None of these

9 If $A = \begin{bmatrix} 1 & 2 \\ -4 & -1 \end{bmatrix}$, then A^{-1} is

a. $\frac{1}{7}\begin{bmatrix} 1 & 2 \\ -4 & -1 \end{bmatrix}$ b. $\begin{bmatrix} -1 & -2 \\ 4 & 1 \end{bmatrix}$

c. $\frac{1}{9}\begin{bmatrix} 1 & 2 \\ 4 & 1 \end{bmatrix}$ d. $\frac{1}{7}\begin{bmatrix} -1 & -2 \\ 4 & 1 \end{bmatrix}$

10 For two suitable order matrices A and B, correct statement is

a. $(AB)^{-1} = A^{-1}B^{-1}$ b. $(AB)^{-1} = B^{-1}A^{-1}$

c. $(AB)^{-1} = (BA)^{-1}$ d. None of these

11 If $A = \begin{bmatrix} 2 & 0 & 0 \\ 0 & 2 & 0 \\ 0 & 0 & 2 \end{bmatrix}$, then the value of A^{-1} is

a. $\begin{bmatrix} 1 & 0 & 0 \\ 0 & 1 & 0 \\ 0 & 0 & 1 \end{bmatrix}$ b. $\begin{bmatrix} 1/2 & 0 & 0 \\ 0 & 1/2 & 0 \\ 0 & 0 & 1/2 \end{bmatrix}$

c. $\begin{bmatrix} -2 & 0 & 0 \\ 0 & -2 & 0 \\ 0 & 0 & -2 \end{bmatrix}$ d. None of these

12 Inverse of the matrix $\begin{bmatrix} 3 & -2 & -1 \\ -4 & 1 & -1 \\ 2 & 0 & 1 \end{bmatrix}$ is

a. $\begin{bmatrix} 1 & 2 & 7 \\ 3 & 3 & 3 \\ -2 & -4 & -5 \end{bmatrix}$ b. $\begin{bmatrix} 1 & -3 & 5 \\ 7 & 4 & 6 \\ 4 & 2 & 7 \end{bmatrix}$

c. $\begin{bmatrix} 1 & 2 & 3 \\ 2 & 5 & 7 \\ -2 & -4 & -5 \end{bmatrix}$ d. $\begin{bmatrix} 1 & 2 & -4 \\ 8 & -4 & -5 \\ 3 & 5 & 2 \end{bmatrix}$

13 If $A = \begin{bmatrix} 1 & 0 & -k \\ 2 & 1 & 3 \\ k & 0 & 1 \end{bmatrix}$ is invertible, then

a. $k = 1$ b. $k = -1$

c. $k = 0$ d. all real values of k

14 Adjoint of the matrix $N = \begin{bmatrix} -4 & -3 & -3 \\ 1 & 0 & 1 \\ 4 & 4 & 3 \end{bmatrix}$, is

a. N b. $2N$

c. $-N$ d. None of these

Working Space

15 Suppose A is a square matrix of order 3 and $B = |A| A^{-1}$. If $|A| = -5$, then $|B|$ is equal to

a. 1 b. −5
c. −1 d. 25

16 Let A, B, C be $n \times n$ matrices. Which one of the following is a correct statement?

a. If $A^2 = 0$, then $A = 0$
b. If $A^3 + 2A^2 + 3A + 5I = 0$; then A is invertible
c. If $AB = AC$, then $B = C$
d. None of the above

17 If K is a scalar and I is a unit matrix of order 3. Then, adj.(KI) is

a. $K^3 I$ b. $K^2 I$
c. $-K^3 I$ d. $-K^2 I$

18 If A and B are square matrices of the same order such that $(A+B)(A-B) = A^2 - B^2$, then $(ABA^{-1})^2$ is

a. $A^2 B^2$ b. A^2
c. B^2 d. I

19 Which of the following statements is/are incorrect?

I. adjoint of symmetric matrix is symmetric.
II. adjoint of a unit matrix is a unit matrix.
III. $A(\text{adj } A) = (\text{adj } A) = |A| I$
IV. adjoint of a diagonal matrix is a diagonal matrix.

a. Only I
b. Only II
c. Both III and IV
d. None of the above

20 Which of the following statements is true?

a. non-singular square matrix does not have a unique inverse
b. determinant of a non-singular matrix is zero
c. If $A^T = A$, then A is a square matrix
d. If $|A| \neq 0$, then $|A \cdot \text{adj } A| = |A|^{n-1}$, where $A = [a_{ij}]_{n \times n}$

21 Which of the following statement is not true?

a. Every skew-symmetric matrix of odd order is non-singular
b. If determinant of a square matrix is non-zero, then it is non-singular
c. adjoint of symmetric matrix is symmetric
d. adjoint of diagonal matrix is diagonal

22 If A, B are two $n \times n$ non-singular matrices, then

a. AB is non-singular
b. AB is singular
c. $(AB)^{-1} = A^{-1}B^{-1}$
d. $(AB)^{-1}$ does not exists

23 If $A = \begin{bmatrix} 1 & -1 & 1 \\ 2 & 1 & -3 \\ 1 & 1 & 1 \end{bmatrix}$ and $10B = \begin{bmatrix} 4 & 2 & 2 \\ -5 & 0 & \alpha \\ 1 & -2 & 3 \end{bmatrix}$ where $B = A^{-1}$, then α is equal to **[AIEEE 2004]**

a. 2 b. −1
c. −2 d. 5

Working Space

24 If $A = \begin{bmatrix} 1 & 0 & 0 \\ 0 & 1 & 1 \\ 0 & -2 & 4 \end{bmatrix}$ and $I = \begin{bmatrix} 1 & 0 & 0 \\ 0 & 1 & 0 \\ 0 & 0 & 1 \end{bmatrix}$ and $A^{-1} = \frac{1}{6}(A^2 + cA + dI)$, then the values of c and d are **[IIT JEE 2005]**

a. $(-6, -11)$
b. $(6, 11)$
c. $(-6, 11)$
d. $(6, -11)$

25 If $A^2 - A + I = 0$, then inverse of A is

a. $I - A$
b. $A - I$
c. A
d. $A + I$ **[AIEEE 2005]**

One or More than One Options Correct Type

Direction (Q. No. 26) This section contains 1 multiple choice questions. Each question has four choices (a), (b), (c) and (d), out of which ONE or MORE THAN ONE are correct.

26 If the adjoint of a 3×3 matrix P is $\begin{bmatrix} 1 & 4 & 4 \\ 2 & 1 & 7 \\ 1 & 1 & 3 \end{bmatrix}$, then the possible value(s) of the determinant of P is **[IIT JEE 2012]**

a. -2
b. -1
c. 1
d. $+2$

Statement Type

Direction (Q. Nos. 27-31) This section is based on Statement I and Statement II. Select the correct answer from the codes given below.

a. Both Statement I and Statement II are correct and Statement II is the correct explanation of Statement I
b. Both Statement I and Statement II are correct but Statement II is not the correct explanation of Statement I
c. Statement I is correct but Statement II is incorrect
d. Statement II is correct but Statement I is incorrect

27 Let A be 2×2 matrix with real entries. Let I be the 2×2 matrix. Denoted by $t_r(A)$, the sum of diagonal entries of A. Assume that $A^2 = I$. **[AIEEE 2008]**

Statement I If $A \neq I$ and $A \neq -I$, then $|A| = -1$

Statement II If $A \neq I$ and $A \neq -I$, then $t_r(A) \neq 0$.

28 Let A be 2×2 matrix. **[AIEEE 2009]**

Statement I adj (adj A) = A.

Statement II $|\text{adj } A| = |A|$

29 **Statement I** Determinant of a skew-symmetric matrix of order 3 is zero.

Statement II For any matrix A, $|A^T| = |A|$ and $|-A| = -|A|$.

30 Let A and B be real matrices of the form $\begin{bmatrix} \alpha & 0 \\ 0 & \beta \end{bmatrix}$ and $\begin{bmatrix} 0 & \gamma \\ \gamma & 0 \end{bmatrix}$, respectively. **[AIEEE 2012]**

Statement I $AB - BA$ is always invertible matrix.

Statement II $AB - BA$ is never an identity matrix.

Working Space

31 **Statement I** If $A = \begin{bmatrix} a^2+x^2 & ab-cx & ac+bx \\ ab+cx & b^2+x^2 & bc-ax \\ ac-bx & bc+ax & c^2+x^2 \end{bmatrix}$ and $B = \begin{bmatrix} x & c & -b \\ -c & x & a \\ b & -a & x \end{bmatrix}$, then $|A| = |B|^2$

Statement II If A^c is cofactor of a square matrix A of order n, then $|A^c| = |A|^{n-1}$.

Comprehension Type

Direction (Q. Nos. 32-37) *This section contains 2 paragraphs, each describing theory, experiments, data, etc. Six questions related to the paragraphs have been given. Each question has only one correct answer among the four given options (a), (b), (c) and (d).*

Passage I

If A and B are two matrices of same order 3×3, where

$$A = \begin{bmatrix} 1 & 2 & 3 \\ 2 & 3 & 4 \\ 5 & 6 & 8 \end{bmatrix} \text{ and } B = \begin{bmatrix} 3 & 2 & 5 \\ 2 & 3 & 8 \\ 7 & 2 & 9 \end{bmatrix}$$

32 The value of adj(adj A) is equal to

a. $-A$ b. A
c. $8A$ d. $16A$

33 The value of $|\text{adj}(\text{adj}A)|$ is equal to

a. 9 b. 1
c. -1 d. 81

34 The value of $|\text{adj}\, B|$ is equal to

a. 24 b. $(24)^2$
c. $(24)^3$ d. $(8)^2$

35 The value of $|\text{adj}(AB)|$ is equal to

a. 24 b. $(24)^2$
c. $(24)^3$ d. (65)

Passage II

Let A be a square matrix of order 2 or 3 and I be the identity matrix of the same order, then the matrix $A - \lambda I$ is called characteristic matrix of the matrix A, where λ is some complex number. The determinant of the characteristic matrix is called characteristic determinant of the matrix A which will of course be a polynomial of degree 3 in λ.

The equation $|A - \lambda I| = 0$ is called the characteristic equation of the matrix A and its roots are called characteristic roots or eigenvalues. It is known that every square matrix satisfies the characteristic equation.

36 If one of the eigenvalues of a square matrix A of order 3×3 is zero, then

a. det A must be non-zero
b. det A must be zero
c. adj A must be a zero matrix
d. None of the above

37 If A' denotes transpose of matrix A, $A'A = I$ and det $A = 1$, then $|\det(A - I)|$ must be equal to

a. 0 b. -1
c. 1 d. None of these

Working Space

One Integer Value Correct Type

Direction *(Q. Nos. 38-40) This section contains 3 questions. When worked out will result in an integer from 0 to 9 (both inclusive).*

38 Suppose A is a matrix such that $A^2 = A$ and $(I + A)^{10} = I + kA$, where $k = 2 \cdot 2^{\lambda} - 1$. Then, the value of λ is

39 For any 2×2 matrix A, if $A(\text{adj } A) = \begin{bmatrix} 10 & 0 \\ 0 & 10 \end{bmatrix}$, then the value of $|A|$ is

40 Let k be a positive real number and **[IIT JEE 2010]**

$$A = \begin{bmatrix} 2k-1 & 2\sqrt{k} & 2\sqrt{k} \\ 2\sqrt{k} & 1 & -2k \\ -2\sqrt{k} & 2k & -1 \end{bmatrix}$$

$$B = \begin{bmatrix} 0 & 2k-1 & \sqrt{k} \\ 1-2k & 0 & 2\sqrt{k} \\ -\sqrt{k} & -2\sqrt{k} & 0 \end{bmatrix}$$

If $\det(\text{adj } A) + \det(\text{adj } B) = 10^6$, then the value of $[k]$ is

Note *adj M denotes the adjoint of a square matrix M and [k] denotes the largest integer less than or equal to k.*

DPP-5 Solving of Equations

Only One Option Correct Type

Working Space

Direction *(Q. Nos. 1-19) This section contains 19 multiple choice questions. Each question has four choices (a), (b), (c) and (d), out of which ONLY ONE is correct.*

1 The equation $x + 2y + 3z = 1$, $2x + y + 3z = 2$ and $5x + 5y + 9z = 4$ have

a. unique solution
b. infinitely many solutions
c. inconsistent
d. None of the above

2 The number of solutions of the following equations $x_2 - x_3 = 1, -x_1 + 2x_3 = -2$ and $x_1 - 2x_2 = 3$ is

a. 0 b. 1
c. 2 d. infinite

3 The number of solutions of the system of equations is $x - y + z = 2$, $2x + y - z = 5$ and $4x - y + z = 10$ is

a. infinite b. 1
c. 2 d. 0

4 The following system of equations $3x - 2y + z = 0$, $\lambda x - 14y + 15z = 0$, $x + 2y - 3z = 0$ has a solution other then $x = y = z = 0$ for λ equals

a. 1 b. 2
c. 3 d. 5

5 The real value of k for which the system of equations $2kx - 2y + 3z = 0$, $x + ky + 2z = 0$, $2x + kz = 0$, has non-trivial solution is

a. 2 b. −2
c. 3 d. −3

6 If $x + ky - z = 0$, $3x - ky - z = 0$ and $x - 3y + z = 0$ has non-zero solution, then k is

a. −1 b. 0
c. 1 d. 2

7 The system of equations $\alpha x + y + z = \alpha - 1$, $x + \alpha y + z = \alpha - 1$, $x + y + \alpha z = \alpha - 1$ has no solution, if α is

a. −2 b. 1
c. −2 d. either −2 or 1

8 The number of solutions of the equations $x + 4y - z = 0$, $3x - 4y - z = 0$, $x - 3y + z = 0$ is

a. 0 b. 1
c. 2 d. infinite

Working Space

9 If $a_1x + b_1y + c_1z = 0$, $a_2x + b_2y + c_2z = 0$ and $a_3x + b_3y + c_3z = 0$ and $\begin{vmatrix} a_1 & b_1 & c_1 \\ a_2 & b_2 & c_2 \\ a_3 & b_3 & c_3 \end{vmatrix} = 0$, then the given system has

a. one trivial and one non-trivial solution
b. no solution
c. one solution
d. infinite solution

10 The value of k for which the set of equations $x + ky + 3z = 0$, $3x + ky - 2z = 0$, $2x + 3y - 4z = 0$ has a non-trivial solution over the set of rationals is

a. 15 b. 31/2 c. 16 d. 33/2

11 Set of equations $a + b - 2c = 0$, $2a - 3b + c = 0$ and $a - 5b + 4c = \alpha$ is consistent for α equal to

a. 1 b. 0 c. −1 d. 2

12 The system of linear equations $x + y + z = 2$, $2x + y - z = 3$, $3x + 2y + kz = 4$ has a unique solution if

a. $k \neq 0$
b. $-1 < k < 1$
c. $-2 < k < 2$
d. $k = 0$

13 The number of values of k for which the system of equations $(k + 1)x + 8y = 4k$, $kx + (k + 3)y = (3k - 1)$ has infinitely many solutions, is

a. 0
b. 1
c. 2
d. infinite

14 The existence of the unique solution of the system $x + y + z = \lambda$, $5x - y + \mu z = 10$, $2x + 3y - z = 6$ depends on

a. μ only
b. λ only
c. λ and μ both
d. Neither λ nor μ

15 The value of λ for which the system of equations $2x - y - z = 12$, $x - 2y + z = 4$, $x + y + \lambda z = 4$ has no solution is **[IIT JEE 2004]**

a. 3 b. −3 c. 2 d. −2

16 Let M and N be two 3×3 non-singular skew-symmetric matrices such that $MN = NM$. If P^T denotes the transpose of P, then $M^2N^2(M^TN)^{-1}(MN^{-1})^T$ is equal to **[IIT JEE 2011]**

a. M^2 b. $-N^2$ c. $-M^2$ d. MN

17 Let $P = [a_{ij}]$ be a 3×3 matrix and $Q = [b_{ij}]$, where $b_{ij} = 2^{i+j}a_{ij}$ for $1 \le i, j \le 3$. If the determinant of P is 2, then the determinant of the matrix Q is **[IIT JEE 2012]**

a. 2^{10} b. 2^{11} c. 2^{12} d. 2^{13}

18 If P is 3×3 matrix such that $P^T = 2P + I$, where P^T is the transpose of P and I is the 3×3 identity matrix, then there exists a column matrix $X = \begin{bmatrix} x \\ y \\ z \end{bmatrix} \neq \begin{bmatrix} 0 \\ 0 \\ 0 \end{bmatrix}$ such that **[IIT JEE 2012]**

a. $PX = \begin{bmatrix} 0 \\ 0 \\ 0 \end{bmatrix}$
b. $PX = X$
c. $PX = 2X$
d. $PX = -X$

Working Space

19 The number of values of k, for which system of equations $(k+1)x+8y=4k,\ kx+(k+3)y=(3k-1)$ has no solution,is **[IIT JEE 2013]**

a. 1
b. 2
c. 3
d. Infinite

One or More than One Options Correct Type

Direction (Q. No. 20) This section contains 1 multiple choice question. Each question has four choices (a), (b), (c) and (d), out of which ONE or MORE THAN ONE are correct.

20 If a system of the equation $(\alpha+1)^3x+(\alpha+2)^3y-(\alpha+3)^3=0$ and $(\alpha+1)x+(\alpha+2)y-(\alpha+3)=0,\ x+y-1=0$ is consistent, what is the value of α?

a. 1
b. 0
c. −3
d. −2

Comprehension Type

Direction (Q. Nos. 21-30) This section contains 4 paragraphs, each describing theory, experiments, data, etc. Ten questions related to the paragraphs have been given. Each question has only one correct answer among the four given options (a), (b), (c) and (d).

Passage I

$A=\begin{bmatrix}1&0&0\\2&1&0\\3&2&1\end{bmatrix}$, if U_1, U_2 and U_3 are columns matrices satisfying $AU_1=\begin{bmatrix}1\\0\\0\end{bmatrix}$, $AU_2=\begin{bmatrix}2\\3\\0\end{bmatrix}$ and $AU_3=\begin{bmatrix}2\\3\\1\end{bmatrix}$. If U is 3×3 matrix whose columns are U_1, U_2, U_3. **[IIT JEE 2006]**

21 The value of $|U|$ is

a. 3
b. −3
c. 3/2
d. 2

22 The sum of the elements of U^{-1} is

a. −1
b. 0
c. 1
d. 3

23 The value of $[3\ \ 2\ \ 0]\ U\begin{bmatrix}3\\2\\0\end{bmatrix}$ is

a. [5]
b. [5/2]
c. [4]
d. $\begin{bmatrix}\frac{3}{2}\end{bmatrix}$

Passage II

If A is a square matrix, then the equation in λ obtained by $|A-\lambda I|=0$ is called as characteristic equation. Every matrix satisfies its own characteristic equation.

Given that $A=\begin{bmatrix}1&2&1\\0&1&-1\\3&-1&1\end{bmatrix}$. Then,

24 If $A^3+kA^2+lA+mI=0$. Then, $(k+m+l)$ is

a. 0
b. 9
c. 5
d. 6

Working Space

25 Inverse of A is

a. $A^2 + 3A + I$

b. $A^2 - 3A - I$

c. $-1/9\,(A^2 - 3A - I)$

d. None of these

Passage III

Consider the equations $x + y - z = b_1$, $2x + z = b_2$, $x - y + az = b_3$

26 If this system is denoted by $AX = B$. Then, for $a = 1$, the value of A^{-1}, is

a. $\begin{bmatrix} 1/2 & 0 & 1/2 \\ -1/2 & 1 & 3/2 \\ -1 & -1 & 1 \end{bmatrix}$

b. $\begin{bmatrix} 1/2 & 0 & 1/2 \\ -1/2 & 1 & 3/2 \\ -1 & 1 & 1 \end{bmatrix}$

c. $\begin{bmatrix} 1/2 & 0 & 1/2 \\ -1/2 & 1 & -3/2 \\ 1 & 1 & -1 \end{bmatrix}$

d. $\begin{bmatrix} 1/2 & 0 & 1/2 \\ -1/2 & 1 & -3/2 \\ -1 & 1 & -1 \end{bmatrix}$

27 If $B = 0$, then the value of a for which $AX = B$ have non-trivial solution, is

a. 2

b. 3

c. 1

d. None of these

Passage IV

Let A be the set of all 3×3 symmetric matrices all of whose entries are either 0 or 1. Five of these entries are 1 and four of them are 0. **[IIT JEE 2009]**

28 The number of matrices in A, is

a. 12

b. 6

c. 9

d. 3

29 The number of matrices of A for which the system of linear equations $A\begin{bmatrix} x \\ y \\ z \end{bmatrix} = \begin{bmatrix} 1 \\ 0 \\ 0 \end{bmatrix}$ has a unique solution, is

a. less than 4

b. atleast 4 but less than 7

c. atleast 7 but less than 10

d. atleast 10

30 The number of matrices of A for which the system of linear equations $A\begin{bmatrix} x \\ y \\ z \end{bmatrix} = \begin{bmatrix} 1 \\ 0 \\ 0 \end{bmatrix}$ inconsistent, is

a. 0

b. more than 2

c. 2

d. 1

DPP-6 Rank of Matrix, Differentiation and Product of Determinants

Subjective Questions

Working Space

Direction *(Q. Nos. 1-8) These questions are subjective in nature, need to be solved completely on notebook.*

1 Show $\begin{vmatrix} (a-x)^2 & (a-y)^2 & (a-z)^2 \\ (b-x)^2 & (b-y)^2 & (b-z)^2 \\ (c-x)^2 & (c-y)^2 & (c-z)^2 \end{vmatrix}$ can be expressed as product of two determinants.

2 Express the determinant $\Delta = \begin{vmatrix} 1 & \cos(\beta-\alpha) & \cos(\gamma-\alpha) \\ \cos(\alpha-\beta) & 1 & \cos(\gamma-\beta) \\ \cos(\alpha-\gamma) & \cos(\beta-\gamma) & 1 \end{vmatrix}$ as the product of two determinants and hence show that $\Delta = 0$.

3 If $S_r = \alpha^r + \beta^r + \gamma^r$. Show that $\begin{vmatrix} S_0 & S_1 & S_2 \\ S_1 & S_2 & S_3 \\ S_2 & S_3 & S_4 \end{vmatrix} = (\alpha-\beta)^2(\beta-\gamma)^2(\gamma-\alpha)^2$

4 If α, β, γ are the roots of the equation $ax^2 + bx + c = 0$ and $S_n = \alpha^n + \beta^n$, then find $\begin{vmatrix} 3 & 1+S_1 & 1+S_2 \\ 1+S_1 & 1+S_2 & 1+S_3 \\ 1+S_2 & 1+S_3 & 1+S_4 \end{vmatrix}$.

5 If a, b, c are distinct real numbers, then show that $\begin{vmatrix} a^2+b^2+c^2 & bc+ca+ab & bc+ca+ab \\ bc+ca+ab & a^2+b^2+c^2 & bc+ca+ab \\ bc+ca+ab & bc+ca+ab & a^2+b^2+c^2 \end{vmatrix}$ is always positive.

6 If f, g, h are differentiable functions of x and $\Delta = \begin{vmatrix} f & g & h \\ (xf)' & (xg)' & (xh)' \\ (x^2f)'' & (x^2g)'' & (x^2h)'' \end{vmatrix}$, then prove that $\Delta' = \begin{vmatrix} f & g & h \\ f' & g' & h' \\ (x^2f'')' & (x^2g'')' & (x^2h'')' \end{vmatrix}$

Working Space

7 Show that the determinant $\Delta(x)$ given by

$$\Delta(x) = \begin{vmatrix} \sin(x+\alpha) & \cos(x+\alpha) & a + x\sin\alpha \\ \sin(x+\beta) & \cos(x+\beta) & b + x\sin\beta \\ \sin(x+\gamma) & \cos(x+\gamma) & c + x\sin\gamma \end{vmatrix}$$, is independent of x.

8 If $f(x) = \begin{vmatrix} x^n & \sin x & \cos x \\ n! & \sin\frac{n\pi}{2} & \cos\frac{n\pi}{2} \\ a & a^2 & a^3 \end{vmatrix}$, then show that $\frac{d^n}{dx^n}\{f(x)\} = 0$ at $x = 0$.

Only One Option Correct Type

Direction (*Q. Nos.* 9 *and* 10) *This section contains* 2 *multiple choice questions. Each question has four choices (a), (b), (c) and (d), out of which ONLY ONE is correct.*

9 If the matrix $A = \begin{bmatrix} y+a & b & c \\ a & y+b & c \\ a & b & y+c \end{bmatrix}$ has rank 3, then

a. $y \neq (a+b+c)$
b. $y \neq 1$
c. $y = 0$
d. $y \neq -(a+b+c)$ and $y \neq 0$

10 If A is a non-zero column matrix of order $m \times 1$ and B is a non-zero row matrix of order $1 \times n$, then rank of AB is equal to

a. n
b. m
c. 1
d. None of these

SET - 1

Revisal Problems for JEE Main

Only One Option Correct Type

Working Space

Direction *(Q. Nos. 1-36) This section contains 36 multiple choice questions. Each question has four choices (a), (b), (c) and (d), out of which ONLY ONE option is correct.*

1 If $A=\begin{bmatrix}3 & 4\\ 2 & 4\end{bmatrix}$, $B=\begin{bmatrix}-2 & -2\\ 0 & -1\end{bmatrix}$, then $(A+B)^{-1}$

a. is a symmetric matrix
b. $A^{-1}+B^{-1}$
c. does not exists
d. None of these

2 If $AB=A$ and $BA=B$, where A and B are square matrices, then

a. $B^2 \neq B$ and $A^2=A$
b. $B^2=B$ and $A^2=A$
c. $A^2 \neq A$ and $B^2=B$
d. None of these

3 If A and B are two matrices such that $AB=B$ and $BA=A$, then A^2+B^2 is

a. $2AB$
b. $2BA$
c. $A+B$
d. AB

4 If A is a singular matrix, then (adj A) is

a. non-singular
b. singular
c. symmetric
d. Not defined

5 If $A=\begin{bmatrix}a & 0 & 0\\ 0 & a & 0\\ 0 & 0 & a\end{bmatrix}$, then $|\text{adj } A|$ is

a. a^{27}
b. a^9
c. a^6
d. a^2

6 For any 2×2 matrix, if $A\cdot(\text{adj } A)=\begin{bmatrix}10 & 0\\ 0 & 10\end{bmatrix}$. Then $|A|$, is

a. 20
b. 100
c. 10
d. 0

7 If $A=\begin{bmatrix}1 & a\\ 0 & 1\end{bmatrix}$, then A^n (where $n\in N$) equals

a. $\begin{bmatrix}1 & na\\ 0 & 1\end{bmatrix}$
b. $\begin{bmatrix}1 & n^2a\\ 0 & 1\end{bmatrix}$
c. $\begin{bmatrix}1 & na\\ 0 & 0\end{bmatrix}$
d. $\begin{bmatrix}n & na\\ 0 & n\end{bmatrix}$

8 If $A^5=0$, such that $A^n \neq I$, for $1\le n\le 4$, then $(I-A)^{-1}$ is

a. A^4
b. A^3
c. $I+A$
d. None of these

9 If $A=\begin{bmatrix}\cos\theta & -\sin\theta\\ \sin\theta & \cos\theta\end{bmatrix}$, then $A^T+A=I_2$, if

a. $\theta=n\pi, n\in Z$
b. $\theta=(2n+1)\pi/2, n\in Z$
c. $\theta=2n\pi+\pi/3, n\in Z$
d. None of the above

Working Space

10 If $A=\begin{bmatrix}1 & a\\0 & 1\end{bmatrix}$, then $\lim_{n\to\infty}\frac{1}{n}A^n$ is

a. $\begin{bmatrix}0 & a\\0 & 0\end{bmatrix}$ b. $\begin{bmatrix}0 & 0\\0 & 0\end{bmatrix}$

c. $\begin{bmatrix}0 & 1\\0 & 0\end{bmatrix}$ d. None of these

11 If $A=\begin{bmatrix}1 & -1\\2 & -1\end{bmatrix}$, $B=\begin{bmatrix}a & 1\\b & -1\end{bmatrix}$ and $(A+B)^2=A^2+B^2$, values of a and b are

a. $a=4, b=1$ b. $a=1, b=4$

c. $a=0, b=4$ d. $a=2, b=4$

12 The matrix $A=\begin{bmatrix}1 & -3 & -4\\-1 & 3 & 4\\1 & -3 & -4\end{bmatrix}$ is nilpotent of index

a. 2 b. 3

c. 4 d. 6

13 For what value of λ, the system of equations $x+y+z=6$, $x+2y+3z=10$, $x+2y+\lambda z=12$ is inconsistent?

a. $\lambda=1$ b. $\lambda=2$

c. $\lambda=-2$ d. $\lambda=3$

14 The solution of equations $x+y+z=6$, $x+2y+3z=10$, $x+2y+\lambda z=\mu$ has no solutions for

a. $\lambda\neq 3, \mu=10$ b. $\lambda=3, \mu\neq 10$

c. $\lambda\neq 3, \mu\neq 10$ d. None of these

15 A square non-singular matrix A satisfies $A^2-A+2I=0$, then A^{-1} is equal to

a. $I-A$ b. $\frac{1}{2}(I-A)$

c. $I+A$ d. $\frac{1}{2}(I+A)$

16 If $A^3=0$, then $I+A+A^2$ equals

a. $I-A$ b. $(I-A)^{-1}$

c. $(I+A)^{-1}$ d. None of these

17 If A and B are square matrices of order 3×3 such that A is an orthogonal matrix and B is a skew-symmetric matrix, then which of the following statements is true?

a. $|AB|=1$ b. $|AB|=0$

c. $|AB|=-1$ d. None of these

18 If $\begin{bmatrix}a & b^3\\2 & 0\end{bmatrix}=\begin{bmatrix}1 & 8\\2 & 0\end{bmatrix}$, then $\begin{bmatrix}a & b\\2 & 0\end{bmatrix}^{-1}$ is

a. $\begin{bmatrix}0 & -2\\-2 & 1\end{bmatrix}$ b. $\begin{bmatrix}1 & 0\\0 & 1\end{bmatrix}$

c. $\begin{bmatrix}0 & 1/2\\1/2 & -1/4\end{bmatrix}$ d. None of these

19 If $A=\begin{bmatrix}0 & 1 & -1\\2 & 1 & 3\\3 & 2 & 1\end{bmatrix}$, then $(A(\text{adj}\,A)\cdot A^{-1})A$ is equal to

a. $2\begin{bmatrix}3 & 0 & 0\\0 & 3 & 0\\0 & 0 & 3\end{bmatrix}$ b. $\begin{bmatrix}-6 & 0 & 0\\0 & -6 & 0\\0 & 0 & -6\end{bmatrix}$

c. $\begin{bmatrix}0 & 1/6 & -1/6\\2/6 & 1/6 & 3/6\\3/6 & 2/6 & 1/6\end{bmatrix}$ d. None of these

Working Space

20 If the value of determinant $\begin{vmatrix} a & 1 & 1 \\ 1 & b & 1 \\ 1 & 1 & c \end{vmatrix}$ is positive, then

a. $abc > 1$
b. $abc > -8$
c. $abc < -8$
d. $abc > -2$

21 The value of the determinant $\begin{vmatrix} b+c & a-b & a \\ c+a & b-c & b \\ a+b & c-a & c \end{vmatrix}$ is

a. $a^3 + b^3 + c^3 - 3abc$
b. $3abc - a^3 - b^3 - c^3$
c. $3abc + a^3 + b^3 + c^3$
d. None of the above

22 If $\Delta = \begin{vmatrix} \cos\alpha & -\sin\alpha & 1 \\ \sin\alpha & \cos\alpha & 1 \\ \cos(\alpha+\beta) & -\sin(\alpha+\beta) & 1 \end{vmatrix}$, then

a. $\Delta \in [1-\sqrt{2}, 1+\sqrt{2}]$
b. $\Delta \in [-1, 1]$
c. $\Delta \in [-\sqrt{2}, \sqrt{2}]$
d. None of the above

23 The largest value of a third order determinant whose elements are 0 and 1, is

a. 1
b. 0
c. 2
d. 3

24 If α, β and γ are the roots of the equation $x^3 + px + q = 0$ (with $p \neq 0$ and $q \neq 0$), the value of determinant $\begin{vmatrix} \alpha & \beta & \gamma \\ \beta & \gamma & \alpha \\ \gamma & \alpha & \beta \end{vmatrix}$ is

a. p
b. q
c. $p^2 - 2q$
d. None of these

25 If $\begin{vmatrix} x^n & x^{n+2} & x^{n+3} \\ y^n & y^{n+2} & y^{n+3} \\ z^n & z^{n+2} & z^{n+3} \end{vmatrix} = (x-y)(y-z)(z-x)\cdot\left(\frac{1}{x}+\frac{1}{y}+\frac{1}{z}\right)$. Then, n equals

a. 1
b. -1
c. 2
d. -2

26 If $f(x) = \begin{vmatrix} 1 & x & x+1 \\ 2x & x(x-1) & (x+1)x \\ 3x(x-1) & (x-1)(x-2) & (x+1)x(x-1) \end{vmatrix}$, then $f(100)$ is

a. 0
b. 1
c. 100
d. -100

27 Let $\{D_1, D_2, D_3, \cdots, D_n\}$ be the set of all third order determinants that can be made with the distinct non-zero real numbers $a_1, a_2, a_3, \cdots, a_9$. Then,

a. $\sum_{i=1}^{n} D_i = 1$
b. $\sum_{i=1}^{n} D_i = 0$
c. $D_i = D_j \ \forall \ i, j$
d. None of these

28 The coefficient of x in $f(x) = \begin{vmatrix} x & (1+\sin x)^3 & \cos x \\ 1 & \log(1+x) & 2 \\ x^2 & (1+x)^2 & 0 \end{vmatrix}$ is

a. 0
b. 1
c. -2
d. None of these

Working Space

29 If $D_K = \begin{vmatrix} 1 & n & n \\ 2k & n^2+n+2 & n^2+n \\ 2k-1 & n^2 & n^2+n+2 \end{vmatrix}$ and $\sum_{k=1}^{n} D_k = 48$, then n is

a. 4
b. 6
c. 8
d. None of the above

30 Let $\begin{vmatrix} x^2+3x & x-1 & x+3 \\ x+1 & -2x & x-4 \\ x-3 & x+4 & 3x \end{vmatrix} = ax^4 + bx^3 + cx^2 + dx + e$ be an identity in x, where a, b, c, d, e are independent of x. Then, the value of e is

a. 4
b. 3
c. 1
d. None of the above

31 If $I_n = \begin{vmatrix} 1 & k & k \\ 2n & k^2+k+1 & k^2+k \\ 2n-1 & k^2 & k^2+k+1 \end{vmatrix}$ and $\sum_{n=1}^{k} I_n = 72$, then k is equal to

a. 8
b. 9
c. 6
d. None of the above

32 If $\alpha + \beta + \gamma = \pi$, then the value of the determinant $\begin{vmatrix} e^{2i\alpha} & e^{-i\gamma} & e^{-i\beta} \\ e^{-i\gamma} & e^{2i\beta} & e^{-i\alpha} \\ e^{-i\beta} & e^{-i\alpha} & e^{2i\gamma} \end{vmatrix}$ is

a. 4
b. −4
c. 0
d. None of the above

33 If w is a non-real cube root of unity, then $\Delta = \begin{vmatrix} a_1+b_1\omega & a_1\omega^2+b_1 & a_1+b_1\omega+c_1\omega^2 \\ a_2+b_2\omega & a_2\omega^2+b_2 & a_2+b_2\omega+c_2\omega^2 \\ a_3+b_3\omega & a_3\omega^2+b_3 & a_3+b_3\omega+c_3\omega^2 \end{vmatrix}$ is

a. −1
b. 0
c. $-\omega^2$
d. None of these

34 If $\Delta_r = \begin{vmatrix} 2^{r-1} & \dfrac{(r+1)!}{1+\frac{1}{r}} & 2r \\ a & b & c \\ 2^n-1 & (n+1)!-1 & n(n+1) \end{vmatrix}$, then $\sum_{r=1}^{n} \Delta_r$ is

a. 0
b. $(n+3)!$
c. $a(n!)+b$
d. None of the above

35 If $a > b > c$ and the system of equations $ax + by + cz = 0$, $bx + cy + az = 0$ and $cx + ay + bz = 0$ has non-trivial solution, then the quadratic equation $ax^2 + bx + c = 0$ has

a. atleast one positive root
b. roots of opposite sign
c. positive roots
d. imaginary roots

Working Space

36 Let A be a square matrix all of whose entries are integers. Then, which one of the following is true? **[AIEEE 2008]**

a. If $|A| = \pm 1$, then A^{-1} exists but all its entries are not necessarily integers
b. If $|A| \neq \pm 1$, then A^{-1} exists and all its entries are non-integers
c. If $|A| = \pm 1$, then A^{-1} exists and all its entries are integers
d. If $|A| = \pm 1$, then A^{-1} need not exists

Statement Type

Direction *(Q. Nos. 37-40) This section is based on Statement I and Statement II. Select the correct answer from the codes given below.*

a. Both Statement I and Statement II are correct and Statement II is the correct explanation of Statement I
b. Both Statement I and Statement II are correct but Statement II is not the correct explanation of Statement I
c. Statement I is correct but Statement II is incorrect
d. Statement II is correct but Statement I is incorrect

37 **Statement I** If dot product of any two different row vectors in matrix A is zero, then A is orthogonal.

Statement II In an orthogonal matrix, dot product of any two different row vectors is zero.

38 **Statement I** If $A^2 - 3A + 2I = 0$, then A can be equal to I or $2I$ or $\begin{bmatrix} 3 & -2 \\ 1 & 0 \end{bmatrix}$ or $\begin{bmatrix} 3 & 1 \\ -2 & 0 \end{bmatrix}$.

Statement II For a second order square matrix A, if $\det|A - xI| = ax^2 + bx + c$, then $aA^2 + bA + cI = 0$.

39 **Statement I** If $A = \begin{bmatrix} 2 & 1+2i \\ 1-2i & 7 \end{bmatrix}$, then $\det(A)$ is real.

Statement II If $A = \begin{bmatrix} a_{11} & a_{12} \\ a_{21} & a_{22} \end{bmatrix}$, a_{ij} being complex numbers, then $|A|$ is always real.

40 Let A and B be two symmetric matrices of order 3. **[AIEEE 2011]**

Statement I $A(BA)$ and $(AB)A$ are symmetric matrices.

Statement II AB is symmetric matrix, if matrix multiplication of A with B is commutative.

SET - 2

Revisal Problems
for JEE Advanced

Only One Option Correct Type

Working Space

***Direction** (Q. Nos. 1-30) This section contains 30 multiple choice questions. Each question has four choices (a), (b), (c) and (d), out of which ONLY ONE option is correct.*

1 If A is involuntary matrix given by $A = \begin{bmatrix} 0 & 1 & -1 \\ 4 & -3 & 4 \\ 3 & -3 & 4 \end{bmatrix}$, then the inverse of $A/2$ will be

a. $2A$
b. $\frac{A^{-1}}{2}$
c. $\frac{A}{2}$
d. A^2

2 If an idempotent matrix is also skew-symmetric, then it must be

a. an involuntary matrix
b. an identity matrix
c. an orthogonal matrix
d. a null matrix

3 If A and B are invertible matrices, which one of the following statement is not correct?

a. $\text{adj} \cdot A = |A| A^{-1}$
b. $\det(A^{-1}) = |\det(A)|^{-1}$
c. $(A + B)^{-1} = B^{-1} + A^{-1}$
d. $(AB)^{-1} = B^{-1}A^{-1}$

4 $\text{tr}(A)$ of a 3×3 matrix $A = (a_{ij})$ is defined by the relation $\text{tr}(A) = a_{11} + a_{22} + a_{33}$ (i.e. $tr(A)$ is sum of the main diagonals elements). Which of the following statement cannot always hold?

a. $\text{tr}(KA) = K\,\text{tr}(A)$, ($K$ is scalar)
b. $\text{tr}(A + B) = \text{tr}(A) + \text{tr}(B)$
c. $\text{tr}(I_3) = 3$
d. $\text{tr}(A^2) = [\text{tr}(A)]^2$

5 If A and B are different matrices satisfying $A^3 = B^3$ and $A^2B = B^2A$, then

a. $|A^2 + B^2| = 0$
b. $|A - B| = 0$
c. $|A^2 + B^2|$ as well as $|A - B|$ must be zero
d. atleast one of $|A^2 + B^2|$ or $|A - B|$ must be zero

6 Which of the following statements is incorrect for a square matrix $A(|A| \neq 0)$?

a. If A is a diagonal matrix, A^{-1} will also be a diagonal matrix
b. If A is a symmetric matrix, A^{-1} will also be a symmetric matrix
c. If $A^{-1} = A \Rightarrow A$ is an idempotent matrix
d. If $A^{-1} = A \Rightarrow A$ is an involuntary matrix

7 Let $\Delta_0 = \begin{vmatrix} a_{11} & a_{12} & a_{13} \\ a_{21} & a_{22} & a_{23} \\ a_{31} & a_{32} & a_{33} \end{vmatrix}$ and Δ_1 denote the determinant formed by the cofactors of elements of Δ_0 and Δ_2 denote the determinant formed by the cofactors at Δ_1 and so on Δ_n denotes the determinant formed by the cofactors at Δ_{n-1}, then the determinant value of Δ_n denotes the determinant formed by the cofactors at Δ_{n-1}, then the determinant value of Δ_n is

a. Δ_0^{2n}
b. $\Delta_0^{2^n}$
c. $\Delta_0^{n^2}$
d. Δ_0^{2}

Working Space

8 The determinant $\begin{vmatrix} 1+a+x & a+y & a+z \\ b+x & 1+b+y & b+z \\ c+x & c+y & 1+c+z \end{vmatrix}$ is equal to

a. $(1+a+b+c)(1+x+y+z)-3(ax+by+cz)$
b. $a(x+y)+b(y+z)+c(z+x)-(xy+yz+zx)$
c. $x(a+b)+y(b+c)+z(c+a)-(ab+bc+ca)$
d. None of the above

9 Let $A+2B=\begin{bmatrix} 1 & 2 & 0 \\ 6 & -3 & 3 \\ -5 & 3 & 1 \end{bmatrix}$ and $2A-B=\begin{bmatrix} 2 & -1 & 5 \\ 2 & -1 & 6 \\ 0 & 1 & 2 \end{bmatrix}$, then $T_r(A)-T_r(B)$ has the value equal to

a. 0
b. 1
c. 2
d. None of these

10 The number of solutions of the matrix $X^2=I$ other than I, is (where, I is the 2×2 matrix)

a. 0
b. 1
c. 2
d. more than 2

11 Let A, B, C, D be (not necessarily square) real matrices such that

$$A^T=BCD,\ B^T=CDA,\ C^T=DAB,\ D^T=ABC$$

for the matrix $S=ABCD$. Consider the two statements.

I. $S^3=S$

II. $S^2=S^4$

a. II is true but not I
b. I is true but not II
c. Both I and II are true
d. Both I and II are false

12 Number of triplets of a, b and c for which the system of equations,

$$ax-by=2a-b \text{ and } (c+1)x+cy=10-a+3b$$

has infinitely many solutions and $x=1, y=3$ is one of the solutions, is

a. exactly one
b. exactly two
c. exactly three
d. infinitely many

13 There are two possible values of A in the solution of the matrix equation

$\begin{bmatrix} 2A+1 & -5 \\ -4 & A \end{bmatrix}^{-1}\begin{bmatrix} A-5 & B \\ 2A-2 & C \end{bmatrix}=\begin{bmatrix} 14 & D \\ E & F \end{bmatrix}$, where A, B, C, D, E, F are real numbers. The absolute value of the difference of these two solutions, is

a. 8/3
b. 11/3
c. 1/3
d. 19/3

14 If every element of a square non-singular matrix A is multiplied by k and the new matrix is denoted by B, then $|A^{-1}|$ and $|B^{-1}|$ are related as

a. $|A^{-1}|=k|B^{-1}|$
b. $|A^{-1}|=\frac{1}{k}|B^{-1}|$
c. $|A^{-1}|=k^n|B^{-1}|$
d. None of these

15 Matrix A satisfies $A^2=2A-I$, where I is the identity matrix, then for $n\geq 2$, A^n is equal to (where, $n\in N$)

a. $nA-I$
b. $2^{n-1}A-(n-1)I$
c. $nA-(n-1)I$
d. $2^{n-1}A-I$

16 Let three matrices $A=\begin{bmatrix} 2 & 1 \\ 4 & 1 \end{bmatrix}, B=\begin{bmatrix} 3 & 4 \\ 2 & 3 \end{bmatrix}$ and $C=\begin{bmatrix} 3 & -4 \\ -2 & 3 \end{bmatrix}$, then

$$\text{tr}(A)+\text{tr}\left(\frac{ABC}{2}\right)+\text{tr}\left(\frac{A(BC)^2}{4}\right)+\text{tr}\left(\frac{A(BC)^3}{8}\right)+\cdots\infty \text{ is equal to}$$

a. 6
b. 9
c. 12
d. None of these

Working Space

17 The equation $\begin{vmatrix} (1+x)^2 & (1-x)^2 & -(2+x^2) \\ 2x+1 & 3x & 1-5x \\ x+1 & 2x & 2-3x \end{vmatrix} + \begin{vmatrix} (1+x)^2 & 2x+1 & x+1 \\ (1-x)^2 & 3x & 2x \\ 1-2x & 3x-2 & 2x-3 \end{vmatrix} = 0$

a. has no real solution
b. has 4 real solutions
c. has two real and two non-real solutions
d. has infinite number of solutions, real or non-real

18 In a square matrix A of order 3, the elements, a_{ii}'s are the sum of the roots of the equation $x^2 - (a+b)x + ab = 0$; $a_{i,\,i+1}$'s are the product of the roots, $a_{i,\,i-1}$'s are all unity and the rest of the elements are all zero. The value of the det (A) is equal to

a. 0
b. $(a+b)^3$
c. $a^3 - b^3$
d. $(a^2+b^2)(a+b)$

19 Let D_k be the $k \times k$ matrix with 0's in the main diagonal, unity as the element of Ist row and $(f(k))$th column and k for all entries. If $f(x) = x - \{x\}$. (where, $\{x\}$ denotes the fractional part function). Then, the value of $\det(D_2) + \det(D_3)$ equals

a. 32
b. 34
c. 36
d. None of these

20 For a matrix $A = \begin{bmatrix} 1 & 2r-1 \\ 0 & 1 \end{bmatrix}$, the value of $\prod_{r=1}^{50} \begin{bmatrix} 1 & 2r-1 \\ 0 & 1 \end{bmatrix}$ is equal to

a. $\begin{bmatrix} 1 & 100 \\ 0 & 1 \end{bmatrix}$
b. $\begin{bmatrix} 1 & 4950 \\ 0 & 1 \end{bmatrix}$
c. $\begin{bmatrix} 1 & 5050 \\ 0 & 1 \end{bmatrix}$
d. $\begin{bmatrix} 1 & 2500 \\ 0 & 1 \end{bmatrix}$

21 Let $A = \begin{bmatrix} 1+x^2-y^2-z^2 & 2(xy+z) & 2(zx-y) \\ 2(xy-z) & 1+y^2-z^2-x^2 & 2(zy+x) \\ 2(zx+y) & 2(yz-x) & 1+z^2-x^2-y^2 \end{bmatrix}$, then $\det(A)$ is equal to

a. $(1+xy+yz+zx)^3$
b. $(1+x^2+y^2+z^2)^3$
c. $(xy+yz+zx)^3$
d. $(1+x^3+y^3+z^3)^2$

22 Let $A = \{1^2, 3^2, 5^2, \cdots\}$. 9 elements are selected from the set A to make 3×3 matrix, then det (A) will be divisible by

a. 9
b. 36
c. 8
d. 64

23 A and B are two square matrices such that $A^2B = BA$ and if $(AB)^{10} = A^k B^{10}$. Then, k is

a. 1001
b. 1023
c. 1042
d. None of these

24 If α, β, γ are the roots of the equation $x^3 + ax^2 + bx + c = 0$ and if the system of equations $\alpha u + \beta v + \gamma w = 0$, $\beta u + \gamma v + \alpha w = 0$ and $\gamma u + \alpha v + \beta w = 0$ posses indeterminable solutions, then $a^3 + 6c$ equals

a. ab
b. $2ab$
c. $3ab$
d. None of these

25 If A and B are two non-singular matrices which commute, then $(A(A+B)^{-1}B)^{-1}(AB)$ is

a. $A+B$
b. $A^{-1}+B^{-1}$
c. $A^{-1}+B$
d. None of these

26 If A and B are two square matrices such that $AB = B$ and $BA = A$, then for $n \in N$ $(A+B)^n$ is

a. $2^n(A+B)$
b. $2^{n-1}(A+B)$
c. $2^{n+1}(A+B)$
d. $2^{n/2}(A+B)$

27 If $\begin{vmatrix} a & b & 1 \\ b & c & 1 \\ c & a & 1 \end{vmatrix} = 2010$ and $\begin{vmatrix} c-a & c-b & ab \\ a-b & a-c & bc \\ b-c & b-a & ca \end{vmatrix} - \begin{vmatrix} c-a & c-b & c^2 \\ a-b & a-c & a^2 \\ b-c & b-a & b^2 \end{vmatrix} = p$, then the number of positive divisors of the number P is

a. 36
b. 49
c. 64
d. 81

Working Space

28 Let $A=\begin{bmatrix} a & b \\ c & d \end{bmatrix}$ be a matrix such that $a+d=1=ad-bc$, then A^{2010} is

a. I_2 b. $-I_2$ c. A d. null matrix

29 Let A, B, are square matrices of same order satisfying $AB = A$ and $BA = B$, then $(A^{2010}+B^{2010})^{2011}$ equals

a. $A+B$ b. $2010(A+B)$ c. $2011(A+B)$ d. $2^{2011}(A+B)$

30 Let A be a 3×3 matrix given by $A=(a_{ij})_{3\times 3}$. If for every column vector X satisfies $X^T AX=0$ and $a_{12}=2008$, $a_{13}=2010$ and $a_{23}=-2012$. Then, the value of $a_{21}+a_{31}+a_{32}$ is

a. −6 b. 2006 c. −2006 d. 0

One or More than One Options Correct Type

Direction (Q. Nos. 31-36) *This section contains 6 multiple choice questions. Each question has four choices (a), (b), (c) and (d), out of which ONE or MORE THAN ONE are correct.*

31 Let A and B are two square idempotent matrices such that $AB \pm BA$ is a null matrix, then the value of the $\det|A-B|$ can be equal to

a. −1 b. 1 c. 0 d. 2

32 A square matrix A with elements from the set of real numbers is said to be orthogonal, if $A' = A^{-1}$. If A is an orthogonal matrix, then

a. A^T is orthogonal b. A^{-1} is orthogonal c. $\text{adj}(A)=A^T$ d. $|A^{-1}|=1$

33 D is a 3×3 diagonal matrix. Which of the following statement is not true?

a. $D^T=D$

b. $AD=DA$ for every matrix A of order 3×3

c. D^{-1} if exists is a scalar matrix

d. None of the above

34 If A and B are two 3×3 matrices such that their product AB is a null matrix, then

a. $\det A \neq 0 \Rightarrow B$ must be a null matrix

b. $\det B \neq 0 \Rightarrow A$ must be a null matrix

c. If none of A and B are null matrices, then atleast one of the two matrices must be singular

d. If neither $\det A$ nor $\det B$ is zero, then the given statement is not possible

35 If there are three square matrices A, B, C of same order satisfying the equation $A^2 = A^{-1}$ and $B=A^{2^n}$ and $C=A^{2^{(n-2)}}$, then which of the following statements are true?

a. $|B-C|=0$ b. $(B+C)(B-C)=0$

c. B must be equal to C d. None of these

36 If p, q, r, s are in an AP and $f(x)=\begin{vmatrix} p+\sin x & q+\sin x & p-r+\sin x \\ q+\sin x & r+\sin x & -1+\sin x \\ r+\sin x & s+\sin x & s-q+\sin x \end{vmatrix}$ such that $\int_0^2 f(x)\,dx=-4$, then the common difference of then AP can be

a. −1 b. 1/2 c. 1 d. 2

Comprehension Type

Direction (Q. Nos. 37-45) *This section contains 3 paragraphs, each describing theory, experiments, data, etc. Nine questions related to the paragraphs have been given. Each question has only one correct answer among the four given options (a), (b), (c) and (d).*

Passage I

If $A_0=\begin{bmatrix} 2 & -2 & -4 \\ -1 & 3 & 4 \\ 1 & -2 & -3 \end{bmatrix}$ and $B_0=\begin{bmatrix} -4 & -3 & -3 \\ 1 & 0 & 1 \\ 4 & 4 & 3 \end{bmatrix}$, $B_n=\text{adj}(B_{n-1})$, $n\in N$ and I is an identity matrix of order 3, then.

Working Space

37 $\det[A_0 + A_0^2B_0^2 + A_0^3 + A_0^4B_0^4 + \cdots 10 \text{ terms}]$ is equal to

a. 1000 b. −800
c. 0 d. −8000

38 $B_1 + B_2 + \cdots + B_{49}$ is equal to

a. B_0 b. $7B_0$
c. $49B_0$ d. $49I$

39 For a variable matrix X, the equation $A_0X = B_0$ will have

a. a unique solution b. infinite solution
c. finitely many solutions d. no solution

Passage II

If A is a symmetric and B is a skew-symmetric matrix and $A + B$ is non-singular and $C = (A+B)^{-1}(A-B)$, then

40 $C^T(A+B)C$, is

a. $A+B$ b. $A-B$
c. A d. B

41 $C^T(A-B)C$, is

a. $A+B$ b. $A-B$
c. A d. B

42 C^TAC, is

a. $A+B$ b. $A-B$
c. A d. B

Passage III

3×3 matrices are formed using the elements from the set $\{-3,-2,-1,0,1,2,3\}$, then

43 Number of matrices that are skew-symmetric matrices (or) scalar matrices is

a. $7^7 - 7^4$ b. $7^7 - 7^3$
c. $7^3 + 6$ d. $7^4(7^3 - 6)$

44 Number of matrices that are lower diagonal (or) upper diagonal is

a. $7^6 - 7^3$ b. $7^6 - 2(7^3)$
c. $7^3(2\times7^3 - 1)$ d. $7^9 - 7^6$

45 Number of matrices each of whose trace is atleast 7 is

a. $7^9 - 7^3$ b. $5(7^6)$
c. $10(7^6)$ d. 7^7

One Integer Value Correct Type

Direction *(Q. Nos. 46-50) This section contains 5 questions. When worked out will result in an integer from 0 to 9 (both inclusive).*

46 A square matrix P satisfies $P^2 = I - P$, where I is an identity matrix. If $P^n = 5I - 8P$, then n is

47 Let A be an idempotent matrix satisfying $(I - 0.4A)^{-1} = I - \lambda A$, where I is unit matrix of the same order as A. Then, the value of $|3\lambda|$ is

48 The number of 2×2 matrices A, that are there with the elements as real numbers satisfying $A + A^T = I$ and $AA^T = I$ is

49 The number of orthogonal matrices of order 2×2 whose entries are 0, 1 is

50 Matrix A is given by $A = \begin{bmatrix} 3 & 11 \\ 2 & 8 \end{bmatrix}$, then the determinant of $A^{2011} - 5A^{2010}$ is $|\lambda \cdot 2^{2012}|$, then λ is

JEE Main & AIEEE Archive

(Compilation of Last 13 Years' Questions)

1 If $\alpha, \beta \neq 0$, $f(n) = \alpha^n + \beta^n$ and

$$\begin{vmatrix} 3 & 1+f(1) & 1+f(2) \\ 1+f(1) & 1+f(2) & 1+f(3) \\ 1+f(2) & 1+f(3) & 1+f(4) \end{vmatrix}$$

$= K(1-\alpha)^2 (1-\beta)^2 (\alpha-\beta)^2$, then K is equal to **(JEE Main 2014)**

a. $\alpha\beta$ b. $\frac{1}{\alpha\beta}$

c. 1 d. -1

Two determinants can be multiplied row-to-row or row-to-column.

$f(n) = \alpha^n + \beta^n$

$f(1) = \alpha + \beta$,

$f(2) = \alpha^2 + \beta^2$, $f(3) = \alpha^3 + \beta^3$, $f(4) = \alpha^4 + \beta^4$

2 If A, an 3×3 non-singular matrix such that $AA' = A'A$ and $B = A^{-1}A'$, then BB' is equal to **(JEE Main 2014)**

a. $I + B$ b. I

c. B^{-1} d. $(B^{-1})'$

$(AB)^T = B^T A^T$ and $A^{-1}A = I$

(where, $A^T = A' =$ Transpose of A)

3 The number of values of k, for which the system of equations

$$(k+1)x + 8y = 4k$$
$$kx + (k+3)y = 3k - 1$$

has no solution, is **(JEE Main 2013)**

a. ∞ b. 1

c. 2 d. 3

4 If $P = \begin{bmatrix} 1 & \alpha & 3 \\ 1 & 3 & 3 \\ 2 & 4 & 4 \end{bmatrix}$ is the adjoint of a 3×3 matrix A and $|A| = 4$, then α is equal to **(JEE Main 2013)**

a. 4 b. 11

c. 5 d. 0

5 Let $A = \begin{bmatrix} 1 & 0 & 0 \\ 2 & 1 & 0 \\ 3 & 2 & 1 \end{bmatrix}$. If U_1 and U_2 are column matrices such that $AU_1 = \begin{bmatrix} 1 \\ 0 \\ 0 \end{bmatrix}$ and $AU_2 = \begin{bmatrix} 0 \\ 1 \\ 0 \end{bmatrix}$, then $U_1 + U_2$ is equal to **(AIEEE 2012)**

a. $\begin{bmatrix} -1 \\ 1 \\ 0 \end{bmatrix}$ b. $\begin{bmatrix} -1 \\ 1 \\ -1 \end{bmatrix}$

c. $\begin{bmatrix} -1 \\ -1 \\ 0 \end{bmatrix}$ d. $\begin{bmatrix} 1 \\ -1 \\ -1 \end{bmatrix}$

6 Let P and Q be 3×3 matrices $P \neq Q$. If $P^3 = Q^3$ and $P^2Q = Q^2P$, then determinant of $(P^2 + Q^2)$ is equal to **(AIEEE 2012)**

a. -2 b. 1

c. 0 d. -1

7 The number of values of k for which the linear equations $4x + ky + 2z = 0$, $kx + 4y + z = 0$ and $2x + 2y + z = 0$ posses a non-zero solution, is **(AIEEE 2011)**

a. 2 b. 1

c. 0 d. 3

8 Let A and B be two symmetric matrices of order 3.

Statement I $A(BA)$ and $(AB)A$ are symmetric matrices.

Statement II AB is symmetric matrix, if matrix multiplication of A with B is commutative. **(AIEEE 2011)**

a. Both Statement I and Statement II are correct and Statement II is the correct explanation of Statement I

b. Both Statement I and Statement II are correct but Statement II is not the correct explanation of Statement I

c. Statement I is correct but Statement II is incorrect

d. Statement II is correct but Statement I is incorrect

9 If the trivial solution is the only solution of the system of equations

$$x - ky + z = 0,$$
$$kx + 3y - kz = 0$$

and $$3x + y - z = 0$$

Then, the set of all values of k is **(AIEEE 2011)**

a. $\{2, -3\}$ b. $R - \{2, -3\}$

c. $R - \{2\}$ d. $R - \{-3\}$

10 **Statement I** Determinant of a skew- symmetric matrix of order 3 is zero.

Statement II For any matrix A,

$$\det(A^T) = \det(A)$$

and $$\det(-A) = -\det(A).$$

Then, **(AIEEE 2011)**

a. Both Statement I and Statement II are correct and Statement II is the correct explanation of Statement I

b. Both Statement I and Statement II are correct but Statement II is not the correct explanation of Statement I

c. Statement I is correct but Statement II is incorrect

d. Statement II is correct but Statement I is incorrect

11 If $\omega \neq 1$ is the complex cube root of unity and matrix $H = \begin{bmatrix} \omega & 0 \\ 0 & \omega \end{bmatrix}$, then H^{70} is equal to

a. H b. 0 **(AIEEE 2011)**
c. $-H$ d. H^2

12 Consider the system of linear equations

$$x_1 + 2x_2 + x_3 = 3, \quad 2x_1 + 3x_2 + x_3 = 3$$

and $3x_1 + 5x_2 + 2x_3 = 1$

The system has **(AIEEE 2010)**

a. infinite number of solutions
b. exactly 3 solutions
c. a unique solution
d. no solution

13 The number of 3×3 non-singular matrices, with four entries as 1 and all other entries as 0, is

a. less than 4 b. 5 **(AIEEE 2010)**
c. 6 d. atleast 7

14 Let A be 2×2 matrix with non-zero entries and $A^2 = I$, where I is 2×2 identity matrix.

Define tr (A) = Sum of diagonal elements of A and $|A|$ = Determinant of matrix A.

Statement I tr $(A) = 0$

Statement II $|A| = 1$ **(AIEEE 2010)**

a. Both Statement I and Statement II are correct and Statement II is the correct explanation of Statement I
b. Both Statement I and Statement II are correct but Statement II is not the correct explanation of Statement I
c. Statement I is correct but Statement II is incorrect
d. Statement II is correct but Statement I is incorrect

15 Let a, b and c be such that $(b + c) \neq 0$.

If $\begin{vmatrix} a & a+1 & a-1 \\ -b & b+1 & b-1 \\ c & c-1 & c+1 \end{vmatrix} + \begin{vmatrix} a+1 & b+1 & c-1 \\ a-1 & b-1 & c+1 \\ (-1)^{n+2}a & (-1)^{n+1}b & (-1)^n c \end{vmatrix} = 0,$

then the value of n is **(AIEEE 2009)**

a. zero b. any even integer
c. any odd integer d. any integer

16 Let A be a 2×2 matrix.

Statement I adj (adj A) = A

Statement II $|\text{adj} A| = A$ **(AIEEE 2009)**

a. Both Statement I and Statement II are correct and Statement II is the correct explanation of Statement I
b. Both Statement I and Statement II are correct and Statement II is not the correct explanation of Statement I
c. Statement I is correct but Statement II is incorrect
d. Statement II is correct but Statement I is incorrect

17 Let A be a 2×2 matrix with real entries. Let I be the 2×2 identity matrix. Denote by tr (A), the sum of diagonal entries of A. Assume that $A^2 = I$.

Statement I If $A \neq I$ and $A \neq -I$, then det $(A) = -1$.

Statement II If $A \neq I$ and $A \neq -I$, then tr $(A) \neq 0$. **(AIEEE 2008)**

a. Both Statement I and Statement II are correct and Statement II is the correct explanation of Statement I
b. Both Statement I and Statement II are correct but Statement II is not the correct explanation of Statement I
c. Statement I is correct but Statement II is incorrect
d. Statement II is correct but Statement I is incorrect

18 Let a, b and c be any real numbers. Suppose that there are real numbers x, y, z not all zero such that $x = cy + bz$, $y = az + cx$ and $z = bx + ay$. Then, $a^2 + b^2 + c^2 + 2abc$ is equal to **(AIEEE 2008)**

a. 1 b. 2
c. -1 d. 0

19 Let A be a square matrix, all of whose entries are integers. Then, which one of the following is true? **(AIEEE 2008)**

a. If det $(A) = \pm 1$, then A^{-1} need not exist
b. If det $(A) = \pm 1$, then A^{-1} exists but all its entries are not necessarily integers
c. If det $(A) \neq \pm 1$, then A^{-1} exists and all its entries are non-integers
d. If det $(A) = \pm 1$, then A^{-1} exists and all its entries are integers

20 If $D = \begin{vmatrix} 1 & 1 & 1 \\ 1 & 1+x & 1 \\ 1 & 1 & 1+y \end{vmatrix}$ for $x \neq 0$, $y \neq 0$, then D is **(AIEEE 2007)**

a. divisible by neither x nor y
b. divisible by both x and y
c. divisible by x but not y
d. divisible by y but not x

21 Let $A = \begin{bmatrix} 5 & 5\alpha & \alpha \\ 0 & \alpha & 5\alpha \\ 0 & 0 & 5 \end{bmatrix}$. If $|A^2| = 25$, then $|\alpha|$ is equal to **(AIEEE 2007)**

a. 5^2 b. 1
c. $\frac{1}{5}$ d. 5

22 If A and B are square matrices of size $n \times n$ such that $A^2 - B^2 = (A - B)(A + B)$, then which of the following will be always true? **(AIEEE 2006)**

a. $AB = BA$
b. Either of A or B is a zero matrix
c. Either of A or B is an identity matrix
d. $A = B$

23 Let $A=\begin{bmatrix}1 & 2\\ 3 & 4\end{bmatrix}$ and $B=\begin{bmatrix}a & 0\\ 0 & b\end{bmatrix}$; $a, b \in N$. Then, **(AIEEE 2006)**

a. there exists more than one but finite number of B's such that $AB = BA$
b. there exists exactly one B such that $AB = BA$
c. there exist infinitely many B's such that $AB = BA$
d. there cannot exist any B such that $AB = BA$

24 If $A^2 - A + I = O$, then the inverse of A is **(AIEEE 2005)**

a. $I - A$ b. $A - I$
c. A d. $A + I$

25 If $a^2 + b^2 + c^2 = -2$ and

$$f(x)=\begin{vmatrix}1+a^2x & (1+b^2)x & (1+c^2)x\\ (1+a^2)x & 1+b^2x & (1+c^2)x\\ (1+a^2)x & (1+b^2)x & 1+c^2x\end{vmatrix},$$

then $f(x)$ is a polynomial of degree **(AIEEE 2005)**

a. 2   b. 3
c. 0   d. 1

26 The system of equations

$$\alpha x + y + z = \alpha - 1,$$
$$x + \alpha y + z = \alpha - 1$$
and $$x + y + \alpha z = \alpha - 1$$

has no solution, if α is **(AIEEE 2005)**

a. 1
b. not -2
c. either -2 or 1
d. -2

27 If $a_1, a_2, \ldots, a_n, \ldots$ are in GP, then the determinant

$$\Delta=\begin{vmatrix}\log a_n & \log a_{n+1} & \log a_{n+2}\\ \log a_{n+3} & \log a_{n+4} & \log a_{n+5}\\ \log a_{n+6} & \log a_{n+7} & \log a_{n+8}\end{vmatrix}$$

is equal to **(AIEEE 2005, 2004)**

a. 2   b. 4
c. 0   d. 1

28 Let $A=\begin{bmatrix}0 & 0 & -1\\ 0 & -1 & 0\\ -1 & 0 & 0\end{bmatrix}$. The only correct statement about the matrix A is **(AIEEE 2004)**

a. A is a zero matrix
b. $A = (-1)I$, where I is a unit matrix
c. A^{-1} does not exist
d. $A^2 = I$

29 Let $A=\begin{bmatrix}1 & -1 & 1\\ 2 & 1 & -3\\ 1 & 1 & 1\end{bmatrix}$ and $10B=\begin{bmatrix}4 & 2 & 2\\ -5 & 0 & \alpha\\ 1 & -2 & 3\end{bmatrix}$.

If B is the inverse of matrix A, then α is equal to **(AIEEE 2004)**

a. -2 b. 1
c. 2   d. 5

30 If the system of linear equations

$$x + 2ay + az = 0,\ x + 3by + bz = 0$$
and $$x + 4cy + cz = 0$$

has a non-zero solution, then a, b and c **(AIEEE 2003)**

a. are in AP
b. are in GP
c. are in HP
d. satisfy $a + 2b + 3c = 0$

31 If $A=\begin{bmatrix}a & b\\ b & a\end{bmatrix}$ and $A^2=\begin{bmatrix}\alpha & \beta\\ \beta & \alpha\end{bmatrix}$, then **(AIEEE 2003)**

a. $\alpha = a^2 + b^2$ and $\beta = ab$
b. $\alpha = a^2 + b^2$ and $\beta = 2ab$
c. $\alpha = a^2 + b^2$ and $\beta = a^2 - b^2$
d. $\alpha = 2ab$ and $\beta = a^2 + b^2$

32 If 1, ω and ω^2 are the cube roots of unity, then

$$\Delta=\begin{vmatrix}1 & \omega^n & \omega^{2n}\\ \omega^n & \omega^{2n} & 1\\ \omega^{2n} & 1 & \omega^n\end{vmatrix}$$ is equal to **(AIEEE 2003)**

a. 0   b. 1
c. ω d. ω^2

33 If l, m and n are the pth, qth and rth terms of a GP and all positive, then $\begin{vmatrix}\log l & p & 1\\ \log m & q & 1\\ \log n & r & 1\end{vmatrix}$ is equal to **(AIEEE 2002)**

a. 3   b. 2
c. 1   d. 0

34 If $\omega (\neq 1)$ is a cubic root of unity, then

$$\begin{vmatrix}1 & 1+i+\omega^2 & \omega^2\\ 1-i & -1 & \omega^2-1\\ -i & -1+\omega-i & -1\end{vmatrix}$$ is equal to **(AIEEE 2002)**

a. 0   b. 1
c. i d. ω

JEE Advanced & IIT JEE Archive

(Compilation of Last 10 Years' Questions)

1 Let X and Y be two arbitrary, 3×3, non-zero skew-symmetic matrices and Z be an arbitrary 3×3 non-zero symmetric matrix. Then, which of the folloiwng matrices is (are) skew-symmetric?

(2015, One or More than One Options Correct Type)

a. $Y^3 Z^4 - Z^4 Y^3$ b. $X^{44} + Y^{44}$
c. $X^4 Z^3 - Z^3 X^4$ d. $X^{23} + Y^{23}$

2 Which of the following values of α satisfy the equation **(2015, One or More than One Options Correct Type)**

$$\begin{vmatrix} (1+\alpha)^2 & (1+2\alpha)^2 & (1+3\alpha)^2 \\ (2+\alpha)^2 & (2+2\alpha)^2 & (2+3\alpha)^2 \\ (3+\alpha)^2 & (3+2\alpha)^2 & (3+3\alpha)^2 \end{vmatrix} = -648\alpha ?$$

a. -4 b. 9
c. -9 d. 4

3 Let M be a 2×2 symmetric matrix with integer entries. Then, M is invertible, if

(2014 Adv., One or More than One Options Correct Type)

a. the first column of M is the transpose of the second row of M
b. the second row of M is the transpose of the first column of M
c. M is a diagonal matrix with non-zero entries in the main diagonal
d. the product of entries in the main diagonal of M is not the square of an integer

4 Let M and N be two 3×3 matrices such that $MN = NM$. Further, if $M \neq N^2$ and $M^2 = N^4$, then

(2014 Adv., One or More than One Options Correct Type)

a. determinant of $(M^2 + MN^2)$ is 0
b. there is a 3×3 non-zero matrix U such that $(M^2 + MN^2)U$ is zero matrix
c. determinant of $(M^2 + MN^2) \geq 1$
d. for a 3×3 matrix U, if $(M^2 + MN^2)U$ equals the zero matrix, then U is the zero matrix

5 Let ω be a complex cube root of unity with $\omega \neq 1$ and $P = [p_{ij}]$ be a $n \times n$ matrix with $p_{ij} = \omega^{i+j}$. Then, $P^2 \neq 0$, when n is equal to

(2013 Adv., One or More than One Options Correct Type)

a. 57 b. 55
c. 58 d. 56

6 For 3×3 matrices M and N, which of the following statement(s) is (are) not correct ?

(2013 Adv., One or More than One Options Correct Type)

a. $N^T M N$ is symmetric or skew-symmetric, according as M is symmetric or skew-symmetric
b. $MN - NM$ is symmetric for all symmetric matrices M and N
c. $M N$ is symmetric for all symmetric matrices M and N
d. (adj M) (adj N) = adj(MN) for all invertible matrices M and N

7 If the adjoint of a 3×3 matrix P is $\begin{bmatrix} 1 & 4 & 4 \\ 2 & 1 & 7 \\ 1 & 1 & 3 \end{bmatrix}$, then the possible value(s) of the determinant of P is/are **(2012, One or More than One Options Correct Type)**

a. -2 b. -1
c. 1 d. 2

8 If P is a 3×3 matrix such that $P^T = 2P + I$, where P^T is the transpose of P and I is the 3×3 identity matrix, then there exists a column matrix,

$$X = \begin{bmatrix} x \\ y \\ z \end{bmatrix} \neq \begin{bmatrix} 0 \\ 0 \\ 0 \end{bmatrix} \text{ such that}$$

(2012, Only One Option Correct Type)

a. $PX = \begin{bmatrix} 0 \\ 0 \\ 0 \end{bmatrix}$
b. $PX = X$
c. $PX = 2X$
d. $PX = -X$

9 Let $P = [a_{ij}]$ be a 3×3 matrix and let $Q = [b_{ij}]$, where $b_{ij} = 2^{i+j} a_{ij}$ for $1 \leq i, j \leq 3$. If the determinant of P is 2, then the determinant of the matrix Q is

(2012, Only One Option Correct Type)

a. 2^{10} b. 2^{11}
c. 2^{12} d. 2^{13}

10 Let M and N be two 3×3 non-singular skew-symmetric matrices such that $MN = NM$. If P^T denotes the transpose of P, then $M^2N^2(M^T N)^{-1}(MN^{-1})^T$ is equal to

(2011, One or More than One Options Correct Type)

a. M^2 b. $-N^2$
c. $-M^2$ d. MN

11 Let $\omega \neq 1$ be a cube root of unity and S be the set of all non-singular matrices of the form $\begin{bmatrix} 1 & a & b \\ \omega & 1 & c \\ \omega^2 & \omega & 1 \end{bmatrix}$, where each of a, b and c is either ω or ω^2. Then, the number of distinct matrices in the set S is **(2011, Only One Option Correct Type)**

a. 2 b. 6
c. 4 d. 8

12 Let M be a 3×3 matrix satisfying

$$M\begin{bmatrix}0\\1\\0\end{bmatrix}=\begin{bmatrix}-1\\2\\3\end{bmatrix}, M\begin{bmatrix}1\\-1\\0\end{bmatrix}=\begin{bmatrix}1\\1\\-1\end{bmatrix} \text{ and } M\begin{bmatrix}1\\1\\1\end{bmatrix}=\begin{bmatrix}0\\0\\12\end{bmatrix}.$$

Then, the sum of the diagonal entries of M is

(2011, Integer Type)

13 Let k be a positive real number and let

$$A=\begin{bmatrix}2k-1 & 2\sqrt{k} & 2\sqrt{k}\\ 2\sqrt{k} & 1 & -2k\\ -2\sqrt{k} & 2k & -1\end{bmatrix}$$

and $$B=\begin{bmatrix}0 & 2k-1 & \sqrt{k}\\ 1-2k & 0 & 2\sqrt{k}\\ -\sqrt{k} & -2\sqrt{k} & 0\end{bmatrix}$$

If $\det(\text{adj } A)+\det(\text{adj } B)=10^6$, then $[k]$ is equal to

(2010, Integer Type)

Note adj M denotes the adjoint of a square matrix M and $[k]$ denotes the largest integer less than or equal to k.

Passage for Q. Nos. (14-16)

Let p be an odd prime number and T_p be the following set of 2×2 matrices

$$T_p=\left\{A=\begin{bmatrix}a & b\\ c & a\end{bmatrix}; a, b, c \in \{0, 1, 2, \ldots p-1\}\right\}$$

(2010, Comprehension Type)

14 The number of A in T_p such that A is either symmetric or skew-symmetric or both and det A is divisible by p is

a. $(p-1)^2$
b. $2(p-1)$
c. $(p-1)^2+1$
d. $2p-1$

15 The number of A in T_p such that the trace of A is not divisible by p but det (A) is divisible by p is

[Note The trace of a matrix is the sum of its diagonal entries.]

a. $(p-1)(p^2-p+1)$
b. $p^3-(p-1)^2$
c. $(p-1)^2$
d. $(p-1)(p^2-2)$

16 The number of A in T_p such that det (A) is not divisible by p, is

a. $2p^2$ b. p^3-5p
c. p^3-3p d. p^3-p^2

17 The number of 3×3 matrices A whose entries are either 0 or 1 and for which the system $A\begin{bmatrix}x\\y\\z\end{bmatrix}=\begin{bmatrix}1\\0\\0\end{bmatrix}$ has exactly two distinct solutions, is

(2010, Only One Option Correct Type)

a. 0 b. 2^9-1
c. 168 d. 2

18 Consider the system of equations

$$x-2y+3z=-1,$$
$$x-3y+4z=1 \text{ and } -x+y-2z=k$$

(2008, Statement Type)

Statement I The system of equations has no solution for $k \neq 3$.

Statement II The determinant

$$\begin{vmatrix}1 & 3 & -1\\ -1 & -2 & k\\ 1 & 4 & 1\end{vmatrix} \neq 0, \text{ for } k \neq 0.$$

a. Both Statement I and Statement II are correct and Statement II is the correct explanation of Statement I
b. Both Statement I and Statement II are correct but Statement II is not the correct explanation of Statement I
c. Statement I is correct but Statement II is incorrect
d. Statement II is correct but Statement I is incorrect

19 If $A=\begin{bmatrix}1 & 0 & 0\\ 0 & 1 & 1\\ 0 & -2 & 4\end{bmatrix}$, $6A^{-1}=A^2+cA+dI$, then (c, d) is

(2005, Only One Option Correct Type)

a. $(-6, 11)$ b. $(-11, 6)$
c. $(11, 6)$ d. $(6, 11)$

20 If $P=\begin{bmatrix}\sqrt{3}/2 & 1/2\\ -1/2 & \sqrt{3}/2\end{bmatrix}$, $A=\begin{bmatrix}1 & 1\\ 0 & 1\end{bmatrix}$ and $Q=PAP^T$, then $P^TQ^{2005}P$ is

(2005, Only One Option Correct Type)

a. $\begin{bmatrix}1 & 2005\\ 0 & 1\end{bmatrix}$ b. $\begin{bmatrix}1 & 2005\\ 2005 & 1\end{bmatrix}$
c. $\begin{bmatrix}1 & 0\\ 2005 & 1\end{bmatrix}$ d. $\begin{bmatrix}1 & 0\\ 0 & 1\end{bmatrix}$

Answers with Explanations

Answers

1. Inverse Trigonometric Functions

DPP-1 Domain and Range of ITF, Trig. $(\text{Trig}^{-1}(x)) = x, \text{Trig}^{-1}(\text{Trig}(x)) = x$ and Interconversion between ITF's

1. (i) Domain $=[-1,1]$ and range $=\left[\frac{-\pi}{2},\frac{\pi}{2}\right]$ or $\left[\frac{\pi}{2},\frac{3\pi}{2}\right]$ or $\left[\frac{3\pi}{2},\frac{5\pi}{2}\right]$ (ii) Domain $=[-1,1]$ and range $=[0,\pi]$

(iii) Domain $=R$ and range $=\left(\frac{-\pi}{2},\frac{\pi}{2}\right)$ (iv) Domain $=(-\infty,-1]\cup[1,\infty]$ and range $=\left[\frac{-\pi}{2},\frac{\pi}{2}\right]-\{0\}$

(v) Domain $=(-\infty,1]\cup[1,\infty)$ and range $=[0,\pi]-\left\{\frac{\pi}{2}\right\}$ (vi) Domain $=R$ and range $=(0,\pi)$

2. (i) $\frac{-\pi}{3}$ (ii) 1 (iii) $\frac{-\pi}{4}$ (iv) -1 (v) $\frac{1}{\sqrt{3}}$

3. (i) $x\in\left(\frac{1}{2},1\right]$ (ii) $-\frac{1}{2}<x\le 1$ (iii) $x\in(1,2]$ (iv) $x\in[\sin 1,1]$

6. (i) $\frac{\pi}{3}$ (ii) $\frac{2\pi}{3}$ (iii) $\frac{\pi}{4}$ (iv) $\frac{\pi}{3}$ (v) $\frac{5\pi}{6}$ (vi) $\frac{-\pi}{4}$ (vii) -1

7. (i) $\frac{4}{5}$ (ii) $\frac{3}{5}$ (iii) $\frac{4}{5}$ (iv) $\frac{15}{8}$ (v) $\frac{13}{5}$ **8.** *d* **9.** *d* **10.** *c* **11.** *b* **12.** *b* **13.** *a*

14. *c* **15.** *b* **16.** *c* **17.** *d* **18.** *a* **19.** *c* **20.** *c*

DPP-2 Identities on Addition and Subtraction

2. (i) $1, \frac{-1}{6}$ (ii) $\frac{1}{4}$ (iii) $\sqrt{\frac{56}{224}}, \sqrt{\frac{24}{224}}$ (iv) ± 1 **3.** *d* **4.** *a* **5.** *b* **6.** *a* **7.** *c* **8.** *c*

9. *c* **10.** *c* **11.** *d* **12.** *a* **13.** *b* **14.** *d* **15.** *b* **16.** *a* **17.** *c* **18.** *b*

19. *a* **20.** *c* **21.** *b* **22.** *a* **23.** *a,b* **24.** *a,b,c* **25.** *b,c* **26.** *b,d* **27.** *b* **28.** *d*

29. 4 **30.** 4 **31.** 4 **32.** 2 **33.** 2 **34.** 2 **35.** 0

DPP-3 Special Formulae

1. *a* **2.** *b* **3.** *a* **4.** *d* **5.** *a* **6.** *c* **7.** *c* **8.** *d* **9.** *b* **10.** *b*

11. *b* **12.** *d* **13.** *c* **14.** *c* **15.** *c* **16.** *b* **17.** *c* **18.** *b,c* **19.** *a,b,c*

20. *b,c,d* **21.** *b* **22.** *c* **23.** *c* **24.** *b* **25.** *a* **26.** *c* **27.** *a* **28.** 2 **29.** 2

30. 0

Revisal Problems (JEE Main)

1. *c* **2.** *d* **3.** *c* **4.** *a* **5.** *b* **6.** *d* **7.** *c* **8.** *c* **9.** *c* **10.** *b*

11. *a* **12.** *a* **13.** *a* **14.** *a* **15.** *c* **16.** *d* **17.** *c* **18.** *c* **19.** *b* **20.** *d*

21. *c* **22.** *c* **23.** *d* **24.** *b* **25.** *a* **26.** *a* **27.** *b* **28.** *a* **29.** *b* **30.** *b*

31. *c* **32.** *a* **33.** *a* **34.** *c* **35.** *b* **36.** *a* **37.** *c* **38.** *b* **39.** *a* **40.** *b*

Revisal Problems (JEE Advanced)

1. *c* **2.** *c* **3.** *b* **4.** *a* **5.** *b* **6.** *c* **7.** *a* **8.** *a* **9.** *b* **10.** *c*

11. *c* **12.** *d* **13.** *d* **14.** *c* **15.** *c* **16.** *a* **17.** *a* **18.** *b* **19.** *a* **20.** *b*

21. *b* **22.** *a* **23.** *c* **24.** *c* **25.** *c* **26.** *a* **27.** *a* **28.** *c* **29.** *a,c* **30.** *a,d*

31. *a,b,c* **32.** *a* **33.** *a* **34.** *b* **35.** *d* **36.** *c* **37.** *a* **38.** *c* **39.** *d* **40.** *d*

41. *c* **42.** *d* **43.** *a* **44.** *c* **45.** *b* **46.** 5 **47.** 0 **48.** 1 **49.** 2 **50.** 6

JEE Main & AIEEE Archive

1. *a* **2.** *b* **3.** *b* **4.** *b* **5.** *b* **6.** *c* **7.** *a* **8.** *d*

JEE Advanced & IIT JEE Archive

1. *b,c,d* **2.** 3 **3.** *b* **4.** *c* **5.** *b*

2. Relations and Functions

DPP-1 Relations

1. *b*	2. *a*	3. *a*	4. *c*	5. *a*	6. *d*	7. *c*	8. *d*	9. *d*	10. *a*
11. *d*	12. *d*	13. *b*	14. *c*	15. *d*	16. *d*	17. *d*	18. *d*	19. *b*	20. *b*
21. *b*	22. *c*	23. *d*	24. *a*	25. *d*	26. *c*	27. *c*	28. *d*	29. *b*	30. *d*
31. *b*	32. *d*	33. *a*	34. *a*	35. *d*					

DPP-2 Definition of Functions

2. (i), (iii), (iv) and (vi) are functions. 4. *f* is a function 5. *c* 6. *d* 7. *d* 8. *a*

9. *c* 10. *c* 11. *d* 12. *d* 13. *b* 14. *b* 15. *c* 16. 4 17. 7 18. 1

19. 5 20. 1

DPP-3 Domain

1. (i) $R-\{-1,1\}$ (ii) $[-1,0)\cup(0,1]$ (iii) $(0,\infty)$ (iv) R (v) $[-2,0)\cup(0,1)$ (vi) $x\in\left(-\frac{1}{3},\frac{1}{2}\right]$ (vii) ϕ (viii) $\left(\frac{1}{2},1\right)\cup\left(1,\frac{3}{2}\right)$

2. (i) $[0,1]$ (ii) $[-1,1]$ (iii) R (iv) ϕ (v) $\bigcup_{n\in I}\left[n\pi, n\pi+\frac{\pi}{4}\right]$ (vi) $R-\{2n\pi\}, n\in I$ (vii) $(0,1]\cup[4,5)$ (viii) $(2,3)$

3. *c*	4. *b*	5. *b*	6. *a*	7. *d*	8. *b*	9. *c*	10. *c*	11. *d*	12. *c*
13. *a*	14. *d*	15. *c*	16. *a*	17. *d*	18. *b*	19. *a,b,d*	20. *a,c*	21. *a*	22. *b*
23. *c*	24. 1	25. 3							

DPP-4 Range of Functions

1. (i) $[0,\infty)$ (ii) $\left[-\frac{1}{2},\frac{1}{2}\right]$ (iii) $[0,4]$ (iv) $\{1,1\}$ (v) $[5-\sqrt{2},5+\sqrt{2}]$ (vi) $(0,1]$

2. (i) Domain $=R$, Range $=\left[\frac{1}{\sqrt{7}},1\right]$

(ii) Domain $=N\cup\{0\}$, Range $=n!, n\in N\cup\{0\}$

(iii) Domain $=R-\{3\}$, Range $=R-\{6\}$ (iv) Domain $=R$, Range $=\{1\}$

3. *a*	4. *c*	5. *d*	6. *b*	7. *b*	8. *b*	9. *b*	10. *d*		
11. *b*	12. *b*	13. *d*	14. *c*	15. *b*	16. *a*	17. *b*	18. *b*	19. *c*	20. *d*
21. *a,b,c*	22. *a,d*	23. *b*	24. *a*	25. *a*	26. 2	27. 2	28. 1	29. 0	30. 2

DPP-5 Identical, Even and Odd Functions

1. *b* 2. *b,c,d* 3. *b* 4. *d*

5. (i) Odd (ii) Even (iii) Even (iv) Neither even nor odd
(v) Odd (vi) Even (vii) Odd (viii) Neither even nor Odd
(ix) Even (x) Neither even nor odd
(xi) Even (xii) Odd (xiii) Even (xiv) Odd

DPP-6 Periodicity of Functions

1. *c*	2. *b*	3. *b*	4. *b*	5. *a*	6. *a*	7. *c*	8. *d*	9. *a*	10. *b*
11. *c*	12. *d*	13. *a*	14. *c*	15. *a*	16. *a*	17. *a,b,d*	18. *c*	19. *a,b,c*	
20. *a,b,c*	21. *a*	22. *c*	23. *a*	24. *a*	25. 4	26. 2	27. 8	28. 2	29. 2

DPP-7 Composition of Functions

3. (i) $fog=x,\ x>0, gof=x, x\in R$

(ii) $fog=|\sin x|, gof=\sin|x|$

(iii) $fog=\sin^{-1}x^2, gof=(\sin^{-1}x)^2$

(iv) $fog=\frac{3x^2-4x+2}{(1-x)^2}, gof=\frac{x^2+2}{x^2+1}$

4. *d*	5. *a*	6. *b*	7. *a*	8. *a*	9. *d*	10. *a*	11. *b*	12. *b*	13. *d*
14. *b*	15. *c*	16. *c*	17. *a*	18. *c*	19. 0	20. 6			

DPP-8 Type of Functions

1. *c* 2. *a* 3. *d* 4. *b* 5. *d* 6. *c* 7. *c* 8. *b* 9. *a* 10. *c*
11. *a* 12. *c* 13. *b* 14. *d* 15. *d* 16. *d* 17. *b* 18. *a* 19. *b* 20. *a*
21. *b* 22. *c* 23. *a* 24. *a* 25. 1 26. 6 27. 1 28. 6 29. 3 30. 5

DPP-9 Inverse of Functions

1. *b* 2. *c* 3. *c* 4. *b* 5. *a* 6. *a* 7. *c* 8. *b* 9. *d* 10. *a*
11. *a* 12. *c* 13. *b* 14. 4 15. 3

Revisal Problems (JEE Main)

1. *d* 2. *c* 3. *c* 4. *c* 5. *d* 6. *a* 7. *b* 8. *b* 9. *a* 10. *a*
11. *a* 12. *a* 13. *d* 14. *c* 15. *d* 16. *b* 17. *b* 18. *c* 19. *a* 20. *d*
21. *a* 22. *c* 23. *b* 24. *b* 25. *c* 26. *b* 27. *c* 28. *b* 29. *b* 30. *c*
31. *a* 32. *a* 33. *b* 34. *a* 35. *d* 36. *b* 37. *b* 38. *b* 39. *c* 40. *a*
41. *b* 42. *a* 43. *a* 44. *d* 45. *a* 46. *d* 47. *c* 48. *b* 49. *c* 50. *b*
51. *b* 52. *c* 53. *b* 54. *a* 55. *d* 56. *b* 57. *d* 58. *a* 59. *b* 60. *d*

Revisal Problems (JEE Advanced)

1. *b* 2. *c* 3. *b* 4. *c* 5. *a* 6. *c* 7. *b* 8. *d* 9. *d* 10. *d*
11. *d* 12. *d* 13. *a* 14. *b* 15. *d* 16. *c* 17. *b* 18. *c* 19. *c* 20. *a*
21. *c* 22. *c* 23. *d* 24. *a* 25. *a* 26. *a,b,c,d* 27. *a,b* 28. *a,b,c* 29. *a,b*
30. *c* 31. *a* 32. *c* 33. *a* 34. *b* 35. *c* 36. *c* 37. *c* 38. *d* 39. *a*
40. *c* 41. 1 42. 1 43. 9 44. 2

JEE Main & AIEEE Archive

1. *d* 2. *d* 3. *c* 4. *b* 5. *a* 6. *b* 7. *a* 8. *c* 9. *c* 10. *c*
11. *a* 12. *a* 13. *d* 14. *a* 15. *d* 16. *a* 17. *d* 18. *b* 19. *b* 20. *c*
21. *d* 22. *d* 23. *b* 24. *b* 25. *a* 26. *b* 27. *a*

JEE Advanced & IIT JEE Archive

1. *a,b,c* 2. *a,b,c* 3. *a,d* 4. *d* 5. *b,d* 6. *b* 7. *a* 8. *b* 9. *d* 10. *b*
11. *d* 12. *b* 13. *a* 14. *d* 15. *c* 16. Range $=\left[\frac{-\pi}{2},\frac{-\pi}{10}\right]\cup\left[\frac{3\pi}{10},\frac{\pi}{2}\right]$

3. Matrices and Determinants

DPP-1 Types of Matrices, Addition and Subtraction of Matrices

1. *c* 2. *b* 3. *a* 4. *c* 5. *a* 6. *a* 7. *a* 8. *a* 9. *d* 10. *c*
11. *b* 12. *b* 13. *b* 14. *a* 15. *c*

DPP-2 Multiplication and Transpose of a Matrix

1. *d* 2. *b* 3. *a* 4. *a* 5. *b* 6. *b* 7. *c* 8. *d* 9. *a* 10. *d*
11. *a* 12. *a* 13. *a* 14. *d* 15. *a* 16. *d* 17. *c* 18. *a* 19. *a* 20. *d*
21. *d* 22. *c* 23. *a* 24. *b* 25. *b* 26. *a,b,c,d* 27. *b* 28. *a,c,d* 29. *b,c*
30. *b* 31. 4 32. 4 33. 2 34. 1 35. 5

DPP-3 Properties of Determinants

1. *c* 2. *c* 3. *b* 4. *b* 5. *b* 6. *d* 7. *d* 8. *c* 9. *b* 10. *b*
11. *b* 12. *a* 13. *a* 14. *c* 15. *a* 16. *b* 17. *c* 18. *c* 19. *b* 20. *a*
21. *b* 22. *c* 23. *b* 24. *a* 25. *a* 26. *c* 27. *b* 28. *a,d* 29. *a,b,c* 30. *b,d*
31. *a,b* 32. *c,d* 33. *d* 34. *c* 35. 2 36. 2 37. 1 38. 6 39. 2 40. 4

DPP-4 Adjoint and Inverse of a Matrix

1. *c* 2. *b* 3. *b* 4. *a* 5. *a* 6. *a* 7. *d* 8. *a* 9. *a* 10. *b*
11. *b* 12. *c* 13. *d* 14. *a* 15. *b* 16. *b* 17. *b* 18. *c* 19. *d* 20. *c*
21. *a* 22. *a* 23. *d* 24. *c* 25. *a* 26. *a,d* 27. *c* 28. *b* 29. *c* 30. *d*
31. *a* 32. *a* 33. *b* 34. *b* 35. *b* 36. *b* 37. *d* 38. 9 39. 10 40. 4

DPP-5 Solving of Equations

1. *a* 2. *a* 3. *d* 4. *d* 5. *a* 6. *c* 7. *c* 8. *b* 9. *d* 10. *d*
11. *b* 12. *a* 13. *b* 14. *a* 15. *d* 16. *c* 17. *d* 18. *d* 19. *c* 20. *b,c*
21. *a* 22. *b* 23. *a* 24. *c* 25. *c* 26. *d* 27. *a* 28. *a* 29. *b* 30. *b*

DPP-6 Rank of Matrix, Differentiation and Product of Determinants

4. $\dfrac{(b^2-4ac)(a+b+c)^2}{a^4}$
9. *d*
10. *c*

Revisal Problems (JEE Main)

1. *a* 2. *b* 3. *c* 4. *b* 5. *c* 6. *c* 7. *a* 8. *d* 9. *c* 10. *a*
11. *b* 12. *a* 13. *d* 14. *b* 15. *b* 16. *b* 17. *b* 18. *c* 19. *a* 20. *b*
21. *b* 22. *a* 23. *c* 24. *d* 25. *b* 26. *a* 27. *b* 28. *c* 29. *a* 30. *d*
31. *a* 32. *b* 33. *b* 34. *a* 35. *a* 36. *c* 37. *d* 38. *b* 39. *c* 40. *b*

Revisal Problems (JEE Advanced)

1. *a* 2. *d* 3. *c* 4. *d* 5. *d* 6. *c* 7. *b* 8. *a* 9. *c* 10. *d*
11. *c* 12. *b* 13. *d* 14. *c* 15. *c* 16. *a* 17. *d* 18. *d* 19. *b* 20. *d*
21. *b* 22. *c* 23. *b* 24. *d* 25. *a* 26. *b* 27. *d* 28. *a* 29. *d* 30. *c*
31. *a,b,c* 32. *a,b* 33. *a,b* 34. *a,b,c,d* 35. *a,b,c* 36. *a,c* 37. *c* 38. *c* 39. *d*
40. *a* 41. *b* 42. *c* 43. *c* 44. *c* 45. *c* 46. 6 47. 2 48. 2 49. 2
50. 7

JEE Main & AIEEE Archive

1. *c* 2. *b* 3. *b* 4. *b* 5. *d* 6. *c* 7. *a* 8. *b* 9. *b* 10. *c*
11. *a* 12. *d* 13. *d* 14. *c* 15. *c* 16. *b* 17. *c* 18. *a* 19. *d* 20. *b*
21. *c* 22. *a* 23. *c* 24. *a* 25. *a* 26. *d* 27. *c* 28. *d* 29. *d* 30. *c*
31. *b* 32. *a* 33. *d* 34. *a*

JEE Advanced & IIT JEE Archive

1. *c,d* 2. *b,c* 3. *c,d* 4. *a,b* 5. *b,c,d* 6. *c,d* 7. *a,d* 8. *d* 9. *d* 10. *c*
11. *a* 12. 9 13. 4 14. *d* 15. *c* 16. *d* 17. *a* 18. *a* 19. *a* 20. *a*

1. Inverse Trigonometric Functions

DPP-1 Domain and Range of ITF, Trig. $(\text{Trig}^{-1}(x)) = x$, $\text{Trig}^{-1}(\text{Trig}(x)) = x$ and Interconversion between ITF's

1. (i) $\sin^{-1} x$ is a function with domain $[-1, 1]$ and range $\left[\frac{-\pi}{2}, \frac{\pi}{2}\right]$.

By considering $\sin^{-1} x$ as a function with domain $[-1, 1]$ and range $\left[\frac{-\pi}{2}, \frac{\pi}{2}\right]$

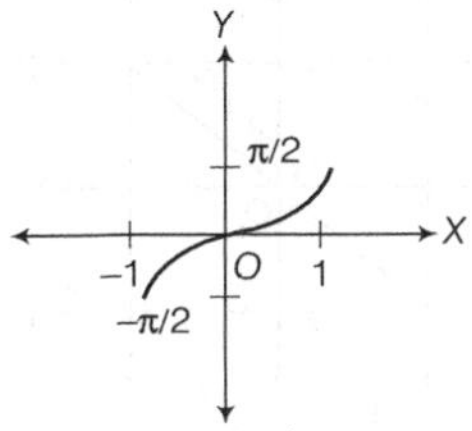

$y=\sin^{-1}x$

(ii) $\cos^{-1} x$ is a function with domain $[-1, 1]$ and range $[0, \pi]$. The graph of $\cos^{-1} x$ is obtained from the graph of cosine function by interchanging the coordinate axes.

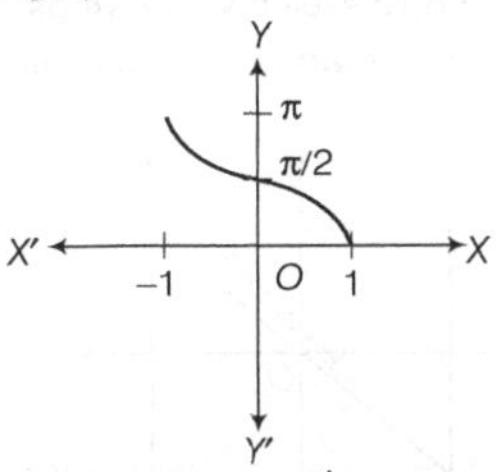

$y=\cos^{-1} x$

(iii) We have, $y = \tan^{-1} x$

Domain $= R$

Range $= \left(-\frac{\pi}{2}, \frac{\pi}{2}\right)$

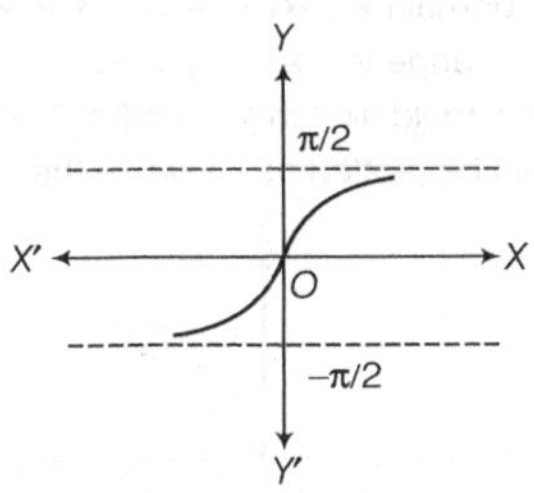

(iv) We have, $y = \operatorname{cosec}^{-1} x$

Domain $= (-\infty, -1] \cup [1, \infty)$

Range $= \left[-\frac{\pi}{2}, \frac{\pi}{2}\right] - \{0\}$

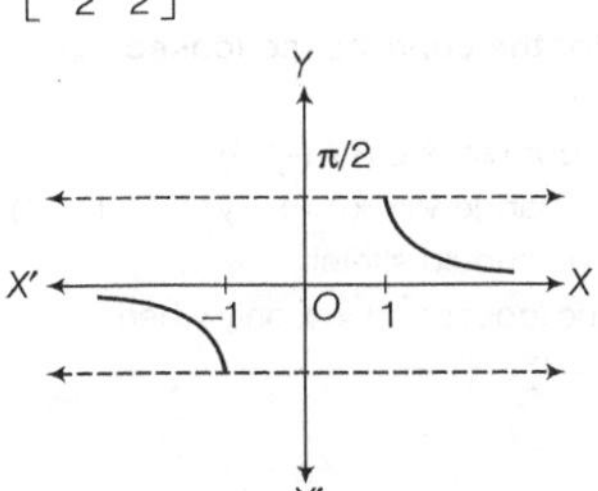

(v) We have,

$$y = \sec^{-1} x$$

Domain $= (-\infty, 1] \cup [1, \infty)$

Range $= [0, \pi] - \left\{\frac{\pi}{2}\right\}$

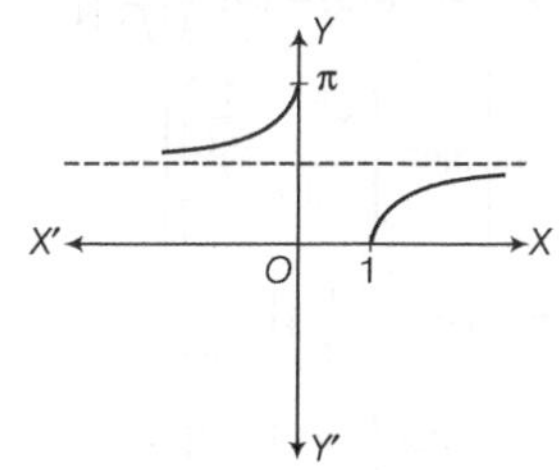

(vi)

$$y = \cot^{-1} x$$

Domain $= R$

Range $= (0, \pi)$

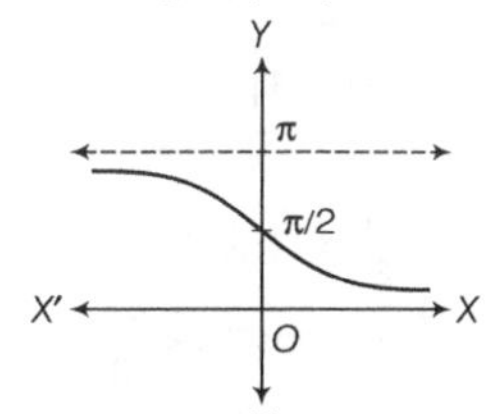

2. (i) We have,

$$\sin^{-1}\left[\sin\left(-\frac{2\pi}{3}\right)\right] = \sin^{-1}\left(-\sin\frac{2\pi}{3}\right) \quad [\because \sin(-\theta) = -\sin\theta]$$

$$= -\sin^{-1}\left(\sin\frac{2\pi}{3}\right) \quad [\because \sin^{-1}(-\theta) = -\sin^{-1}\theta]$$

$$= -\sin^{-1}\left[\sin\left(\pi - \frac{\pi}{3}\right)\right]$$

$$= -\sin^{-1}\left(\sin\frac{\pi}{3}\right) = -\frac{\pi}{3} \quad [\because \sin^{-1}(\sin\theta) = \theta]$$

(ii) We have,

$$\sin\left[\frac{\pi}{3} - \sin^{-1}\left(\frac{-1}{2}\right)\right] = \sin\left[\frac{\pi}{3} + \sin^{-1}\left(\frac{1}{2}\right)\right]$$

$$= \sin\left(\frac{\pi}{3} + \frac{\pi}{6}\right) \quad \left[\because \sin^{-1}\left(\frac{1}{2}\right) = \frac{\pi}{6}\right]$$

$$= \sin\left(\frac{3\pi}{6}\right)$$

$$= \sin\left(\frac{\pi}{2}\right) = 1$$

(iii) We have,

$$\tan^{-1}\left(\tan\frac{3\pi}{4}\right) = \tan^{-1}\left[\tan\left(\pi - \frac{\pi}{4}\right)\right]$$

$$= \tan^{-1}\left(-\tan\frac{\pi}{4}\right)$$

$$= -\tan^{-1}\left(\tan\frac{\pi}{4}\right) \quad [\tan^{-1}(-\theta) = -\tan^{-1}\theta]$$

$$= -\frac{\pi}{4}$$

(iv) We have,

$\cot[\sec^{-1}(-\sqrt{2}) + \cos^{-1}(-1)]$

$$= \cot\left(\frac{3\pi}{4} + \pi\right) = \cot\left(\frac{7\pi}{4}\right) = -1$$

(v) We have,

$$\tan\left[\cos^{-1}\left(\frac{1}{2}\right) + \tan^{-1}\left(-\frac{1}{\sqrt{3}}\right)\right]$$

$$= \tan\left[\tan^{-1}(\sqrt{3}) + \tan^{-1}\left(-\frac{1}{\sqrt{3}}\right)\right]$$

$$= \tan\left[\tan^{-1}(\sqrt{3}) - \tan^{-1}\left(\frac{1}{\sqrt{3}}\right)\right]$$

$$= \tan\left[\tan^{-1}\left(\frac{\sqrt{3} - \frac{1}{\sqrt{3}}}{1 + 1}\right)\right]$$

$$= \tan\left[\tan^{-1}\left(\frac{1}{\sqrt{3}}\right)\right] = \frac{1}{\sqrt{3}}$$

3. (i) We have, $\sin^{-1} x > \frac{\pi}{6}$; $x \in [-1, 1]$

$\Rightarrow \quad x > \sin\frac{\pi}{6}$; $x \in [-1, 1]$

$\Rightarrow \quad x > \frac{1}{2}$; $x \in [-1, 1]$

$\Rightarrow \quad x \in \left(\frac{1}{2}, 1\right]$

(ii) We have,

$\frac{-\pi}{4} < \cos^{-1} x < \frac{2\pi}{3}$; $x \in [-1, 1]$

$\Rightarrow \quad 0 \le \cos^{-1} x < \frac{2\pi}{3}$

$\Rightarrow \quad \cos 0 \ge x > \cos\frac{2\pi}{3}$

$\therefore \quad 1 \ge x > -\frac{1}{2}$

(iii) We have,

$\log_{\frac{\pi}{4}}[\tan^{-1}(x-1)] \ge 1$

$\Rightarrow \quad 0 < \tan^{-1}(x-1) \le \frac{\pi}{4}$

$\Rightarrow \quad 0 < x - 1 \le \tan\frac{\pi}{4}$

$\Rightarrow \quad 0 < x - 1 \le 1$

$\Rightarrow \quad 1 < x \le 2$

$\therefore \quad x \in (1, 2]$

(iv) We have,

$(\sin^{-1} x)^2 - 3(\sin^{-1} x) + 2 \le 0$

Put $\quad \sin^{-1} x = t$

$\Rightarrow \quad t^2 - 3t + 2 \le 0$

$\Rightarrow \quad t^2 - 2t - t + 2 \le 0$

$\Rightarrow \quad t(t-2) - 1(t-2) \le 0$

$\Rightarrow \quad (t-2)(t-1) \le 0$

$\Rightarrow \quad 1 \le t \le 2$

But $\sin^{-1} x \le \frac{\pi}{2}$

$\Rightarrow \quad 1 \le \sin^{-1} x \le \frac{\pi}{2}$

$\therefore \quad \sin 1 \le x \le 1$

4. As we know, all the mentioned six curves are non-periodic but have restricted domain and range. So, we shall first define each curve for its domain and range and then sketch the curves.

(i) **Graph for the curve $\sin(\sin^{-1}x)$**

We know, domain $x \in [-1, 1]$

and $\quad$ range $y \in [-1, 1]$

Hence, we should sketch $y = \sin(\sin^{-1} x)$ only when $x \in [-1, 1]$ and $y = x$.

So, its graph could be plotted as shown in the figure.

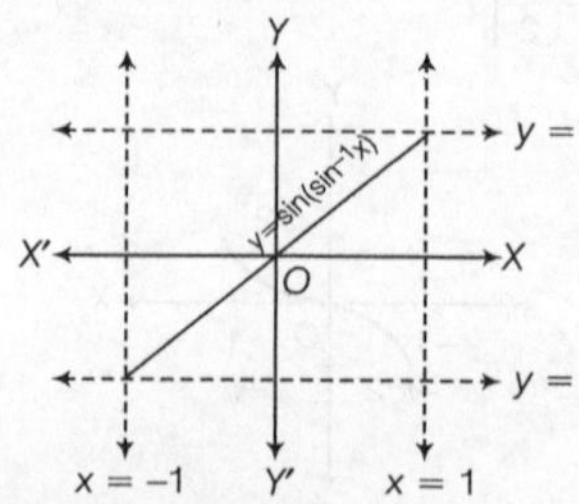

(ii) **Graph for the curve $y = \cos(\cos^{-1} x)$**

We know, domain $x \in [-1, 1]$

and range $y = x \Rightarrow y \in [-1, 1]$

Hence, we should sketch $y = \cos(\cos^{-1} x) = x$ only when $x \in [-1, 1]$. So, its graph could be plotted as shown in the figure.

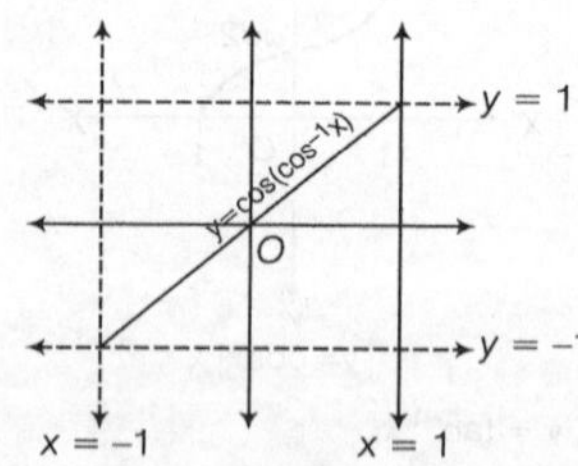

(iii) **Graph for the curve $\tan(\tan^{-1} x)$**

We know,

domain $x \in R$ (i.e. $-\infty < x < \infty$)

and $\quad$ range $y = x \Rightarrow y \in R$.

Hence, we should sketch $y = \tan(\tan^{-1} x) = x$, $\forall x \in R$. So, its graph could be plotted as shown in the figure.

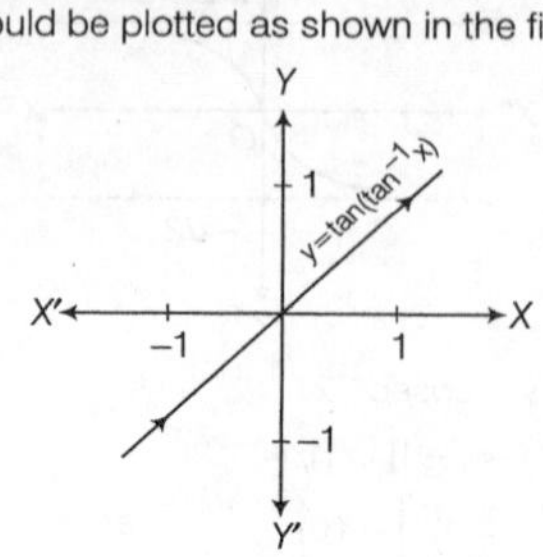

(iv) **Graph for the curve $\text{cosec}(\text{cosec}^{-1}x)$**

We know,

domain $x \in R - (-1, 1)$

and $\quad$ range $y = x \Rightarrow y \in R - (-1, 1)$

Hence, we should sketch

$y = \text{cosec}(\text{cosec}^{-1}x) = x$ only when

$x \in (-\infty, -1] \cup [1, \infty)$.

So, its graph could be plotted as shown in the figure.

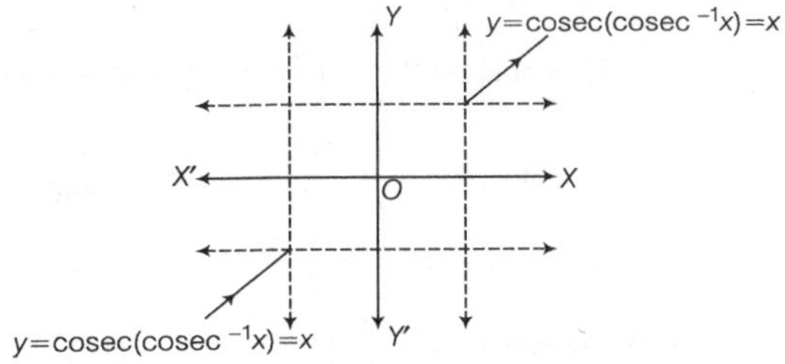

(v) **Graph for the curve sec (sec^{-1} x)**

We know

domain $x \in R - (-1, 1)$

and range $y = x \Rightarrow y \in R - (-1, 1)$

Hence, we should sketch $y = \sec(\sec^{-1} x) = x$ only when $x \in (-\infty, -1] \cup [1, \infty)$.

So, its graph could be plotted as shown in the figure.

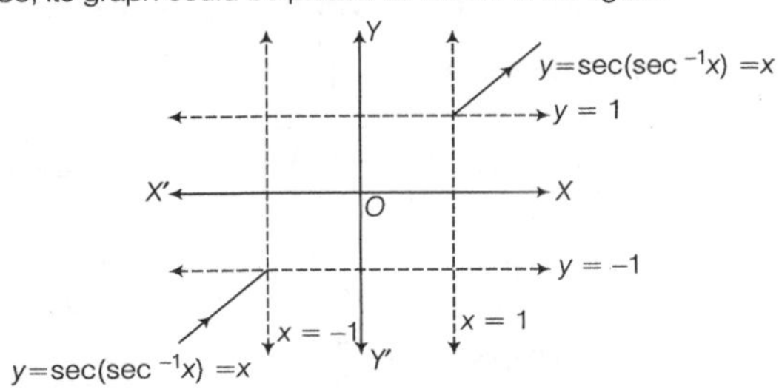

(vi) **Graph for the curve cot (cot^{-1} x)**

We know,

domain $x \in R$ (i.e. $-\infty < x < \infty$)

Range $y = x \Rightarrow y \in R$

Hence, we should sketch

$y = \cot(\cot^{-1} x) = x, \forall x \in R$ as shown in the figure.

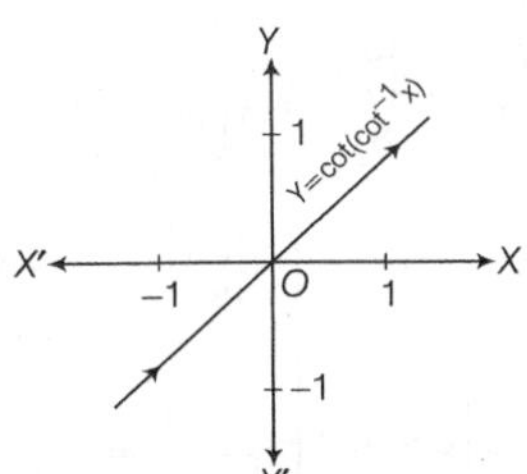

5. (i) As $y = \sin^{-1}(\sin x)$ is periodic with period $= 2\pi$.

$\therefore$ To draw this graph, we should draw the graph for one interval of length 2π and repeat it for entire values of x.

As we know,

$$\sin^{-1}(\sin x) = \begin{cases} x, & \dfrac{-\pi}{2} \le x \le \dfrac{\pi}{2} \\ (\pi - x), & -\dfrac{\pi}{2} \le \pi - x < \dfrac{\pi}{2} \end{cases}$$

$$\left[\text{i.e. } \frac{\pi}{2} \le x \le \frac{3\pi}{2}\right]$$

or $$\sin^{-1}(\sin x) = \begin{cases} x, & -\dfrac{\pi}{2} \le x \le \dfrac{\pi}{2} \\ \pi - x, & \dfrac{\pi}{2} \le x \le \dfrac{3\pi}{2} \end{cases}$$

which is defined for the interval of length 2π plotted as

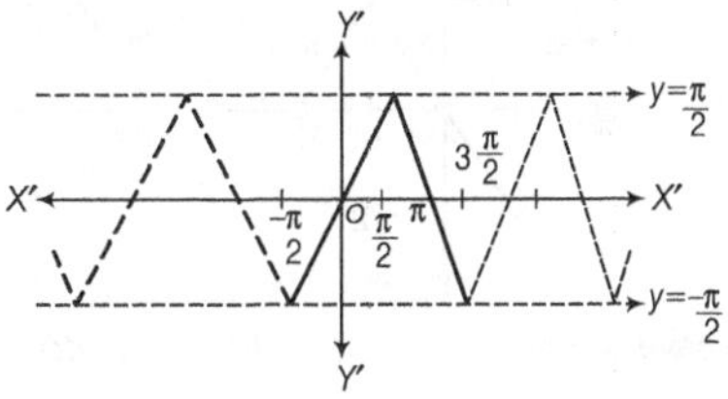

Thus, the graph for $y = \sin^{-1}(\sin x)$ is a straight line up and a straight line down with slope 1 and –1 respectively, lying between $\left[-\dfrac{\pi}{2}, \dfrac{\pi}{2}\right]$.

(ii) As $y = \cos^{-1}(\cos x)$ is periodic with a period $= 2\pi$.

$\therefore$ To draw this graph, we should draw the graph for one interval of length 2π and repeat it for entire values of x of length 2π.

As we know,

$$\cos^{-1}(\cos x) = \begin{cases} x, & 0 \le x \le \pi \\ 2\pi - x, & 0 \le 2\pi - x \le \pi \end{cases}$$

or $$\cos^{-1}(\cos x) = \begin{cases} x, & 0 \le x \le \pi \\ 2\pi - x, & \pi \le x \le 2\pi \end{cases}$$

Thus, it has been define for $0 < x < 2\pi$ that has length 2π.

So, its graph could be plotted as :

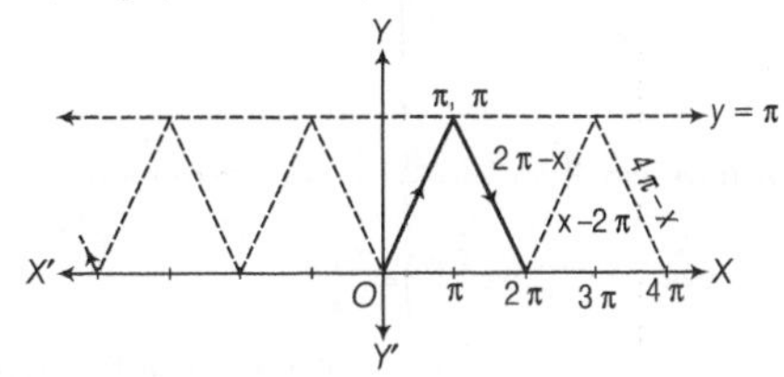

(iii) As $y = \tan^{-1}(\tan x)$ is periodic with period π.

$\therefore$ To draw this graph, we should draw the graph for one interval of length π and repeat it for entire values of x.

As we know,

$$\tan^{-1}(\tan x) = \left\{x, \frac{-\pi}{2} < x < \frac{\pi}{2}\right\}$$

Thus, it has been defined for $\dfrac{-\pi}{2} < x < \dfrac{\pi}{2}$

that has length π. So, its graph could be plotted as :

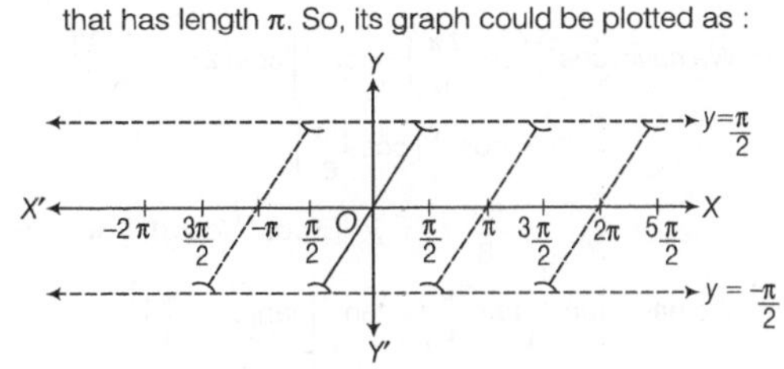

Thus, the curve for $y = \tan^{-1}(\tan x)$, where y is not defined for $x \in (2n + 1)\dfrac{\pi}{2}$.

(iv) We have, $y = \operatorname{cosec}^{-1}(\operatorname{cosec} x)$

as range of $y = \operatorname{cosec}^{-1}(\operatorname{cosec} x)$ is $\left[-\frac{\pi}{2}, \frac{\pi}{2}\right] - \{0\}$.

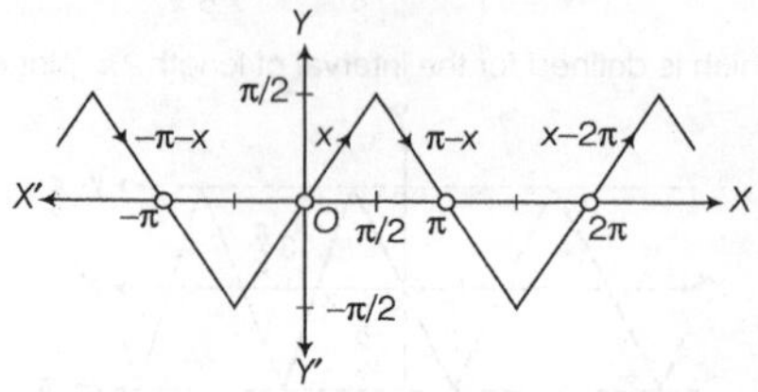

(v) We have, $y = \sec^{-1}(\sec x)$, $y \in [0, \pi] - \{\pi/2\}$

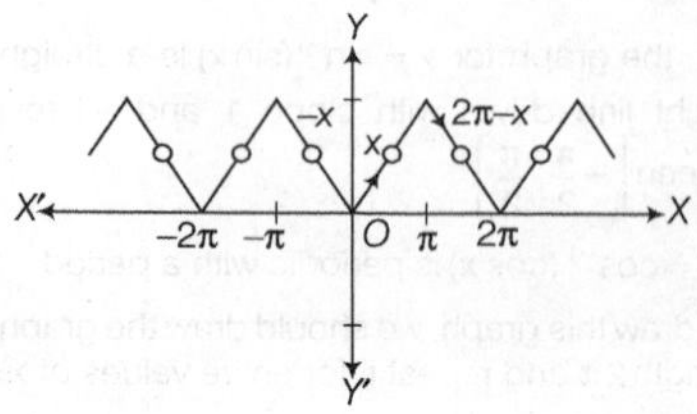

(vi) We have, $y = \cot^{-1}(\cot x)$, $y \in (0, \pi)$

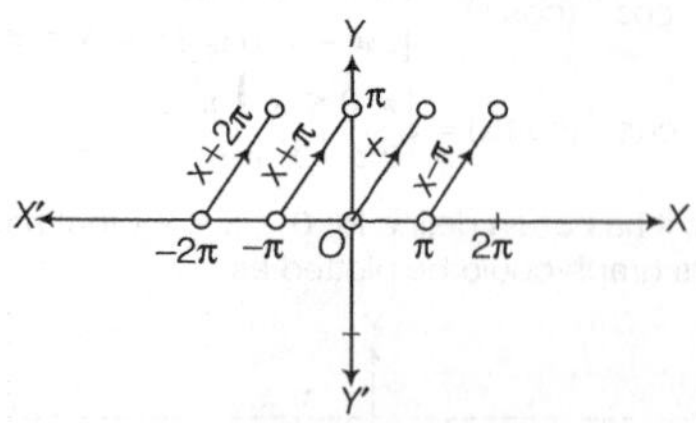

6. (i) We have, $\sin^{-1}\left(\sin\frac{\pi}{3}\right) = \frac{\pi}{3}$ $\quad \left[\because \sin^{-1}(\sin\theta) = \theta, \forall \theta \in \left(\frac{-\pi}{2}, \frac{\pi}{2}\right)\right]$

(ii) We have, $\cos^{-1}\left(\cos\frac{2\pi}{3}\right) = \frac{2\pi}{3}$

$$[\because \cos^{-1}(\cos\theta) = \theta, \forall \theta \in [0, \pi]]$$

(iii) We have, $\tan^{-1}\left(\tan\frac{\pi}{4}\right) = \frac{\pi}{4}$

$$\left[\because \tan^{-1}(\tan\theta) = \theta, \forall \theta \in \left(-\frac{\pi}{2}, \frac{\pi}{2}\right)\right]$$

(iv) We have, $\sin^{-1}\left(\sin\frac{2\pi}{3}\right) = \sin^{-1}\left[\sin\left(\pi - \frac{\pi}{3}\right)\right] = \sin^{-1}\left(\sin\frac{\pi}{3}\right)$

$$= \frac{\pi}{3} \qquad \left[\because \sin^{-1}(\sin\theta) = \theta, \forall \theta \in \left(\frac{-\pi}{2}, \frac{\pi}{2}\right)\right]$$

(v) We have, $\cos^{-1}\left(\cos\frac{7\pi}{6}\right) = \cos^{-1}\left[\cos\left(2\pi - \frac{5\pi}{6}\right)\right]$

$$= \cos^{-1}\left(\cos\frac{5\pi}{6}\right)$$

$$= \frac{5\pi}{6} \qquad [\because \cos^{-1}(\cos\theta) = \theta, \forall \theta \in [0, \pi]]$$

(vi) We have, $\tan^{-1}\left(\tan\frac{3\pi}{4}\right) = \tan^{-1}\left[\tan\left(\pi - \frac{\pi}{4}\right)\right]$

$$= \tan^{-1}\left(-\tan\frac{\pi}{4}\right) = -\tan^{-1}\left(\tan\frac{\pi}{4}\right) = -\frac{\pi}{4}$$

(vii) We have, $\cos\left[\cos^{-1}\left(\frac{-\sqrt{3}}{2}\right) + \frac{\pi}{6}\right]$

$$= \cos\left\{\pi - \cos^{-1}\left(\frac{\sqrt{3}}{2}\right) + \frac{\pi}{6}\right\}$$

$$= \cos\left\{\pi - \frac{\pi}{6} + \frac{\pi}{6}\right\} = \cos\pi = -1$$

7. (i) We have, $\cos\left(\sin^{-1}\frac{3}{5}\right)$

$$= \cos\left(\cos^{-1}\sqrt{1 - \left(\frac{3}{5}\right)^2}\right) \qquad [\because \sin^{-1} x = \cos^{-1}\sqrt{1 - x^2}]$$

$$= \cos\left(\cos^{-1}\sqrt{1 - \frac{9}{25}}\right) = \cos\left(\cos^{-1}\sqrt{\frac{16}{25}}\right)$$

$$= \cos\left(\cos^{-1}\frac{4}{5}\right) = \frac{4}{5} \qquad [\cos^{-1}(\cos\theta) = \theta]$$

(ii) We have, $\sin\left[\cos^{-1}\left(\frac{4}{5}\right)\right]$

$$= \sin\left[\sin^{-1}\sqrt{1 - \left(\frac{4}{5}\right)^2}\right] \qquad [\because \cos^{-1} x = \sin^{-1}\sqrt{1 - x^2}]$$

$$= \sin\left(\sin^{-1}\sqrt{1 - \frac{16}{25}}\right)$$

$$= \sin\left[\sin^{-1}\left(\sqrt{\frac{9}{25}}\right)\right] = \sin\left(\sin^{-1}\frac{3}{5}\right) = \frac{3}{5}$$

(iii) We have, $\cos\left[\sin^{-1}\left(\frac{-3}{5}\right)\right]$

$$= \cos\left[\cos^{-1}\sqrt{1 - \left(\frac{-3}{5}\right)^2}\right] \qquad [\because \sin^{-1} x = \cos^{-1}\sqrt{1 - x^2}]$$

$$= \cos\left(\cos^{-1}\sqrt{1 - \frac{9}{25}}\right) = \cos\left(\cos^{-1}\sqrt{\frac{16}{25}}\right)$$

$$= \cos\left(\cos^{-1}\frac{4}{5}\right) = \frac{4}{5}$$

(iv) We have, $\tan\left[\cos^{-1}\left(\frac{8}{17}\right)\right]$

$$= \tan\left(\tan^{-1}\frac{\sqrt{1 - \left(\frac{8}{17}\right)^2}}{\frac{8}{17}}\right) \qquad \left[\because \cos^{-1} x = \tan^{-1}\frac{\sqrt{1 - x^2}}{x}\right]$$

$$= \tan\left(\tan^{-1}\frac{\frac{15}{17}}{\frac{8}{17}}\right) = \tan\left(\tan^{-1}\frac{15}{8}\right) = \frac{15}{8}$$

(v) We have, $\operatorname{cosec}\left[\cos^{-1}\left(-\frac{12}{13}\right)\right]$

$$= \operatorname{cosec}\left[\operatorname{cosec}^{-1}\left(\frac{1}{\sqrt{1 - \left(\frac{-12}{13}\right)^2}}\right)\right]$$

$$\left[\because \cos^{-1} x = \operatorname{cosec}^{-1}\left(\frac{1}{\sqrt{1 - x^2}}\right)\right]$$

$$= \operatorname{cosec}\left[\operatorname{cosec}^{-1}\left(\frac{13}{5}\right)\right] = \frac{13}{5}$$

8. (d) We have, $\cot\left[\cos^{-1}\left(\frac{7}{25}\right)\right]$

$$= \cot\left(\cot^{-1}\frac{\frac{7}{25}}{\sqrt{1 - \left(\frac{7}{25}\right)^2}}\right) \qquad \left[\because \cos^{-1} x = \cot^{-1}\left(\frac{x}{\sqrt{1 - x^2}}\right)\right]$$

$$= \cot\left[\cot^{-1}\left(\frac{\frac{7}{25}}{\frac{24}{25}}\right)\right] = \cot\left(\cot^{-1}\frac{7}{24}\right) = \frac{7}{24}$$

9. (d) We have, $\sin(\cot^{-1} x)$

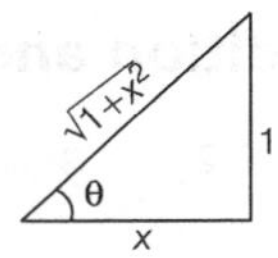

Let $\quad \cot^{-1} x = \theta$

$\cot\theta = x$

Now, $\quad \sin\theta = \dfrac{1}{\sqrt{1+x^2}}$

$\sin(\cot^{-1} x) = \dfrac{1}{\sqrt{1+x^2}} \quad [\because \theta = \cot^{-1} x]$

$= (1+x^2)^{-1/2}$

10. (c) We have, $1 + \cot^2(\sin^{-1} x) = 1 + [\cot(\sin^{-1} x)]^2$

$$= 1 + \left\{\cot\left(\cot^{-1}\frac{\sqrt{1-x^2}}{x}\right)\right\}^2 \quad \left[\because \sin^{-1} x = \cot^{-1}\left(\frac{\sqrt{1-x^2}}{x}\right)\right]$$

$$= 1 + \left(\frac{\sqrt{1-x^2}}{x}\right)^2 = 1 + \frac{1-x^2}{x^2} = 1 + \frac{1}{x^2} - 1 = \frac{1}{x^2}$$

11. (b) We have, $\cos(\tan^{-1} x)$

$$= \cos\left[\cos^{-1}\left(\frac{1}{\sqrt{1+x^2}}\right)\right] \quad \left[\because \tan^{-1} x = \cos^{-1}\left(\frac{1}{\sqrt{1+x^2}}\right)\right]$$

$$= \frac{1}{\sqrt{1+x^2}}$$

12. (b) We have, $\tan(\sec^{-1}\sqrt{1+x^2})$

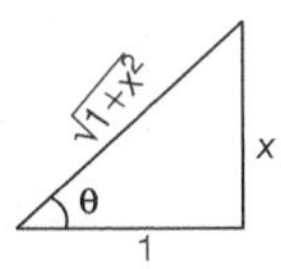

Let, $\quad \sec^{-1}\sqrt{1+x^2} = \theta$

$\sec\theta = \sqrt{1+x^2}$

Now, $\quad \tan\theta = \dfrac{x}{1}$

$\tan(\sec^{-1}\sqrt{1+x^2}) = x \quad [\because \theta = \sec^{-1}\sqrt{1+x^2}]$

13. (a) We have, $\sin[\cot^{-1}\{\tan(\cos^{-1} x)\}]$

$$= \sin\left[\cot^{-1}\left\{\tan\left(\tan^{-1}\frac{\sqrt{1-x^2}}{x}\right)\right\}\right]$$

$$\left[\because \cos^{-1} x = \tan^{-1}\frac{\sqrt{1-x^2}}{x}\right]$$

$$= \sin\left[\left(\cot^{-1}\frac{\sqrt{1-x^2}}{x}\right)\right] \quad [\because \tan(\tan^{-1}\theta) = \theta]$$

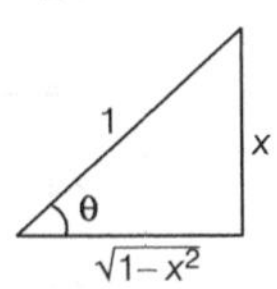

Let $\cot^{-1}\dfrac{\sqrt{1-x^2}}{x} = \theta$

$\Rightarrow \quad \cot\theta = \dfrac{\sqrt{1-x^2}}{x}$

Now, $\quad \sin\theta = x$

$\Rightarrow \quad \sin\left(\cot^{-1}\dfrac{\sqrt{1-x^2}}{x}\right) = x \quad \left[\because \theta = \cot^{-1}\dfrac{\sqrt{1-x^2}}{x}\right]$

14. (c) Let $\sin^{-1}\left(\dfrac{\sqrt{x}}{\sqrt{x+a}}\right) = \theta$

$\Rightarrow \quad \sin\theta = \dfrac{\sqrt{x}}{\sqrt{x+a}}$

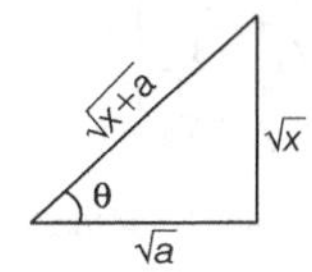

Now, $\quad \tan\theta = \dfrac{\sqrt{x}}{\sqrt{a}}$

$\Rightarrow \quad \theta = \tan^{-1}\sqrt{\dfrac{x}{a}}$

$\Rightarrow \quad \sin^{-1}\left(\dfrac{\sqrt{x}}{\sqrt{x+a}}\right) = \tan^{-1}\sqrt{\dfrac{x}{a}}$

15. (b) Given, $\quad \sin^{-1} x = \theta + \beta \Rightarrow \sin(\theta + \beta) = x \quad \ldots(i)$

and $\quad \sin^{-1} y = \theta - \beta \Rightarrow \sin(\theta - \beta) = y \quad \ldots(ii)$

Now, $\quad 1 + xy = 1 + \sin(\theta + \beta)\sin(\theta - \beta)$

[using Eqs. (i) and (ii)]

$= 1 + \sin^2\theta - \sin^2\beta$

$= 1 - \sin^2\beta + \sin^2\theta$

$= \cos^2\beta + \sin^2\theta$

16. (c) We have, $\sec^2(\tan^{-1} 2) + \operatorname{cosec}^2(\cot^{-1} 3)$

$= \{\sec(\tan^{-1} 2)\}^2 + \{\operatorname{cosec}(\cot^{-1} 3)\}^2$

$= \{\sec(\sec^{-1}(\sqrt{1+(2)^2}\}^2$

$+ \{\operatorname{cosec}(\operatorname{cosec}^{-1}(\sqrt{1+3^2})\}^2$

$= (\sqrt{5})^2 + (\sqrt{10})^2 = 5 + 10 = 15$

17. (d) It is evident from the graph of Q. 5 (ii)

$\cos^{-1}(\cos x) = 2\pi - x, \pi \le x \le 2\pi$

18. (a) Given, $\tan^{-1} x = \sin^{-1}\left(\dfrac{3}{\sqrt{10}}\right)$

$$\Rightarrow \quad \tan^{-1} x = \tan^{-1}\left(\frac{\frac{3}{\sqrt{10}}}{\sqrt{1-\left(\frac{3}{\sqrt{10}}\right)^2}}\right)$$

$$\Rightarrow \quad \tan^{-1} x = \tan^{-1}\left(\frac{\frac{3}{\sqrt{10}}}{\frac{1}{\sqrt{10}}}\right)$$

$\Rightarrow \quad \tan^{-1} x = \tan^{-1}(3)$

$\therefore \quad x = 3$

19. (c) Given, $\quad \cos^{-1} x + \cos^{-1} y + \cos^{-1} z = 3\pi$

and $\quad 0 \le \cos^{-1} x \le \pi$

$\therefore \quad \cos^{-1} x = \cos^{-1} y = \cos^{-1} z = \pi$

$\Rightarrow \quad x = y = z = -1$

$\therefore \quad xy + yz + zx = 1 + 1 + 1 = 3$

20. (c) Given, $\cos^{-1} x + \cos^{-1} y + \cos^{-1} z + \cos^{-1} t = 4\pi$

$\Rightarrow \quad \cos^{-1} x = \cos^{-1} y = \cos^{-1} z = \cos^{-1} t = \pi \quad [\because 0 \le \cos^{-1} x \le \pi]$

$\Rightarrow \quad x = y = z = t = -1$

$\Rightarrow \quad x^2 + y^2 + z^2 + t^2 = 1 + 1 + 1 + 1 = 4$

DPP 2 Identities on Addition and Subtraction

1. (i) LHS $= \tan^{-1}\left(\frac{1}{7}\right) + \tan^{-1}\left(\frac{1}{13}\right)$

$= \tan^{-1}\left(\dfrac{\frac{1}{7}+\frac{1}{13}}{1-\frac{1}{7}\times\frac{1}{13}}\right)$

$\left[\because \tan^{-1}x + \tan^{-1}y = \tan^{-1}\left(\frac{x+y}{1-xy}\right), xy < 1\right]$

$= \tan^{-1}\left(\dfrac{\frac{13+7}{91}}{\frac{90}{91}}\right) = \tan^{-1}\left(\frac{20}{90}\right) = \tan^{-1}\left(\frac{2}{9}\right)$

(ii) LHS $= \tan^{-1}(2) + \tan^{-1}(3)$

$= \pi + \tan^{-1}\left(\frac{2+3}{1-6}\right) = \pi + \tan^{-1}\left(\frac{5}{-5}\right)$

$= \pi + \tan^{-1}(-1) = \pi - \frac{\pi}{4} = \frac{3\pi}{4}$

(iii) LHS $= \tan^{-1}\left(\frac{1}{2}\right) + \tan^{-1}\left(\frac{1}{5}\right) + \tan^{-1}\left(\frac{1}{8}\right)$

$= \tan^{-1}\left(\dfrac{\frac{1}{2}+\frac{1}{5}}{1-\frac{1}{10}}\right) + \tan^{-1}\left(\frac{1}{8}\right)$

$= \tan^{-1}\left(\frac{7}{9}\right) + \tan^{-1}\left(\frac{1}{8}\right)$

$= \tan^{-1}\left(\dfrac{\frac{7}{9}+\frac{1}{8}}{1-\frac{7}{9}\times\frac{1}{8}}\right) = \tan^{-1}\left(\dfrac{\frac{56+9}{72}}{\frac{65}{72}}\right)$

$= \tan^{-1}(1)$

$= \frac{\pi}{4} =$ RHS

(iv) LHS $= \sin^{-1}\left(\frac{3}{5}\right) + \sin^{-1}\left(\frac{8}{17}\right)$

$= \sin^{-1}\left[\frac{3}{5}\sqrt{1-\left(\frac{8}{17}\right)^2} + \frac{8}{17}\sqrt{1-\left(\frac{3}{5}\right)^2}\right]$

$[\because \sin^{-1}x + \sin^{-1}y = \sin^{-1}(x\sqrt{1-y^2} + y\sqrt{1-x^2}]$

$= \sin^{-1}\left(\frac{3}{5}\times\frac{15}{17} + \frac{8}{17}\times\frac{4}{5}\right)$

$= \sin^{-1}\left(\frac{45}{85} + \frac{32}{85}\right)$

$= \sin^{-1}\left(\frac{77}{85}\right) =$ RHS

(v) LHS $= \sin^{-1}\left(\frac{4}{5}\right) + \sin^{-1}\left(\frac{5}{13}\right) + \sin^{-1}\left(\frac{16}{65}\right)$

$= \sin^{-1}\left[\frac{4}{5}\sqrt{1-\left(\frac{5}{13}\right)^2} + \frac{5}{13}\sqrt{1-\left(\frac{4}{5}\right)^2}\right] + \sin^{-1}\left(\frac{16}{65}\right)$

$= \sin^{-1}\left(\frac{4}{5}\times\frac{12}{13} + \frac{5}{13}\times\frac{3}{5}\right) + \sin^{-1}\left(\frac{16}{65}\right)$

$= \sin^{-1}\left(\frac{63}{65}\right) + \sin^{-1}\left(\frac{16}{65}\right)$

$= \sin^{-1}\left(\frac{63}{65}\times\frac{63}{65} + \frac{16}{65}\times\frac{16}{65}\right)$

$= \sin^{-1}\left(\frac{4225}{65\times 65}\right)$

$= \sin^{-1}(1) = \frac{\pi}{2}$

2. (i) Given, $\tan^{-1}(2x) + \tan^{-1}(3x) = n\pi + \frac{3\pi}{4}$

$\Rightarrow \tan^{-1}\left(\frac{2x+3x}{1-6x^2}\right) = n\pi + \frac{3\pi}{4}$

$\Rightarrow \tan^{-1}\left(\frac{5x}{1-6x^2}\right) = n\pi + \frac{3\pi}{4}$

$\Rightarrow \frac{5x}{1-6x^2} = \tan\left(n\pi + \frac{3\pi}{4}\right)$

$\Rightarrow \frac{5x}{1-6x^2} = \tan\left(\frac{3\pi}{4}\right)$

$\Rightarrow \frac{5x}{1-6x^2} = \tan\left(\pi - \frac{\pi}{4}\right)$

$\Rightarrow \frac{5x}{1-6x^2} = -\tan\frac{\pi}{4}$

$\Rightarrow \frac{5x}{1-6x^2} = -1$

$\Rightarrow 5x = -1 + 6x^2$

$\Rightarrow 6x^2 - 5x - 1 = 0$

$\Rightarrow x = \frac{5 \pm \sqrt{25+24}}{12}$

$\Rightarrow x = \frac{5 \pm \sqrt{49}}{12}$

$\Rightarrow x = \frac{5 \pm 7}{12}$

$\therefore x = 1, -\frac{1}{6}$

(ii) Given, $\tan^{-1}(x+1) + \tan^{-1}(x-1) = \tan^{-1}\left(\frac{8}{31}\right)$

$\Rightarrow \tan^{-1}\left[\frac{x+1+x-1}{1-(x+1)(x-1)}\right] = \tan^{-1}\left(\frac{8}{31}\right)$

$\Rightarrow \tan^{-1}\left(\frac{2x}{1-x^2+1}\right) = \tan^{-1}\left(\frac{8}{31}\right)$

$\Rightarrow \tan^{-1}\left(\frac{2x}{2-x^2}\right) = \tan^{-1}\left(\frac{8}{31}\right)$

$\Rightarrow \frac{2x}{2-x^2} = \frac{8}{31}$

$\Rightarrow 62x = 16 - 8x^2$

$\Rightarrow 8x^2 + 62x - 16 = 0$

$\Rightarrow 4x^2 + 31x - 8 = 0$

$\Rightarrow x = \frac{-31 \pm \sqrt{961+128}}{8}$

$\Rightarrow x = \frac{-31 \pm \sqrt{1089}}{8}$

$\Rightarrow x = \frac{-31 \pm 33}{8}$

$\Rightarrow x = \frac{1}{4}, -8$

But $x = -8$ [not possible]

$\therefore x = \frac{1}{4}$

(iii) Given, $\sin^{-1}x + \sin^{-1}2x = \frac{\pi}{3}$

$\Rightarrow \sin^{-1}[x\sqrt{1-4x^2} + 2x\sqrt{1-x^2}] = \frac{\pi}{3}$

$\Rightarrow x\sqrt{1-4x^2} + 2x\sqrt{1-x^2} = \sin\frac{\pi}{3}$

$$\Rightarrow \quad x\sqrt{1-4x^2} + 2x\sqrt{1-x^2} = \frac{\sqrt{3}}{2}$$

$$\Rightarrow \quad x\sqrt{1-4x^2} = \frac{\sqrt{3}}{2} - 2x\sqrt{1-x^2}$$

$$\Rightarrow \quad x^2(1-4x^2) = \frac{3}{4} + 4x^2(1-x^2) - 2 \times \frac{\sqrt{3}}{2} \times 2x\sqrt{1-x^2}$$

$$\Rightarrow x^2(1-4x^2) = \frac{3}{4} + 4x^2 - 4x^4 - 2\sqrt{3}x\sqrt{1-x^2}$$

$$\Rightarrow \quad x^2 - 4x^4 = \frac{3}{4} + 4x^2 - 4x^4 - 2\sqrt{3}x\sqrt{1-x^2}$$

$$\Rightarrow \quad x^2 = \frac{3}{4} + 4x^2 - 2\sqrt{3}x\sqrt{1-x^2}$$

$$\Rightarrow \quad -3x^2 - \frac{3}{4} = -2\sqrt{3}x\sqrt{1-x^2}$$

$$\Rightarrow \quad 3x^2 + \frac{3}{4} = 2\sqrt{3}x\sqrt{1-x^2}$$

$$\Rightarrow \quad 12x^2 + 3 = 8\sqrt{3}x\sqrt{1-x^2}$$

$$\Rightarrow \quad (12x^2+3)^2 = 64 \times 3x^2(1-x^2)$$

$$\Rightarrow 144x^4 + 9 + 72x^2 = 192x^2 - 192x^4$$

$$\Rightarrow \quad 336x^4 - 120x^2 + 9 = 0$$

$$\Rightarrow \quad 112x^4 - 40x^2 + 3 = 0$$

$$\Rightarrow \quad x^2 = \frac{40 \pm \sqrt{1600-1344}}{224}$$

$$\Rightarrow \quad x^2 = \frac{40 \pm \sqrt{256}}{224}$$

$$x^2 = \frac{40 \pm 16}{224} = \frac{56}{224}, \frac{24}{224}$$

$$x^2 = \frac{56}{224} \text{ or } x^2 = \frac{24}{224}$$

$$x = \sqrt{\frac{56}{224}} \text{ or } x = \sqrt{\frac{24}{224}}$$

(iv) Given, $\cos^{-1} x + \sin^{-1}\frac{x}{2} = \frac{\pi}{6}$

$$\Rightarrow \quad \sin^{-1}\sqrt{1-x^2} + \sin^{-1}\frac{x}{2} = \frac{\pi}{6}$$

$$\Rightarrow \sin^{-1}\left(\sqrt{1-x^2}\sqrt{1-\frac{x^2}{4}} + \frac{x}{2}\sqrt{1-1+x^2}\right) = \frac{\pi}{6}$$

$$\Rightarrow \quad \sin^{-1}\left(\sqrt{1-x^2}\,\frac{\sqrt{4-x^2}}{2} + \frac{x^2}{2}\right) = \frac{\pi}{6}$$

$$\Rightarrow \quad \sin^{-1}\left(\frac{\sqrt{x^4-5x^2+4}}{2} + \frac{x^2}{2}\right) = \frac{\pi}{6}$$

$$\Rightarrow \quad \frac{\sqrt{x^4-5x^2+4}}{2} + \frac{x^2}{2} = \frac{1}{2}$$

$$\Rightarrow \quad \sqrt{x^4-5x^2+4} + x^2 = 1$$

$$\Rightarrow \quad \sqrt{x^4-5x^2+4} = 1 - x^2$$

$$\Rightarrow \quad x^4 - 5x^2 + 4 = (1-x^2)^2$$

$$\Rightarrow \quad x^4 - 5x^2 + 4 = 1 + x^4 - 2x^2$$

$$\Rightarrow \quad 3x^2 - 3 = 0$$

$$\Rightarrow \quad x^2 = 1$$

$$\therefore \quad x = \pm 1$$

3. (d) Given, $\sin\left(\sin^{-1}\frac{1}{5} + \cos^{-1}x\right) = 1$

$$\Rightarrow \quad \sin^{-1}\left(\frac{1}{5}\right) + \cos^{-1}x = \frac{\pi}{2}$$

$$\therefore \quad x = \frac{1}{5}, \quad \left[\text{using } \sin^{-1}x + \cos^{-1}x = \frac{\pi}{2}\right]$$

4. (a) Given, $\sin\left(\sin^{-1}\frac{1}{3} + \sec^{-1}3\right) + \cos\left(\tan^{-1}\frac{1}{2} + \tan^{-1}2\right)$

$$= \sin\left(\sin^{-1}\frac{1}{3} + \sin^{-1}\frac{\sqrt{8}}{3}\right) + \cos\left[\tan^{-1}\left(\frac{\frac{1}{2}+2}{1-1}\right)\right]$$

$$= \sin\left[\sin^{-1}\left(\frac{1}{3}\times\frac{1}{3} + \frac{\sqrt{8}}{3}\times\frac{\sqrt{8}}{3}\right)\right] + \cos\left(\frac{\pi}{2}\right)$$

$$= \sin\left[\sin^{-1}\left(\frac{1}{9}+\frac{8}{9}\right)\right] + \cos\frac{\pi}{2}$$

$$= \sin[\sin^{-1}(1)] + \cos\frac{\pi}{2} = \sin\frac{\pi}{2} + \cos\frac{\pi}{2}$$

$$= 1 + 0 = 1$$

5. (b) We have, $\cos\left[\cos^{-1}\left(-\frac{1}{7}\right) + \sin^{-1}\left(-\frac{1}{7}\right)\right]$

$$= \cos\left(\frac{\pi}{2}\right) = 0$$

6. (a) Given, $4\sin^{-1}x + \cos^{-1}x = \frac{\pi}{2}$

$$\Rightarrow \quad 4\sin^{-1}x = \frac{\pi}{2} - \cos^{-1}x$$

$$\Rightarrow \quad 4\sin^{-1}x = \sin^{-1}x$$

$$\Rightarrow \quad 3\sin^{-1}x = 0$$

$$\Rightarrow \quad \sin^{-1}x = 0$$

$$\therefore \quad x = 0$$

7. (c) $\cot^{-1}x + \tan^{-1}3 = \frac{\pi}{2}$

$$\therefore \quad x = 3 \quad \left[\text{using } \tan^{-1}x + \cot^{-1}x = \frac{\pi}{2}\right]$$

8. (c) We have, $\cot^{-1}9 + \operatorname{cosec}^{-1}\frac{\sqrt{41}}{4}$

$$\Rightarrow \quad \tan^{-1}\left(\frac{1}{9}\right) + \tan^{-1}\left(\frac{4}{5}\right)$$

$$\Rightarrow \quad \tan^{-1}\left(\frac{\frac{1}{9}+\frac{4}{5}}{1-\frac{4}{45}}\right) = \tan^{-1}\left[\frac{\frac{5+36}{45}}{\frac{45-4}{45}}\right] = \tan^{-1}\left[\frac{41}{41}\right] = \frac{\pi}{4}$$

9. (c) Given, $\sin^{-1}\left(\frac{1}{3}\right) + \sin^{-1}\left(\frac{2}{3}\right) = \sin^{-1}x$

$$\Rightarrow \quad \sin^{-1}\left(\frac{1}{3}\sqrt{1-\frac{4}{9}} + \frac{2}{3}\sqrt{1-\frac{1}{9}}\right) = \sin^{-1}x$$

$$\Rightarrow \quad \sin^{-1}\left(\frac{1}{3}\times\frac{\sqrt{5}}{3} + \frac{2}{3}\times\frac{\sqrt{8}}{3}\right) = \sin^{-1}x$$

$$\Rightarrow \quad \frac{\sqrt{5}}{9} + \frac{2\sqrt{8}}{9} = x$$

$$\therefore \quad x = \frac{\sqrt{5}+4\sqrt{2}}{9}$$

10. (c) Given, $\tan^{-1}x + 2\cot^{-1}x = \frac{2\pi}{3}$

$$\Rightarrow \quad \tan^{-1}x + 2\left(\frac{\pi}{2} - \tan^{-1}x\right) = \frac{2\pi}{3}$$

$$\Rightarrow \quad \tan^{-1}x + \pi - 2\tan^{-1}x = \frac{2\pi}{3}$$

$$\Rightarrow \quad -\tan^{-1}x + \pi = \frac{2\pi}{3}$$

$$\Rightarrow \quad -\tan^{-1}x = \frac{2\pi}{3} - \pi$$

$$\Rightarrow \quad -\tan^{-1}x = -\frac{\pi}{3}$$

$$\Rightarrow \quad \tan^{-1}x = \frac{\pi}{3}$$

$$\Rightarrow \quad x = \tan\frac{\pi}{3}$$

$$\therefore \quad x = \sqrt{3}$$

11. (d) We have,

$$\cos^{-1}\left(\frac{15}{17}\right) + 2\tan^{-1}\left(\frac{1}{5}\right)$$

$$= \cos^{-1}\left(\frac{15}{17}\right) + \cos^{-1}\left(\frac{1-\frac{1}{25}}{1+\frac{1}{25}}\right)$$

$$\left[\because 2\tan^{-1}x = \cos^{-1}\frac{1-x^2}{1+x^2}, \text{ if } 0 \le x < \infty\right]$$

$$= \cos^{-1}\left(\frac{15}{17}\right) + \cos^{-1}\left(\frac{24}{26}\right)$$

$$= \cos^{-1}\left(\frac{15}{17} \times \frac{24}{26} - \sqrt{1-\left(\frac{15}{17}\right)^2}\sqrt{1-\left(\frac{24}{26}\right)^2}\right)$$

$$= \cos^{-1}\left(\frac{15}{17} \times \frac{12}{13} - \frac{8 \times 10}{17 \times 26}\right)$$

$$= \cos^{-1}\left(\frac{180}{221} - \frac{40}{221}\right) = \cos^{-1}\left(\frac{140}{221}\right)$$

12. (a) We have, $\sin^{-1}\left(\frac{3}{5}\right) + \tan^{-1}\left(\frac{1}{7}\right)$

$$= \tan^{-1}\left(\frac{\frac{3}{5}}{\sqrt{1-\frac{9}{25}}}\right) + \tan^{-1}\left(\frac{1}{7}\right)$$

$$= \tan^{-1}\left(\frac{3}{4}\right) + \tan^{-1}\left(\frac{1}{7}\right)$$

$$= \tan^{-1}\left(\frac{\frac{3}{4}+\frac{1}{7}}{1-\frac{3}{4}\times\frac{1}{7}}\right)$$

$$= \tan^{-1}\left(\frac{\frac{21+4}{28}}{\frac{25}{28}}\right) = \tan^{-1}\left(\frac{\frac{25}{28}}{\frac{25}{28}}\right)$$

$$= \tan^{-1}(1) = \frac{\pi}{4}$$

13. (b) We have,

$$\tan^{-1}\left(\frac{a-b}{1+ab}\right) + \tan^{-1}\left(\frac{b-c}{1+bc}\right)$$

$$= \tan^{-1}a - \tan^{-1}b + \tan^{-1}b - \tan^{-1}c$$

$$\because \tan^{-1}x - \tan^{-1}y = \tan^{-1}\left(\frac{x-y}{1+xy}\right)$$

$$= \tan^{-1}a - \tan^{-1}c$$

14. (d) Given, $\tan^{-1}x + \tan^{-1}y + \tan^{-1}z = \frac{\pi}{2}$

$$\Rightarrow \quad \tan^{-1}x + \tan^{-1}y = \frac{\pi}{2} - \tan^{-1}z$$

$$\Rightarrow \quad \tan^{-1}\left(\frac{x+y}{1-xy}\right) = \frac{\pi}{2} - \tan^{-1}z$$

$$\Rightarrow \quad \tan^{-1}\left(\frac{x+y}{1-xy}\right) = \cot^{-1}z$$

$$\Rightarrow \quad \tan^{-1}\left(\frac{x+y}{1-xy}\right) = \tan^{-1}\left(\frac{1}{z}\right)$$

$$\Rightarrow \quad \frac{x+y}{1-xy} = \frac{1}{z}$$

$$\Rightarrow \quad xz + yz = 1 - xy$$

$$\Rightarrow \quad xy + yz + zx = 1$$

$$\therefore \quad xy + yz + zx - 1 = 0$$

15. (b) Let $\sin^{-1}(x) = A, \sin^{-1}y = B, \sin^{-1}z = C$

$$\therefore \quad A + B + C = \frac{\pi}{2} \quad \ldots(i)$$

and $\quad x^2 + y^2 + z^2 + 2xyz$

Now $\sin^2 A + \sin^2 B + \sin^2 C + 2\sin A \sin B \sin C$

$= 1 - \cos^2 A + \sin^2 B + \sin^2 C + 2\sin A \sin B \sin C$

$= 1 - [\cos(A+B)\cdot\cos(A-B)] + \sin^2 C + 2\sin A \sin B \sin C$

$= 1 - \cos\left(\frac{\pi}{2} - C\right)\cdot\cos(A-B) + \sin^2 C + 2\sin A \sin B \sin C$

[using Eq. (i)]

$= 1 - \sin C\,[\cos(A-B) - \sin C] + 2\sin A \sin B \sin C$

$= 1 - \sin C\,[\cos(A-B) - \cos(A+B)] + 2\sin A \sin B \sin C$

$= 1 - \sin C\,[2\sin A \sin B] + 2\sin A \sin B \sin C$

$\left[\text{using } A + B = \frac{\pi}{2} - C\right]$

$\Rightarrow 1$

16. (a) Given, $\quad \sin^{-1}x = 2\sin^{-1}a$

$$\because \quad -\frac{\pi}{2} \le \sin^{-1}x \le \frac{\pi}{2}$$

$$\Rightarrow \quad -\frac{\pi}{2} \le 2\sin^{-1}a \le \frac{\pi}{2}$$

$$\Rightarrow \quad \frac{-\pi}{4} \le \sin^{-1}a \le \frac{\pi}{4}$$

$$\Rightarrow \quad \sin\left(-\frac{\pi}{4}\right) \le a \le \sin\frac{\pi}{4}$$

$$\Rightarrow \quad -\frac{1}{\sqrt{2}} \le a \le \frac{1}{\sqrt{2}}$$

$$\therefore \quad |a| \le \frac{1}{\sqrt{2}}$$

17. (c) Given, $\quad \cos^{-1}x - \cos^{-1}\frac{y}{2} = \alpha$

$$\Rightarrow \quad \cos^{-1}\left(\frac{xy}{2} + \sqrt{1-x^2}\sqrt{1-\frac{y^2}{4}}\right) = \alpha$$

$$\Rightarrow \quad \frac{xy}{2} + \sqrt{1-x^2}\sqrt{1-\frac{y^2}{4}} = \cos\alpha$$

$$\Rightarrow \quad 2\sqrt{1-x^2}\sqrt{1-\frac{y^2}{4}} = 2\cos\alpha - xy$$

On squaring both sides, we get

$$4\frac{(1-x^2)(4-y^2)}{4} = 4\cos^2\alpha + x^2y^2 - 4xy\cos\alpha$$

$$\Rightarrow \quad 4 - 4x^2 - y^2 + x^2y^2 = 4\cos^2\alpha + x^2y^2 - 4xy\cos\alpha$$

$$\therefore \quad 4x^2 - 4xy\cos\alpha + y^2 = 4\sin^2\alpha$$

18. (b) Given, $\sin^{-1}\left(\frac{x}{5}\right) + \text{cosec}^{-1}\left(\frac{5}{4}\right) = \frac{\pi}{2}$

$$\Rightarrow \quad \sin^{-1}\left(\frac{x}{5}\right) + \sin^{-1}\left(\frac{4}{5}\right) = \frac{\pi}{2}$$

$$\Rightarrow \quad \sin^{-1}\left(\frac{x}{5}\right) = \frac{\pi}{2} - \sin^{-1}\left(\frac{4}{5}\right)$$

$$\Rightarrow \quad \sin^{-1}\left(\frac{x}{5}\right) = \cos^{-1}\left(\frac{4}{5}\right)$$

$$\Rightarrow \quad \sin^{-1}\left(\frac{x}{5}\right) = \sin^{-1}\left(\frac{3}{5}\right) \Rightarrow \frac{x}{5} = \frac{3}{5}$$

$$\therefore \quad x = 3$$

19. (a) Since, $\text{cosec}^{-1}\left(\frac{5}{3}\right) = \tan^{-1}\left(\frac{3}{4}\right)$

$$\therefore \quad = \cot\left(\tan^{-1}\frac{3}{4} + \tan^{-1}\frac{2}{3}\right)$$

$$= \cot\left[\tan^{-1}\left(\frac{\frac{3}{4}+\frac{2}{3}}{1-\frac{1}{2}}\right)\right]$$

$$= \cot\left[\tan^{-1}\left(\frac{\frac{17}{12}}{\frac{1}{2}}\right)\right] = \cot\left[\tan^{-1}\left(\frac{17}{6}\right)\right]$$

$$= \cot\left[\cot^{-1}\left(\frac{6}{17}\right)\right] = \frac{6}{17}$$

20. (c) $\tan^{-1}\sqrt{x(x+1)}$ and $\sin^{-1}\sqrt{x^2+x+1}$ are defined if

$$x(x+1) \ge 0 \text{ and } x^2 + x + 1 \le 1$$

$$\Rightarrow \quad x^2 + x \ge 0 \text{ and } x^2 + x \le 0$$

$$\Rightarrow \quad x^2 + x = 0$$

$$\Rightarrow \quad x = 0, -1$$

Thus, LHS of the given equation is meaningful for $x = 0, -1$.

For $x = 0$, we have

$$\text{LHS} = \tan^{-1}(0) + \sin^{-1}(1) = \frac{\pi}{2} = \text{RHS}$$

For $x = -1$, we have

$$\text{LHS} = \tan^{-1}(0) + \sin^{-1}(1) = \frac{\pi}{2} = \text{RHS}$$

Hence, these are two real solutions of the given equations.

21. (b) We have,

$$\sin^{-1}\left(x - \frac{x^2}{2} + \frac{x^3}{4} - \frac{x^4}{8} + \dots\right) + \cos^{-1}\left(x^2 - \frac{x^4}{2} + \frac{x^6}{4} + \dots\right) = \frac{\pi}{2}$$

$$\Rightarrow \quad \sin^{-1}\left(\frac{x}{1+\frac{x}{2}}\right) + \cos^{-1}\left(\frac{x^2}{1+\frac{x^2}{2}}\right) = \frac{\pi}{2}$$

$$\Rightarrow \quad \sin^{-1}\left(\frac{2x}{2+x}\right) + \cos^{-1}\left(\frac{2x^2}{2+x^2}\right) = \frac{\pi}{2}$$

$$\Rightarrow \quad \sin^{-1}\left(\frac{2x}{2+x}\right) = \sin^{-1}\left(\frac{2x^2}{2+x^2}\right)$$

$$\Rightarrow \quad \frac{2x}{2+x} = \frac{2x^2}{2+x^2}$$

$$\Rightarrow \quad \frac{x}{x+2} = \frac{x^2}{x^2+2}$$

$$\Rightarrow \quad 1 + \frac{2}{x} = 1 + \frac{2}{x^2} \Rightarrow \frac{1}{x} = \frac{1}{x^2}$$

$$\Rightarrow \quad x = x^2$$

$$\therefore \quad x = 1 \quad [\because x > 0]$$

22. (a) Since, x, y, z are in an AP.

$$\therefore \quad 2y = x + z$$

Also, $\tan^{-1}x$, $\tan^{-1}y$ and $\tan^{-1}z$ are in an AP.

$$2\tan^{-1}y = \tan^{-1}x + \tan^{-1}z$$

$$\Rightarrow \quad \tan^{-1}\left(\frac{2y}{1-y^2}\right) = \tan^{-1}\left(\frac{x+z}{1-xz}\right)$$

$$\Rightarrow \quad \frac{2y}{1-y^2} = \frac{x+z}{1-xz}$$

$$\Rightarrow \quad \frac{x+z}{1-y^2} = \frac{x+z}{1-xz}$$

$$\Rightarrow \quad y^2 = xz$$

Since, x, y, z are in an AP as well as in GP.

$$\therefore \quad x = y = z$$

23. (a,b) $\tan^{-1}x + \sin^{-1}x = \tan^{-1}2x$

$$\Rightarrow \quad \tan^{-1}x + \tan^{-1}\left(\frac{x}{\sqrt{1-x^2}}\right) = \tan^{-1}2x$$

$$\Rightarrow \quad 1 - y^2 = 1 - xy$$

$$\Rightarrow \quad \tan^{-1}\left(\frac{x + \frac{x}{\sqrt{1-x^2}}}{1 - \frac{x^2}{\sqrt{1-x^2}}}\right) = \tan^{-1}2x$$

$$\Rightarrow \quad \tan^{-1}\left(\frac{x\sqrt{1-x^2}+x}{\sqrt{1-x^2}-x^2}\right) = \tan^{-1}2x$$

$$\Rightarrow \quad \frac{x\sqrt{1-x^2}+x}{\sqrt{1-x^2}-x^2} = 2x$$

$$\Rightarrow \quad \sqrt{1-x^2} + 1 = 2\sqrt{1-x^2} - 2x^2$$

$$\Rightarrow \quad -\sqrt{1-x^2} + 1 = -2x^2$$

$$\Rightarrow \quad \sqrt{1-x^2} = 1 + 2x^2$$

$$\Rightarrow \quad 1 - x^2 = (1 + 2x^2)^2$$

$$\Rightarrow \quad 1 - x^2 = 1 + 4x^4 + 4x^2$$

$$\Rightarrow \quad 4x^4 + 5x^2 = 0$$

$$\Rightarrow \quad x^2(4x^2 + 5) = 0$$

$$\Rightarrow \quad x^2 = 0 \text{ or } x^2 = -\frac{5}{4} \quad \text{[not possible]}$$

$$\therefore \quad x = 0$$

24. (a,b,c) Given, $\sin^{-1}(x^2 - 4x + 5) + \cos^{-1}(y^2 - 2y + 2) = \frac{\pi}{2}$

as $\quad x^2 - 4x + 5 = (x-2)^2 + 1 \ge 1$

But $\quad |x^2 - 4x + 5| \le 1$

$\therefore \quad x^2 - 4x + 5 = 1$

$\Rightarrow \quad x = 2 \quad \dots$(i)

Again, $\quad y^2 - 2y + 2 = (y-1)^2 + 1 \ge 1$

But $\quad |y^2 - 2y + 2| \le 1$

$\therefore \quad y^2 - 2y + 2 = 1$

$\Rightarrow \quad y = 1 \quad \dots$(ii)

Hence, $\quad x = 2$ and $y = 1$

Now, from options

(a) $x + y = 2 + 1 = 3$ [correct]

(b) $xy = 2 \times 1 = 2$ [correct]

(c) $x - y = 2 - 1 = 1$ [correct]

(d) $x - y = 2 - 1 \ne 0$ [incorrect]

25. (b,c) Given, $\tan^{-1} x + \cot^{-1} y = \tan^{-1} 3$

$\Rightarrow \quad \tan^{-1}(1/y) = \tan^{-1} 3 - \tan^{-1} x$

$\Rightarrow \quad \tan^{-1}(1/y) = \tan^{-1}\left(\frac{3-x}{1+3x}\right)$

or $\quad y = \frac{1+3x}{3-x}$ [as $x, y \in$ natural numbers]

$\therefore$ x can be only 1 or 2.

When $x = 1 \Rightarrow y = 2$ and when $x = 2 \Rightarrow y = 7$

26. (b,d) Given, $\cos^{-1}\frac{7}{|x|} = \frac{\pi}{2} - \cos^{-1}\frac{4\sqrt{15}}{|x|} = \sin^{-1}\frac{4\sqrt{15}}{|x|}$

$\Rightarrow \quad \cos^{-1}\frac{7}{|x|} = \cos^{-1}\left(\sqrt{1 - \frac{240}{|x|^2}}\right)$

$\Rightarrow \quad \frac{49}{|x|^2} = 1 - \frac{240}{|x|^2}$

$\Rightarrow \quad 289 = |x|^2$

$\therefore \quad x = \pm 17$

Explanations for Q. Nos. (27-28)

Given,

$$\cos^{-1} x + (\sin^{-1} y)^2 = \frac{p\pi^2}{4} \quad \ldots\text{(i)}$$

$$\cos^{-1} x \cdot (\sin^{-1} y)^2 = \frac{\pi^2}{16} \quad \ldots\text{(ii)}$$

From Eqs. (i) and (ii), we get

$$\cos^{-1} x + \frac{\pi^2}{16} \times \frac{1}{\cos^{-1} x} = \frac{p\pi^2}{4}$$

$$16(\cos^{-1} x)^2 + \pi^2 = 16\frac{p\pi^2}{4}\cos^{-1} x$$

$$16(\cos^{-1} x)^2 + \pi^2 = 4p\pi^2 \cos^{-1} x$$

Put $\cos^{-1} x = t$

$$16t^2 + \pi^2 = 4p\pi^2 t$$

$\Rightarrow \quad 16t^2 - 4p\pi^2 t + \pi^2 = 0$

$\Rightarrow \quad t = \frac{4p\pi^2 \pm \sqrt{16p^2\pi^4 - 16\pi^2}}{32}$

The value of t exists, if

$$16p^2\pi^4 - 64\pi^2 \geq 0$$

$\Rightarrow \quad p^2\pi^4 \geq 4\pi^2$

$\Rightarrow \quad p^2\pi^2 \geq 4$

$\Rightarrow \quad p^2 \geq \frac{4}{\pi^2}$

$\therefore \quad p \geq \pm\frac{2}{\pi}$

27. (b) Let $\cos^{-1} x = a$

$\Rightarrow \quad a \in (0, \pi]$

and $\quad \sin^{-1} y = b$

$\Rightarrow \quad b \in \left[\frac{-\pi}{2}, \frac{\pi}{2}\right]$

We have,

$$a + b^2 = \frac{p\pi^2}{4} \quad \ldots\text{(i)}$$

and $$ab^2 = \frac{\pi^4}{16} \quad \ldots\text{(ii)}$$

Since, $b^2 \in \left[0, \frac{\pi^2}{4}\right]$, we get

$$a + b^2 \in \left[0, \pi + \frac{\pi^2}{4}\right]$$

So, from Eq. (i), we get

$$0 \leq \frac{p\pi^2}{4} \leq \pi + \frac{\pi^2}{4}$$

i.e. $\quad 0 \leq p \leq \frac{4}{\pi} + 1$

Since, $p \in z$, so $p = 0$, 1 or 2.

Substituting the value of b^2 from Eq. (i) in Eq. (ii), we get

$$a\left(\frac{p\pi^2}{4} - a\right) = \frac{\pi^4}{16}$$

$\Rightarrow \quad 16a^2 - 4p\pi^2 a + \pi^4 = 0 \quad \ldots\text{(iii)}$

Since, $\quad a \in R$

$\Rightarrow \quad \beta \geq 0$

i.e. $\quad 16p^2\pi^4 - 64\pi^4 \geq 0$

$\Rightarrow \quad p^2 \geq 4$

$\Rightarrow \quad p \geq 2$

$\therefore \quad p = 2$

28. (d) Substituting $p = 2$ in Eq. (iii), we get

$$16a^2 - 8\pi^2 a + \pi^4 = 0$$

$\Rightarrow \quad (4a - \pi^2)^2 = 0$

$\Rightarrow \quad a = \frac{\pi^2}{4} = \cos^{-1} x$

$\therefore \quad x = \cos\frac{\pi^2}{4}$

29. (4) We have,

$$\tan^{-1}\left(\frac{1}{1+1+1^2}\right) + \tan^{-1}\left(\frac{1}{1+2+2^2}\right) + \tan^{-1}\left(\frac{1}{1+3+3^2}\right) + \ldots \infty$$

$$= \lim_{n\to\infty}\sum_{r=1}^{n}\tan^{-1}\frac{1}{r^2+r+1}$$

$$= \lim_{n\to\infty}\sum_{r=1}^{n}\tan^{-1}\left\{\frac{1}{1+r(r+1)}\right\}$$

$$= \lim_{n\to\infty}\sum_{r=1}^{n}\tan^{-1}\left\{\frac{(r+1)-r}{1+r(r+1)}\right\}$$

$$= \lim_{n\to\infty}\sum_{r=1}^{n}\{\tan^{-1}(r+1) - \tan^{-1} r\}$$

$$= \lim_{n\to\infty}\{\tan^{-1}(n+1) - \tan^{-1} 1\}$$

$$= \tan^{-1}\infty - \tan^{-1} 1 = \frac{\pi}{2} - \frac{\pi}{4} = \frac{\pi}{4}$$

$\therefore \quad k = 4$

30. (4) Given series,

$$\tan^{-1}\left(\frac{1}{2\cdot 1^2}\right) + \tan^{-1}\left(\frac{1}{2\cdot 2^2}\right) + \tan^{-1}\left(\frac{1}{2\cdot 3^2}\right) + \ldots\infty = \frac{\pi}{k}$$

$\Rightarrow \quad \sum_{r=1}^{\infty}\tan^{-1}\left(\frac{1}{2r^2}\right) = \frac{\pi}{k}$

$\Rightarrow \quad \sum_{r=1}^{\infty}\tan^{-1}\left(\frac{2}{1+4r^2-1}\right) = \frac{\pi}{k}$

$\Rightarrow \quad \sum_{r=1}^{\infty}\tan^{-1}\left(\frac{(2r+1)-(2r-1)}{1+(2r+1)(2r-1)}\right) = \frac{\pi}{k}$

$\Rightarrow \quad \sum_{r=1}^{\infty}\{\tan^{-1}(2r+1) - \tan^{-1}(2r-1)\} = \frac{\pi}{k}$

$\Rightarrow \quad \lim_{n\to\infty}\sum_{r=1}^{n}\{\tan^{-1}(2r+1) - \tan^{-1}(2r-1)\} = \frac{\pi}{k}$

$\Rightarrow \quad \lim_{n\to\infty}\{(\tan^{-1} 3 - \tan^{-1} 1) + (\tan^{-1} 5 - \tan^{-1} 3) + \ldots + \tan^{-1}(2n+1) - \tan^{-1}(2n-1)\} = \frac{\pi}{k}$

$$\Rightarrow \lim_{n\to\infty}(\tan^{-1}(2n+1)-\tan^{-1}1)=\frac{\pi}{k}$$

$$\Rightarrow \frac{\pi}{2}-\frac{\pi}{4}=\frac{\pi}{k}$$

$$\Rightarrow \frac{\pi}{4}=\frac{\pi}{k}$$

$$\therefore k=4$$

31. (4) $\sum_{m=1}^{n}\tan^{-1}\left(\frac{2m}{m^4+m^2+2}\right)=\tan^{-1}\left(\frac{n^2+n}{an^2+bn+c}\right)$

$$\Rightarrow \sum_{m=1}^{n}\tan^{-1}\left(\frac{2m}{1+(m^2+m+1)(m^2-m+1)}\right)=\tan^{-1}\left(\frac{n^2+n}{an^2+bn+c}\right)$$

$$\Rightarrow \sum_{m=1}^{n}\tan^{-1}\left(\frac{(m^2+m+1)-(m^2-m+1)}{1+(m^2+m+1)(m^2-m+1)}\right)=\tan^{-1}\left(\frac{n^2+n}{an^2+bn+c}\right)$$

$$\Rightarrow \sum_{m=1}^{n}[\tan^{-1}(m^2+m+1)-\tan^{-1}(m^2-m+1)]$$

$$=\tan^{-1}\left(\frac{n^2+n}{an^2+bn+c}\right)$$

$$\Rightarrow (\tan^{-1}3-\tan^{-1}1)+(\tan^{-1}7-\tan^{-1}3)$$
$$+(\tan^{-1}13-\tan^{-1}7)+\ldots+(\tan^{-1}(n^2+n+1)-\tan^{-1}(n^2-n+1))$$

$$=\tan^{-1}\left(\frac{n^2+n}{an^2+bn+c}\right)$$

$$\Rightarrow \tan^{-1}(n^2+n+1)-\tan^{-1}(1)=\tan^{-1}\left(\frac{n^2+n}{an^2+bn+c}\right)$$

$$\Rightarrow \tan^{-1}\left(\frac{n^2+n}{2+n^2+n}\right)=\tan^{-1}\left(\frac{n^2+n}{an^2+bn+c}\right)$$

$$\Rightarrow a=1,\quad b=1,\quad c=2$$

$$\therefore a+b+c=1+1+2=4$$

32. (2) We have,

$$\cot^{-1}\left(2^2+\frac{1}{2}\right)+\cot^{-1}\left(2^3+\frac{1}{2^2}\right)+\cot^{-1}\left(2^4+\frac{1}{2^3}\right)+\cdots\infty$$

$$\Rightarrow \lim_{n\to\infty}\sum_{r=1}^{n}\cot^{-1}\left(2^{r+1}+\frac{1}{2^r}\right)$$

$$\Rightarrow \lim_{n\to\infty}\sum_{r=1}^{n}\tan^{-1}\left(\frac{2^r}{2^{2r+1}+1}\right)$$

$$\Rightarrow \lim_{n\to\infty}\sum_{r=1}^{n}\tan^{-1}\left(\frac{2^{r+1}-2^r}{1+2^{r+1}\cdot 2^r}\right)$$

$$\lim_{n\to\infty}\sum_{r=1}^{n}(\tan^{-1}2^{r+1}-\tan^{-1}2^r)$$

$$\Rightarrow \lim_{n\to\infty}\{(\tan^{-1}2^2-\tan^{-1}2)+(\tan^{-1}2^3-\tan^{-1}2^2)+\cdots+(\tan^{-1}2^{n+1}-\tan^{-1}2^n)\}$$

$$\Rightarrow \lim_{n\to\infty}(\tan^{-1}2^{n+1}-\tan^{-1}2)$$

$$\Rightarrow \frac{\pi}{2}-\tan^{-1}2=\cot^{-1}2$$

$$\therefore k=2$$

33. (2) $S_n=\cot^{-1}\left(1+\frac{3}{4}\right)+\cot^{-1}\left(2^2+\frac{3}{4}\right)+\cot^{-1}\left(3^2+\frac{3}{4}\right)+\cdots+\cot^{-1}\left(n^2+\frac{3}{4}\right)$

$$=\sum_{r=1}^{n}\cot^{-1}\left(r^2+\frac{3}{4}\right)=\sum_{r=1}^{n}\tan^{-1}\left(\frac{4}{3+4r^2}\right)$$

$$=\sum_{r=1}^{n}\tan^{-1}\left(\frac{1}{\frac{3}{4}+r^2}\right)$$

$$=\sum_{r=1}^{n}\tan^{-1}\left[\frac{1}{1+r^2-\frac{1}{4}}\right]$$

$$=\sum_{r=1}^{n}\tan^{-1}\left[\frac{\left(r+\frac{1}{2}\right)-\left(r-\frac{1}{2}\right)}{1+\left(r+\frac{1}{2}\right)\left(r-\frac{1}{2}\right)}\right]$$

$$=\sum_{r=1}^{n}\left\{\tan^{-1}\left(r+\frac{1}{2}\right)-\tan^{-1}\left(r-\frac{1}{2}\right)\right\}$$

$$=\left(\tan^{-1}\frac{3}{2}-\tan^{-1}\frac{1}{2}\right)+\left(\tan^{-1}\frac{5}{2}-\tan^{-1}\frac{3}{2}\right)+\cdots+\left(\tan^{-1}\left(n+\frac{1}{2}\right)-\tan^{-1}\left(n-\frac{1}{2}\right)\right)$$

$$=\tan^{-1}\left(n+\frac{1}{2}\right)-\tan^{-1}\left(\frac{1}{2}\right)$$

$\therefore$ Required sum

$$=\lim_{n\to\infty}S_n=\lim_{n\to\infty}\tan^{-1}\left(n+\frac{1}{2}\right)-\tan^{-1}\left(\frac{1}{2}\right)$$

$$=\frac{\pi}{2}-\tan^{-1}\left(\frac{1}{2}\right)$$

$$=\cot^{-1}\left(\frac{1}{2}\right)=\tan^{-1}(2)$$

$$\therefore k=2$$

34. (2) $\sin^{-1}\left(\frac{1}{\sqrt{2}}\right)+\sin^{-1}\left(\frac{\sqrt{2}-1}{\sqrt{6}}\right)+\ldots+\sin^{-1}\left(\frac{\sqrt{n}-\sqrt{n-1}}{\sqrt{n(n+1)}}\right)+\ldots\infty=\frac{\pi}{k}$

$$\Rightarrow \sum_{r=1}^{\infty}\sin^{-1}\left(\frac{\sqrt{r}-\sqrt{r-1}}{\sqrt{r(r+1)}}\right)=\frac{\pi}{k}$$

$$\Rightarrow \sum_{r=1}^{\infty}\tan^{-1}\left(\frac{\sqrt{r}-\sqrt{r-1}}{1+\sqrt{r(r-1)}}\right)=\frac{\pi}{k}$$

$$\Rightarrow \lim_{n\to\infty}\sum_{r=1}^{n}(\tan^{-1}(\sqrt{r})-\tan^{-1}\sqrt{r-1})=\frac{\pi}{k}$$

$$\Rightarrow \lim_{n\to\infty}\tan^{-1}\sqrt{n}=\frac{\pi}{k}$$

$$\Rightarrow \frac{\pi}{2}=\frac{\pi}{k}$$

$$\therefore k=2$$

35. (0) We have,

$$x=\text{cosec}\,[\tan^{-1}\{\cos(\cot^{-1}(\sec(\sin^{-1}a)))\}]$$

$$\Rightarrow x=\text{cosec}\left[\tan^{-1}\left\{\cos\left(\cot^{-1}\left(\frac{1}{\sqrt{1-a^2}}\right)\right)\right\}\right]$$

$$\Rightarrow x=\text{cosec}\left[\tan^{-1}\left(\frac{1}{\sqrt{2-a^2}}\right)\right]$$

$$\Rightarrow x=\sqrt{3-a^2} \quad \ldots\text{(i)}$$

and $y=\sec[\cot^{-1}\{\sin(\tan^{-1}(\text{cosec}(\cos^{-1}a)))\}]$

$$\Rightarrow y=\sec\left[\cot^{-1}\left\{\sin\left(\tan^{-1}\left(\frac{1}{\sqrt{1-a^2}}\right)\right)\right\}\right]$$

$$\Rightarrow y=\sec\left[\cot^{-1}\left(\frac{1}{\sqrt{2-a^2}}\right)\right]$$

$$\Rightarrow y=\sqrt{3-a^2}$$

From Eqs. (i) and (ii), we get

$$x=y=\sqrt{3-a^2}$$

$$\therefore x^2-y^2=(\sqrt{3-a^2})^2-(\sqrt{3-a^2})^2=0$$

DPP-3 Special Formulae

1. (a) Graph of $\sin^{-1}(2x\sqrt{1-x^2})$ is shown below :

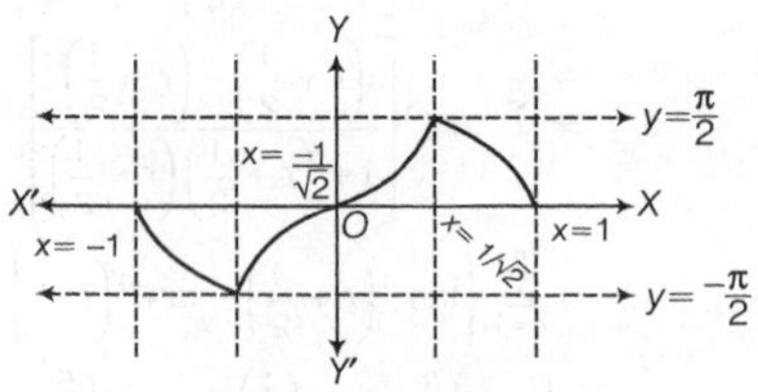

2. (b) Defining the curve let $x = \cos\theta$.

$$\therefore \quad \cos^{-1}(2x^2-1) = \begin{cases} 2\cos^{-1}x, & 0 \le x \le 1 \\ 2\pi - 2\cos^{-1}x, & -1 \le x < 0 \end{cases}$$

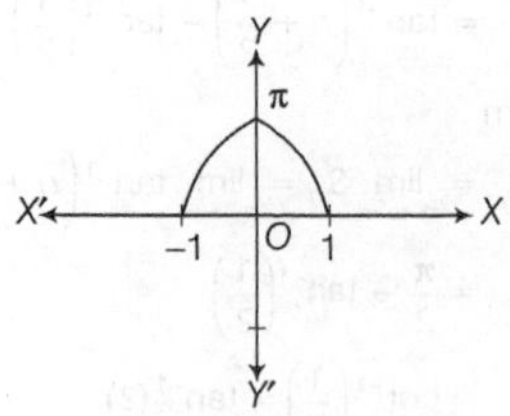

3. (a) Graph for the curve $y = \sin^{-1}\left(\frac{2x}{1+x^2}\right)$

Hence, for domain $\quad \frac{2x}{1+x^2} \le 1$

$$\Rightarrow \quad 2|x| \le 1 + x^2 \qquad [\because 1 + x^2 > 0, \forall x]$$

$$\Rightarrow \quad |x|^2 - 2|x| + 1 \ge 0$$

$$\Rightarrow \quad \{|x| - 1\}^2 \ge 0$$

$$\Rightarrow \quad x \in R$$

For range $y = \sin^{-1}\left(\frac{2x}{1+x^2}\right) \Rightarrow y \in \left[-\frac{\pi}{2}, \frac{\pi}{2}\right]$

Defining the curve let $x = \tan\theta$

$$\Rightarrow \quad y = \sin^{-1}(\sin 2\theta) = \begin{cases} \pi - 2\theta; & 2\theta > \frac{\pi}{2} \\ 2\theta; & -\frac{\pi}{2} \le 2\theta \le \frac{\pi}{2} \\ -\pi - 2\theta; & 2\theta < -\frac{\pi}{2} \end{cases}$$

$$\text{or} \quad y = \begin{cases} \pi - 2\tan^{-1}x, & \tan^{-1}x > \frac{\pi}{4} \\ 2\tan^{-1}x, & \frac{-\pi}{4} \le \tan^{-1}x \le \frac{\pi}{4} \\ -\pi - 2\tan^{-1}x, & \tan^{-1}x < \frac{\pi}{4} \end{cases}$$

$$\text{or} \quad y = \begin{cases} \pi - 2\tan^{-1}x, & x \ge 1 \\ 2\tan^{-1}x, & -1 \le x \le 1 \\ -\pi - 2\tan^{-1}x, & x < -1 \end{cases} \qquad \ldots\text{(i)}$$

Thus, $y = \sin^{-1}\left(\frac{2x}{1+x^2}\right)$ is defined for $x \in R$, where $y \in \left[-\frac{\pi}{2}, \frac{\pi}{2}\right]$,

so the graph for Eq. (i) could be shown in the figure given below.

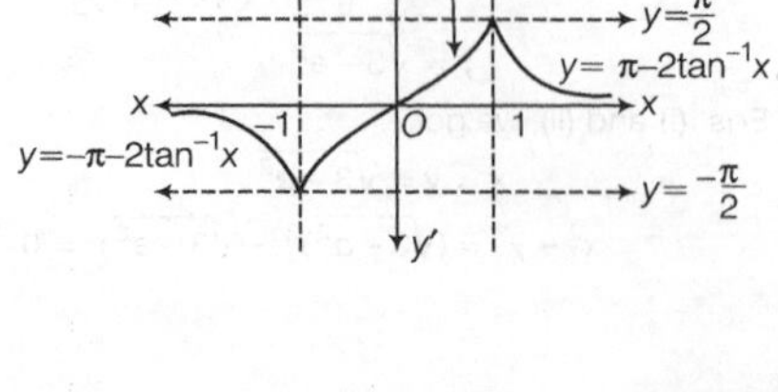

4. (d) Graph for $y = \cos^{-1}\left(\frac{1-x^2}{1+x^2}\right)$

Here, for domain $\left|\frac{1-x^2}{1+x^2}\right| \le 1$

$$\Rightarrow \quad |1 - x^2| \le 1 + x^2 \qquad [\because 1 + x^2 > 0, \forall x \in R]$$

which is true for all x; as $1 + x^2 > 1 - x^2$

$$\therefore \quad x \in R$$

For range $y = \cos^{-1}\left(\frac{1-x^2}{1+x^2}\right) \Rightarrow y \in (0, \pi)$

Define the curve, let $x = \tan\theta$

$$\therefore \quad y = \cos^{-1}\left(\frac{1-\tan^2\theta}{1+\tan^2\theta}\right) = \cos^{-1}(\cos 2\theta) = \begin{cases} 2\theta, & 2\theta \ge 0 \\ -2\theta, & 2\theta < 0 \end{cases}$$

$$\Rightarrow \quad y = \begin{cases} 2\tan^{-1}x, & \tan^{-1}x \ge 0 \\ -2\tan^{-1}x, & \tan^{-1}x < 0 \end{cases}$$

So, the graph of $y = \cos^{-1}\left(\frac{1-x^2}{1+x^2}\right) = \begin{cases} 2\tan^{-1}x, & x \ge 0 \\ -2\tan^{-1}x, & x < 0 \end{cases}$

is shown as,

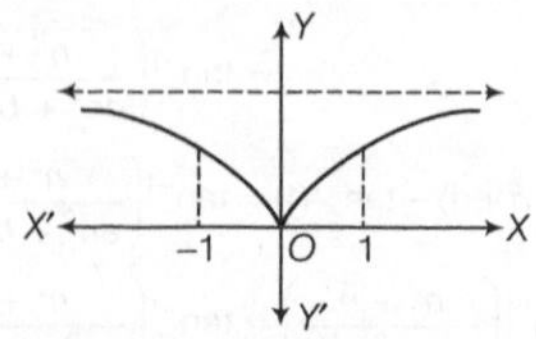

5. (a) Graph for the curve

$$y = \tan^{-1}\left(\frac{2x}{1-x^2}\right)$$

Here, for domain $\frac{2x}{1-x^2} \in R$ except $1 - x^2 = 0$

i.e. $\quad x = \pm 1$ or $x \in R - \{1, -1\}$

For range $\quad y = \tan^{-1}\left(\frac{2x}{1-x^2}\right) \Rightarrow y \in \left(\frac{-\pi}{2}, \frac{\pi}{2}\right)$

Defining the curve, let $x = \tan\theta$

$$\Rightarrow \quad y = \tan^{-1}\left(\frac{2\tan\theta}{1-\tan^2\theta}\right) = \tan^{-1}(\tan 2\theta)$$

$$= \begin{cases} \pi + 2\theta, & 2\theta < -\frac{\pi}{2} \\ 2\theta, & -\frac{\pi}{2} < 2\theta < \frac{\pi}{2} \\ -\pi + 2\theta, & 2\theta > \frac{\pi}{2} \end{cases} = \begin{cases} \pi + 2\tan^{-1}x, & \tan^{-1}x < \frac{-\pi}{4} \\ 2\tan^{-1}x, & -\frac{\pi}{4} < \tan^{-1}x < \frac{\pi}{4} \\ -\pi + 2\tan^{-1}x, & \tan^{-1}x > \frac{\pi}{4} \end{cases}$$

$$= \begin{cases} \pi + 2\tan^{-1}x, & x < -1 \\ 2\tan^{-1}x, & -1 < x < 1 \\ -\pi + 2\tan^{-1}x, & x > 1 \end{cases}$$

So, the graph of

$$y = \tan^{-1}\left(\frac{2x}{1-x^2}\right) = \begin{cases} \pi + 2\tan^{-1}x, & x < -1 \\ 2\tan^{-1}x, & -1 < x < 1 \\ -\pi + 2\tan^{-1}x, & x > 1 \end{cases}$$

is shown as :

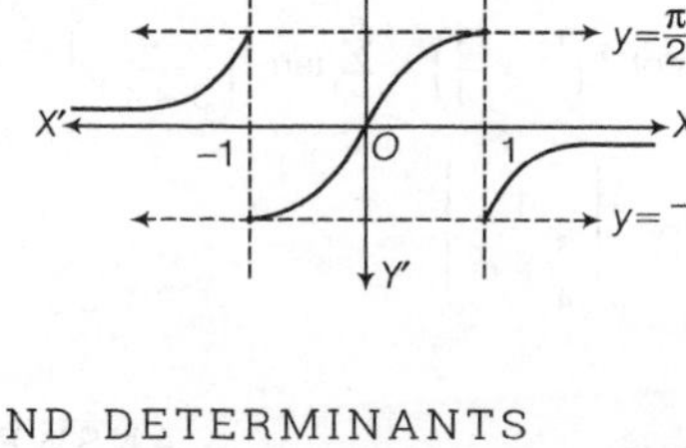

6. (c) Graph for $y = \cos^{-1}(4x^3 - 3x)$

Here, domain $\in[-1, 1]$, range $\in[0, \pi]$.

Now, define the curve

Let $x = \cos\theta$

$$y = \cos^{-1}(\cos 3\theta) = \begin{cases} 2\pi - 3\theta, & \pi \le 3\theta \le 2\pi \\ 3\theta & 0 \le 3\theta \le \pi \\ -2\pi + 3\theta & -\pi \le 3\theta \le 0 \end{cases}$$

$$= \begin{cases} 2\pi - 3\cos^{-1} x, & \frac{\pi}{3} \le \cos^{-1} x \le \frac{2\pi}{3} \\ 3\cos^{-1} x, & 0 < \cos^{-1} x \le \frac{\pi}{3} \\ -2\pi + 3\cos^{-1} x, & -\frac{\pi}{3} \le \cos^{-1} x \le 0 \end{cases}$$

$$= \begin{cases} 2\pi - 3\cos^{-1} x, & -\frac{1}{2} \le x \le \frac{1}{2} \\ 3\cos^{-1} x, & \frac{1}{2} \le x \le 1 \\ -2\pi + 3\cos^{-1} x & -1 \le x \le -\frac{1}{2} \end{cases}$$

$\because$ If $0 \le \theta \le \frac{\pi}{3} \Rightarrow \cos\frac{\pi}{3} \le \cos\theta \le \cos 0$

$\Rightarrow \frac{1}{2} \le \cos\theta \le 1$

Hence, the interval is changed, since $\cos x$ is decreasing in $[0, \pi]$.

So, the graph of

$$y = \cos^{-1}(4x^3 - 3x) = \begin{cases} 2\pi - 3\cos^{-1} x, & -\frac{1}{2} \le x \le \frac{1}{2} \\ 3\cos^{-1} x, & \frac{1}{2} \le x < 1 \\ -2\pi + 3\cos^{-1} x, & -1 \le x \le -\frac{1}{2} \end{cases}$$

is shown as:

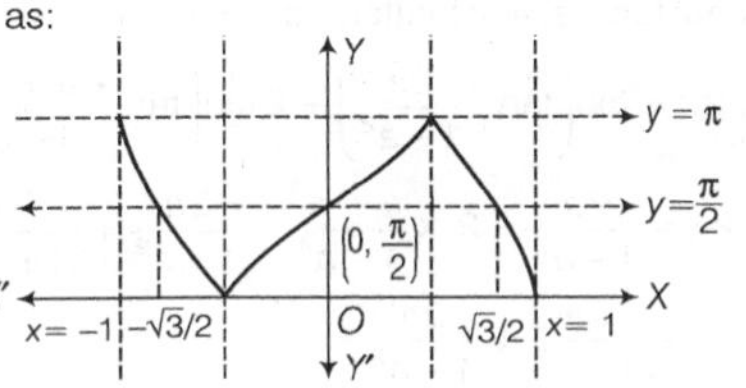

7. (c) Graph for $y = \sin^{-1}(3x - 4x^3)$

Defining the curve

Let $x = \sin\theta$

$$\Rightarrow \quad y = \sin^{-1}(\sin 3\theta) = \begin{cases} \pi - 3\theta, & \frac{\pi}{2} \le 3\theta \le \frac{3\pi}{2} \\ 3\theta, & -\frac{\pi}{2} \le 3\theta \le \frac{\pi}{2} \\ -\pi - 3\theta & -\frac{3\pi}{2} \le 3\theta \le -\frac{\pi}{2} \end{cases}$$

$$= \begin{cases} \pi - 3\sin^{-1} x, & \frac{\pi}{6} \le \sin^{-1} x \le \frac{\pi}{2} \\ 3\sin^{-1} x, & \frac{-\pi}{6} \le \sin^{-1} x < \frac{\pi}{6} \\ -\pi - 3\sin^{-1} x, & -\frac{\pi}{2} \le \sin^{-1} x \le -\frac{\pi}{6} \end{cases}$$

$\therefore \quad y = \sin^{-1}(3x - 4x^3)$

$$= \begin{cases} \pi - 3\sin^{-1} x, & \frac{1}{2} \le x \le 1 \\ 3\sin^{-1} x, & -\frac{1}{2} \le x \le \frac{1}{2} \\ -\pi - 3\sin^{-1} x, & -1 \le x \le -\frac{1}{2} \end{cases}$$

For domain, $y = \sin^{-1}(3x - 4x^3)$

$\Rightarrow x \in [-1, 1]$

For range, $y = \sin^{-1}(3x - 4x^3)$

$\Rightarrow y \in \left[-\frac{\pi}{2}, \frac{\pi}{2}\right]$

So, the graph of

$$y = \sin^{-1}(3x - 4x^3) = \begin{cases} \pi - 3\sin^{-1} x, & \frac{1}{2} \le x \le 1 \\ 3\sin^{-1} x, & -\frac{1}{2} \le x \le \frac{1}{2} \\ -\pi - 3\sin^{-1} x, & -1 \le x \le -\frac{1}{2} \end{cases}$$

is shown as :

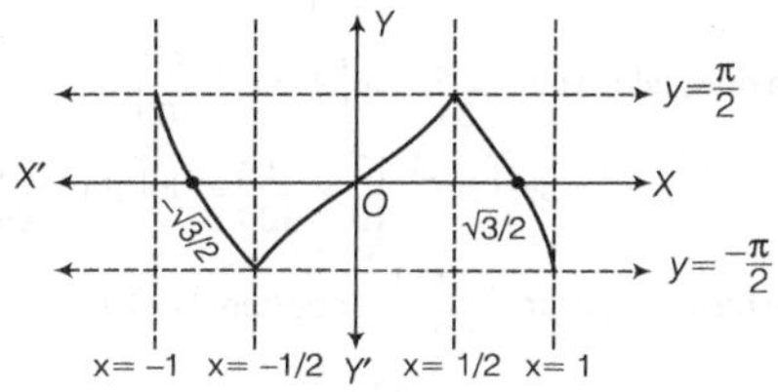

8. (d) Graph of $y = \tan^{-1}\left(\frac{3x - x^3}{1 - 3x^2}\right)$

Here, for domain $y = \tan^{-1}\left(\frac{3x - x^3}{1 - 3x^2}\right)$

$\Rightarrow x \in R$, except $1 - 3x^2 = 0$

$\Rightarrow x \ne \pm \frac{1}{\sqrt{3}}$

$\therefore x \in R - \left\{\pm \frac{1}{\sqrt{3}}\right\}$

For range, $y = \tan^{-1}\left(\frac{3x - x^3}{1 - 3x^2}\right) \Rightarrow y \in \left(-\frac{\pi}{2}, \frac{\pi}{2}\right)$

Defining the curve, let $x = \tan\theta$

$\Rightarrow y = \tan^{-1}(\tan 3\theta)$

$$= \begin{cases} \pi + 3\theta, & 3\theta < -\frac{\pi}{2} \\ 3\theta, & -\frac{\pi}{2} < 3\theta < \frac{\pi}{2} \\ -\pi + 3\theta & 3\theta > \frac{\pi}{2} \end{cases}$$

$$= \begin{cases} \pi + 3\tan^{-1} x, & \tan^{-1} x < -\frac{\pi}{6} \\ 3\tan^{-1} x, & -\frac{\pi}{6} < \tan^{-1} x < \frac{\pi}{6} \\ -\pi + 3\tan^{-1} x, & \tan^{-1} x > \frac{\pi}{6} \end{cases}$$

$$= \begin{cases} \pi + 3\tan^{-1} x, & x < -\frac{1}{\sqrt{3}} \\ 3\tan^{-1} x, & -\frac{1}{\sqrt{3}} < x < \frac{1}{\sqrt{3}} \\ -\pi + 3\tan^{-1} x, & x > \frac{1}{\sqrt{3}} \end{cases}$$

So, the graph of

$$y = \tan^{-1}\left(\frac{3x - x^3}{1 - 3x^2}\right) = \begin{cases} \pi + 3\tan^{-1} x, & x < -\frac{1}{\sqrt{3}} \\ 3\tan^{-1} x, & -\frac{1}{\sqrt{3}} < x < \frac{1}{\sqrt{3}} \\ -\pi + 3\tan^{-1} x, & x > \frac{1}{\sqrt{3}} \end{cases}$$

is shown as :

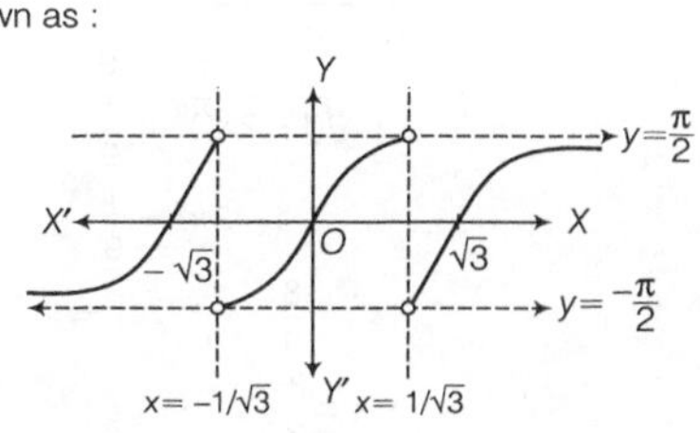

9. (b) We have,

$$\cos\left(2\cos^{-1}\frac{1}{5}+\sin^{-1}\frac{1}{5}\right)=\cos\left(2\cos^{-1}\frac{1}{5}+\frac{\pi}{2}-\cos^{-1}\frac{1}{5}\right)$$

$$=\cos\left(\cos^{-1}\frac{1}{5}+\frac{\pi}{2}\right)=-\sin\left(\cos^{-1}\left(\frac{1}{5}\right)\right)$$

$$=-\sin^{-1}\left(\sin^{-1}\frac{2\sqrt{6}}{5}\right)=\frac{-2\sqrt{6}}{5}$$

10. (b) We have, $\sin\left(4\tan^{-1}\frac{1}{3}\right)=\sin\left(2\cdot\sin^{-1}\frac{6}{10}\right)$

$$=\sin\left(\sin^{-1}\frac{12}{10}\times\frac{8}{10}\right)=\sin\left(\sin^{-1}\frac{24}{25}\right)=\frac{24}{25}$$

11. (b) We have, $\sin\left[2\tan^{-1}\left(\frac{1}{3}\right)\right]+\cos(\tan^{-1}2\sqrt{2})$

$$=\sin\left[\sin^{-1}\left(\frac{2\times\frac{1}{3}}{1+\frac{1}{9}}\right)\right]+\cos\left(\cos^{-1}\frac{1}{3}\right)$$

$$=\frac{2\times\frac{1}{3}}{\frac{10}{9}}+\frac{1}{3}=\frac{2}{3}\times\frac{9}{10}+\frac{1}{3}$$

$$=\frac{6}{10}+\frac{1}{3}=\frac{18+10}{30}=\frac{28}{30}=\frac{14}{15}$$

12. (d) We have,

$$\tan\left[2\tan^{-1}\left(\frac{1}{5}\right)-\frac{\pi}{4}\right]=\tan\left(\tan^{-1}\frac{\frac{2}{5}}{1-\frac{1}{25}}-\tan^{-1}(1)\right)$$

$$=\tan\left(\tan^{-1}\frac{2}{5}\times\frac{25}{24}-\tan^{-1}(1)\right)$$

$$=\tan\left[\tan^{-1}\left(\frac{10}{24}\right)-\tan^{-1}(1)\right]$$

$$=\tan\left[\tan^{-1}\left(\frac{\frac{10}{24}-1}{1+\frac{10}{24}}\right)\right]$$

$$=\tan\left[\tan^{-1}\left(\frac{-14}{34}\right)\right]=-\frac{14}{34}=-\frac{7}{17}$$

13. (c) We have,

$$\tan\left(2\cos^{-1}\frac{3}{5}\right)=\tan\left[\cos^{-1}\left(2\times\frac{9}{25}-1\right)\right]$$

$$=\tan\left[\cos^{-1}\left(\frac{-7}{25}\right)\right]=\tan\left[\tan^{-1}\left(-\frac{24}{7}\right)\right]=\frac{-24}{7}$$

14. (c) Given, $\cos(2\sin^{-1}x)=\frac{1}{9}$

$$\cos\left[\sin^{-1}(2x\sqrt{1-x^2})\right]=\frac{1}{9}$$

$$\cos\left[\cos^{-1}\sqrt{1-4x^2(1-x^2)}\right]=\frac{1}{9}$$

$$\cos(\cos^{-1}\sqrt{1-4x^2+4x^4})=\frac{1}{9}$$

$\Rightarrow$ $\sqrt{1-4x^2+4x^4}=\frac{1}{9}$

$\Rightarrow$ $\sqrt{(1-2x^2)^2}=\frac{1}{9}$

$\Rightarrow$ $1-2x^2=\frac{1}{9}$

$\Rightarrow$ $2x^2=\frac{8}{9}$ $\Rightarrow$ $2x^2=\frac{4}{9}$

$\therefore$ $x=\pm\frac{2}{3}$

15. (c) Given, $\sin^{-1}x=2\tan^{-1}x$

$$\sin^{-1}x=\sin^{-1}\left(\frac{2x}{1+x^2}\right)$$

$$x=\frac{2x}{1+x^2}$$

$$x+x^3=2x$$

$\Rightarrow$ $x^3-x=0$

$\Rightarrow$ $x(x^2-1)=0$

$\therefore$ $x=0,-1,1$

16. (b) Given, $2\sin^{-1}x=\sin^{-1}(2x\sqrt{1-x^2})$ is only possible, when

$$-\frac{\pi}{2}\le 2\sin^{-1}x\le\frac{\pi}{2}$$

$$-\frac{\pi}{4}\le\sin^{-1}x\le\frac{\pi}{4}$$

$\therefore$ $-\frac{1}{2}\le x\le\frac{1}{\sqrt{2}}$

17. (c) As $\sin^{-1}\left(\frac{2x}{1+x^2}\right)=\begin{cases}-\pi-2\tan^{-1}x, & x\le-1\\ 2\tan^{-1}x, & -1\le x\le 1\\ \pi-2\tan^{-1}x, & x\ge 1\end{cases}$

$\therefore$ $2\tan^{-1}x+\sin^{-1}\left(\frac{2x}{1+x^2}\right)$ is independent of x.

when $x\le-1$ and $x\ge 1$,

i.e. $x\in(-\infty,-1]\cup[1,\infty)$

18. (b,c) Given, equation

$2x=\tan(2\tan^{-1}a)+2\tan(\tan^{-1}a+\tan^{-1}a^3)$

$$2x=\tan\left(\tan^{-1}\frac{2a}{1-a^2}\right)+2\tan\left(\tan^{-1}\frac{a+a^3}{1-a^4}\right)$$

$$2x=\frac{2a}{1-a^2}+2\times\frac{a+a^3}{1-a^4}=\frac{2a}{1-a^2}+\frac{2a(1+a^2)}{(1+a^2)(1-a^2)}$$

$$=\frac{2a}{1-a^2}+\frac{2a}{1-a^2}$$

$$2x=\frac{4a}{1-a^2}\Rightarrow x=\frac{2a}{1-a^2}\Rightarrow a\ne 1$$

Now, $x=\frac{2a}{1-a^2}$

$\Rightarrow$ $x-xa^2=2a$

$\Rightarrow$ $a^2x+2a=x$

19. (a,b,c) We have,

$2\tan(\tan^{-1}x+\tan^{-1}x^3)$

$$=2\tan\left(\tan^{-1}\frac{x+x^3}{1-x^4}\right)$$

$$=2\tan\left(\tan^{-1}\frac{x}{1-x^2}\right)=\frac{2x}{1-x^2}$$

$$=\tan(2\tan^{-1}x)$$

Also, $\tan(\cot^{-1}(-x)-\cot^{-1}x)$

$$=\tan\left[\tan^{-1}\left(-\frac{1}{x}\right)-\tan^{-1}\left(\frac{1}{x}\right)\right]$$

$$=\tan\left[\tan^{-1}\left(\frac{-\frac{1}{x}-\frac{1}{x}}{1-\frac{1}{x^2}}\right)\right]=\tan\left[\tan^{-1}\left(\frac{-\frac{2}{x}}{\frac{x^2-1}{x^2}}\right)\right]$$

$$=\frac{-2}{x}\times\frac{x^2}{x^2-1}=\frac{-2x}{x^2-1}=\frac{2x}{1-x^2}$$

20. (b,c,d) We have, $\cos\left[\frac{1}{2}\cos^{-1}\left\{\cos\left(-\frac{14}{5}\pi\right)\right\}\right]$

$\Rightarrow \quad \cos\left[\frac{1}{2}\cos^{-1}\left(\cos\frac{14\pi}{5}\right)\right] \Rightarrow \cos\left[\frac{1}{2}\cos^{-1}\cos\left(3\pi-\frac{\pi}{5}\right)\right]$

$\Rightarrow \quad \cos\left[\frac{1}{2}\cos^{-1}\left(-\cos\frac{\pi}{5}\right)\right] = \cos\left[\frac{1}{2}\cdot\left(\pi-\frac{\pi}{5}\right)\right]$

$\therefore \quad \cos\left(\frac{2\pi}{5}\right) = \sin\left(\frac{\pi}{10}\right) = -\cos\left(\frac{3\pi}{5}\right)$

21. (b)

(i) $0 \le \cos^{-1}(8x^4 - 8x^2 + 1) \le \pi$

$\Rightarrow \quad 0 \le 4\cos^{-1}x \le \pi$

$\Rightarrow \quad 0 \le \cos^{-1}x \le \frac{\pi}{4}$

$\Rightarrow \quad \frac{1}{\sqrt{2}} \le x \le 1$

(ii) $0 \le \cos^{-1}(8x^4 - 8x^2 + 1) \le \pi$

$\Rightarrow \quad 0 \le 4\cos^{-1}x - 2\pi \le \pi$

$\Rightarrow \quad 2\pi \le 4\cos^{-1}x \le 3\pi$

$\Rightarrow \quad \frac{\pi}{2} \le \cos^{-1}x \le \frac{3\pi}{4}$

$\Rightarrow \quad -\frac{1}{\sqrt{2}} \le x \le 0$

(iii) $0 \le 2\pi - 4\cos^{-1}x \le \pi$

$\Rightarrow \quad -2\pi \le -4\cos^{-1}x \le -\pi$

$\Rightarrow \quad 2\pi \ge 4\cos^{-1}x \ge \pi$

$\Rightarrow \quad \frac{\pi}{2} \ge \cos^{-1}x \ge \frac{\pi}{4}$

$\Rightarrow \quad 0 \le x \le \frac{1}{\sqrt{2}}$

(iv) $0 \le 4\pi - 4\cos^{-1}x \le \pi$

$\Rightarrow \quad -4\pi \le -4\cos^{-1}x \le -3\pi$

$\Rightarrow \quad 4\pi \ge 4\cos^{-1}x \ge 3\pi$

$\Rightarrow \quad \pi \ge \cos^{-1}x \ge \frac{3\pi}{4}$

$\Rightarrow \quad -1 \le x \le -\frac{1}{\sqrt{2}}$

(i) → (q), (ii) → (s), (iii) → (p), (iv) → (r)

22. (c) (i) $\cos^{-1}(4x^3 - 3x) = 3\cos^{-1}x$

$\Rightarrow \quad 0 \le \cos^{-1}(4x^3 - 3x) \le \pi$

$\Rightarrow \quad 0 \le 3\cos^{-1}x \le \pi$

$\Rightarrow \quad 0 \le \cos^{-1}x \le \frac{\pi}{3}$

$\Rightarrow \quad 1 \le x \le \frac{1}{2}$

(ii) $\sin^{-1}(3x - 4x^3) = 3\sin^{-1}x$

$\Rightarrow \quad \frac{-\pi}{2} \le \sin^{-1}(3x - 4x^3) \le \frac{\pi}{2}$

$\Rightarrow \quad \frac{-\pi}{2} \le 3\sin^{-1}x \le \frac{\pi}{2}$

$\Rightarrow \quad \frac{-\pi}{6} \le \sin^{-1}x \le \frac{\pi}{6}$

$\Rightarrow \quad -\frac{1}{2} \le \sin^{-1}x \le \frac{1}{2}$

(iii) $\cos^{-1}(4x^3 - 3x) = 3\sin^{-1}x$

$\Rightarrow \quad 0 \le \cos^{-1}(4x^3 - 3x) \le \pi$

$\Rightarrow \quad 0 \le 3\sin^{-1}x \le \pi$

$\Rightarrow \quad 0 \le \sin^{-1}x \le \frac{\pi}{3}$

$\Rightarrow \quad 0 \le x \le \frac{\sqrt{3}}{2}$

(iv) $\sin^{-1}(3x - 4x^3) = 3\cos^{-1}x$

$\Rightarrow \quad \frac{-\pi}{2} \le \sin^{-1}(3x - 4x^3) \le \frac{\pi}{2}$

$\Rightarrow \quad \frac{-\pi}{2} \le 3\cos^{-1}x \le \frac{\pi}{2}$

$\Rightarrow \quad \frac{-\pi}{6} \le \cos^{-1}x \le \frac{\pi}{6}$

$\Rightarrow \quad 0 \le \cos^{-1}x \le \frac{\pi}{6}$

$\Rightarrow \quad 1 \ge x \ge \frac{\sqrt{3}}{2}$

(i) → (p), (ii) → (q,s), (iii) → (r,s), (iv) → (r,s)

23. (c) When $\frac{\pi}{2} \le \sin^{-1}x \le \frac{3\pi}{2} \Rightarrow \sin^{-1}(-x) = 2\pi - \sin^{-1}x$

24. (b) We have, $f(-x) = 3\sin^{-1}(-x) - 2\cos^{-1}(-x)$

$= 3(2\pi - \sin^{-1}x) - 2(3\pi - \cos^{-1}x)$

$= 6\pi - 3\sin^{-1}x - 6\pi + 2\cos^{-1}x$

$f(-x) = -(3\sin^{-1}x - 2\cos^{-1}x) = -f(x)$

$\therefore$ $f(x)$ is an odd function.

25. (a) $(\sin^{-1}x)^3 - (\cos^{-1}x)^3$, minimum occurs

when $\quad \sin^{-1}x = \frac{\pi}{2}$ and $\cos^{-1}x = 2\pi$

$\Rightarrow \quad \frac{\pi^3}{8} - 8\pi^3 = \frac{\pi^3 - 64\pi^3}{8} = \frac{-63\pi^3}{8}$

26. (c) We have, $\sin^{-1}x + \cos^{-1}x = \frac{3\pi}{2} + \pi = \frac{5\pi}{2}$

27. (a) Given, $\quad \sin^{-1} + \cos^{-1}x = \frac{5\pi}{2}$

$\Rightarrow \quad \frac{-3\pi}{2} + \cos^{-1}x = \frac{5\pi}{2}$

$\Rightarrow \quad \cos^{-1}x = \frac{5\pi}{2} + \frac{3\pi}{2}$

$\Rightarrow \quad \cos^{-1}x = \frac{8\pi}{2}$

$\therefore$ Codomain of $\cos^{-1}x$ is $\left[\frac{8\pi}{2}, 5\pi\right]$ i.e. $[4\pi, 5\pi]$

28. (2) Given, $\cos^{-1}(4x^3 - 3x) = a + b\cos^{-1}x, -1 < x < -1/2$

But $\quad \cos^{-1}(4x^3 - 3x) = -2\pi + 3\cos^{-1}x, -1 < x < -1/2$

$\Rightarrow \quad a = -2\pi, b = 3$

Now, $\quad |(a + b + 2)| = |[-2\pi + 3 + 2]| = |[-2\pi + 5]| = 2$

29. (2) $D \ge 0 \Rightarrow (\tan^{-1}\alpha)^3 - 8(\cot^{-1}\alpha)^3 \ge 0$

$\Rightarrow \quad \tan^{-1}\alpha \ge 2\cot^{-1}\alpha$

$\Rightarrow \quad \tan^{-1}\alpha \ge \pi - 2\tan^{-1}\alpha$

$\Rightarrow \quad \tan^{-1}\alpha \ge \frac{\pi}{3} \Rightarrow \alpha \ge \sqrt{3}$

$\frac{-b}{2\alpha} \ge 0 \Rightarrow \frac{\tan^{-1}\alpha}{2\cot^{-1}\alpha} > 0$

$\Rightarrow \quad \tan^{-1}\alpha > 0 \Rightarrow \alpha > 0$

$af(0) > 0 \Rightarrow 2(\cot^{-1}\alpha)^3 > 0 \Rightarrow \alpha \in R$

On comparing, we get $\alpha \ge \sqrt{3}$.

$\therefore$ Minimum integral value is 2.

30. (0) $\frac{\pi}{4} < A < \frac{\pi}{2} \Rightarrow \cot A < 1$

$\Rightarrow \quad \tan^{-1}(\cot A) + \tan^{-1}(\cot^3 A) = \tan^{-1}\left(-\frac{1}{2}\tan 2A\right)$

$\Rightarrow \quad k = \tan^{-1}\left(\frac{1}{2}\tan 2A\right) - \tan^{-1}\left(\frac{1}{2}\tan 2A\right) = 0$

$\therefore \quad [k] = 0$

Revisal Problems (JEE Main)

1. (c) We have, $1 + \cot^2 (\sin^{-1} x) = 1 + \{\cot (\sin^{-1} x)\}^2$

$$= 1 + \left\{\cot\left(\cot^{-1} \frac{\sqrt{1-x^2}}{x}\right)\right\}^2$$

$$= 1 + \left\{\frac{\sqrt{1-x^2}}{x}\right\}^2 = 1 + \frac{1-x^2}{x^2}$$

$$= \frac{x^2 + 1 - x^2}{x^2} = \frac{1}{x^2}$$

2. (d) We have, $\sec^2 (\tan^{-1} 2) + \text{cosec}^2 (\cot^{-1} 3)$

$$= \{1 + \tan^2(\tan^{-1} 2)\} + \{1 + \cot^2 (\cot^{-1} 3)\}$$

$$= \{1 + \tan\{\tan^{-1} 2\}^2\} + \{1 + \cot\{\cot^{-1} 3\}^2\}$$

$$= (1 + 2^2) + (1 + 3^2) = 5 + 10 = 15$$

3. (c) We have,

$\tan^2 (\sec^{-1} 2) + \cot^2 (\text{cosec}^{-1} 3)$

$$= [\sec^2 (\sec^{-1} 2) - 1] + [\text{cosec}^2(\text{cosec}^{-1} 3) - 1]$$

$$= [\{\sec (\sec^{-1} 2)\}^2 - 1][\text{cosec} (\text{cosec}^{-1} 3\}^2 - 1]$$

$$= \{2^2 - 1\} + \{3^2 - 1\}$$

$$= 3 + 8 = 11$$

4. (a) We have,

$\sin\left[\frac{1}{2} \cos^{-1}\left(\frac{4}{5}\right)\right]$

Let $\frac{1}{2} \cos^{-1}\left(\frac{4}{5}\right) = x$

$\Rightarrow \quad \cos 2x = \frac{4}{5}$

$\Rightarrow \quad 1 - 2 \sin^2 x = \frac{4}{5}$

$\Rightarrow \quad 2 \sin^2 x = 1 - \frac{4}{5}$

$\Rightarrow \quad \sin^2 x = \frac{1}{10}$

$\Rightarrow \quad \sin x = \frac{1}{\sqrt{10}}$

$\Rightarrow \quad \sin\left[\frac{1}{2} \cos^{-1}\left(\frac{4}{5}\right)\right] = \frac{1}{\sqrt{10}} \qquad \left[\because x = \frac{1}{2} \cos^{-1}\left(\frac{4}{5}\right)\right]$

5. (b) Given, $\sin^{-1} x + \sin^{-1} y = \frac{2\pi}{3}$

$\Rightarrow \quad \frac{\pi}{2} - \cos^{-1} x + \frac{\pi}{2} - \cos^{-1} y = \frac{2\pi}{3}$

$\Rightarrow \quad \cos^{-1} x + \cos^{-1} y = \pi - \frac{2\pi}{3} = \frac{\pi}{3}$

6. (d) We have,

$$2 \tan^{-1} (\cos x) = \tan^{-1} (\text{cosec}^2 x)$$

$\Rightarrow \quad \tan [2 \tan^{-1}(\cos x)] = \text{cosec}^2 x$

$\Rightarrow \quad \tan\left(\tan^{-1} \frac{2 \cos x}{1 - \cos^2 x}\right) = \text{cosec}^2 x$

$\Rightarrow \quad \frac{2 \cos x}{\sin^2 x} = \text{cosec}^2 x$

$\Rightarrow \quad 2 \cos x = 1 \qquad \Rightarrow \cos x = \frac{1}{2}$

$\Rightarrow \quad x = \cos^{-1}\left(\frac{1}{2}\right)$

$\therefore \quad x = \frac{\pi}{3}$

7. (c) We have, $\tan^{-1} (x - 1) + \tan^{-1} x + \tan^{-1}(x + 1) = \tan^{-1}(3x)$

$\Rightarrow \tan^{-1}(x - 1) + \tan^{-1} (x + 1) + \tan^{-1} x = \tan^{-1}(3x)$

$\Rightarrow \quad \tan^{-1}\left(\frac{x - 1 + x + 1}{1 - x^2 + 1}\right) + \tan^{-1} x = \tan^{-1}(3x)$

$\Rightarrow \quad \tan^{-1}\left(\frac{2x}{2 - x^2}\right) + \tan^{-1} x = \tan^{-1} 3x$

$\Rightarrow \quad \tan^{-1}\left(\frac{\frac{2x}{2 - x^2} + x}{1 - \frac{2x^2}{2 - x^2}}\right) = \tan^{-1} 3x$

$\Rightarrow \quad \frac{2x + 2x - x^3}{2 - x^2 - 2x^2} = 3x \quad \Rightarrow \quad \frac{4x - x^3}{2 - 3x^2} = 3x$

$\Rightarrow \quad 4x - x^3 = 6x - 9x^3$

$\Rightarrow \quad 8x^3 - 2x = 0$

$\Rightarrow \quad x (8x^2 - 2) = 0$

$\Rightarrow \quad x = 0 \text{ or } x^2 = \frac{2}{8} \Rightarrow x = 0 \text{ or } x = \pm\frac{1}{2}$

$\therefore$ Three solutions are possible.

8. (c) We know that,

$\sin^{-1} x$ and $\cos^{-1} x$ exist for $x \in [-1, 1]$

Now, $\quad \sin^{-1} x > \cos^{-1} x$

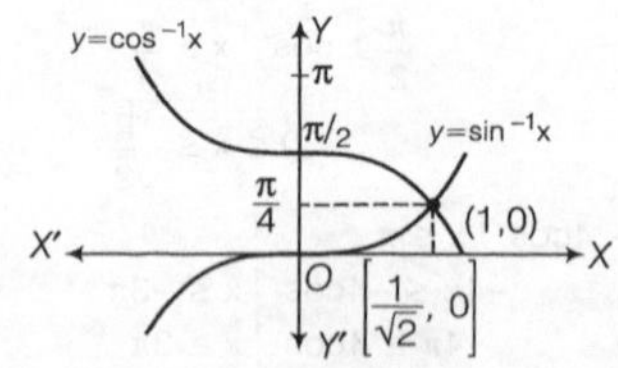

$\Rightarrow \quad \sin^{-1} x > \frac{\pi}{2} - \sin^{-1} x$

$\Rightarrow \quad 2 \sin^{-1} x > \frac{\pi}{2}$

$\Rightarrow \quad \sin^{-1} x > \frac{\pi}{4}$

$\Rightarrow \quad x > \sin \frac{\pi}{4} \Rightarrow x > \frac{1}{\sqrt{2}}$

$\Rightarrow \quad x \in \left[\frac{1}{\sqrt{2}}, 1\right] \qquad [\because -1 \le x \le 1]$

9. (c) We have, $3 \cos^{-1}\left(x^2 - 7x + \frac{25}{2}\right) = \pi$

$\Rightarrow \quad x^2 - 7x + \frac{25}{2} = \cos \frac{\pi}{3}$

$\Rightarrow \quad x^2 - 7x + \frac{25}{2} = \frac{1}{2}$

$\Rightarrow \quad x^2 - 7x + 12 = 0$

$\Rightarrow \quad x = \frac{7 \pm \sqrt{49 - 48}}{2} \Rightarrow x = \frac{7 \pm 1}{2}$

$\therefore \quad x = 4, 3$

10. (b) We have, $\sin^{-1} x + \cot^{-1} \frac{1}{2} = \frac{\pi}{2}$

$\sin^{-1} x = \frac{\pi}{2} - \cot^{-1} \frac{1}{2}$

$\Rightarrow \quad \sin^{-1} x = \tan^{-1}\left(\frac{1}{2}\right)$

$\Rightarrow \quad \sin^{-1} x = \sin^{-1}\left(\frac{1}{\sqrt{5}}\right) \Rightarrow x = \frac{1}{\sqrt{5}}$

11. (a) We have, $\sin^{-1} x + \sin^{-1} y + \sin^{-1} z = \dfrac{3\pi}{2}$

This is only possible when $\sin^{-1} x = \sin^{-1} y = \sin^{-1} z = \dfrac{\pi}{2}$

$\Rightarrow \quad x = y = z = 1$

Now, $x^{100} + y^{100} + z^{100} - \dfrac{9}{x^{100} + y^{100} + z^{100}}$

$= (1)^{100} + (1)^{100} + (1)^{100} - \dfrac{9}{(1)^{100} + (1)^{100} + (1)^{100}}$

$= 3 - \dfrac{9}{3} = 3 - 3 = 0$

12. (a) We have, $\tan^{-1} x + \cos^{-1} \dfrac{y}{\sqrt{1-y^2}} = \sin^{-1} \dfrac{3}{\sqrt{10}}$

$\Rightarrow \quad \tan^{-1} x + \tan^{-1}\left(\dfrac{1}{y}\right) = \tan^{-1}(3)$

$\Rightarrow \quad \tan^{-1}\left(\dfrac{1}{y}\right) = \tan^{-1} 3 - \tan^{-1} x$

$\Rightarrow \quad \tan^{-1}\left(\dfrac{1}{y}\right) = \tan^{-1}\left(\dfrac{3-x}{1+3x}\right) \quad \Rightarrow \quad y = \dfrac{1+3x}{3-x}$

As x, y are positive integers, $x = 1, 2$ and corresponding values of y are 2, 7.

$\therefore$ The solutions are $(x, y) = (1, 2)$ and $(2, 7)$

13. (a) We have,

$\cos^{-1}\left(\cos \dfrac{5\pi}{3}\right) + \sin^{-1}\left(\sin \dfrac{5\pi}{3}\right)$

$= \cos^{-1}\left[\cos\left(2\pi - \dfrac{\pi}{3}\right)\right] + \sin^{-1}\left[\sin\left(2\pi - \dfrac{\pi}{3}\right)\right]$

$= \cos^{-1}\left(\cos \dfrac{\pi}{3}\right) + \sin^{-1}\left[\sin\left(-\dfrac{\pi}{3}\right)\right]$

$= \dfrac{\pi}{3} + \left(-\dfrac{\pi}{3}\right) = 0$

14. (a) Given, $\cot^{-1}(\sqrt{\cos\alpha}) - \tan^{-1}(\sqrt{\cos\alpha}) = x \quad \ldots(i)$

We know that,

$\cot^{-1}(\sqrt{\cos\alpha}) + \tan^{-1}(\sqrt{\cos\alpha}) = \dfrac{\pi}{2}$

On adding Eqs. (i) and (ii), we get

$2\cot^{-1}(\sqrt{\cos\alpha}) = \dfrac{\pi}{2} + x$

$\Rightarrow \quad \sqrt{\cos\alpha} = \cot\left(\dfrac{\pi}{4} + \dfrac{\pi}{2}\right)$

$\Rightarrow \quad \sqrt{\cos\alpha} = \dfrac{\cot\dfrac{x}{2} - 1}{1 + \cot\dfrac{x}{2}}$

$\Rightarrow \quad \sqrt{\cos\alpha} = \dfrac{\cos\dfrac{x}{2} - \sin\dfrac{x}{2}}{\cos\dfrac{x}{2} + \sin\dfrac{x}{2}}$

On squaring both sides, we get

$\cos\alpha = \dfrac{\cos^2\dfrac{x}{2} + \sin^2\dfrac{x}{2} - 2\sin\dfrac{x}{2}\cos\dfrac{x}{2}}{\cos^2\dfrac{x}{2} + \sin^2\dfrac{x}{2} + 2\sin\dfrac{x}{2}\cos\dfrac{x}{2}}$

$\Rightarrow \quad \cos\alpha = \dfrac{1 - \sin x}{1 + \sin x}$

$\Rightarrow \quad \dfrac{1 - \tan^2\alpha}{1 + \tan^2\alpha} = \dfrac{1 - \sin x}{1 + \sin x}$

Applying componendo and dividendo, we get

$\sin x = \tan^2\left(\dfrac{\alpha}{2}\right)$

15. (c) Given, $\sin^{-1}\left(\tan\dfrac{\pi}{4}\right) - \sin^{-1}\left(\dfrac{\sqrt{3}}{x}\right) = \dfrac{\pi}{6}$

$\Rightarrow \quad \sin^{-1}(1) - \sin^{-1}\left(\dfrac{\sqrt{3}}{x}\right) = \dfrac{\pi}{6}$

$\Rightarrow \quad \dfrac{\pi}{2} - \sin^{-1}\left(\sqrt{\dfrac{3}{x}}\right) = \dfrac{\pi}{6}$

$\Rightarrow \quad \sin^{-1}\left(\dfrac{\sqrt{3}}{\sqrt{x}}\right) = \dfrac{\pi}{2} - \dfrac{\pi}{6}$

$\Rightarrow \quad \sin^{-1}\sqrt{\dfrac{3}{x}} = \dfrac{3\pi - \pi}{6}$

$\Rightarrow \quad \sin^{-1}\sqrt{\dfrac{3}{x}} = \dfrac{2\pi}{6}$

$\Rightarrow \quad \sin^{-1}\sqrt{\dfrac{3}{x}} = \dfrac{\pi}{3}$

$\Rightarrow \quad \sqrt{\dfrac{3}{x}} = \dfrac{\sqrt{3}}{2}$

$\Rightarrow \quad \dfrac{3}{x} = \dfrac{3}{4}$

$\therefore \quad x = 4$

16. (d) We have

$\sin^{-1}[\cos\{\cos^{-1}(\cos x) + \sin^{-1}(\sin x)\}]$

$= \sin^{-1}[\cos\{x + \pi - x\}]$

$= \sin^{-1}[\cos\pi]$

$= \sin^{-1}(-1) = -\dfrac{\pi}{2}$

17. (c) We have,

$\cos[\tan^{-1}(\tan 2)]$

$= \cos)(2 - \pi) = \cos 2 \quad [\because \tan^{-1}(\tan\theta) = \theta - \pi]$

18. (c) We have,

$\sin^{-1}(\sin 10) = \sin^{-1}[\sin(3\pi - 10)]$

$= 3\pi - 10$

19. (b) We have,

$\cos^{-1}(\cos 12) - \sin^{-1}(\sin 14)$

$= \cos^{-1}[\cos(4\pi - 12)] - \sin^{-1}[\sin(14 - 4\pi)]$

$= 4\pi - 12 - (14 - 4\pi)$

$= 4\pi - 12 - 14 + 4\pi$

$= 8\pi - 26$

20. (d) We have,

$\cos^{-1}\sqrt{\dfrac{2}{3}} - \cos^{-1}\left(\dfrac{\sqrt{6}+1}{2\sqrt{3}}\right)$

$= \tan^{-1}\dfrac{1}{\sqrt{2}} - \tan^{-1}\left(\dfrac{\sqrt{3}-\sqrt{2}}{1+\sqrt{6}}\right)$

$= \tan^{-1}\left(\dfrac{1}{\sqrt{2}}\right) - [\tan^{-1}\sqrt{3} - \tan^{-1}\sqrt{2}]$

$= \left(\tan^{-1}\dfrac{1}{\sqrt{2}} + \tan^{-1}\sqrt{2}\right) - \tan^{-1}\sqrt{3}$

$= \dfrac{\pi}{2} - \dfrac{\pi}{3} = \dfrac{\pi}{6}$

21. (c) We have, $\cos^{-1}[\cos\{2\cot^{-1}(\sqrt{2} - 1\}]$

$= \cos^{-1}\left[\cos\left(2 \times \dfrac{3\pi}{8}\right)\right] \quad \left[\because \cot\dfrac{3\pi}{8} = \sqrt{2} - 1\right]$

$= \cos^{-1}\left[\cos\left(\dfrac{3\pi}{4}\right)\right] = \dfrac{3\pi}{4}$

22. (c) Given, $x_1 = 2\tan^{-1}\left(\frac{1+x}{1-x}\right)$

and $x_2 = \sin^{-1}\left(\frac{1-x^2}{1+x^2}\right) = \tan^{-1}\left(\frac{1-x^2}{2x}\right)$

Now, $\frac{1+x}{1-x} > 1$

$$\Rightarrow \quad x_1 = \pi + \tan^{-1}\left(\frac{2\left(\frac{1+x}{1-x}\right)}{1-\left(\frac{1+x}{1-x}\right)^2}\right)$$

$$= \pi + \tan^{-1}\left(\frac{1-x^2}{-2x}\right)$$

$$= \pi - \tan^{-1}\left(\frac{1-x^2}{2x}\right)$$

$$x_1 + x_2 = \pi - \tan^{-1}\left(\frac{1-x^2}{2x}\right) + \sin^{-1}\left(\frac{1-x^2}{1+x^2}\right)$$

$$= \pi - \tan^{-1}\left(\frac{1-x^2}{2x}\right) + \tan^{-1}\left(\frac{1-x^2}{2x^2}\right)$$

$\therefore \quad x_1 + x_2 = \pi$

23. (d) We have,

$$\tan^{-1}\left(\frac{3\sin 2x}{5+3\cos 2x}\right) + \tan^{-1}\left(\frac{\tan x}{4}\right)$$

$$= \tan^{-1}\left(\frac{6\tan x}{8+2\tan^2 x}\right) + \tan^{-1}\left(\frac{\tan x}{4}\right)$$

$$= \tan^{-1}\left(\frac{\frac{3\tan x}{4+\tan^2 x} + \frac{\tan x}{4}}{1 - \frac{3\tan^2 x}{16+4\tan^2 x}}\right) \quad \left[\because \frac{3\tan^2\alpha}{16+4\tan^2\alpha} < 1\right]$$

$$= \tan^{-1}\left(\frac{12\tan x + 4\tan x + \tan^3 x}{16+\tan^2 x}\right)$$

$$= \tan^{-1}(\tan x) = x$$

24. (b) Let $\tan^{-1} x = A$, $\tan^{-1} y = B$ and $\tan^{-1} z = C$

$\Rightarrow \quad x = \tan A,\ y = \tan B,\ z = \tan C$

$\therefore \quad x + y + z = \tan A + \tan B + \tan C,$

since $\quad A + B + C = \pi$

$\Rightarrow \quad \tan A + \tan B + \tan C = \tan A \cdot \tan B \cdot \tan C$

$\therefore \quad x + y + z = \tan A \cdot \tan B \cdot \tan C$

$= xyz$

25. (a) Here,

$$\sin^{-1}(3x - 4x^3) = 3\sin^{-1} x - \pi$$

$\therefore \quad \frac{-\pi}{2} \le 3\sin^{-1} x - \pi \le \frac{\pi}{2}, \left\{\text{as, } \frac{-\pi}{2} \le \sin^{-1}(3x-4x^3) \le \frac{\pi}{2}\right\}$

$\Rightarrow \quad \frac{\pi}{2} \le 3\sin^{-1} x \le \frac{3\pi}{2}$

$\Rightarrow \quad \frac{\pi}{6} \le \sin^{-1} x \le \frac{\pi}{2}$

$\therefore \quad \frac{1}{2} \le x \le 1$

26. (a) Given, $\sin^{-1}(\sin 5) > x^2 - 4x$

$\Rightarrow \quad \sin^{-1}[\sin(5 - 2\pi)] > x^2 - 4x$

$\Rightarrow \quad 5 - 2\pi > x^2 - 4x$

$\Rightarrow \quad x^2 - 4x - 5 + 2\pi < 0$

$\Rightarrow \quad 2 - \sqrt{9 - 2\pi} < x < 2 + \sqrt{9 - 2\pi}$

$\Rightarrow \quad |x - 2| < \sqrt{9 - 2\pi}$

27. (b) We have,

$$\left(\sqrt{\frac{a\lambda}{bc}}\right)\left(\sqrt{\frac{b\lambda}{ca}}\right) = \frac{\lambda}{c} = \frac{a+b+c}{c} = 1 + \frac{b}{c} + \frac{a}{c} > 1$$

$$\Rightarrow \quad \left[\tan^{-1}\sqrt{\frac{a\lambda}{bc}} + \tan^{-1}\sqrt{\frac{b\lambda}{ca}}\right] + \tan^{-1}\left(\sqrt{\frac{c\lambda}{ab}}\right)$$

$$= \pi + \tan^{-1}\left(\frac{\sqrt{\frac{a\lambda}{bc}} + \sqrt{\frac{b\lambda}{ca}}}{1 - \sqrt{\frac{a\lambda}{bc}} \cdot \sqrt{\frac{b\lambda}{ca}}}\right) + \tan^{-1}\left(\frac{\sqrt{c\lambda}}{ab}\right)$$

$$= \pi + \tan^{-1}\frac{\sqrt{\frac{\lambda}{c}}\left[\frac{a+b}{\sqrt{ab}}\right]}{1 - \frac{\lambda}{c}} + \tan^{-1}\sqrt{\frac{c\lambda}{ab}}$$

$$= \pi + \tan^{-1}\left(\frac{\sqrt{\frac{\lambda c}{ab}}[a+b]}{c - \lambda}\right) + \tan^{-1}\left(\frac{\sqrt{c\lambda}}{ab}\right)$$

$$= \pi + \tan^{-1}\left(\frac{\sqrt{\frac{\lambda c}{ab}}(\lambda - c)}{c - \lambda}\right) + \tan^{-1}\sqrt{\frac{c\lambda}{ab}}$$

$$= \pi + \tan^{-1}\left(\frac{-\sqrt{\lambda c}}{ab}\right) + \tan^{-1}\left(\frac{\sqrt{c\lambda}}{ab}\right) = \pi$$

28. (a) Given, $\tan^2(\sin^{-1} x) > 1$

$\Rightarrow \quad |\tan(\sin^{-1} x)| > 1$

$\Rightarrow \quad |\sin^{-1} x| > \frac{\pi}{4}$

$\Rightarrow \quad |x| > \frac{1}{\sqrt{2}}$

$\Rightarrow \quad x \in \left(-1, -\frac{1}{2}\right) \cup \left(\frac{1}{\sqrt{2}}, 1\right)$

29. (b) We have, $\sin^{-1}[\cos\{\cos^{-1}(\cos x) + \sin^{-1}(\sin x)\}]$

$$\sin^{-1}\{\cos\{2\pi - x - 2\pi + x\}\} = \sin^{-1}[\cos(0)]$$

$$= \sin^{-1}(1) = \frac{\pi}{2}$$

30. (b) Since, α, β, γ are the roots of the equation

$$x^3 + mx^2 + 3x + m = 0$$

$\therefore \quad \alpha + \beta + \gamma = -m$

$\alpha\beta + \beta\gamma + \gamma\alpha = 3$

$\alpha\beta\gamma = -m$

Now, $\tan^{-1}(\alpha) + \tan^{-1}(\beta) + \tan^{-1}(\gamma)$

$$= \tan^{-1}\left(\frac{\alpha+\beta}{1-\alpha\beta}\right) + \tan^{-1}\gamma$$

$$= \tan^{-1}\left(\frac{\frac{\alpha+\beta}{1-\alpha\beta} + \gamma}{1 - \frac{\alpha+\beta}{1-\alpha\beta} \times \gamma}\right)$$

$$= \tan^{-1}\left(\frac{\frac{\alpha+\beta+\gamma-\alpha\beta\gamma}{1-\alpha\beta}}{\frac{1-\alpha\beta-\beta\gamma-\gamma\alpha}{1-\alpha\beta}}\right)$$

$$= \tan^{-1}\left(\frac{\alpha+\beta+\gamma-\alpha\beta\gamma}{1-(\alpha\beta+\beta\gamma+\gamma\alpha)}\right)$$

$$= \tan^{-1}\left[\frac{-m-(-m)}{1-3}\right] = \tan^{-1}\left(\frac{0}{-2}\right)$$

$$= \tan^{-1}(0) = n\pi$$

31. (c) Given, $\alpha = \sin^{-1}[\cos(\sin^{-1} x)]$

$$= \sin^{-1}\left[\cos(\cos^{-1}\sqrt{1-x^2})\right]$$

$$\alpha = \sin^{-1}\left(\sqrt{1-x^2}\right)$$

$$\sin\alpha = \sqrt{1-x^2}$$

$\Rightarrow$ $\cos\alpha = \sqrt{1-(1-x^2)} = x$

and $\beta = \cos^{-1}[\sin(\cos^{-1} x)]$

$$\beta = \cos^{-1}\left[(\sin(\sin^{-1}\sqrt{1-x^2})\right]$$

$$\beta = \cos^{-1}\sqrt{1-x^2}$$

$$\cos\beta = \sqrt{1-x^2}$$

Since, $\sin\alpha = \sqrt{1-x^2}$

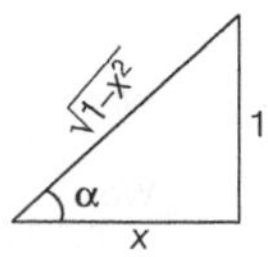

$$\tan\alpha = \frac{\sqrt{1-x^2}}{x}$$

$$\cos\beta = \sqrt{1-x^2} \Rightarrow \sin\beta = x$$

$$\tan\beta = \frac{x}{\sqrt{1-x^2}}, \cot\beta = \frac{\sqrt{1-x^2}}{x}$$

Hence, $\tan\alpha = \cot\beta$

32. (a) As $\frac{2e+4}{3} > e$

and $\log\left(\frac{2e+4}{3}\right) > 1$

$\Rightarrow$ $\cos^{-1}\left[\log\left(\frac{2e+4}{3}\right)\right]$ does not exists.

33. (a) Since, $\cot^{-1}(-x) = \pi - \cot^{-1} x$

$\therefore$ $\cot^{-1}(-1) = \pi - \cot^{-1}(1) = \pi - \frac{\pi}{4} = \frac{3\pi}{4}$

34. (c) **Statement I** $x + \frac{1}{x} = 2$

$$x^2 + 1 = 2x$$

$\Rightarrow$ $x^2 - 2x + 1 = 0$

$$x = 1$$

$\therefore$ $\sin^{-1}(1) = \frac{\pi}{2}$

Statement I is correct.

Statement II

$\sin^{-1}(\sin x) = x, x \in \left[\frac{-\pi}{2}, \frac{\pi}{2}\right]$

$\therefore$ Statement II is incorrect.

35. (b) $\sin^{-1} x + \sin^{-1} y = \pi$, only when $\sin^{-1} x = \sin^{-1} y = \frac{\pi}{2}$

$\therefore$ $x = y = 1$

Also, $\cos^{-1} x + \cos^{-1} y = 0$, only when $x = y = 1$.

$\therefore$ Only one solution, i.e. (1, 1).

It also satisfy $x^2 + y^2 = 2$.

$\Rightarrow$ Statement I is correct, Statement II is true but Statement II is not the correct explanation of Statement I.

36. (a) Given, $6x^2 + 11x + 3 = 0$

$\Rightarrow$ $6x^2 + 9x + 2x + 3 = 0$

$\Rightarrow$ $3x(2x+3) + 1(2x+3) = 0$

$\Rightarrow$ $(2x+3)(3x+1) = 0$

$\Rightarrow$ $x = -\frac{3}{2}, \frac{-1}{3}$

$\therefore$ $\beta = \frac{-3}{2}, \alpha = \frac{-1}{3}$

$$\frac{-1}{3} > \frac{-3}{2}$$

$\therefore$ $\cos^{-1}\left(\frac{-1}{3}\right)$ exists.

[since, domain of $\cos^{-1} x$ is $[-1,1]$]

37. (c) **Statement I**

$$\tan^2(\sec^{-1}2) + \cot^2(\text{cosec}^{-1}3)$$

$$= \sec^2(\sec^{-1}2) - 1 + \text{cosec}^2(\text{cosec}^{-1}3) - 1$$

$$= 2^2 + 3^2 - 2 = 13 - 2 = 11$$

$\therefore$ Statement I is correct.

Statement II

$\tan^2\theta + \sec^2\theta = 1 = \cot^2\theta + \text{cosec}^2\theta$

It is incorrect.

38. (b) **Statement I**

$$\tan^{-1}\left(\frac{a}{x}\right) + \tan^{-1}\left(\frac{b}{x}\right) = \frac{\pi}{2}$$

$\Rightarrow$ $\tan^{-1}\left(\frac{\frac{a}{x}+\frac{b}{x}}{1-\frac{ab}{x^2}}\right) = \frac{\pi}{2}$

$\Rightarrow$ $\frac{\frac{a}{x}+\frac{b}{x}}{1-\frac{ab}{x^2}} = \infty$

$\Rightarrow$ $1 - \frac{ab}{x^2} = 0$

$\Rightarrow$ $x^2 = ab$

$\therefore$ $x = \sqrt{ab}$

Statement II

$$\tan^{-1}\left(\frac{m}{n}\right) + \tan^{-1}\left(\frac{n-m}{n+m}\right)$$

$$\tan^{-1}\left(\frac{\frac{m}{n}+\frac{n-m}{n+m}}{1-\frac{m(n-m)}{n(n+m)}}\right)$$

$$\tan^{-1}\left(\frac{\frac{m}{n}+\frac{n-m}{n+m}}{\frac{n^2+mn-mn+m^2}{n(m+n)}}\right)$$

$\therefore$ $\tan^{-1}\left(\frac{m^2+n^2}{m^2+n^2}\right) = \tan^{-1}(1) = \frac{\pi}{4}$

39. (a) Maximum value of $(\sin x + \cos x)$ is $\sqrt{2}$ while $\sin^{-1} x + \cos^{-1} x = \frac{\pi}{2}$, $\forall$ $x \in [-1, 1]$. So, the equation $\sin x + \cos x = \sin^{-1} x + \cos^{-1} x$ has no solution.

40. (b) $\because \tan^{-1} x + \tan^{-1} y + \tan^{-1} z = \frac{\pi}{4}$

Let $x = \tan A, y = \tan B, z = \tan C$

$\therefore$ $\tan(A+B+C) = \frac{x+y+z-xyz}{1-(xy+yz+zx)} = 1$

$\Rightarrow$ $x + y + z - xyz = 1 - (xy + yz + zx)$

$\Rightarrow$ $(x-1)(y-1)(z-1) = 0$

One of x, y, z is equal to 1.

If $z = 1, x + y = 0$

$\therefore$ $(x)^{\text{odd}} + (-x)^{\text{odd}} + 1^{\text{odd}} = 1$

$\Rightarrow$ AM of odd powers of x, y, z is equal to $\left(\frac{1}{3}\right)$.

Revisal Problems (JEE Advanced)

1. (c) $\sqrt{(x^2+1)^2-4x^2} = \sqrt{(x^2-1)^2} = |x^2-1|$

$\therefore \qquad |\tan^{-1}|x|| = |x^2-1|$

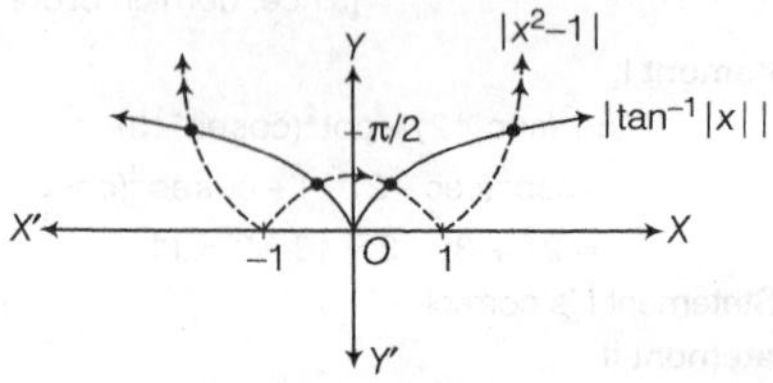

Clearly, the two graphs cut at 4 points.

$\therefore$ Number of solutions is 4.

2. (c) We have, $\sin^{-1}(\cos^{-1} x) < 1$

$\Rightarrow \qquad \cos^{-1} x < \sin(1)$

$\therefore \qquad x > \cos(\sin(1)) \qquad \ldots(i)$

Also, $\qquad \cos^{-1}(\cos^{-1} x) < 1$

$\Rightarrow \qquad \cos^{-1} x > \cos 1$

$\Rightarrow \qquad x < \cos(\cos 1) \qquad \ldots(ii)$

From Eqs. (i) and (ii), we get

$$\cos(\sin 1) < x < \cos(\cos 1)$$

3. (b) We have,

$$\tan\left[\sin^{-1}\left(\frac{x}{\sqrt{2}} + \sqrt{\frac{1-x^2}{2}}\right) - \sin^{-1} x\right]$$

$$= \tan\left[\sin^{-1}\frac{x+\sqrt{1-x^2}}{\sqrt{2}} - \sin^{-1} x\right]$$

Let $x = \sin\theta$, where $0 \le x \le \frac{1}{\sqrt{2}} \Rightarrow 0 \le \theta \le \frac{\pi}{4}$

$$= \tan\left[\sin^{-1}\left(\frac{\sin\theta+\cos\theta}{\sqrt{2}}\right) - \theta\right]$$

$$= \tan\left[\sin^{-1}\left\{\sin\left(\theta+\frac{\pi}{4}\right)\right\} - \theta\right]$$

$$= \tan\left[\theta + \frac{\pi}{4} - \theta\right] = \tan\frac{\pi}{4} = 1$$

4. (a) Let $f(x) = x^3 + bx^2 + cx + 1$

$f(0) = 1 > 0$ and $f(-1) = b - c < 0 \qquad [\because b < c]$

$\Rightarrow \alpha$ lies between -1 and 0.

$\Rightarrow \qquad \alpha < 0$

$\Rightarrow \qquad \tan^{-1}\left(\frac{1}{\alpha}\right) = -\pi + \cot^{-1}\alpha$

$\Rightarrow \qquad \tan^{-1}\alpha + \tan^{-1}\left(\frac{1}{\alpha}\right) = -\pi + \tan^{-1}\alpha + \cot^{-1}\alpha$

$\Rightarrow \qquad \tan^{-1}\alpha + \tan^{-1}\frac{1}{\alpha} = -\pi + \frac{\pi}{2} = -\frac{\pi}{2}$

5. (b) Given, $\cos^{-1}\left(\frac{x^2-1}{x^2+1}\right) + \tan^{-1}\left(\frac{2x}{x^2-1}\right) = \frac{2\pi}{3}, x > 1$

Put $x = \cot\theta$

$$\cos^{-1}\left(\frac{\cot^2\theta-1}{\cot^2\theta+1}\right) + \tan^{-1}\left(\frac{2\cot\theta}{\cot^2\theta-1}\right) = \frac{2\pi}{3}$$

$$\Rightarrow \cos^{-1}\left(\frac{1-\tan^2\theta}{1+\tan^2\theta}\right) + \tan^{-1}\left(\frac{2\tan\theta}{1-\tan^2\theta}\right) = \frac{2\pi}{3}$$

$\Rightarrow \qquad \cos^{-1}(\cos 2\theta) + \tan^{-1}(\tan 2\theta) = \frac{2\pi}{3}$

$\Rightarrow \qquad 2\theta + 2\theta = \frac{2\pi}{3}$

$\Rightarrow \qquad 4\theta = \frac{2\pi}{3}$

$\Rightarrow \qquad 4\cot^{-1} x = \frac{2\pi}{3}$

$\Rightarrow \qquad \cot^{-1} x = \frac{\pi}{6}$

$\therefore \qquad x = \cot\frac{\pi}{6} = \sqrt{3}$

6. (c) Given, $\sqrt{|\sin^{-1}|\cos x|| + |\cos^{-1}|\sin x||}$

$$= \sin^{-1}|\cos x| - \cos^{-1}|\sin x|$$

When $x \in \left[0, \frac{\pi}{2}\right], 0 \le \frac{\pi}{2} - x \le \frac{\pi}{2}$

We have,

$$\sqrt{\sin^{-1}\left[\sin\left(\frac{\pi}{2}-x\right)\right] + \cos^{-1}\left[\cos\left(\frac{\pi}{2}-x\right)\right]}$$

$$= \sin^{-1}\left[\sin\left(\frac{\pi}{2}-x\right)\right] - \cos^{-1}\left[\cos\left(\frac{\pi}{2}-x\right)\right]$$

$\Rightarrow \qquad \sqrt{2\left(\frac{\pi}{2}-x\right)} = \frac{\pi}{2} - x - \frac{\pi}{2} + x$

$\therefore \qquad x = \frac{\pi}{2}$

Similarly, one solution exists for each of the interval $(\pi, 2\pi)$, $(2\pi, 3\pi)$ and $(3\pi, 4\pi)$.

Hence, four solutions are possible.

7. (a) $\sin^{-1}\left|\frac{1+x^2}{2x}\right|$, exists only when $\left|\frac{1+x^2}{2x}\right| \le 1 \qquad \ldots(i)$

But $\qquad \left|\frac{1+x^2}{2x}\right| \ge 1, \forall x \in R - \{0\} \qquad \ldots(ii)$

From Eqs. (i) and (ii), we get

$$\frac{x^2+1}{2x} = \pm 1 \text{ or } x^2 + 1 = \pm 2x \text{ or } (x \pm 1)^2 = 0$$

$\therefore \qquad x = \pm 1$

But $x = 1$ only satisfy the equation.

$$\sin^{-1}\left(\frac{1+x^2}{2x}\right) = \frac{\pi}{2} \cdot \sec(x-1)$$

$\therefore$ Only one solution.

8. (a) Let $\sqrt{\tan\alpha} = \tan x$

Then, $\qquad u = \cot^{-1}(\tan x) - \tan^{-1}(\tan x)$

$$= \frac{\pi}{2} - x - x = \frac{\pi}{2} - 2x$$

$\Rightarrow \qquad 2x = \frac{\pi}{2} - u \Rightarrow \frac{\pi}{4} - \frac{u}{2} = x$

$\Rightarrow \qquad \tan x = \tan\left(\frac{\pi}{4} - \frac{u}{2}\right)$

$\Rightarrow \qquad \sqrt{\tan\alpha} = \tan\left(\frac{\pi}{4} - \frac{u}{2}\right)$

9. (b) Given, $\sin^{-1} x + \cos^{-1}(1-x) = -\sin^{-1} x$

$\Rightarrow \qquad 2\sin^{-1} x + \cos^{-1}(1-x) = 0$

has domain, $\qquad -1 \le x \le 1$ and $-1 \le 1 - x \le 1$

or $\qquad (-1 \le x \le 1$ and $0 \le x \le 2)$

$\therefore \qquad 0 \le x \le 1$

For domain, $x \in [0, 1]$, $\sin^{-1} x \ge 0$ and $\cos^{-1}(1-x) \ge 0$

$\therefore \qquad 2\sin^{-1} x + \cos^{-1}(1-x) = 0$

only when $\quad 2\sin^{-1} x = 0$ and $\cos^{-1}(1-x) = 0$

$\Rightarrow \qquad x = 0$ is the only solution.

10. (c) Graph of $|y| = \sin x$ and $y = \cos^{-1}(\cos x)$ are shown in the figure given below.

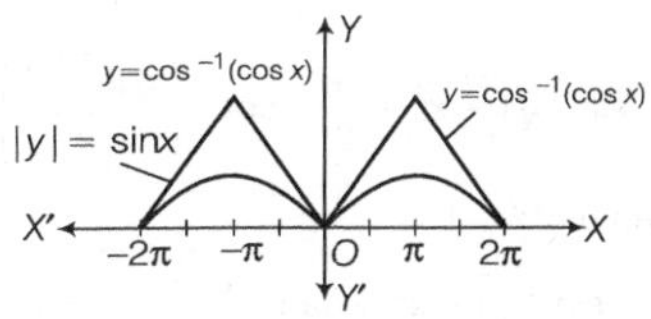

It is evident from the graph, that the curves intersect at (0, 0), $(2\pi, 0)$ and $(-2\pi, 0)$. Hence, there are three solutions.

11. (c) Given, $\cos^{-1}\sqrt{p} + \cos^{-1}\sqrt{1-p} + \cos^{-1}\sqrt{1-q} = \frac{3\pi}{4}$

Clearly, left hand side of the given equation is meaningfull, if $0 \le p \le 1$ and $0 \le q \le 1$. Let $p = \cos^2\theta$, then

$$\cos^{-1}\sqrt{p} + \cos^{-1}\sqrt{1-p} + \cos^{-1}\sqrt{1-q} = \frac{3\pi}{4}$$

$$\Rightarrow \cos^{-1}(\cos\theta) + \cos^{-1}(\sin\theta) + \cos^{-1}\sqrt{1-q} = \frac{3\pi}{4}$$

$$\Rightarrow \theta + \frac{\pi}{2} - \theta + \cos^{-1}\sqrt{1-q} = \frac{3\pi}{4}$$

$$\Rightarrow \cos^{-1}\sqrt{1-q} = \frac{\pi}{4}$$

$$\Rightarrow \sqrt{1-q} = \frac{1}{\sqrt{2}}$$

$$\Rightarrow 1 - q = \frac{1}{2} \Rightarrow q = \frac{1}{2}$$

Alternate Method

Let $\alpha = \cos^{-1}\sqrt{p}$, $\beta = \cos^{-1}\sqrt{1-p}$ and $\gamma = \cos^{-1}\sqrt{1-q}$

$\Rightarrow \cos\alpha = \sqrt{p}$, $\cos\beta = \sqrt{1-p}$ and $\cos\gamma = \sqrt{1-q}$

Therefore,

$\sin\alpha = \sqrt{1-p}$, $\sin\beta = \sqrt{p}$ and $\sin\gamma = \sqrt{q}$

The given equation becomes

$$\alpha + \beta + \gamma = \frac{3\pi}{4}$$

$$\Rightarrow \alpha + \beta = \frac{3\pi}{4} - \gamma$$

$$\Rightarrow \cos(\alpha + \beta) = \cos\left(\frac{3\pi}{4} - \gamma\right)$$

$$\Rightarrow \cos\alpha\cos\beta - \sin\alpha\sin\beta = \cos\left[\pi - \left(\frac{\pi}{4} + \gamma\right)\right]$$

$$= -\cos\left(\frac{\pi}{4} + \gamma\right)$$

$$\Rightarrow \sqrt{p}\sqrt{1-p} - \sqrt{1-p}\sqrt{p} = -\left(\frac{1}{\sqrt{2}}\sqrt{1-q} - \frac{1}{\sqrt{2}}\sqrt{q}\right)$$

$$\Rightarrow 0 = \sqrt{1-q} - \sqrt{q}$$

$$\Rightarrow 1 - q = q$$

$$q = \frac{1}{2} \text{ and } 0 \le p \le 1$$

12. (d) Given,

$$x^2 + ax + \sin^{-1}(x^2 - 4x + 5) + \cos^{-1}(x^2 - 4x + 5) = 0$$

We have, $x^2 - 4x + 5 = (x-2)^2 + 1 \ge 1$

So, the given equation is meaningful for $x = 2$. Putting $x = 2$, we have

$$4 + 2a + \sin^{-1}(-1) + \cos^{-1}(-1) = 0$$

$$\Rightarrow 4 + 2a + \frac{-\pi}{2} + \pi = 0$$

$$\Rightarrow 2a = -\frac{\pi}{2} - 4$$

$$\therefore a = -\frac{\pi}{4} - 2$$

13. (d) We have, $\sin^{-1}(\sin 4) = \sin^{-1}[\sin(\pi - 4)] = \pi - 4$

$$\therefore x^2 - kx + \sin^{-1}(\sin 4) > 0, \forall x \in R$$

$$\Rightarrow x^2 - kx + (\pi - 4) > 0, \forall x \in R$$

$$\Rightarrow k^2 - 4(\pi - 4) < 0 \Rightarrow k^2 + 16 - 4\pi < 0$$

But $k^2 + 16 - 4\pi > 0, \forall k \in R$.

So, there is no value of k for which the given equation hold true.

14. (c) Here, $\cos^{-1}\left\{\frac{x^2}{2} + \sqrt{1-x^2}\sqrt{1-\frac{x^2}{4}}\right\} = \cos^{-1}\frac{x}{2} - \cos^{-1}x$

Only when $\frac{x}{2} \le x$ and $-1 \le x \le 1$

$\Rightarrow x \le 2x$ and $-1 \le x \le 1$

$\Rightarrow x \ge 0$ and $-1 \le x \le 1$

$\therefore 0 \le x \le 1$

15. (c) Given, $[\cot^{-1}x] + [\cos^{-1}x] = 0$

As $\cos^{-1}x, \cot^{-1}x \ge 0$

$\Rightarrow [\cot^{-1}x] = [\cos^{-1}x] = 0$

$[\cot^{-1}x] = 0$

$\Rightarrow x \in (\cot 1, \infty)$ ...(i)

$[\cos^{-1}x] = 0$

$\Rightarrow x \in (\cos 1, 1]$ ...(ii)

Hence, from Eqs. (i) and (ii), we get

$x \in (\cot 1, 1]$

16. (a) $[\cot^{-1}x]^2 - 6[\cot^{-1}x] + 9 \le 0$

Let $[\cot^{-1}x] = t$

$t^2 - 6t + 9 \le 0$

$\Rightarrow (t-3)^2 \le 0$

$\Rightarrow t = 3, 3$

$[\cot^{-1}x] = 3$

$\Rightarrow 3 \le \cot^{-1}x < \pi$

$\therefore x \in (-\infty, \cot 3]$

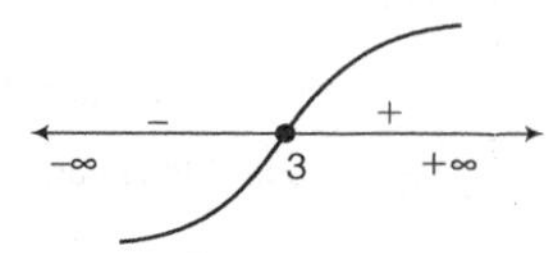

17. (a) As $\tan^{-1}x + \cot^{-1}x = \frac{\pi}{2}, \forall x \in R$

So, the given equation can be written as

$$(\sin^{-1}x)^3 + (\cos^{-1}x)^3 = 7\left(\frac{\pi^3}{8}\right)$$

$$\Rightarrow (\sin^{-1}x + \cos^{-1}x)^3 - 3(\sin^{-1}x)(\cos^{-1}x)$$

$$(\sin^{-1}x + \cos^{-1}x) = 7\left(\frac{\pi}{2}\right)^3$$

$$\Rightarrow \left(\frac{\pi}{2}\right)^3 - 3(\sin^{-1}x)(\cos^{-1}x)\left(\frac{\pi}{2}\right) = 7\left(\frac{\pi}{2}\right)^3$$

$$\Rightarrow (\sin^{-1}x)(\cos^{-1}x) = -\frac{\pi^2}{2}$$

Now, since maximum value of $\cos^{-1}x$ is π.

$\therefore \cos^{-1}x \ge 0$

and minimum value of $\sin^{-1}x$ is $\frac{-\pi}{2}$.

This happen at the same value of x,

i.e. $x = -1$

So, the minimum value of $(\sin^{-1}x)(\cos^{-1}x)$

$$= \left(-\frac{\pi}{2}\right)(\pi) = -\frac{\pi^2}{2}$$

So, if $(\sin^{-1}x)(\cos^{-1}x) = -\frac{\pi^2}{2}$, then $x = -1$.

18. (b) Given, $(\cot^{-1} x)(\tan^{-1} x) + \left(2 - \frac{\pi}{2}\right)\cot^{-1} x$

$$- 3\tan^{-1} x - 3\left(2 - \frac{\pi}{2}\right) > 0$$

$\Rightarrow \quad \cot^{-1} x > 0$

$\Rightarrow \quad (\cot^{-1} x - 3)(2 - \cot^{-1} x) > 0$

$\Rightarrow \quad (\cot^{-1} x - 3)(\cot^{-1} x - 2) < 0$

$\Rightarrow \quad 2 < \cot^{-1} x < 3$

$\Rightarrow \quad \cot 3 < x < \cot 2$

[as $\cot^{-1} x$ is a decreasing function]

Hence, $x \in (\cot 3, \cot 2)$.

19. (a) Given, $\quad 2\cos^{-1} x = \cot^{-1}\left(\frac{2x^2 - 1}{2x\sqrt{1 - x^2}}\right)$

Put $x = \cos\theta$

LHS $= 2\theta$; $0 \le \theta \le \pi$ and $-1 \le x \le 1$ ...(i)

RHS $= \cot^{-1}\left(\frac{\cos 2\theta}{2\cos\theta|\sin\theta|}\right) = \cot^{-1}(\cot 2\theta)$

If $\quad 0 < 2\theta < \pi$ ...(ii)

From Eqs. (i) and (ii), we get

$$0 < \theta < \frac{\pi}{2}$$

$\therefore \quad x \in (0, 1)$

20. (b) Given, $\sin[\cos^{-1}\{\cos(\tan^{-1} x)\}] = p$

For $x \in R$, $\quad \tan^{-1} x \in \left(-\frac{\pi}{2}, \frac{\pi}{2}\right)$

$$\cos(\tan^{-1} x) \in (0, 1]$$

$$\cos^{-1}[\cos(\tan^{-1} x)] \in \left[0, \frac{\pi}{2}\right)$$

$$\sin[\cos^{-1}\{\cos(\tan^{-1} x)\}] \in [0, 1)$$

21. (b) Given, $\quad \sin^{-1}\left(\frac{5}{x}\right) + \sin^{-1}\left(\frac{12}{x}\right) = \frac{\pi}{2}$

Put $\quad \sin^{-1}\frac{5}{x} = A \quad \Rightarrow \quad \frac{5}{x} = \sin A$

$\sin^{-1}\frac{12}{x} = B \quad \Rightarrow \quad \frac{12}{x} = \sin B$

$\Rightarrow \quad A + B = \frac{\pi}{2} \quad \Rightarrow \quad \sin A = \sin\left(\frac{\pi}{2} - B\right)$

$\sin A = \cos B$

$\Rightarrow \quad \frac{5}{x} = \sqrt{1 - \frac{144}{x^2}} \quad \Rightarrow \quad \frac{169}{x^2} = 1$

$\Rightarrow \quad x^2 = 169$

$\therefore \quad x = 13$

[$\because x = -13$ does not satisfy the given equation]

22. (a) Given, $\quad \sin^{-1} x = \cos^{-1} x + \sin^{-1}(3x - 2)$

$\Rightarrow \quad \sin^{-1} x - \cos^{-1} x = \sin^{-1}(3x - 2)$

$\Rightarrow \quad \frac{\pi}{2} - \cos^{-1} x - \cos^{-1} x = \frac{\pi}{2} - \cos^{-1}(3x - 2)$

$\Rightarrow \quad 2\cos^{-1} x = \cos^{-1}(3x - 2)$, also $x \in [-1, 1]$

$\Rightarrow \quad \cos^{-1}(2x^2 - 1) = \cos^{-1}(3x - 2)$

and $\quad 3x - 2 \in [-1, 1]$ i.e. $-1 \le 3x - 2 \le 1$

$\Rightarrow \quad 2x^2 - 1 = 3x - 2$

Hence, $x \in \left[\frac{1}{3}, 1\right]$

$\Rightarrow \quad 2x^2 - 3x + 1 = 0$

$\therefore \quad x = 1$ or $\frac{1}{2}$

23. (c) Given, $\sin^{-1}\sqrt{1 - x^2} + \cos^{-1} x = \cot^{-1}\left(\frac{\sqrt{1 - x^2}}{x}\right) - \sin^{-1} x$

$\Rightarrow \quad \sin^{-1}\sqrt{1 - x^2} + \frac{\pi}{2} = \cot^{-1}\left(\frac{\sqrt{1 - x^2}}{x}\right)$ ...(i)

Case I When $0 < x \le 1$

$$\sin^{-1}\sqrt{1 - x^2} + \tan^{-1}\frac{\sqrt{1 - x^2}}{x} = 0$$

$\Rightarrow \quad \sin^{-1}\sqrt{1 - x^2} + \sin^{-1}\sqrt{1 - x^2} = 0$

$\therefore \quad \sin^{-1}\sqrt{1 - x^2} = 0$

$x = \pm 1$ but $x > 0$

$\Rightarrow \quad x = 1$ ...(ii)

Case II When $-1 \le x < 0$

$$\sin^{-1}\sqrt{1 - x^2} + \frac{\pi}{2} = \pi - \tan^{-1}\left(\frac{x}{\sqrt{1 - x^2}}\right)$$

$\Rightarrow \quad \sin^{-1}\sqrt{1 - x^2} = \frac{\pi}{2} - \tan^{-1}\left(\frac{x}{\sqrt{1 - x^2}}\right)$

$\Rightarrow \quad \sin^{-1}\sqrt{1 - x^2} = \cot^{-1}\left(\frac{x}{\sqrt{1 - x^2}}\right)$

$\therefore \quad \sin^{-1}\sqrt{1 - x^2} = \sin^{-1}\sqrt{1 - x^2}$

Always true

$\therefore \quad x \in [-1, 0)$ ...(iii)

From Eqs. (ii) and (iii), we get

$$x \in [-1, 0) \cup \{1\}$$

24. (c) Given, $\quad f(x) = (1 + \sec^{-1} x)(1 + \cos^{-1} x)$

The given expression is defined only for $x = 1$ and $x = -1$.

$\because \quad f(1) = 1$ and $f(-1) = (1 + \pi)(1 + \pi) = (1 + \pi)^2$

Hence, range $= \{1, (1 + \pi)^2\}$

25. (c) Given,

$\cos^{-1} x + \cos^{-1} 2x + \cos^{-1} 3x = \pi$

$\Rightarrow \quad \cos^{-1} 2x + \cos^{-1} 3x = \pi - \cos^{-1} x$

$\Rightarrow \quad \cos^{-1} 2x + \cos^{-1} 3x = \cos^{-1}(-x)$

$\Rightarrow \quad \cos^{-1}[(2x)(3x) - \sqrt{1 - 4x^2}\sqrt{1 - 9x^2}] = \cos^{-1}(-x)$

$\Rightarrow \quad 6x^2 - \sqrt{1 - 4x^2}\sqrt{1 - 9x^2} = -x$

$\Rightarrow \quad (6x^2 + x)^2 = (1 - 4x^2)(1 - 9x^2)$

$\Rightarrow \quad x^2 + 12x^3 = 1 - 13x^2$

$\Rightarrow \quad 12x^3 + 14x^2 - 1 = 0$

$\Rightarrow \quad a = 12, b = 14$, and $c = 0$

$\therefore \quad a + b + c = 12 + 14 + 0 = 26$

26. (a) Given,

$$\sin^{-1}\left[\cos\left(\frac{2x^2 + 10|x| + 4}{x^2 + 5|x| + 3}\right)\right] = \cot\left(\cot^{-1}\frac{2 - 8|x|}{9|x|}\right) + \frac{\pi}{2}$$

$\Rightarrow \quad \frac{\pi}{2} - \cos^{-1}\left(\cos\left(\frac{2(x^2 + 5|x| + 3) - 2}{x^2 + 5|x| + 3}\right)\right)$

$$= \cot\left(\cot^{-1}\left(\frac{2}{9|x|} - 2\right)\right) + \frac{\pi}{2}$$

$\Rightarrow \quad \frac{\pi}{2} - 2 + \frac{2}{x^2 + 5|x| + 3} = \frac{2}{9|x|} - 2 + \frac{\pi}{2}$

$\Rightarrow \quad |x|^2 - 4|x| + 3 = 0 \quad \Rightarrow \quad |x| = 1, 3$

$\Rightarrow \quad x = \pm 1, \pm 3$

$\therefore \quad$ Product $= (1)(3)(-1)(-3) = 9$

27. (a) Domain of f is $[-1, 1]$;

$f(x) = \sin x + \cos x + \tan x + \sin^{-1} x + \cos^{-1} x + \tan^{-1} x$

$$f'(x) = \cos x - \sin x + \sec^2 x + 0 + \frac{1}{1+x^2}$$

Here, $\sec^2 x \ge 1$ and $\frac{1}{1+x^2} \in \left[\frac{1}{2}, 1\right]$

Hence, $f'(x) > 0 \Rightarrow f$ is increasing.

$\Rightarrow$ Range is $[f(-1), f(1)]$

$$\therefore \quad [f(x)]_{\min} = f(-1) = -\sin 1 + \cos 1 - \tan 1 - \frac{\pi}{2} + \pi - \frac{\pi}{4}$$

$$= \frac{\pi}{4} + \cos 1 - \sin 1 - \tan 1$$

$$[f(x)]_{\max} = f(1) = \sin 1 + \cos 1 + \tan 1 + \frac{\pi}{2} + \frac{\pi}{4}$$

$$= \frac{3\pi}{4} + \cos 1 + \sin 1 + \tan 1$$

$$= \frac{M+m}{2} = \frac{\pi}{2} + \cos 1$$

28. (c) Given, $\cos(\tan^{-1} x) = x$

Let $\tan^{-1} x = \theta \Rightarrow x = \tan\theta$

$$\Rightarrow \quad \cos\theta = x \quad \Rightarrow \quad \frac{1}{\sqrt{1+x^2}} = x$$

$$\Rightarrow \quad x^2(1+x^2) = 1 \quad \Rightarrow \quad x^2 = \frac{-1 \pm \sqrt{5}}{2}$$

$$\Rightarrow \quad x^2 = \frac{\sqrt{5}-1}{2} \quad \Rightarrow \quad \frac{x^2}{2} = \frac{\sqrt{5}-1}{4}$$

Now, $\cos^{-1}\left(\frac{x^2}{2}\right) = \cos^{-1}\left(\frac{\sqrt{5}-1}{4}\right)$

$$= \cos^{-1}\left(\sin\frac{\pi}{10}\right) = \cos^{-1}\left(\sin\frac{2\pi}{5}\right) = \frac{2\pi}{5}$$

29. (a,c) Given, $\cos^{-1} x = \tan^{-1} x$

$$\Rightarrow \quad \cos^{-1} x = \cos^{-1}\left(\frac{1}{\sqrt{1+x^2}}\right)$$

$$\therefore \quad x = \frac{1}{\sqrt{1+x^2}} \text{ or } x^2(1+x^2) = 1$$

$$\Rightarrow \quad x^4 + x^2 - 1 = 0$$

$$\therefore \quad x^2 = \frac{-1 \pm \sqrt{5}}{2} \Rightarrow x^2 = \frac{\sqrt{5}-1}{2},$$

neglecting $x^2 = \frac{-\sqrt{5}-1}{2}$

Also, $\sin(\cos^{-1} x) = \sin(\sin^{-1}\sqrt{1-x^2}) = \sqrt{1-x^2}$

$$= x^2 = \frac{\sqrt{5}-1}{2} \quad \left[\because 1 - x^2 = x^4 \; \therefore x^2 = \sqrt{1-x^2}\right]$$

30. (a,d) We have,

$$f(x) = \cos^{-1} x + \cos^{-1}\left\{\frac{x}{2} + \frac{\sqrt{3}}{2}\sqrt{1-x^2}\right\}$$

$$= \begin{cases} \cos^{-1} x + \cos^{-1} x - \cos^{-1}\frac{1}{2}, & \text{when } -1 \le x \le \frac{1}{2} \\ \cos^{-1} x + \cos^{-1}\frac{1}{2} - \cos^{-1} x, & \text{when } \frac{1}{2} \le x \le 1 \end{cases}$$

$$= \begin{cases} 2\cos^{-1} x - \frac{\pi}{3}, & \text{when } -1 \le x \le \frac{1}{2} \\ \frac{\pi}{3}, & \text{when } \frac{1}{2} \le x \le 1 \end{cases}$$

$$\therefore \quad f\left(\frac{2}{3}\right) = \frac{\pi}{3}$$

and $$f\left(\frac{1}{3}\right) = 2\cos^{-1}\left(\frac{1}{3}\right) - \frac{\pi}{3}$$

31. (a,b,c) $\because \cos(\cos^{-1} 1) = 1 \Rightarrow \cos(\cos(\cos^{-1} 1)) = \cos 1$

$$\sin^{-1}[\sin(\pi - 1)] = \pi - (\pi - 1) = 1$$

$$\Rightarrow \quad \sin[\sin^{-1}\sin(\pi - 1)] = \sin 1$$

$$\cos^{-1}[\cos(2\pi - 2)] = \cos^{-1}(\cos 2) = 2$$

$$\Rightarrow \quad \sin[\cos^{-1}(\cos(2\pi - 2))] = \sin 2$$

We know, $\cos 1 < \sin 1 < \sin 2$

$\cos^{-1}(\cos x)$ is periodic and even.

$$\cos^{-1}(\cos(2t\pi - 1)) = \cos^{-1}(\cos 1) = 1 \quad [t \in I]$$

$$\sum_{t=1}^{5000} \cos^{-1}[\cot(2t\pi - 1)] = 5000$$

Now, $\cot^{-1}[\cot(t\pi + 2)] = 2$

$$\therefore \quad \sum_{t=1}^{2500} \cot^{-1}[\cot(t\pi + 2)] = 5000$$

32. (a) Let $\tan^{-1}\frac{a}{x} = \alpha \Rightarrow \tan\alpha = \frac{a}{x}$

$$\therefore \quad \alpha + \beta + \gamma + \delta = \frac{\pi}{2}$$

$$\tan(\alpha + \beta + \gamma + \delta) = \tan\frac{\pi}{2}$$

$$\Rightarrow \quad \frac{S_2 - S_3}{1 - S_2 + S_4} = \infty$$

$$\Rightarrow \quad 1 - S_2 + S_4 = 0$$

Now, $S_4 = \tan\alpha \cdot \tan\beta \cdot \tan\gamma \cdot \tan\delta = \frac{abcd}{x^4}$

$$S_2 = \sum \tan\alpha \cdot \tan\beta = \frac{\sum ab}{x^2}$$

$$\therefore \quad \frac{abcd}{x^4} - \frac{\sum ab}{x^2} + 1 = 0$$

$\Rightarrow \quad x^4 - (\sum ab)x^2 + abcd = 0$, has roots x_1, x_2, x_3, x_4

$$\therefore \quad x_1 + x_2 + x_3 + x_4 = 0$$

$$\sum x_1x_2x_3 = 0$$

$$\sum x_1x_2x_3 = \underbrace{x_1x_2x_3x_4}_{\text{non-zero}}\left[\frac{1}{x_1} + \frac{1}{x_2} + \frac{1}{x_3} + \frac{1}{x_4}\right] = 0$$

$$\Rightarrow \quad \frac{1}{x_1} + \frac{1}{x_2} + \frac{1}{x_3} + \frac{1}{x_4} = 0$$

$$\Rightarrow \quad \sum_{i=1}^{4} \frac{1}{x_i} = 0$$

and $x_1x_2x_3x_4 = abcd$

or $(-x_2 - x_3 - x_4)(-x_1 - x_3 - x_4)(-x_1 - x_2 - x_4)(-x_1 - x_2 - x_3) = abcd$

or $(x_1 + x_2 + x_3)(x_2 + x_3 + x_4)(x_1 + x_3 + x_4)(x_1 + x_2 + x_4) = abcd$

33. (a) Given, $\sin^{-1}(ax) + \cos^{-1} y + \cos^{-1}(bxy) = \frac{\pi}{2}$

$$\Rightarrow \quad \cos^{-1} y + \cos^{-1} bxy = \frac{\pi}{2} - \sin^{-1} ax = \cos^{-1} ax$$

Let $\cos^{-1} y = \alpha$, $\cos^{-1}(bxy) = \beta$, $\cos^{-1}(ax) = \gamma$

$$\Rightarrow \quad y = \cos\alpha,\ bxy = \cos\beta,\ ax = \cos\gamma$$

$$\therefore \quad \alpha + \beta = \gamma$$

$$\Rightarrow \quad \cos(\gamma - \alpha) = bxy$$

$$\Rightarrow \quad \cos\gamma\cos\alpha + \sin\gamma\sin\alpha = bxy$$

$$\Rightarrow \quad (a - b)xy = -\sin\alpha\sin\gamma$$

$$\Rightarrow \quad (a - b)^2x^2y^2 = \sin^2\alpha\sin^2\gamma = (1 - \cos^2\alpha)(1 - \cos^2\gamma)$$

$$\Rightarrow \quad (a - b)^2x^2y^2 = (1 - a^2x^2)(1 - y^2) \quad \ldots(\text{i})$$

(a) For $a = 1$, $b = 0$, Eq. (i) reduces to

$$x^2y^2 = (1 - x^2)(1 - y^2)$$

$$\Rightarrow \quad x^2 + y^2 = 1$$

(b) For $a = 1, b = 1$, Eq. (i) becomes

$$(1 - x^2)(1 - y^2) = 0$$

$$\Rightarrow \quad (x^2 - 1)(y^2 - 1) = 0$$

(c) For $a = 1, b = 2$, Eq. (i) becomes

$$x^2y^2 = (1 - x^2)(1 - y^2)$$

$$\Rightarrow \quad x^2 + y^2 = 1$$

(d) For $a = 2, b = 2$, Eq. (i) reduces to

$$(1 - 4x^2)(1 - y^2)$$

$$\Rightarrow \quad (4x^2 - 1)(y^2 - 1) = 0$$

(i) → (p), (ii) → (q), (iii) → (p), (iv)→ (s)

34. (b)

35. (d)

(i) When $x \in (0, \cos 1)$, then $\sin^{-1} x \in \left(0, \frac{\pi}{2} - 1\right)$ and $\cos^{-1} x \in \left(1, \frac{\pi}{2}\right)$.

(ii) When $x \in \left(\cos 1, \frac{1}{\sqrt{2}}\right)$, then $\sin^{-1} x \in \left(\frac{\pi}{2} - 1, \frac{\pi}{4}\right)$ and $\cos^{-1} x \in \left(\frac{\pi}{4}, 1\right)$.

(iii) When $x \in \left(\frac{1}{\sqrt{2}}, \sin 1\right)$, then $\sin^{-1} x \in \left(\frac{1}{\sqrt{2}}, 1\right)$ and $\cos^{-1} x \in \left(\frac{\pi}{2} - 1, \frac{\pi}{4}\right)$.

(iv) When $x \in (\sin 1, 1)$, then $\sin^{-1} x \in \left(1, \frac{\pi}{2}\right)$ and $\cos^{-1} x \in \left(0, \frac{\pi}{2} - 1\right)$.

So, (i) → (q), (ii) → (s), (iii) → (p), (iv) → (p)

36. (c) Range for $\frac{1}{\pi^2}[(\cos^{-1} x)^2 - (\sin^{-1} x)^2] \in \left[\frac{-1}{4}, \frac{3}{4}\right] \therefore b - a = 1$

Range for $\frac{1}{\pi^2}[(\cos^{-1} x)^2 + (\sin^{-1} x)^2] \in \left[\frac{1}{8}, \frac{5}{4}\right] \therefore \; b - a = \frac{9}{8}$

Range for $\frac{1}{\pi^2}[(\cos^{-1} x)^3 - (\sin^{-1} x)^3] \in \left[\frac{-1}{8}, \frac{9}{8}\right] \therefore b - a = \frac{5}{4}$

Range for $\frac{1}{\pi^2}[(\cos^{-1} x)^3 + (\sin^{-1} x)^3] \in \left[\frac{1}{32}, \frac{7}{8}\right] \therefore b - a = \frac{27}{32}$

So, (i) → q, (ii) → r, (iii) → s, (iv) → p.

Explanations for Q. Nos. (37–39)

$f(x) = 3x^3 - 13x^2 + 14x - 2$ has roots α, β, γ. Also, $\alpha < \beta < \gamma$

Here, $f(0) = -1, f(1) = 2, f(2) = -2, f(3) = 4$

$$\therefore \quad f(0) \cdot f(1) = -2 \Rightarrow \alpha \in (0, 1)$$

$$f(1) \cdot f(2) = -4 \Rightarrow \beta \in (1, 2)$$

$$f(2) \cdot f(3) = -8 \Rightarrow \gamma \in (2, 3)$$

$$\therefore \quad [\alpha] = 0, [\beta] = 1, [\gamma] = 2$$

37. (a) $[\alpha]\,[\beta]\,[\gamma] = 0, 1, 2$ are in an AP.

38. (c) $\lim_{n \to \infty} (\alpha)^{n!} \cdot (\beta)^{\frac{1}{n}}$

as $n \to \infty, (\alpha)^{n!} \to 0$ and $(\beta)^{\frac{1}{n}} \to 1$

$$\therefore \quad \lim_{n \to \infty} (\alpha)^{n!} (\beta)^{\frac{1}{n}} \to 0$$

39. (d) $\because \tan^{-1} \alpha + \tan^{-1} \beta + \tan^{-1} \gamma = \pi + \tan^{-1}\left(\frac{\alpha + \beta + \gamma - \alpha\beta\gamma}{1 - (\alpha\beta + \beta\gamma + \gamma\alpha)}\right)$

$$= \pi + \tan^{-1}(-1)$$

$$= \pi - \frac{\pi}{4} = \frac{3\pi}{4}$$

40. (d) Let $\cos^{-1} x = \theta$

Then, $x = \cos\theta, 0 \le \theta \le \pi$

$$\therefore \quad \sin^{-1}\sqrt{1 - x^2} = \sin^{-1}(\sin\theta)$$

$$= \begin{cases} \theta, & \text{if } 0 \le \theta \le \frac{\pi}{2} \\ \pi - \theta, & \text{if } \frac{\pi}{2} < \theta \le \pi \end{cases}$$

$$= \begin{cases} \cos^{-1} x, & \text{if } 0 \le x \le 1 \\ \pi - \cos^{-1} x, & \text{if } -1 \le x < 0 \end{cases}$$

$\therefore \quad \cos^{-1} x = \sin^{-1}\sqrt{1 - x^2}$, if $0 < x < 1$ is true.

41. (c) Let $\sin^{-1} x = \theta$

Then, $x = \sin\theta$ and $\frac{-\pi}{2} \le \theta \le \frac{\pi}{2}$

$$\therefore \quad \cos^{-1}\sqrt{1 - x^2} = \cos^{-1}(\cos\theta)$$

$$= \begin{cases} -\theta, & \frac{\pi}{2} \le \theta \le 0 \\ \theta, & 0 \le \theta \le \frac{\pi}{2} \end{cases}$$

$$= \begin{cases} -\sin^{-1} x, & -1 \le x \le 0 \\ \sin^{-1} x, & 0 \le x \le 1 \end{cases}$$

42. (d) Let $\cos^{-1} x = \theta$, then $x = \cos\theta$ and $0 \le \theta \le \pi$.

$$\therefore \quad \tan^{-1}\frac{\sqrt{1 - x^2}}{x} = \tan^{-1}(\tan\theta)$$

$$= \begin{cases} \theta, & 0 \le \theta \le \frac{\pi}{2} \\ \theta - \pi, & \frac{\pi}{2} < \theta \le \pi \end{cases}$$

$$= \begin{cases} \cos^{-1} x, & 0 < x \le 1 \\ -\pi + \cos^{-1} x, & -1 \le x < 0 \end{cases}$$

i.e. $\cos^{-1} x = \pi + \tan^{-1}\left(\frac{\sqrt{1 - x^2}}{x}\right), -1 < x < 0$

43. (a) $\because a = 20 - 6\pi, b = 10\pi - 30, c = 10 - 3\pi$

So, $a + b + c = \pi$

Let $f(x) = \sin[2(a + b + c)x] + \cos^2[(a + b + c)x]$

$$= \sin(2\pi x) + \cos^2(\pi x)$$

$$= \sin(2\pi x) + \frac{1 + \cos 2\pi x}{2}$$

$\therefore \quad$ Maximum value of $f(x) = \frac{\sqrt{5} + 1}{2}$.

44. (c) We have,

$$\sin^{-1}(\sin x) \ge |x - \pi| \Rightarrow x \in \left[\frac{\pi}{2}, \pi\right]$$

So, largest integer $x = 3$.

45. (b) We have, $5\sec^{-1} x + 10\sin^{-1} y = 10\pi$

$$\Rightarrow \quad \sec^{-1} x = \pi \text{ and } \sin^{-1} y = \frac{\pi}{2}$$

$$\Rightarrow \quad x = -1 \text{ and } y = 1$$

So, $\tan^{-1}(-1) + \cos^{-1}(0) = \frac{-\pi}{4} + \frac{\pi}{2} = \frac{\pi}{4}$

46. (5) We have, $(n - 2)x^2 + 8x + (n + 4) > 0, \forall x \in R$

$$\Rightarrow \quad (n - 2) > 0 \Rightarrow n - 2 \ge 1$$

$$\Rightarrow \quad n \ge 3 \qquad \text{...(i)}$$

and $D = 64 - 4(n - 2)(n + 4) < 0$

$$\Rightarrow \quad n > 4 \text{ or } n \ge 5 \qquad \text{...(ii)}$$

From Eqs. (i) and (ii), $n = 5$

47. (0) $\sin^{-1} x \in \left[-\frac{\pi}{2}, \frac{\pi}{2}\right]$

$$\cos^{-1} y \in [0, \pi]$$

$$\sec^{-1} z \in \left[0, \frac{\pi}{2}\right) \cup \left(\frac{\pi}{2}, \pi\right]$$

$$\Rightarrow \sin^{-1} x + \cos^{-1} y + \sec^{-1} z \le \frac{\pi}{2} + \pi + \pi = \frac{5\pi}{2}$$

Also, $t^2 - \sqrt{2\pi}\, t + 3\pi = t^2 - 2\sqrt{\frac{\pi}{2}}\, t + \frac{\pi}{2} - \frac{\pi}{2} + 3\pi$

$$= \left(t - \sqrt{\frac{\pi}{2}}\right)^2 + \frac{5\pi}{2} \ge \frac{5\pi}{2}$$

The given inequation holds if equality holds, i.e. LHS = RHS = $\frac{5\pi}{2}$

$\Rightarrow$ $x = 1,\ y = -1,\ z = -1$ and $t = \sqrt{\frac{\pi}{2}}$

$\therefore$ $\tan^{-1} x + \tan^{-1} y + \tan^{-1} z + \tan^{-1}\left(\sqrt{\frac{2}{\pi}}\, t\right)$

$$= \tan^{-1}(1) + \tan^{-1}(-1) + \tan^{-1}(-1) + \tan^{-1}(1)$$

$$= \tan^{-1}(1) - \tan^{-1}(1) - \tan^{-1}(1) + \tan^{-1}(1) = 0$$

48. (1) Domain $\in [-1, 1]$

$$f_{\max}(x) = f(1) = \frac{1}{\pi}\left[\frac{\pi}{2} + \frac{\pi}{4}\right] + \frac{2}{13}$$

$$= \frac{47}{52} = M$$

$\therefore$ $\left[\frac{1}{M}\right] = 1$

49. (2) Domain = $\{-1, 1\}$

$\therefore$ Only 2 solutions.

50. (6) $\because \sin^{-1} x + \sin^{-1} y + \sin^{-1} z = \frac{3\pi}{2}$

$\therefore$ $x = y = z = 1$

$$A = x^2 + y^2 + z^2 = 3$$

$$(\sin^{-1} x)^2 + (\sin^{-1} y)^2 + (\sin^{-1} z)^2 = \frac{3\pi^2}{4}$$

If $x = y = z = -1$, $(x + y + z)$ is minimum.

$B = |(x + y + z)_{\min}| = 3,$

$\therefore$ $A + B = 6$

JEE Main & AIEEE Archive

1. (a) Since, x, y and z are in an AP.

$\therefore$ $2y = x + z$

Also, $\tan^{-1} x$, $\tan^{-1} y$ and $\tan^{-1} z$ are in an AP.

$\therefore$ $2\tan^{-1} y = \tan^{-1} x + \tan^{-1} z$

$$\Rightarrow \tan^{-1}\left(\frac{2y}{1 - y^2}\right) = \tan^{-1}\left(\frac{x + z}{1 - xz}\right)$$

$$\Rightarrow \frac{x + z}{1 - y^2} = \frac{x + z}{1 - xz}$$

$$\Rightarrow y^2 = xz$$

Since x, y and z are in an AP as well as in GP.

$\therefore$ $x = y = z$

2. (b) In difference series, whenever, we have summation involving more than 3 terms, we should always convert into differences.

e.g. $\sum_{r=1}^{5} \frac{1}{r(r+1)}$

$$= \frac{1}{1\cdot(2)} + \frac{1}{2\cdot(3)} + \frac{1}{3\cdot 4} + \frac{1}{4\cdot 5} + \frac{1}{5\cdot 6}$$

$$= \frac{2-1}{1\cdot(2)} + \frac{3-2}{2\cdot(3)} + \frac{4-3}{3\cdot(4)} + \frac{5-4}{4\cdot(5)} + \frac{6-5}{5\cdot(6)}$$

$$= \left(1 - \frac{1}{2}\right) + \left(\frac{1}{2} - \frac{1}{3}\right) + \left(\frac{1}{3} - \frac{1}{4}\right) + \left(\frac{1}{4} - \frac{1}{5}\right) + \left(\frac{1}{5} - \frac{1}{6}\right)$$

$$= 1 - \frac{1}{6} = \frac{5}{6}$$

Situation Analysis

Convert into $\tan^{-1} x - \tan^{-1} y = \tan^{-1}\left(\frac{x - y}{1 + xy}\right)$

$$\cot\left(\sum_{n=1}^{23} \cot^{-1}\left(1 + \sum_{k=1}^{n} 2k\right)\right)$$

$$\Rightarrow \cot\left(\sum_{n=1}^{23} \cot^{-1}(1 + 2 + 4 + 6 + 8 + \dots + 2n)\right)$$

$$\Rightarrow \cot\left(\sum_{n=1}^{23} \cot^{-1}(1 + n(n+1))\right)$$

$$\Rightarrow \cot\left(\sum_{n=1}^{23} \tan^{-1}\frac{1}{1 + n(n+1)}\right)$$

$$\Rightarrow \cot\left(\sum_{n=1}^{23} \tan^{-1}\left(\frac{(n+1) - n}{1 + n(n+1)}\right)\right)$$

$$\Rightarrow \cot\left\{\sum_{n=1}^{23} (\tan^{-1}(n+1) - \tan^{-1} n)\right\}$$

$$\Rightarrow \cot\{\tan^{-1} 2 - \tan^{-1} 1) + (\tan^{-1} 3 - \tan^{-1} 2) + (\tan^{-1} 4 - \tan^{-1} 3) + \dots + (\tan^{-1} 24 - \tan^{-1} 23)\}$$

$$\Rightarrow \cot\{\tan^{-1} 24 - \tan^{-1} 1\}$$

$$\Rightarrow \cot\left(\tan^{-1}\frac{24 - 1}{1 + 24\cdot(1)}\right) = \cot\left(\tan^{-1}\frac{23}{25}\right)$$

$$= \cot\left(\cot^{-1}\frac{25}{23}\right)$$

$$= \frac{25}{23}$$

3. (b) Since, $\operatorname{cosec}^{-1}\left(\frac{5}{3}\right) = \tan^{-1}\left(\frac{3}{4}\right)$

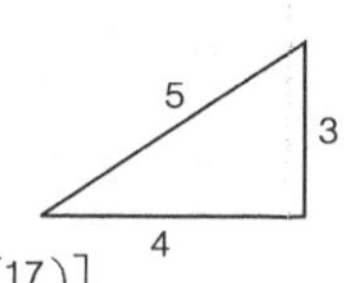

$\therefore$ $\cot\left(\tan^{-1}\frac{3}{4} + \tan^{-1}\frac{2}{3}\right)$

$$= \cot \tan^{-1}\left[\frac{\frac{3}{4} + \frac{2}{3}}{1 - \frac{1}{2}}\right] = \cot \tan^{-1}\left[\frac{\left(\frac{17}{12}\right)}{\left(\frac{1}{2}\right)}\right]$$

$$= \cot\left[\tan^{-1}\left(\frac{17}{6}\right)\right] = \frac{6}{17}$$

4. (b) $\because \sin^{-1}\left(\frac{x}{5}\right)+\text{cosec}^{-1}\left(\frac{5}{4}\right)=\frac{\pi}{2}$

$$\Rightarrow \sin^{-1}\left(\frac{x}{5}\right)+\sin^{-1}\left(\frac{4}{5}\right)=\frac{\pi}{2}$$

$$\Rightarrow \sin^{-1}\left(\frac{x}{5}\right)=\frac{\pi}{2}-\sin^{-1}\left(\frac{4}{5}\right)$$

$$\Rightarrow \sin^{-1}\left(\frac{x}{5}\right)=\cos^{-1}\left(\frac{4}{5}\right)$$

$$\Rightarrow \sin^{-1}\left(\frac{x}{5}\right)=\sin^{-1}\left(\frac{3}{5}\right)$$

$$\therefore \quad x=3$$

5. (b) Given, $\cos^{-1}x-\cos^{-1}\frac{y}{2}=\alpha$

$$\Rightarrow \cos^{-1}\left(\frac{xy}{2}+\sqrt{1-x^2}\sqrt{1-\frac{y^2}{4}}\right)=\alpha$$

$$\Rightarrow \frac{xy}{2}+\sqrt{1-x^2}\sqrt{1-\frac{y^2}{4}}=\cos\alpha$$

$$\Rightarrow 2\sqrt{1-x^2}\sqrt{1-\frac{y^2}{4}}=2\cos\alpha-xy$$

On squaring both sides, we get

$$\frac{4(1-x^2)(4-y^2)}{4}=4\cos^2\alpha+x^2y^2-4xy\cos\alpha$$

$$\Rightarrow 4-4x^2-y^2+x^2y^2=4\cos^2\alpha+x^2y^2-4xy\cos\alpha$$

$$\therefore 4x^2-4xy\cos\alpha+y^2=4\sin^2\alpha$$

6. (c) Given, $\sin^{-1}x=2\sin^{-1}a$

$$\because -\frac{\pi}{2}\le\sin^{-1}x\le\frac{\pi}{2}$$

$$\Rightarrow -\frac{\pi}{2}\le 2\sin^{-1}a\le\frac{\pi}{2}$$

$$\Rightarrow -\frac{\pi}{4}\le\sin^{-1}a\le\frac{\pi}{4}$$

$$\Rightarrow \sin\left(-\frac{\pi}{4}\right)\le a\le\sin\frac{\pi}{4}$$

$$\Rightarrow -\frac{1}{\sqrt{2}}\le a\le\frac{1}{\sqrt{2}}$$

$$\Rightarrow |a|\le\frac{1}{\sqrt{2}}$$

7. (a) Given,

$$\cot^{-1}(\sqrt{\cos\alpha})-\tan^{-1}(\sqrt{\cos\alpha})=x \quad \dots\text{(i)}$$

We know that,

$$\cot^{-1}(\sqrt{\cos\alpha})+\tan^{-1}(\sqrt{\cos\alpha})=\frac{\pi}{2} \quad \dots\text{(ii)}$$

$$\left[\because \cot^{-1}x+\tan^{-1}x=\frac{\pi}{2}\right]$$

On adding Eqs. (i) and (ii), we get

$$2\cot^{-1}(\sqrt{\cos\alpha})=\frac{\pi}{2}+x$$

$$\Rightarrow \sqrt{\cos\alpha}=\cot\left(\frac{\pi}{4}+\frac{x}{2}\right)$$

$$\Rightarrow \sqrt{\cos\alpha}=\frac{\cot\frac{x}{2}-1}{1+\cot\frac{x}{2}}$$

$$\Rightarrow \sqrt{\cos\alpha}=\frac{\cos\frac{x}{2}-\sin\frac{x}{2}}{\cos\frac{x}{2}+\sin\frac{x}{2}}$$

On squaring both sides, we get

$$\Rightarrow \cos\alpha=\frac{\cos^2\frac{x}{2}+\sin^2\frac{x}{2}-2\sin\frac{x}{2}\cos\frac{x}{2}}{\cos^2\frac{x}{2}+\sin^2\frac{x}{2}+2\sin\frac{x}{2}\cos\frac{x}{2}}$$

$$\Rightarrow \cos\alpha=\frac{1-\sin x}{1+\sin x}$$

$$\Rightarrow \frac{1-\tan^2\frac{\alpha}{2}}{1+\tan^2\frac{\alpha}{2}}=\frac{1-\sin x}{1+\sin x}$$

Applying componendo and dividendo rule, we get

$$\sin x=\tan^2\left(\frac{\alpha}{2}\right)$$

8. (d) $\tan^{-1}\left(\frac{1}{4}\right)+\tan^{-1}\left(\frac{2}{9}\right)=\tan^{-1}\left(\frac{\frac{1}{4}+\frac{2}{9}}{1-\frac{1}{4}\times\frac{2}{9}}\right)$

$$=\tan^{-1}\left(\frac{17}{34}\right)=\tan^{-1}\left(\frac{1}{2}\right)$$

JEE Advanced & IIT JEE Archive

1. (b,c,d) Given, $\alpha=3\sin^{-1}\left(\frac{6}{11}\right)$ and $\beta=3\cos^{-1}\left(\frac{4}{9}\right)$

$$\because \frac{6}{11}>\frac{1}{2}$$

$$\Rightarrow \sin^{-1}\left(\frac{6}{11}\right)>\sin^{-1}\left(\frac{1}{2}\right)$$

$$\Rightarrow 3\sin^{-1}\left(\frac{6}{11}\right)>3\cdot\frac{\pi}{6}=\frac{\pi}{2}$$

$$\therefore \alpha>\frac{\pi}{2} \quad \dots\text{(i)}$$

$$\Rightarrow \cos\alpha<0$$

Now, $\beta=3\cos^{-1}\left(\frac{4}{9}\right)$

$$\because \frac{4}{9}<\frac{1}{2}$$

$$\Rightarrow \cos^{-1}\left(\frac{4}{9}\right)>\cos^{-1}\left(\frac{1}{2}\right)$$

$$\Rightarrow 3\cos^{-1}\left(\frac{4}{9}\right)>3\cdot\frac{\pi}{3}$$

$$\Rightarrow \beta>\pi \quad \dots\text{(ii)}$$

$$\Rightarrow \cos\beta<0 \text{ and } \sin\beta<0$$

Since, α is slightly greater than $\frac{\pi}{2}$ and

β is slightly greater than π.

$$\Rightarrow \alpha+\beta>\frac{3\pi}{2}$$

$$\therefore \cos(\alpha+\beta)>0$$

2. (3) **Plan**

(i) Using definition of $f(x)=\cos^{-1}(x)$, we trace the curve $f(x)=\cos^{-1}(\cos x)$.

(ii) The number of solutions of equations involving trigonometric functions and algebraic functions, are found using graphs of the curves.

We know, $\cos^{-1}(\cos x) = \begin{cases} x, \text{ if } x \in [0, \pi] \\ 2\pi - x, \text{ if } x \in [\pi, 2\pi] \\ -2\pi + x, \text{ if } x \in [2\pi, 3\pi] \\ 4\pi - x, \text{ if } x \in [3\pi, 4\pi] \end{cases}$

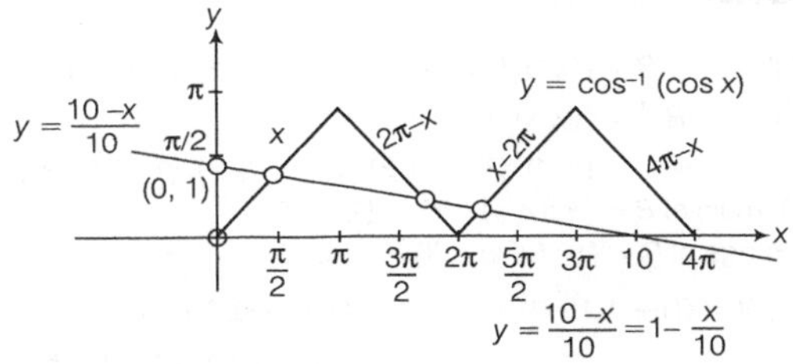

From above figure, it is clear that $y = \frac{10-x}{10}$ and $y = \cos^{-1}(\cos x)$ intersect at three distinct points, so number of solutions is 3.

3. (b)

(i) Here, innermost function is inverse

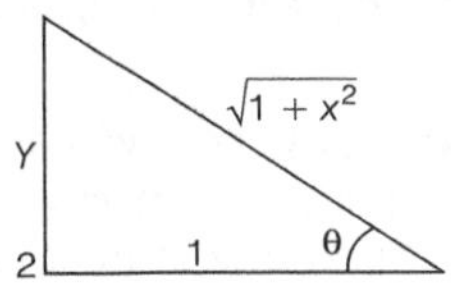

$\therefore$ Put $\tan^{-1} y = \theta$

$\Rightarrow \quad \tan\theta = y$

$$\Rightarrow \left[\frac{1}{y^2}\cdot\left(\frac{\cos(\tan^{-1} y) + y\sin(\tan^{-1} y)}{\cot(\sin^{-1} y) + \tan(\sin^{-1} y)}\right)^2 + y^4\right]^{1/2}$$

$$\Rightarrow \left[\frac{1}{y^2}\left\{\frac{\frac{1}{\sqrt{1+y^2}} + \frac{y^2}{\sqrt{1+y^2}}}{\frac{\sqrt{1-y^2}}{y} + \frac{y}{\sqrt{1-y^2}}}\right\}^2 + y^4\right]^{1/2}$$

$$\Rightarrow \left[\frac{1}{y^2}\cdot y^2(1-y^4) + y^4\right]^{1/2} = 1$$

(ii) Given, $\cos x + \cos y = -\cos z$

$\sin x + \sin y = -\sin z$

On squaring and adding, we get

$\cos^2 x + \sin^2 x + \cos^2 y + \sin^2 y + 2\cos x \cos y + 2\sin x \sin y = 1$

$\Rightarrow \quad 2 + 2(\cos(x-y)) = 1$

$\Rightarrow \quad \cos(x-y) = -\frac{1}{2}$

$\Rightarrow \quad 2\cos^2\left(\frac{x-y}{2}\right) - 1 = -\frac{1}{2}$

$\Rightarrow \quad 2\cos^2\left(\frac{x-y}{2}\right) = \frac{1}{2}$

$\Rightarrow \quad \cos\left(\frac{x-y}{2}\right) = \frac{1}{2}$

(iii) $\cos 2x \cdot \left(\cos\left(\frac{\pi}{4} - x\right) - \cos\left(\frac{\pi}{4} + x\right)\right) + 2\sin^2 x = 2\sin x \cdot \cos x$

$\Rightarrow \quad \cos 2x \cdot (\sqrt{2}\sin x) + 2\sin^2 x = 2\sin x \cdot \cos x$

$\Rightarrow \quad \sqrt{2}\sin x[\cos 2x + \sqrt{2}\sin x - \sqrt{2}\cos x] = 0$

$\Rightarrow \sin x = 0, (\cos x - \sin x)(\cos x + \sin x - \sqrt{2}) = 0$

$\Rightarrow \quad \sec x = 1 \text{ or } \tan x = 1$

$\Rightarrow \quad \sec x = 1 \text{ or } \frac{1}{\sqrt{2}}$

(iv) $\cot(\sin^{-1}\sqrt{1-x^2}) = \sin(\tan^{-1}(x\sqrt{6}))$

$\Rightarrow \quad \frac{x}{\sqrt{1-x^2}} = \frac{x\sqrt{6}}{\sqrt{1+6x^2}}$

$\Rightarrow \quad 1 + 6x^2 = 6 - 6x^2$

$\Rightarrow \quad 12x^2 = 5$

$x = \sqrt{\frac{5}{12}} = \frac{\sqrt{5}}{2\sqrt{3}}$

So, (i)→ (s), (ii) → (r), (iii) → (q,s), (iv) → (p)

4. (c) We have, $0 < x < 1$

Let $\cot^{-1} x = \theta$

$\Rightarrow \quad \cot\theta = x$

$\Rightarrow \quad \sin\theta = \frac{1}{\sqrt{1+x^2}}$

$= \sin(\cot^{-1} x)$

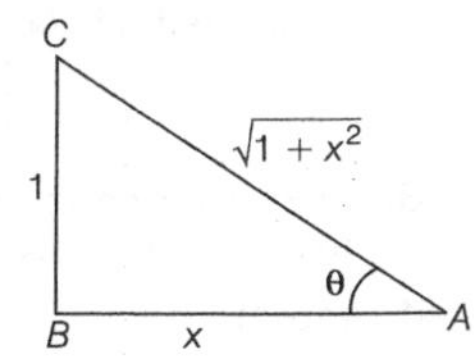

and $\quad \cos\theta = \frac{x}{\sqrt{1+x^2}}$

$= \cos(\cot^{-1} x)$

Now, $\sqrt{1+x^2}\,[\{x\cos(\cot^{-1} x) + \sin(\cot^{-1} x)\}^2 - 1]^{1/2}$

$$= \sqrt{1+x^2}\left[\left(x\frac{x}{\sqrt{1+x^2}} + \frac{1}{\sqrt{1+x^2}}\right)^2 - 1\right]^{1/2}$$

$$= \sqrt{1+x^2}\left[\left(\frac{1+x^2}{\sqrt{1+x^2}}\right)^2 - 1\right]^{1/2}$$

$= \sqrt{1+x^2}[1 + x^2 - 1]^{1/2}$

$= x\sqrt{1+x^2}$

5. (b)

(i) If $a = 1$, $b = 0$, then

$\sin^{-1} x + \cos^{-1} y = 0$

$\Rightarrow \quad \sin^{-1} x = -\cos^{-1} y$

$\Rightarrow \quad x^2 + y^2 = 1$

(ii) If $a = 1$ and $b = 1$, then

$\sin^{-1} x + \cos^{-1} y + \cos^{-1} xy = \frac{\pi}{2}$

$\Rightarrow \quad \cos^{-1} x - \cos^{-1} y = \cos^{-1} xy$

$\Rightarrow \quad xy + \sqrt{1-x^2}\sqrt{1-y^2} = xy$

$\Rightarrow \quad (x^2 - 1)(y^2 - 1) = 0$

(iii) If $a = 1$, $b = 2$, then

$\sin^{-1} x + \cos^{-1} y + \cos^{-1}(2xy) = \frac{\pi}{2}$

$\Rightarrow \quad \cos^{-1} x - \cos^{-1} y = \cos^{-1}(2xy)$

$\Rightarrow \quad xy + \sqrt{1-x^2}\sqrt{1-y^2} = 2xy$

$\Rightarrow \quad x^2 + y^2 = 1$

(iv) If $a = 2$ and $b = 2$, then

$\sin^{-1}(2x) + \cos^{-1}(y) + \cos^{-1}(2xy) = \frac{\pi}{2}$

$\Rightarrow \quad \cos^{-1}(2x) - \cos^{-1}(y) = \cos^{-1}(2xy)$

$\Rightarrow \quad 2xy + \sqrt{1-4x^2}\sqrt{1-y^2} = 2xy$

$\Rightarrow \quad (4x^2 - 1)(y^2 - 1) = 0$

(i)→ (p), (ii) → (q), (iii) → (p), (iv) → (s)

2. Relations and Functions

DPP-1 Relations

1. (b) $o(A \times B) = o(A) \times o(B) = mn$

$\Rightarrow \quad o[P(A \times B)] = 2^{mn}$

$\therefore$ Number of relations from A to B is 2^{mn}.

2. (a) $o(A) = n$. Since, R is reflexive, the ordered pairs (a, a), $a \in A$ are in R.

The ordered pairs (a, a), $a \in A$ are n in number.

$\therefore \quad \{(a, a) : a \in A\} \subseteq R$

$\therefore \quad n \le o(R)$ i.e. $m \ge n$

3. (a) By definition, of cartesian product and relation. If R is relation from a set P to set Q, then $R \subseteq P \times Q$.

4. (c) $o(A) = n$

$\Rightarrow \quad o(A \times A) = n \times n = n^2$

$\Rightarrow \quad o[P(A \times A)] = 2^{n^2}$

$\therefore$ Number of relations in $A = 2^{n^2}$

5. (a) $x \in R$

$\Rightarrow x - x + \sqrt{5} = \sqrt{5}$ is an irrational number.

$\therefore \quad (x, x) \in R \Rightarrow R$ is reflexive

$(\sqrt{5}, 1) \in R$ because $\sqrt{5} - 1 + \sqrt{5} = 2\sqrt{5} - 1$, which is an irrational number.

$\therefore \quad (1, \sqrt{5}) \notin R$

$\Rightarrow \quad R$ is not symmetric.

We have, $(\sqrt{5}, 1), (1, 2\sqrt{5}) \in R$ because $\sqrt{5} - 1 + \sqrt{5} = 2\sqrt{5} - 1$, if $1 - 2\sqrt{5} + \sqrt{5} = 1 - \sqrt{5}$ are irrational numbers.

Also, $(\sqrt{5}, 2\sqrt{5}) \in R$ and $\sqrt{5} - 2\sqrt{5} + \sqrt{5} = 0$, which is not an irrational number.

$\therefore \quad (\sqrt{5}, 2\sqrt{5}) \notin R$

$\therefore$ R is not transitive

6. (d) We have,

$$R = \{(x, y) : x, y \in N, x + 2y = 10, 1 \le x, y \le 10\}$$

$$x + 2y = 10$$

$\Rightarrow \quad y = \dfrac{10 - x}{2}$

$\therefore \quad R = \{(2, 4), (4, 3), (6, 2), (8, 1)\}$

Domain of $R = \{x : (x, y) \in R\} = \{2, 4, 6, 8\}$

Range of $R = \{y : (x, y) \in R\} = \{4, 3, 2, 1\}$

7. (c) We have, $R = [(2, 2), (3, 3), (4, 4), (1, 2)\}$

$1 \in A$ and $(1, 1) \notin R$

$\therefore$ R is not reflexive. $(1, 2) \in R$ but $(2, 1) \notin R$

$\therefore$ R is not symmetric

The relation R is transitive.

8. (d) $\quad x^2 + y^2 = 9$

$\Rightarrow \quad y = 9 - x^2$

$\Rightarrow \quad y = \pm\sqrt{9 - x^2}$

$\therefore \quad R = \{(0, 3), (0, -3), (3, 0), (-3, 0)\}$

Domain of $R = \{x : (x, y) \in R\} = \{0, 3, -3\}$

Range of $R = \{y : (x, y) \in R\} = \{3, -3, 0\}$

9. (d) $\quad R = \{(1, 1), (1, 2), (2, 1), (2, 2), (2, 3), (3, 2), (3, 3)\}$

Now, $R^{-1} = \{(y, x) : (x, y) \in R\}$

$= \{(1, 1), (2, 1), (1, 2), (2, 2), (3, 2), (2, 3), (3, 3)\} = R$

Domain of $R = \{x : (x, y) \in R\} = \{1, 2, 3\}$

Range of $R = \{y : (x, y) \in R\} = \{1, 2, 3\}$

10. (a) $R = \{(1 + 1, 1 + 1), (1 + 2, 1 + 4), (1 + 3, 1 + 9)$
$(1 + 4, 1 + 16), (1 + 5, 1 + 25)\}$

$= \{(2, 3), (3, 5), (4, 10), (5, 17), (6, 26)\}$

Domain of $R = \{x : (x, y) \in R\} = \{2, 3, 4, 5, 6\}$

Range of $R = \{y : (x, y) \in R\} = \{2, 5, 10, 17, 26\}$

11. (d) We have, $R = \{(1, 3), (2, 5) (3, 3)\}$

$\therefore \quad R^{-1}\{(y, x); (x, y) \in R\} = \{(3, 1), (5, 2), (3, 3)\}$

12. (d) We have,

$$R = \{(x, y) : |x^2 - y^2| < 16\}$$

$\therefore R = \{(1, 1), (1, 2), (1, 3), (1, 4), (2, 1), (2, 2), (2, 3),$
$(2, 4), (3, 1), (3, 2), (3, 3), (3, 4), (4, 1), (4, 2), (4, 3),$
$(4, 4), (4, 5), (5, 4), (5, 5)\}$

13. (b) $R = \{(x, y) : x + 2y = 8, x, y \in N\}$

$$x + 2y = 8$$

$\Rightarrow \quad y = \dfrac{8 - x}{2}$

$\therefore \quad R = \{(2, 3), (4, 2), (6, 1)\}$

$\therefore \quad$ Range of $R = \{y : (x, y) \in R\} = \{1, 2, 3\}$

14. (c) We have,

$$R = \{(x, y) : x, y \in Z, x^2 + y^2 \le 4\}$$

$\therefore \quad R = \{(0, 0), (0, -1), (0, 1), (0, -2), (0, 2), (-1, 0),$
$(1, 0), (1, 1), (1, -1), (-1, 1), (-1, -1), (2, 0), (-2, 0)\}$

$\therefore$ Domain of $R = \{x : (x, y) \in R\} = \{0, -1, 1, -2, 2\}$

15. (d) R is not symmetric as $(2, 3) \in R$ but $(3, 2) \notin R$.

16. (d) If $\phi \ne A \in X$, then $A \cap A \ne \phi$

$\therefore \quad A$ is not related to A.

$\therefore \quad R$ is not reflexive.

Let $A, B \in X$ and ARB

$\therefore \quad A \cap B \ne \phi\ B \cap A \ne \phi\backslash$ or BRA.

$\therefore \quad R$ is symmetric.

17. (d) $\quad |i| = \sqrt{0^2 + (1)^2} = \sqrt{1} = 1$

$\therefore \quad iR1$ is correct.

18. (d) Let R denotes the void relation.

Let $x \in A$

$\therefore \quad (x, x) \notin R$ because $R = \phi$

$\therefore \quad R$ is not reflexive,

The relation R is trivially symmetric and transitive.

19. (b) Since, $R \subseteq A \times B$ and $S \subseteq B \times C$

we have $\quad SoR \subseteq A \times C$.

$\therefore$ SoR is a relation from A to C.

20. (b) The relation $R \cup S$ is reflexive, symmetric and not necessarily transitive.

$\therefore$ $R \cup S$ is not equivalence relation but $R \cap S$ is an equivalence relation.

21. (b) We have, $R = \{(4, 5), (1, 4), (4, 6), (7, 6), (3, 7)\}$

$\therefore \quad R^{-1} = \{(5, 4), (4, 1), (6, 4), (6, 7), (7, 3)\}$

$\therefore \quad R^{-1}oR = \{(4, 4), (1, 1), (4, 7), (7, 4), (7, 7), (3, 3)\}$

22. (c) Let $R = \{a, b\} : a + b$ is an even integer $a, b \in Z\}$.

For, $a \in Z, a + a = 2a$ is an even integer.

$\therefore \quad (a, a) \in R, \forall\ a \in Z$

$\therefore$ R is reflexive.

Let $\quad (a, b) \in R \Rightarrow a + b$ is even integer.

$\Rightarrow \quad b + a$ is is even integer $\Rightarrow (b, a) \in R$

$\Rightarrow R$ is symmetric

Let $\quad (a, b), (b, c) \in R$

$\Rightarrow \quad a + b$ and $b + c$ are even integers.

$\Rightarrow \quad (a + b) + (b + c) = a + c + 2b$ is even integer.

$\Rightarrow \quad (a + c + 2b) - 2b = a + c$ is even integer.

$\Rightarrow \quad (a, c) \in R \Rightarrow R$ is transitive

$\therefore$ R is an equivalence relation.

Let $R = \{(a, b) : a - b$ is even integer, $a, b \in Z\}$

For $a \in Z, a - a = 0$ is an even integer.

$\therefore \quad (a - a) \in R\ \forall a \in Z$

$\therefore$ R is reflexive.

Let $(a, b) \in R$. $\Rightarrow a - b$ is even integer.

$\Rightarrow -(a - b)$ is even integer.

$\Rightarrow b - a$ is even integer. $\Rightarrow (b, a) \in R$

$\Rightarrow R$ is symmetric.

Let $(a, b), (b, c) \in R \Rightarrow a - b, b - c$are even integers.

$(a - b) + (b - c)$ is even integer.

$\Rightarrow \quad a - c$ is even integer.

$\Rightarrow \quad (a, c) \in R$

$\Rightarrow \quad R$ is transitive.

$\therefore \quad R$ is an equivalence relation.

Let $\quad R = \{(a, b) : a < b; a, b \in Z\}$

Let $\quad a \in Z, \Rightarrow a < a$ is false.

$\therefore$ R is not reflexive.

$\therefore$ R is not an equivalence relation.

Let $\quad R = \{(a, b) : a = b, a, b \in Z\}$.

It is quite easy to check that R is an equivalence relation.

23. (d) R_1 is not reflexive because $(q, q), (r, r) \notin R_1$.

$\therefore$ R_1 is not an equivalence relation.

R_2 is not reflexive because $(p, p) \notin R_2$.

$\therefore$ R_2 is not an equivalence relation

R_3 is reflexive because $(p, p,), (q, q), (r, r) \in R_3$.

R_3 is not symmetric because $(p, q), \in R_3$ but $(q, p) \notin R_3$.

24. (a) Let $(a, b), (b, c) \in (R \cup S)$.

It is possible that $(a, b) \in R - S$ and $(b, c) \in S - R$.

In such a case, we cannot say that

$(a, c) \in R$ or $(a, c) \in S$

$\therefore \quad (a, c)$ may not be in $R \cup S$.

$\therefore \quad R \cup S$ is not transitive.

(b) Let $(a, b), (b, c) \in R \cap S$.

$\therefore \quad (a, b), (b, c) \in R$ and $(a, b), (b, c) \in S$

$\therefore \quad (a, c) \in R$ and $(a, c) \in S$

$\therefore \quad (a, c) \in R \cap S$

$\therefore$ $R \cap S$ is transitive.

(c) Let $(a, b) \in R \cup S$

$\therefore \quad (a, b) \in R$ or $(a, b) \in S$

Now, $\quad (a, b) \in R \Rightarrow (b, a) \in R \quad (\because R$ is symmetric)

$(a, b) \in S \Rightarrow (b - a) \in S \quad (\because S$ is symmetric)

$\therefore \quad (b, a) \in R \cup S$

$\therefore$ $R \cup S$ is symmetric.

(d) Let $a \in A$.

$\therefore \quad (a, a) \in R$ and $(a, a) \in S$

$\therefore \quad (a, a) \in R \cap S$

$\therefore$ $R \cap S$ is reflexive.

25. (d) Let $(a, b), (c, d) \in N \times N$.

The relation R defined by $(a, b)\ R\ (c, d)$

$\Leftrightarrow a + d = b + c$ is an equivalence relation.

The relation R defined by $(a, b)\ R\ (c, d)$

$\Leftrightarrow ad = bc$ is an equivalence relation.

Let $R = \{[(a, b), (c, d)] : ad\ (b + c)$

$= bc\ (a + d), (a, b), (c, d) \in N \times N\}$

$$ad\ (b + c) = bc\ (a + d) \Leftrightarrow \frac{b + c}{bc} = \frac{a + d}{ad}$$

$$\Leftrightarrow \quad \frac{1}{c} + \frac{1}{b} + \frac{1}{d} + \frac{1}{a} \Leftrightarrow \frac{1}{a} - \frac{1}{b} = \frac{1}{c} - \frac{1}{d}$$

Let $(a, b) \in N \times N$

We have, $\quad \frac{1}{a} - \frac{1}{b} = \frac{1}{a} - \frac{1}{b}$

$\therefore \quad (a, b)\ R\ (a, b)$

$\therefore$ R is reflexive.

Let $\quad (a, b)\ R\ (c, d)$

$\therefore \quad \frac{1}{a} - \frac{1}{b} = \frac{1}{c} - \frac{1}{d}$

$\Rightarrow \quad \frac{1}{c} - \frac{1}{d} = \frac{1}{a} - \frac{1}{b}$

$\Rightarrow \quad (c, d)\ R\ (a, b)$

$\Rightarrow R$ is symmetric.

Let $\quad (a, b)\ R\ (c, d)$ and $(c, d)\ R\ (e, f)$

Let $\quad (a, b)\ R\ (c, d)$ and $(c, d)\ R\ (e, f)$

$\therefore \quad \frac{1}{a} - \frac{1}{b} = \frac{1}{c} - \frac{1}{d}$

and $\quad \frac{1}{c} - \frac{1}{d} = \frac{1}{e} - \frac{1}{f}$

$\Rightarrow \quad \frac{1}{a} - \frac{1}{b} = \frac{1}{e} - \frac{1}{f}$

$\Rightarrow \quad (a, b)\ R\ (e, f)$

$\Rightarrow R$ is transitive.

$\therefore$ R is equivalence relation.

Let $\quad R = \{[(a, b), (c, d)] : bc\ (b + c)$

$= ad\ (a + d), (a, b), (c, d) \in N \times N\}$.

Let $\quad (a, b) \in N \times N$

We have, $ba(b + a) = ab\ (a + b) \Rightarrow R$ is reflexive.

Let $(a, b)\ R\ (c, d)$

$\therefore \quad bc\ (b + c) = ad\ (a + d)$

$\Rightarrow \quad ad\ (a + d) = bc\ (b + c)$

$\Rightarrow \quad da(d + a) = cb\ (c + b)$

$\Rightarrow \quad (c, d)\ R\ (a, b)$

$\Rightarrow \quad R$ is symmetric

Let $(a, b)\ R\ (c, d)$ and $(c, d)\ R\ (e, f)$.

$\therefore \quad bc\ (b + c) = ad\ (a + d)$ and $de\ (d + e) = cf\ (c + f)$

These equations need no imply be $(b + e) = af\ (a + f)$.

$\therefore$ (a, b) may not be R-related to (c, f).

$\therefore$ R is not transitive.

$\therefore$ R is not an equivalence relation.

26. (c) SoR is relation from A to C.

$\therefore$ $(SoR)^{-1}$ is a relation from C to A.

R^{-1} is a relation from B to A.

S^{-1} i a relation from C to B.

$\therefore$ $R^{-1}oS^{-1}$ is a relation from C to A.

Let $(c, a) \in (SoR)^{-1}$

$\therefore$ $(a, c) \in SoR$

$\therefore$ $\exists\, b \in B : (a, b) \in R$ and $(b, c) \in S$

$\therefore$ $(b, a) \in R^{-1}$ and $(c, b) \in S^{-1}$

$\therefore$ $(c, a) \in R^{-1}oS^{-1}$

$\therefore$ $(SoR)^{-1} \subseteq R^{-1}oS^{-1}$

Conversely, let $(c, a) \in R^{-1}oS^{-1}$

$\therefore$ $\exists\, b \in B: (c, b) \in S^{-1}$ and $(b, a) \in R^{-1}$

$\therefore$ $(b, c) \in S$ and $(a, b) \in R$

$\Rightarrow$ $(b, c) \in SoR \Rightarrow (c, a) \in (SoR)^{-1}$

$\therefore$ $R^{-1}oS^{-1} \subseteq (SoR)^{-1}$

Combining, we get

$$(SoR)^{-1} = R^{-1}oS^{-1}$$

27. (c) Let R denotes the relation 'is brother of'. Let $x \in A$. If x is a girl, then we cannot say that x is brother of x.

$\therefore$ $(x, x) \notin R$

$\therefore$ R is not reflexive

Let $(x, y) \in R$

$\therefore$ x is a brother of y.

$\therefore$ y may or may not be a boy.

$\therefore$ We cannot say that $(y, x) \in R$.

$\therefore$ R is not symmetric.

Let $(x, y) \in R$ and $(y, z) \in R$

$\therefore$ x is brother of y and y is brother of z.

$\Rightarrow$ x is brother of $z \Rightarrow (x, z) \in R$

$\therefore$ R is transitive.

28. (d) For $a \in Z$, we have $a \cdot a = a^2 \geq 0$

$\therefore$ $a\,R\,a \,\forall\, a \in Z$

$\therefore$ R is reflexive.

Let $a\,R\,b$.

$\therefore$ $ab \geq 0$ or $ba \geq 0$ or $b\,R\,a$

$\therefore$ R is symmetric.

Let $a\,R\,b$ and $b\,R\,c$

$\therefore$ $ab \geq 0, bc \geq 0$

$\Rightarrow$ $(ab)(bc) \geq 0 \Rightarrow b^2(ac) \geq 0$

$\Rightarrow$ $ac \geq 0$ $(\because b^2 \geq 0)$

$\therefore$ R is transitive.

29. (b) We have, $R = \{(a, b) : a + b \in Z, a, b \in Z \cup \{\sqrt{2}\}$.

$\sqrt{2} \in Z$ and $\sqrt{2} + \sqrt{2} = 2\sqrt{2} \notin Z$

$\therefore$ $(\sqrt{2}, \sqrt{2}) \notin R$

$\therefore$ R is not reflexive.

Let $(a, b) \in R$

$\therefore$ $a + b \in Z$

$\Rightarrow$ $b + a \in Z \Rightarrow (b, a) \in R$

$\Rightarrow R$ is symmetric.

Let $(a, b), (b, c) \in R. \therefore a + b, b + c \in Z$

$\therefore$ None of a, b, c is equal to $\sqrt{2}$.

$\therefore$ $a, b, c \in Z$

$\therefore$ $a + c \in Z$

$\Rightarrow$ $(a, c) \in R \Rightarrow R$ is transitive.

$\therefore$ R is not an equivalence relation.

30. (d) If R be the relation, $x R y \Leftrightarrow x - y$ is divisible by m.

$x R x$ because $x - x$ is divisible by m.

So, R is reflexive.

$x R y \Rightarrow y R x$. So, R is symmetric.

$x R y$ and $y R z$

$\Rightarrow$ $x - y = k_1 m$

$y - z = k_2 m$

$\therefore x - z = (k_1 + k_2)m$. So, R is transitive.

As, R is reflexive, symmetric and transitive, it is an equivalence relation.

31. (b) As $(x, x) \in R \Rightarrow |x - x| \leq 1$

$\Rightarrow$ $0 \leq 1$ (true)

Thus, reflexive.

As, $(x, y) \in R$

$\Rightarrow$ $|x - y| \leq 1$

$\Rightarrow$ $|y - x| \leq 1$

$\Rightarrow$ $(y, x) \in R$

Thus, symmetric.

Again, $(x, y) \in R$ and $(y, z) \in R$

$\Rightarrow$ $|x - y| \leq 1$

and $|y - z| \leq 1 \not\Rightarrow |x - z| \leq 1$

$\therefore$ Not transitive.

32. (d) $(a, -b)\, r\, (a, b)$ because $a + b = b + a$. So, r is reflexive.

$(a, b)\, r\, (c, d)$

$\Rightarrow$ $a + d = b + c$

$\Rightarrow$ $c + b = d + a$

$\Rightarrow$ $(c, d)\, r\, (a, b)$

So, r is symmetric.

$(a, b)\, r\, (c, d)$ and $(c, d)\, r\, (e, f)$

$\Rightarrow$ $a + d = b + c, c + f = d + e$

On adding, $a + d + c + f = b + c + d + e$

$\Rightarrow$ $a + f = b + e$

$\Rightarrow$ $(a, b)\, r\, (e, f)$

$\therefore$ R is transitive.

33. (a) $l_1 R l_1 \Rightarrow l_1$ is not perpendicular to l_1.

$\therefore$ R is not reflexive.

$l_1 R l_2 \Rightarrow l_1 \perp l_2$

$\Rightarrow$ $l_2 \perp l_1 \Rightarrow l_2 R l_1$

Hence, R is symmetric.

$l_1 R l_2$ and $l_2 R l_3 \not\Rightarrow l_1 R l_3$

$\therefore$ R is not transitive.

34. (a) $m\,R\,m$, as m is multiple of m

$\therefore$ R is reflexive.

$mRn \not\Rightarrow nRm$

$\therefore$ $R$is not symmetric.

$m\,R\,n$

and $n\,R\,p \Rightarrow m\,R\,p_2$

$\therefore$ R is transitive.

Hence, reflexive and transitive.

35. (d) $R = \{(1, 1), (2, 2), (3, 3), (4, 4), (1, 2), (2, 1), (3, 1), (1, 3)\}$

As $(1, 1), (2, 2), (3, 3), (4, 4) \in R$

$\therefore$ R is reflexive.

$(1, 2)$ and $(2, 1) \in R$, also $(3, 1)$ and $(1, 3) \in R$.

$\therefore$ R is symmetric

Also, $(1, 2)$ and $(2, 1) \in R \Rightarrow (1, 1) \in R$

Similarly, for other elements, hence R is transitive.

$\therefore$ R is equivalence relation.

DPP-2 Definition of Functions

1. 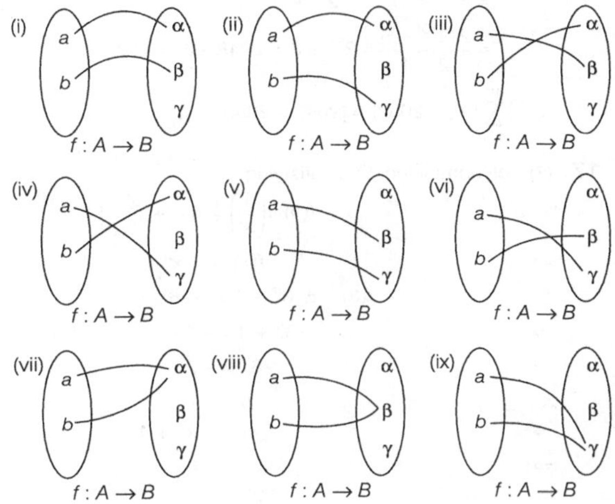

2. To check function draw straight line parallel to Y-axis, if it cuts atmost one point, then it is a function.

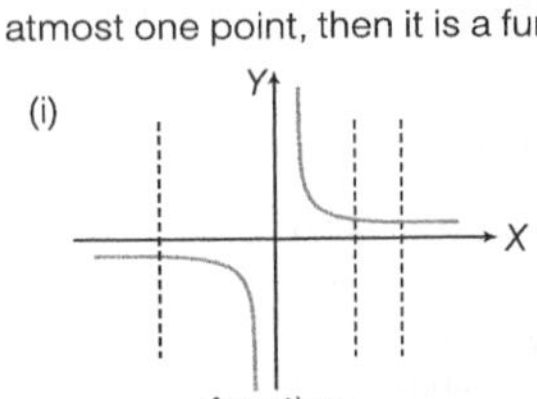

function

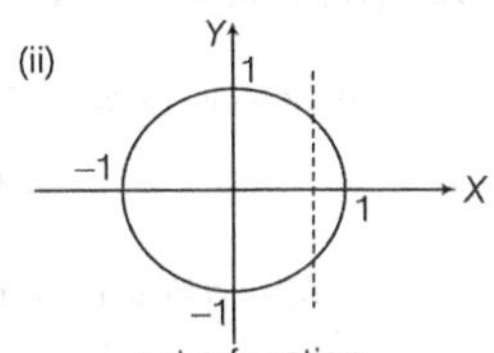

not a function

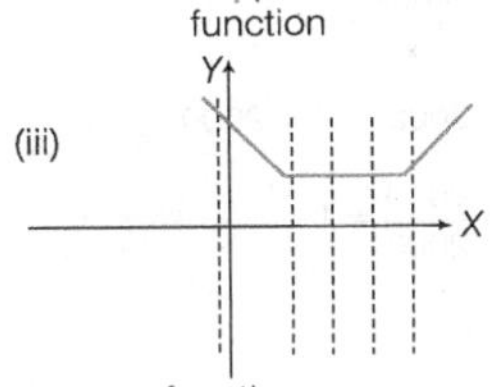

function

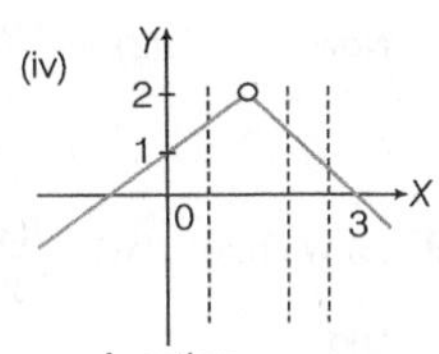

function

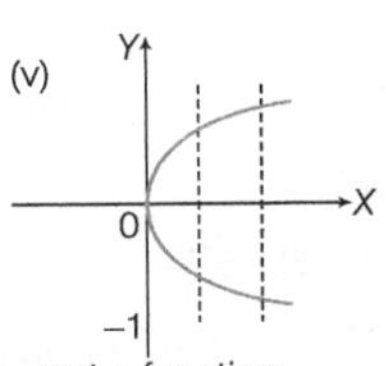

not a function

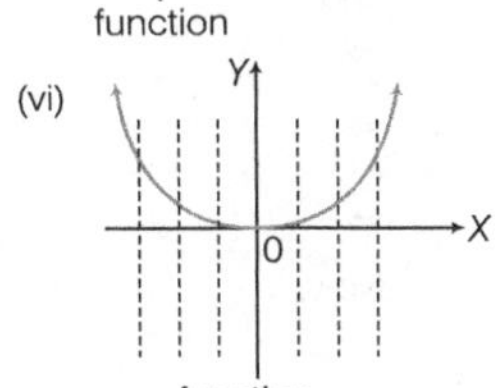

function

3. Given, $f(x) = \dfrac{4^x}{4^x + 2}$

$$\Rightarrow \quad f(1 - x) = \frac{4^{1-x}}{4^{1-x} + 2}$$

Now, $f(x) + f(1 - x) = \dfrac{4^x}{4^x + 2} + \dfrac{4^{1-x}}{4^{1-x} + 2}$

$$= \frac{4^x}{4^x + 2} + \frac{\frac{4}{4^x}}{\frac{4}{4^x} + 2}$$

$$= \frac{4^x}{4^x + 2} + \frac{4}{4 + 2 \cdot 4^x}$$

$$= \frac{4^x}{4^x + 2} + \frac{2}{4^x + 2} = \frac{4^x + 2}{4^x + 2} = 1$$

Hence proved.

4. Given, S = Set of all triangle

and $\quad f : S \to R^+$

$$f(\Delta) = \text{Area of triangle}$$

$\therefore$ f is a function because area of every triangle is positive real number.

5. (c) Given, $f(1) = 1$ and $f(n + 1) = 2f(n) + 1, n \ge 1$

$$f(2) = f(1 + 1) = 2f(1) + 1 = 2 \times 1 + 1 = 3 = 2^2 - 1$$
$$f(3) = f(2 + 1) = 2f(2) + 1 = 2 \times 3 + 1 = 7 = 2^3 - 1$$
$$f(4) = f(3 + 1) = 2f(3) + 1 = 2 \times 7 + 1 = 15 = 2^4 - 1$$
$$f(5) = f(4 + 1) = 2f(4) + 1 = 2 \times 15 + 1 = 31 = 2^5 - 1$$
$$\vdots \qquad \vdots$$
$$f(n) = f((n - 1) + 1) = 2f(n - 1) + 1 = 2^n - 1$$

6. (d) Given,

$$f\left(x + \frac{1}{x}\right) = x^2 + \frac{1}{x^2}, x \ne 0$$

$$\Rightarrow \quad f\left(x + \frac{1}{x}\right) = \left(x + \frac{1}{x}\right)^2 - 2$$

Put $\quad x + \dfrac{1}{x} = t$

$$\Rightarrow \quad f(t) = t^2 - 2$$
$$\Rightarrow \quad f(x) = x^2 - 2$$
$$\therefore \quad |x| \in [2, \infty)$$

7. (d) Given, $f(x) = \cos(\log_e x)$

$$\therefore \quad f(x)\,f(y) - \frac{1}{2}\left\{f\left(\frac{x}{y}\right) + f(xy)\right\}$$

$$= \cos(\log x)\cos(\log y) - \frac{1}{2}\left\{\cos\log\left|\frac{x}{y}\right| + \cos\log(xy)\right\}$$

$$= \cos(\log x)\cos(\log y) - \frac{1}{2}\{\cos(\log x - \log y) + \cos(\log x + \log y)\}$$

$$= \cos(\log x)\cos(\log y) - \frac{1}{2}\{2\cos(\log x)\cos(\log y)\} = 0$$

8. (a) Given,

$$y = \frac{ax - c}{cx - a}$$

$$\Rightarrow \quad cxy - ay = ax - c$$
$$\Rightarrow \quad cxy - ax = ay - c$$
$$\Rightarrow \quad x(cy - a) = ay - c$$
$$\therefore \quad x = \frac{ay - c}{cy - a} = f(y)$$

9. (c) Given,

$$f(x + ay, x - ay) = axy$$

Let $\quad x + ay = t \quad$ ...(i)

and $\quad x - ay = u \quad$ ...(ii)

On solving Eqs. (i) and (ii), we get

$$x = \frac{t + u}{2}$$

and $\quad y = \dfrac{t - u}{2a}$

$$\therefore \quad f(t, u) = \frac{a(t + u)}{2} \cdot \left(\frac{t - u}{2a}\right)$$

$$\Rightarrow \quad f(t, u) = \frac{t^2 - u^2}{4}$$

$$\therefore \quad f(x, y) = \frac{x^2 - y^2}{4}$$

10. (c) Given, $f(x) = \log\left(\frac{1+x}{1-x}\right)$

$$\therefore \quad f\left(\frac{2x}{1+x^2}\right) = \log\left(\frac{1+\frac{2x}{1+x^2}}{1-\frac{2x}{1+x^2}}\right)$$

$$= \log\left(\frac{1+x^2+2x}{1+x^2-2x}\right)$$

$$= \log\left[\frac{(1+x)^2}{(1-x)^2}\right] = \log\left(\frac{1+x}{1-x}\right)^2$$

$$= 2\log\left(\frac{1+x}{1-x}\right) = 2f(x)$$

11. (d) We have,

$$f(x) = 4x^3 + 3x^2 + 3x + 4$$

$$\Rightarrow \quad f\left(\frac{1}{x}\right) = \frac{4}{x^3} + \frac{3}{x^2} + \frac{3}{x} + 4$$

$$\Rightarrow \quad x^3 f\left(\frac{1}{x}\right) = x^3\left(\frac{4}{x^3} + \frac{3}{x^2} + \frac{3}{x} + 4\right)$$

$$= 4 + 3x + 3x^2 + 4x^3$$

$$\therefore \quad x^3 f\left(\frac{1}{x}\right) = f(x)$$

12. (d) Given, $f(x) = \sqrt{x}$

$$\therefore \quad f(25) = \sqrt{25} = 5$$

$$\Rightarrow \quad f(16) = \sqrt{16} = 4$$

$$\Rightarrow \quad f(1) = \sqrt{1} = 1$$

Now, $\frac{f(25)}{f(16)+f(1)} = \frac{5}{4+1} = 1$

Explanations for Q. Nos. (13-15)

We have $f(a, b) = f(a-1, b) + a - 1$

$$= f(a-2, b) + (a-2) + (a-1)$$

$$= f(1, b) + \frac{a(a-1)}{2}$$

$$= f(1, b-1) - (b-1) + \frac{a(a-1)}{2}$$

$$= f(1, 1) - \frac{b(b-1)}{2} + \frac{a(a-1)}{2}$$

$$= \frac{a(a-1)}{2} - \frac{b(b-1)}{2} = 1999$$

$$= (a-b)(a+b-1) = 2 \times 1999$$

Here, 1999 is a prime number and $a - b < a + b - 1$.

Thus, if $a - b = 1$ and $a + b - 1 = 3998$

$\Rightarrow \quad a = 2000$ and $b = 1999$ ... (i)

If $a - b = 2$ and $a + b - 1 = 1999$

$\Rightarrow \quad a = 1001$ and $b = 999$ ... (ii)

$\therefore \quad (a, b) = (2000, 1999)$ or $(1001, 999)$

Thus, number of ordered pairs (a, b) is 2.

The maximum value of $(a + b)$ is 3999.

The minimum value of $(a + b)$ is 2000.

13. (b) **14.** (b) **15.** (c)

16. (4) We have, $f(x+y) = f(x)\,f(y), \forall\ x, y \in N$

and $f(1) = 2$

$$\therefore \sum_{n=1}^{10} f(n) = f(1) + f(2) + f(3) + f(4) + f(5) + f(6) + f(7) + f(8) + f(9) + f(10)$$

$$= f(1) + f(1+1) + f(1+2) + f(1+3) + f(1+4) + f(1+5) + f(1+6) + f(1+7) + f(1+8) + f(1+9)$$

$$= f(1) + f(1)\cdot f(1) + f(1)\cdot f(1)\cdot f(1) + \ldots + \{f(x)\}^{10}$$

$$= f(1) + \{f(1)\}^2 + \{f(1)\}^3 + \{f(1)\}^4 + \ldots + \{f(1)\}^{10}$$

$$= 2 + 2^2 + 2^3 + 2^4 + \cdots + 2^{10}$$

$$= \frac{2(2^{10}-1)}{2-1} = 2^{11} - 2 = 2048 - 2 = 2046$$

$$\therefore \quad \left|\sum_{n=1}^{10} f(n) - 2050\right| = |2046 - 2050| = 4.$$

17. (7) Polynomial function satisfying

$$f(x)\cdot f\left(\frac{1}{x}\right) = f(x) + f\left(\frac{1}{x}\right)$$

$$\Rightarrow \quad f(x) = \pm\, x^n + 1$$

$$\therefore \quad f(3) = \pm\, 3^n + 1 = -26$$

$$\Rightarrow \quad -3^n + 1 = -26$$

$$\Rightarrow \quad -3^n = -27$$

$$\therefore \quad n = 3$$

$$\Rightarrow \quad f(x) = -x^3 + 1$$

$$\Rightarrow \quad f(2) = -8 + 1$$

$$\Rightarrow \quad f(2) = -7$$

Hence, $|f(2)| = 7$

18. (1) We have, $f(x) + 2f\left(\frac{2002}{x}\right) = 3x$ ...(i)

On replacing x by $\frac{2002}{x}$, we get

$$f\left(\frac{2002}{x}\right) + 2f(x) = \frac{6006}{x} \quad \text{...(ii)}$$

On solving Eqs. (i) and (ii), we get

$$f(x) = \frac{4004}{x} - x$$

Now, $f(2) = \frac{4004}{2} - 2 = 2002 - 2 = 2000$

$$\therefore \quad \frac{f(2)}{2000} = \frac{2000}{2000} = 1$$

19. (5) We have, $f(xy) = \frac{f(x)}{y}$

and $f(30) = 20$

$$\Rightarrow \quad f(5 \times 6) = 20$$

$$\Rightarrow \quad \frac{f(5)}{6} = 20 \quad \left[\because f(xy) = \frac{f(x)}{y}\right]$$

$$\Rightarrow \quad f(5) = 120$$

Now, $f(40) = f(5 \times 8) = \frac{f(5)}{8}$

$$= \frac{120}{8} \quad [\because f(5) = 120]$$

$$= 15$$

$$\therefore \quad |f(40) - 10| = |15 - 10| = 5$$

20. (1) Given, $f(x, y) = \max(x, y)^{\min(x, y)}$

and $g(x, y) = \max(x, y) - \min(x, y)$

Now, $g\left(-1, -\frac{3}{2}\right) = -1 - \left(-\frac{3}{2}\right)$

$$= -1 + \frac{3}{2} = \frac{1}{2}$$

Now, $g(-4, -1.75) = -1.75 - (-4)$

$$= -1.75 + 4 = 2.25$$

$$\therefore \quad \left[f\left(g\left(-1, -\frac{3}{2}\right), g(-4, -1.75)\right)\right] = \left[f\left(\frac{1}{2}, 2.25\right)\right]$$

$$= [(2.25)^{1/2}] = [1.125] = 1$$

DPP-3 Domain

1. (i) Given, $f(x) = \dfrac{x^3 - 5x + 3}{x^2 - 1}$

Domain of $f(x)$ is the set of values of x for which the function is defined.

$\Rightarrow \quad x^2 - 1 \neq 0 \quad \Rightarrow \quad x^2 \neq 1$

$\therefore \quad x \neq \pm 1$

Hence, $f(x)$ is not defined at $x = \pm 1$.

$\therefore$ Domain $= R - \{-1, 1\}$

(ii) Given, $f(x) = \dfrac{\sin^{-1} x}{x}$

Now, $f(x)$ is defined, if $x \in [-1, 1]$ and $x \neq 0$

i.e. $\quad x \in [-1, 0) \cup (0, 1]$

(iii) Given, $f(x) = \dfrac{1}{\sqrt{x + |x|}}$

$\therefore \quad x + |x| \begin{cases} x + x, & \text{if } x \geq 0 \\ x - x, & \text{if } x < 0 \end{cases}$

$\Rightarrow \quad x + |x| = \begin{cases} 2x, & x \geq 0 \\ 0, & x < 0 \end{cases} \quad \ldots \text{(i)}$

Now, $f(x)$ assumes real values,

if $\quad x + |x| > 0 \Rightarrow x > 0 \quad$ [using Eq. (i)]

Hence, domain of $f = (0, \infty)$

(iv) Given, $f(x) = e^{x + \sin x}$

Clearly, domain of $f(x)$ is R.

(v) Given, $f(x) = \dfrac{1}{\log_{10}(1 - x)} + \sqrt{x + 2}$

$\Rightarrow \quad 1 - x > 0, x \neq 0 \text{ and } x + 2 \geq 0$

$\Rightarrow \quad x < 1, x \neq 0 \text{ and } x \geq -2$

$\Rightarrow \quad x < 1 \text{ and } x \neq 0 \text{ and } x \geq -2$

$\therefore \quad x \in [-2, 0) \cup (0, 1)$

(vi) Given, $f(x) = \sqrt{1 - 2x} + 3\sin^{-1}\left(\dfrac{3x - 1}{2}\right)$

$\Rightarrow \quad 1 - 2x \geq 0 \text{ and } \left|\dfrac{3x - 1}{2}\right| \leq 1$

$\therefore \quad x \leq \dfrac{1}{2} \text{ and } -1 \leq \dfrac{3x - 1}{2} \leq 1$

$\Rightarrow \quad x \leq \dfrac{1}{2} \text{ and } -1 \leq 3x \leq 3$

$\Rightarrow \quad x \leq \dfrac{1}{2} \text{ and } \dfrac{-1}{3} \leq x \leq 1$

$\therefore \quad x \in \left[\dfrac{-1}{3}, \dfrac{1}{2}\right]$

(vii) Given, $f(x) = 2^{\sin^{-1} x} + \dfrac{1}{\sqrt{x - 2}}$

$\Rightarrow \quad x - 2 \geq 0 \text{ and } x \in [-1, 1]$

$\Rightarrow \quad x \geq 2 \text{ and } x \in [-1, 1]$

$\therefore \quad x \in \phi$

(viii) Given, $f(x) = \log_x \log_2\left(\dfrac{1}{x - \dfrac{1}{2}}\right)$

$\Rightarrow x - \dfrac{1}{2} > 0 \text{ and } \log_2\left(\dfrac{1}{x - \dfrac{1}{2}}\right) > 0, x > 0 - \{1\}$

$\Rightarrow \quad x > \dfrac{1}{2} \text{ and } \dfrac{1}{x - \dfrac{1}{2}} > 2^0 \quad \left[\begin{array}{l}\because \log_y x > z \\ \Rightarrow x > y^z, \text{if } y > 1\end{array}\right]$

$\Rightarrow \quad x > \dfrac{1}{2} \text{ and } x - \dfrac{1}{2} < 1$

$\Rightarrow \quad x > \dfrac{1}{2} \text{ and } x < \dfrac{3}{2}, x > 0 - \{1\}$

$\therefore \quad x \in \left(\dfrac{1}{2}, 1\right) \cup \left(1, \dfrac{3}{2}\right)$

2. (i) Given, $f(x) = \sqrt{3 - 2^x - 2^{1-x}}$

$\Rightarrow \quad 3 - 2^x - 2^{1-x} \geq 0$

$\Rightarrow \quad -2^x - 2^{1-x} \geq -3$

$\Rightarrow \quad 2^x + 2^{1-x} \leq 3$

$\Rightarrow \quad 2^x + \dfrac{2}{2^x} \leq 3$

$\Rightarrow \quad (2^x)^2 + 2 \leq 3 \cdot 2^x$

$\Rightarrow \quad (2^x)^2 - 3 \cdot 2^x + 2 \leq 0$

Let $2^x = t \quad \Rightarrow \quad t^2 - 3t + 2 \leq 0$

$\Rightarrow \quad t^2 - 2t - t + 2 \leq 0$

$\Rightarrow \quad (t - 2)(t - 1) \leq 0$

$\Rightarrow \quad t \in [1, 2] \quad \Rightarrow \quad 2^x \in [1, 2]$

$\therefore \quad x \in [0, 1]$

(ii) Given, $f(x) = \sqrt{1 - \sqrt{1 - x^2}}$

$\Rightarrow \quad 1 - \sqrt{1 - x^2} \geq 0 \text{ and } 1 - x^2 \geq 0$

$\Rightarrow \quad \sqrt{1 - x^2} \leq 1 \text{ and } 1 - x^2 \geq 0$

$\Rightarrow \quad 1 - x^2 \leq 1 \text{ and } x^2 \leq 1$

$\Rightarrow \quad -x^2 \leq 0 \text{ and } x \leq \pm 1$

$\Rightarrow \quad x^2 \geq 0 \text{ and } |x| \leq 1$

$\therefore \quad x \in [-1, 1]$

(iii) Given, $f(x) = (x^2 + x + 1)^{-3/2}$

$\Rightarrow \quad f(x) = \dfrac{1}{(x^2 + x + 1)^{3/2}}$

$\Rightarrow \quad f(x) = \dfrac{1}{\sqrt{(x^2 + x + 1)^3}}$

$\Rightarrow \quad (x^2 + x + 1)^3 > 0$

$\Rightarrow \quad x^2 + x + 1 > 0$

Here, $\quad a > 1$

and $\quad D = b^2 - 4 \cdot ac = -3 < 0$

$\Rightarrow \quad x^2 + x + 1 \geq 0, \forall x \in R$

(iv) Given, $f(x) = \sqrt{\dfrac{x - 2}{x + 2}} + \sqrt{\dfrac{1 - x}{1 + x}}$

We must have,

$\dfrac{x - 2}{x + 2} \geq 0 \text{ and } \dfrac{1 - x}{1 + x} \geq 0$

$\Rightarrow \quad \dfrac{x - 2}{x + 2} \geq 0 \Rightarrow x \geq 2 \text{ or } x < -2 \quad \ldots \text{(i)}$

and $\quad \dfrac{1 - x}{1 + x} \geq 0 \Rightarrow -1 < x \leq 1 \quad \ldots \text{(ii)}$

From Eqs. (i) and (ii), we get $x \in \phi$

(v) Given, $f(x) = \sqrt{\tan x - \tan^2 x}$

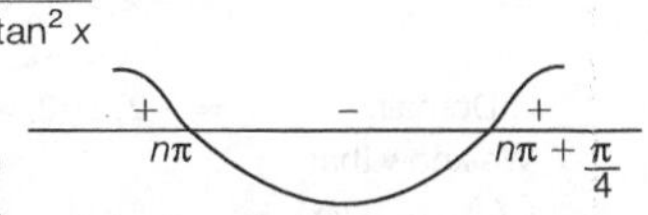

$\Rightarrow \quad \tan x - \tan^2 x \geq 0$

$\Rightarrow \quad \tan^2 x - \tan x \leq 0$

$\Rightarrow \quad \tan x (\tan x - 1) \leq 0$

$\Rightarrow \quad x = n\pi \text{ or } n\pi + \dfrac{\pi}{4}$

$\therefore \quad x \in \bigcup_{n \in I}\left[n\pi, n\pi + \dfrac{\pi}{4}\right]$

(vi) Given, $f(x) = \dfrac{1}{\sqrt{1-\cos x}}$

$\Rightarrow \quad 1 - \cos x > 0$

$\Rightarrow \quad \cos x < 1$, i.e. $\cos x \neq 1 \Rightarrow x \neq 2n\pi$

$\therefore \quad x \in R - \{2n\pi\}, n \in I$

(vii) Given, $f(x) = \sqrt{\log_{\frac{1}{4}}\left(\dfrac{5x - x^2}{4}\right)}$

$\log_{\frac{1}{4}}\left(\dfrac{5x - x^2}{4}\right) \geq 0$ and $\dfrac{5x - x^2}{4} > 0$

$\Rightarrow \quad \dfrac{5x - x^2}{4} \leq \left(\dfrac{1}{4}\right)^0$ and $-x(x-5) > 0$

$\Rightarrow \quad (x-1)(x-4) \geq 0$ and $-x(x-5) > 0$

+   1   −   4   +   and   −   −5   +   0   −

$\therefore \quad x \in (0, 1] \cup [4, 5)$

(viii) We have, $f(x) = \log_{10}[1 - \log_{10}(x^2 - 5x + 16)]$

for domain,

$x^2 - 5x + 16 > 0$

and $\quad 1 - \log_{10}(x^2 - 5x + 16) > 0$

$\Rightarrow \quad x \in R$ and $\log_{10}(x^2 - 5x + 16) < 1$

$\Rightarrow \quad x \in R$ and $x^2 - 5x + 16 < 10$

$\Rightarrow \quad x \in R$ and $x^2 - 5x + 6 < 0$

$\Rightarrow \quad x \in R$ and $(x-2)(x-3) < 0$

+   2   −   3   +

$\Rightarrow \quad x \in R$ and $2 < x < 3$

$\therefore \quad x \in (2, 3)$

3. (c) Given, $f(x) = \sqrt{5 - 4x - x^2} + x^2 \log(x+4)$

$\Rightarrow \quad 5 - 4x - x^2 \geq 0$ and $x + 4 > 0$

$\Rightarrow \quad x^2 + 4x - 5 \leq 0$ and $x > -4$

$\Rightarrow \quad (x+5)(x-1) \leq 0$ and $x > -4$

$\Rightarrow \quad x \in [-5, 1]$ and $x > -4$

$\therefore \quad x \in (-4, 1]$

or $\quad -4 < x \leq 1$

4. (b) Given, $f(x) = \sin^{-1} 5x$

We know that, domain of $\sin^{-1} x$ is $[-1, 1]$.

$\therefore \quad -1 \leq 5x \leq 1 = -1/5 \leq x \leq 1/5$

$\therefore \quad x \in \left[-\dfrac{1}{5}, \dfrac{1}{5}\right]$

5. (b) Given, $f(x) = \dfrac{\sin^{-1}(3-x)}{\log(|x| - 2)}$

Let $\quad g(x) = \sin^{-1}(3-x)$

$\Rightarrow \quad -1 \leq 3 - x \leq 1$

The domain of $g(x)$ is $[2, 4]$.

Let $\quad h(x) = \log(|x| - 2)$

$\Rightarrow \quad |x| - 2 > 0$ or $|x| > 2$

$\Rightarrow \quad x < -2$ or $x > 2$

$\therefore$ Domain $\quad (-\infty, -2) \cup (2, \infty)$

We know that,

$\left(\dfrac{f}{g}\right)(x) = \dfrac{f(x)}{g(x)}, \forall x \in D_1 \cap D_2 - \{x \in R, g(x) = 0\}$

Hence, the domain of $f(x)$ is $(2, 4] - \{3\}$

i.e. $\quad (2, 3) \cup (3, 4]$

6. (a) Let $f(x) = \sin^{-1}\left[\log_3\left(\dfrac{x}{3}\right)\right]$

Since, domain of $\sin^{-1} x$ is $[-1, 1]$.

$\therefore \quad -1 \leq \log_3\left(\dfrac{x}{3}\right) \leq 1 \quad \Rightarrow \quad 3^{-1} \leq \dfrac{x}{3} \leq 3$

$\Rightarrow \quad 1 \leq x \leq 9$

Hence, domain of $\sin^{-1}\left[\log_3\left(\dfrac{x}{3}\right)\right]$ is $[1, 9]$.

7. (d) Given, $f(x) = \dfrac{\log_2(x+3)}{x^2 + 3x + 2} = \dfrac{\log_2(x+3)}{(x+1)(x+2)}$

For numerator $\quad x + 3 > 0$

$\Rightarrow \quad x > -3 \quad \ldots$(i)

and for denominator

$(x+1)(x+2) \neq 0$

$x \neq -1, -2 \quad \ldots$(ii)

From Eqs. (i) and (ii), we have

Domain $(-3, \infty) \sim \{-1, -2\}$

8. (b) Given, $f(x) = \sqrt{\log \dfrac{1}{|\sin x|}}$

$f(x)$ is defined for

$\log\left(\dfrac{1}{|\sin x|}\right) \geq 0 \quad \Rightarrow \quad \dfrac{1}{|\sin x|} \geq 1$ and $|\sin x| \neq 0$

$\Rightarrow \quad |\sin x| \neq 0 \quad \left[\because \dfrac{1}{|\sin x|} \geq 1, \forall x\right]$

$\Rightarrow \quad x \neq n\pi, n \in Z$

Hence, the domain of $f(x)$ is $R - \{n\pi, n \in Z\}$.

9. (c) For domain of $f(x) = \log(\sqrt{x-4} + \sqrt{6-x})$

$x - 4 \geq 0$ and $6 - x \geq 0$

$\Rightarrow \quad x \geq 4$ and $x \leq 6$

$\therefore \quad x \in [4, 6]$

10. (c) Given, $f(x) = \log_{x+3}(x^2 - 1)$

$f(x)$ is to be defined, when $x^2 - 1 > 0$, $3 + x > 0$ and $3 + x \neq 1$.

i.e. $\quad x^2 > 1, \quad x > -3$ and $x \neq -2$

i.e. $\quad x < -1$ or $x > 1, \quad x > -3$ and $x \neq -2$

$\therefore$ Domain $= (-3, -2) \cup (-2, -1) \cup (1, \infty)$

11. (d) Given, $f(x) = \sqrt{x - x^2} + \sqrt{4 + x} + \sqrt{4 - x}$

$\Rightarrow \quad x - x^2 > 0$ and $4 + x > 0$ and $4 - x > 0$

$\Rightarrow \quad x(1 - x) > 0$ and $x > -4$ and $x < 4$

$\Rightarrow \quad x(x-1) < 0, \quad x > -4, x < 4$

$\Rightarrow \quad x \in (0, 1), x > -4, x < 4$

$\therefore \quad x \in [0, 1]$

12. (c) Given, $f(x) = \sqrt{\log(x^2 - 6x + 6)}$

$\Rightarrow \quad \log(x^2 - 6x + 6) \geq 0$ and $x^2 - 6x + 6 > 0$

$\Rightarrow \quad x^2 - 6x + 6 \geq 1$

$\Rightarrow \quad x^2 - 6x + 5 \geq 0$

$\Rightarrow \quad (x-5)(x-1) \geq 0$

$\therefore \quad x \in (-\infty, 1] \cup [5, \infty)$

13. (a) Given, $f(x) = \sin^{-1}(|x - 1| - 2)$

Since, the domain of $\sin^{-1} x$ is $[-1, 1]$, $f(x)$ is defined, if

$-1 \leq |x - 1| - 2 \leq 1$

$\Rightarrow \quad 1 \leq |x - 1| \leq 3$

$\Rightarrow \quad -3 \leq x - 1 \leq -1$ or $1 \leq x - 1 \leq 3$

$\Rightarrow \quad -2 \leq x \leq 0$ or $2 \leq x \leq 4$

$\therefore$ Domain $= [-2, 0] \cup [2, 4]$

14. (d) We have, $f(x) = \sqrt{\dfrac{-\log_{0.3}(x-1)}{x^2+2x+8}}$, for domain

$$\frac{-\log_{0.3}(x-1)}{x^2+2x+8} \geq 0$$

$\Rightarrow$ $\dfrac{\log_{0.3}(x-1)}{x^2+2x+8} \leq 0$, since x^2+2x+8 is always positive, as $x^2+2x+8=(x+1)^2+7$

$\therefore$ $\log_{0.3}(x-1) \leq 0$

$\Rightarrow$ $x-1 \geq (0.3)^0$

$\therefore$ $x \geq 2$

15. (c) We have,

$$f(x) = \cot^{-1}\sqrt{x^2+3x} + \cos^{-1}\sqrt{x^2+3x+1}$$

For domain, $x^2+3x \geq 0$ and $x^2+3x+1 \leq 1$

$\Rightarrow$ $x^2+3x \geq 0$ and $x^2+3x \leq 0$

i.e. only possible, when $x^2+3x=0$.

$\Rightarrow$ $x = 0, -3$

$\therefore$ $x \in \{0, -3\}$

16. (a) Given, $f(x) = \sqrt{\sin^{-1}(2x) + \dfrac{\pi}{6}}$

To find domain, we must have

$$\sin^{-1}(2x) + \frac{\pi}{6} \geq 0 \qquad \left[\text{but } \frac{-\pi}{2} \leq \sin^{-1}\theta \leq \frac{\pi}{2}\right]$$

Now, $-\dfrac{\pi}{6} \leq \sin^{-1}(2x) \leq \dfrac{\pi}{2}$

$\Rightarrow$ $\sin\left(\dfrac{-\pi}{6}\right) \leq 2x \leq \sin\left(\dfrac{\pi}{2}\right)$

$\Rightarrow$ $-\dfrac{1}{2} \leq 2x \leq 1$

$\Rightarrow$ $-\dfrac{1}{4} \leq x \leq \dfrac{1}{2}$

$\therefore$ $x \in \left[-\dfrac{1}{4}, \dfrac{1}{2}\right]$

17. (d) Given that,

$2^x + 2^y = 2, \forall x, y \in R$

$\Rightarrow$ $2^y = 2 - 2^x$

$\Rightarrow$ $y = \log_2(2-x^x)$ for domain, $2 - 2^x > 0$

$\Rightarrow$ $2 > 2x$

$\Rightarrow$ $x < 1$

$\Rightarrow$ $x \in (-\infty, 1)$

18. (b) We have, $f(\cos 4\theta) = \dfrac{2\cos^2\theta}{2\cos^2\theta - 1} = \dfrac{1+\cos 2\theta}{\cos 2\theta}$ ...(i)

when, $\cos 4\theta = \dfrac{1}{3}$

$\Rightarrow$ $2\cos^2 2\theta - 1 = \dfrac{1}{3}$ $\Rightarrow$ $2\cos^2 2\theta = \dfrac{4}{3}$

$\Rightarrow$ $\cos^2 2\theta = \dfrac{2}{3}$

$\therefore$ $\cos 2\theta = \pm\sqrt{\dfrac{2}{3}}$

Thus, $f\left(\dfrac{1}{3}\right) = \dfrac{1 \pm \sqrt{\dfrac{2}{3}}}{\pm\sqrt{\dfrac{2}{3}}} = \dfrac{\sqrt{3}+\sqrt{2}}{\pm\sqrt{2}}$

$\therefore$ $f\left(\dfrac{1}{3}\right) = \sqrt{\dfrac{3}{2}} + 1$ or $-\sqrt{\dfrac{3}{2}} + 1$

19. (a,b,d) A function is said to be transcendental function, if it is not algebraic.

20. (a,c) Here, $[2\tan\pi x] \geq 2$ and $\dfrac{x^2+2x-3}{4x^2-4x-3} > 0$.

$\therefore$ $f(x) = \dfrac{x^2+2x-3}{4x^2-4x-3}$

$\Rightarrow$ $2\tan\pi x \geq 2$ and $\dfrac{(x-1)(x+3)}{(2x+1)(2x-3)} > 0$

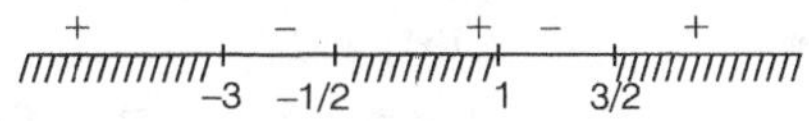

$\Rightarrow$ $\tan\pi x \geq 1$

$\Rightarrow$ $x < -3$ or $-\dfrac{1}{2} < x < 1$ or $x > \dfrac{3}{2}$

$\Rightarrow$ $\pi x \in \left[n\pi + \dfrac{\pi}{4}, n\pi + \dfrac{\pi}{2}\right)$ and

$x \in (-\infty, -3) \cup \left(-\dfrac{1}{2}, 1\right) \cup \left(\dfrac{3}{2}, \infty\right)$

$\therefore$ $x \in \left[n + \dfrac{1}{4}, n + \dfrac{1}{2}\right)$ and

$x \in (-\infty, -3) \cup \left(-\dfrac{1}{2}, 1\right) \cup \left(\dfrac{3}{2}, \infty\right)$

Thus, n cannot be $\{-1, 1\}$

But n can be $\{0, -5\}$.

Explanations for Q. Nos. (21-23)

Here, $\{f(x)\}^2 \cdot f\left(\dfrac{1-x}{1+x}\right) = 64x$ ... (i)

Putting $\dfrac{1-x}{1+x} = y$ and $x = \dfrac{1-y}{1+y}$, we get

$$\left\{f\left(\frac{1-y}{1+y}\right)\right\}^2 \cdot f(y) = 64 \cdot \left(\frac{1-y}{1+y}\right)$$

$\Rightarrow$ $f(x) \cdot \left(f\left(\dfrac{1-x}{1+x}\right)\right)^2 = 64\left(\dfrac{1-x}{1+x}\right)$ ... (ii)

From Eqs. (i) and (ii), we get

$$\frac{\{f(x)\}^4 \cdot \left\{f\left(\frac{1-x}{1+x}\right)\right\}^2}{f(x) \cdot \left\{f\left(\frac{1-x}{1+x}\right)\right\}^2} = \frac{(64x)^2}{64\left(\frac{1-x}{1+x}\right)}$$

$\Rightarrow$ $\{f(x)\}^3 = 64x^2 \cdot \left(\dfrac{1+x}{1-x}\right)$

$\Rightarrow$ $f(x) = 4x^{2/3} \cdot \left(\dfrac{1+x}{1-x}\right)^{1/3}$

where, $x = \dfrac{9}{7} = -4 \cdot \left(\dfrac{9}{7}\right)^{2/3} \cdot (2) = -8\left(\dfrac{9}{7}\right)^{2/3}$

21. (a) **22.** (b) **23.** (c)

24. (1) $f(2) = 2\sin 2 + 2b \cdot \cos 2 + 8$

$\underline{f(-2) = -a\sin 2 - 2b\cos 2 + 8}$

$f(2) + f(-2) = 16$

$\therefore$ $f(-2) = 16 - f(2)$, given $f(2) = 15$

$\Rightarrow$ $f(-2) = 1$

25. (3) $\because f(x) = [x] + \sum_{k=1}^{2015} \dfrac{\{x+r\}}{2015}$

$= [x] + \dfrac{\{x+1\} + \{x+2\} + \ldots + \{x+2015\}}{2015}$

$= [x] + \dfrac{2015\{x\}}{2015} = [x] + \{x\}$

$= x$

$\therefore$ $f(3) = 3$

DPP-4 Range of Functions

1. (i) We have, $f(x) = |x - 3|$

$\therefore$ Range $= f(x) = [0, \infty)$ [$\because$ minimum of modulus function is zero]

(ii) We have, $f(x) = \dfrac{x}{1 + x^2}$

Let $f(x) = y$

$\therefore \quad y = \dfrac{x}{1 + x^2} \Rightarrow x^2 y + y = x$

$\Rightarrow \quad x^2 y - x + y = 0$

$\Rightarrow \quad x = \dfrac{1 \pm \sqrt{1 - 4y^2}}{2}$

$\therefore \quad x \in R,\ 1 - 4y^2 \ge 0$

$\Rightarrow \quad y^2 \le \dfrac{1}{4} \Rightarrow y \in \left[-\dfrac{1}{2}, \dfrac{1}{2}\right]$

$\therefore$ Range $= \left[-\dfrac{1}{2}, \dfrac{1}{2}\right]$

(iii) We have, $f(x) = \sqrt{16 - x^2}$

Let $f(x) = y,\ y \ge 0$ ...(i)

$\Rightarrow \quad y^2 = 16 - x^2 \Rightarrow x^2 = 16 - y^2$

$\Rightarrow \quad x = \sqrt{16 - y^2}$

$\Rightarrow \quad 16 - y^2 \ge 0$

$\Rightarrow \quad y^2 \le 16 \Rightarrow -4 \le y \le 4$ ...(ii)

From Eq. (i) and (ii), we get

$0 \le y \le 4$ [$\because x \in [-4, 4]$]

$\therefore$ Range $= [0, 4]$

(iv) We have, $f(x) = \dfrac{|x - 4|}{x - 4}$

$$\Rightarrow \quad f(x) = \begin{cases} \dfrac{x-4}{x-4}, & \text{if } x > 4 \\ \dfrac{-(x-4)}{x-4}, & \text{if } x < 4 \end{cases}$$

$$\Rightarrow \quad f(x) = \begin{cases} 1, & \text{if } x > 4 \\ -1, & \text{if } x < 4 \end{cases}$$

$\therefore$ Range $= \{-1, 1\}$

(v) We have, $f(x) = 5 + \sin x + \cos x$

$\Rightarrow \quad f(x) = 5 + \sqrt{2}\left[\sin\left(x + \dfrac{\pi}{4}\right)\right]$

$\Rightarrow \quad -1 \le \sin\left(x + \dfrac{\pi}{4}\right) \le 1$

$\therefore$ Range $= [5 - \sqrt{2}, 5 + \sqrt{2}]$

(vi) We have, $f(x) = \dfrac{1}{1 + \sqrt{x}}$

Let $f(x) = y$

$\therefore \quad y = \dfrac{1}{1 + \sqrt{x}} \Rightarrow 1 + \sqrt{x} = \dfrac{1}{y}$

$\Rightarrow \quad \sqrt{x} = \dfrac{1 - y}{y}$

$\Rightarrow \quad \dfrac{1 - y}{y} \ge 0$ [$\because \sqrt{x} \in [0, \infty)$]

$\Rightarrow \quad 0 < y \le 1$ [$\because y \ne 0$]

$\therefore$ Range $= (0, 1]$

2. (i) We have, $f(x) = \dfrac{1}{\sqrt{4 + 3\sin x}}$

$f(x)$ is defined, if $4 + 3\sin x > 0,\ \forall x \in R$ [$\because \sin\theta \in [-1, 1]$]

$\therefore$ Domain of $f(x)$ is R.

Range of $f(x)$ is $\left[\dfrac{1}{\sqrt{4+3}}, \dfrac{1}{\sqrt{4-3}}\right]$. [$\because \sin\theta \in [-1, 1]$]

$\therefore$ Range of $f(x)$ is $\left[\dfrac{1}{\sqrt{7}}, 1\right]$

(ii) We have, $f(x) = x!$
which is the product of n natural numbers and domain $D \in N \cup \{0\}$
Hence, its range is $n!, n \in N \cup \{0\}$.

(iii) We have, $f(x) = \dfrac{x^2 - 9}{x - 3}$

$\Rightarrow \quad x - 3 \ne 0 \Rightarrow x \ne 3$

$\therefore$ Domain is $R - \{3\}$.

Range $= R - \{f(3)\} = R - \{6\}$

(iv) We have, $f(x) = \sin^2 x^3 + \cos^2 x^3$

Its domain is R.

$\therefore f(x) = 1$

Hence, its range is $\{1\}$.

3. (a) We have, $f(x) = \sec\left(\dfrac{\pi}{4}\cos^2 x\right),\ x \in R$

$\therefore \quad 0 \le \cos^2 x \le 1,\ 0 \le \dfrac{\pi}{4}\cos^2 x \le \dfrac{\pi}{4}$

$\sec 0 \le \sec\left(\dfrac{\pi}{4}\cdot\cos^2 x\right) \le \sec\dfrac{\pi}{4}$

$\Rightarrow \quad 1 \le f(x) \le \sqrt{2}$

$\therefore \quad$ Range $= [1, \sqrt{2}]$

4. (c) Let $y = \dfrac{x^2 + x + 2}{x^2 + x + 1}, x \in R$

$\Rightarrow \quad x^2 y + xy + y = x^2 + x + 2$

$\Rightarrow \quad x^2(y - 1) + x(y - 1) + y - 2 = 0$

$\therefore x \in R$, hence $\Delta \ge 0$

$(y - 1)^2 - 4(y - 1)(y - 2) \ge 0$

$\Rightarrow \quad (y - 1)(y - 1 - 4y + 8) \ge 0$

$\Rightarrow \quad (y - 1)(-3y + 7) \ge 0$

$\Rightarrow \quad (y - 1)(3y - 7) \le 0$

$\Rightarrow \quad 1 \le y \le \dfrac{7}{3}$, but $y \ne 1$ as it does not satisfy

$y = \dfrac{x^2 + x + 2}{x^2 + x + 1}$

$\therefore$ Range $= \left(1, \dfrac{7}{3}\right]$

5. (d) We have, $f(x) = a\cos(bx + c) + d$

We know that, $\quad -a \le a\cos(bx + c) \le a$

$\Rightarrow \quad -a + d \le a\cos(bx + c) + d \le a + d$

$\Rightarrow \quad d - a \le f(x) \le d + a$

$\therefore \quad$ Range $= [d - a, d + a]$

6. (b) We have, $f(x) = [x] - x = -\{x\}$

Since, $\quad 0 \le \{x\} < 1 \Rightarrow -1 < -\{x\} \le 0$

$\therefore$ Range $= (-1, 0]$

7. (b) We have, $f(x) = \cos\dfrac{x}{3}$

It range is $[-1, 1]$.

8. (b) We have, $f(x) = \dfrac{x + 2}{|x + 2|}$

$$\therefore \quad f(x) = \begin{cases} 1, & x > -2 \\ -1, & x < -2 \end{cases}$$

$\therefore$ It range is $\{-1, 1\}$.

9. (b) We have, $f(x) = \cos 2x - \sin 2x$

We know that,

$$-\sqrt{a^2+b^2} \le a\sin x \pm b\cos x \le \sqrt{a^2+b^2}$$

$\Rightarrow$ $-\sqrt{2} \le \cos 2x - \sin 2x \le \sqrt{2}$

$\therefore$ Range $= [-\sqrt{2}, \sqrt{2}]$

10. (d) We have, $f(x) = \cos x - \sin x$

$\Rightarrow$ $-\sqrt{2} \le f(x) \le \sqrt{2}$

$\therefore$ Range $= [-\sqrt{2}, \sqrt{2}]$

11. (b) We have, $f(x) = \dfrac{1}{2 - \sin 3x}$

It domain is $x \in R$.

Now, let $y = \dfrac{1}{2-\sin 3x} \Rightarrow 2 - \sin 3x = \dfrac{1}{y} \Rightarrow \sin 3x = 2 - \dfrac{1}{y}$

Since, $-1 \le \sin 3x \le 1 \Rightarrow -1 \le 2 - \dfrac{1}{y} \le 1$

$\Rightarrow$ $-3 \le -\dfrac{1}{y} \le -1 \Rightarrow 1 \le \dfrac{1}{y} \le 3$

$\Rightarrow$ $\dfrac{1}{3} \le y \le 1$

$\therefore$ Range $= \left[\dfrac{1}{3}, 1\right]$

12. (b) Let $y = \dfrac{x^2+14x+9}{x^2+2x+3}$

$\Rightarrow$ $x^2y + 2xy + 3y = x^2 + 14x + 9$

$\Rightarrow$ $x^2(y-1) + x(2y-14) + 3y - 9 = 0$

$\because$ $x \in R$

$\therefore$ $\Delta \ge 0$

$\Rightarrow$ $4(y-7)^2 - 4\cdot(y-1)\,3(y-3) \ge 0$

$\Rightarrow$ $y^2 + 49 - 14y - 3(y^2 - 4y + 3) \ge 0$

$\Rightarrow$ $-2y^2 - 2y + 40 \ge 0$

$\Rightarrow$ $y^2 + y - 20 \le 0$

$\Rightarrow$ $(y+5)(y-4) \le 0$

$\Rightarrow$ $-5 \le y \le 4$

$\therefore$ Range $= [-5, 4]$

13. (d) We have, $f(\theta) = \sec^2\theta + \cos^2\theta$

We know that, AM $\ge$ GM

$$\frac{\sec^2\theta + \cos^2\theta}{2} \ge \sqrt{\sec^2\theta \cdot \cos^2\theta}$$

$\therefore$ $\sec^2\theta + \cos^2\theta \ge 2$

$\therefore$ Its range is $[2, \infty)$.

14. (c) We have, $f(x) = \log_e \sqrt{4 - x^2}$

Here, $-x^2 \le 0,\ 4 - x^2 \le 4$

$0 < \sqrt{4-x^2} \le 2$

$\log_e 0 < \log_e \sqrt{4-x^2} \le \log_e 2$

$\Rightarrow$ $-\infty < \log_e \sqrt{4-x^2} \le \log_e 2$

Range $(-\infty, \log_e 2]$

15. (b) We have, $f(x) = \tan\sqrt{\dfrac{\pi^2}{9} - x^2}$

It is defined, when $\dfrac{\pi^2}{9} - x^2 \ge 0$

$\Rightarrow$ $-\dfrac{\pi}{3} \le x \le \dfrac{\pi}{3}$

Clearly, for $x \in \left[-\dfrac{\pi}{3}, \dfrac{\pi}{3}\right] \Rightarrow \sqrt{\dfrac{\pi^2}{9} - x^2} \in \left[0, \dfrac{\pi}{3}\right]$

Since, $\tan x$ is strictly increasing function on $\left[0, \dfrac{\pi}{3}\right]$.

$$\tan 0 \le \tan\left(\sqrt{\frac{\pi^2}{9} - x^2}\right) \le \tan\frac{\pi}{3}$$

$\Rightarrow$ $0 \le f(x) \le \sqrt{3}$

$\therefore$ Range $= [0, \sqrt{3}]$

16. (a) We have, $f(x) = \log_e(3x^2 - 4x + 5)$

Now, $3x^2 - 4x + 5 = 3\left[x^2 - \dfrac{4}{3}x + \dfrac{5}{3} + \dfrac{4}{9} - \dfrac{4}{9}\right]$

$= 3\left[\left(x - \dfrac{2}{3}\right)^2 + \dfrac{11}{9}\right]$

$= 3\left(x - \dfrac{2}{3}\right)^2 + \dfrac{11}{3}$

$\Rightarrow$ $\dfrac{11}{3} \le 3x^2 - 4x + 5 < \infty$

$\Rightarrow$ $\log_e \dfrac{11}{3} \le \log_e(3x^2 - 4x + 5) < \log_e \infty$

$\Rightarrow$ $\log_e \dfrac{11}{3} \le f(x) < \infty$

$\therefore$ Range $= \left[\log_e \dfrac{11}{3}, \infty\right)$

17. (b) We have, $f(x) = 4^x + 2^x + 1$

Let $y = 2^{2x} + 2^x + 1$

$\Rightarrow 2^{2x} + 2^x + 1 - y = 0$

$\Rightarrow$ $2^x = \dfrac{-1 + \sqrt{1 - 4(1-y)}}{2}$

$\therefore$ $x = \log_2\left(\dfrac{\sqrt{4y-3} - 1}{2}\right)$

which defined, when $4y - 3 \ge 0$

$\Rightarrow$ $y \ge \dfrac{3}{4}$ ...(i)

and $\dfrac{\sqrt{4y-3} - 1}{2} > 0 \Rightarrow \sqrt{4y - 3} > 1$

$\Rightarrow$ $4y - 3 > 1$

$y > 1$ ...(ii)

From Eqs. (i) and (ii), we get

$\therefore$ Range of $f(x) = (1, \infty)$

18. (b) We have, $f(x) = \log_{\sqrt{5}}\{\sqrt{2}(\sin x - \cos x) + 3\}$

Since, $-\sqrt{2} \le \sin x - \cos x < \sqrt{2}$

$\Rightarrow$ $1 \le \sqrt{2}(\sin x - \cos x) + 3 \le 5$

$\Rightarrow \log_{\sqrt{5}} 1 \le \log_{\sqrt{5}}\{\sqrt{2}(\sin x - \cos x) + 3\} \le \log_{\sqrt{5}} 5$

$\Rightarrow$ $0 \le f(x) \le 2$

$\therefore$ Range $= [0, 2]$

19. (c) We have, $f(x) = {}^{7-x}P_{x-3}$

It is defined, when $7 - x \ge 0$

$\Rightarrow$ $x \le 7$ ...(i)

$x - 3 \ge 0 \Rightarrow x \ge 3$ ...(ii)

and $7 - x \ge x - 3$

$10 \ge 2x \Rightarrow x \le 5$ ...(iii)

On combining Eqs. (i), (ii) and (iii), we get

$x = 3, 4, 5$

Now, $f(3) = {}^4P_0 = 1$

$f(4) = {}^3P_1 = 3$

and $f(5) = {}^2P_2 = 2$

$\therefore$ Range $= \{1, 2, 3\}$

20. (d) We have, $f(x) = \dfrac{x}{1+|x|} = \begin{cases} \dfrac{x}{1-x}, & x < 0 \\ \dfrac{x}{1+x}, & x \geq 0 \end{cases}$

Let $y_1 = \dfrac{x}{1-x} \Rightarrow y_1 - xy_1 = x$

$\Rightarrow \quad x(1 + y_1) = y_1$

$\Rightarrow \quad x = \dfrac{y_1}{1 + y_1}$

which is defined, when

$\Rightarrow \quad \dfrac{y_1}{1+y_1} < 0$

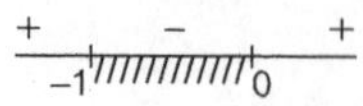

$\Rightarrow \quad y_1 \in (-1, 0)$ ...(i)

Now, $y_2 = \dfrac{x}{1+x} \Rightarrow x = \dfrac{y_2}{1-y_2}$

It is defined, when

$\dfrac{y_2}{1-y_2} \geq 0$

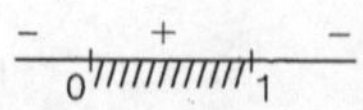

$\Rightarrow \quad y_2 \in [0, 1)$ ...(ii)

From Eqs. (i) and (ii), we get

$y \in (-1, 1)$

21. (a,b,c) We have,

$f(x) = |x-1| + |x-2| + |x-3| + |x-4|$

$f(x) = \begin{cases} -4x + 10; & x < 1 \\ -2x + 8; & 1 \leq x < 2 \\ 4; & 2 \leq x < 3 \\ 2x - 2; & 3 \leq x < 4 \\ 4x - 10; & 4 \leq x < \infty \end{cases}$

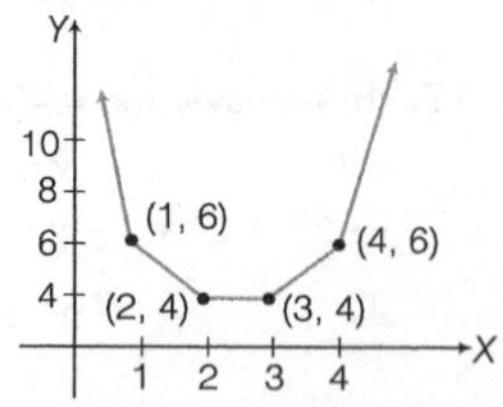

Clearly, we see that $f(x)$ is least i.e. 4, which is not unique and its least value is 4.

22. (a,d) We have, $f(x) = \sec^{-1}[1 + \cos^2 x]$

We know that domain of $\sec^{-1} x$ is $|x| \geq 1$.

$\because \quad |[1 + \cos^2 x]| \geq 0$

which is true for every $x \in R$.

To find the range $0 \leq \cos^2 x \leq 1 \Rightarrow 1 \leq (1 + \cos^2 x) \leq 2$

$\Rightarrow \quad [1 + \cos^2 x] = 1$ and 2

$\therefore$ Range of $f(x)$ is $\{\sec^{-1} 1, \sec^{-1} 2\}$.

Explanations for Q. Nos. (23-25)

From the paragraph, the function is given by

$f(x) = \begin{cases} \pi(2R - x)^2; & 0 \leq x \leq R \\ \pi R^2; & R \leq x \leq 3R \\ \pi(6Rx - x^2 - 8R^2); & 3R \leq x \leq 4R \end{cases}$

We see that, $f(x)$ is not defined outside the interval $[0, 4R]$.

$f(x)$ is decreases on $[0, R]$, so $f(x)$ takes values in $[\pi R^2, 4\pi R^2]$ for $x \in [0, R]$.

We see that on the interval $[3R, 4R]$, $f(x)$ is decreases, so $f(x)$ takes values in $[0, \pi R^2]$. Hence, range of f is $[0, 4\pi R^2]$, since $f(x)$ is strictly decreasing function on $[0, R] \cup [3R, 4R]$, so it is one-one.

On this interval, moreover, the function is constant on $[R, 3R]$, thus it cannot be one-one on $[R, 3R]$ or a set containing $[R, 3R]$.

23. (b) **24.** (a) **25.** (a)

26. (2) We have, $f(x) = \log_e(\sin^{-1} x)$

$f(x)$ is defined, when $\sin^{-1} x > 0$ and $-1 < x < 1$

$\Rightarrow \quad 0 < x < 1$

$\Rightarrow \quad 0 < \sin^{-1} x \leq \dfrac{\pi}{2}$

$\Rightarrow \quad \log_e 0 < \log_e \sin^{-1} x < \log_e \dfrac{\pi}{2}$

$\Rightarrow \quad -\infty < f(x) \leq \log_e \dfrac{\pi}{2}$

But it is $\left(-\infty, \log_e \dfrac{\pi}{2}\right)$.

By comparing, we get $k = 2$.

27. (2) We have, $f(x) = 2 - 3x - 5x^2$

$= -5\left[x^2 + \dfrac{3x}{5} - \dfrac{2}{5}\right] = -5\left[x^2 + \dfrac{3x}{5} + \dfrac{9}{100} - \dfrac{9}{100} - \dfrac{2}{5}\right]$

$= -5\left[\left(x + \dfrac{3}{10}\right)^2 - \dfrac{49}{100}\right] = \dfrac{49}{20} - 50\left(x + \dfrac{3}{10}\right)^2$

$\therefore$ Range of $f(x) = \left(-\infty, \dfrac{49}{20}\right) \Rightarrow \lambda = 2$

By comparing, we get $\lambda = 2$.

28. (1) $\because f(x) = 3|\sin x| - 4|\cos x|$

$f(x)$ is a periodic function period 2π. So, analysis is limited in $[0, 2\pi)$.

$f_{max} = 3 \cdot (1) - 4(0) = 3$ at $x = \dfrac{\pi}{2}$

$f_{min} = 3 \cdot (0) - 4(1) = -4$ at $x = 0$

$\Rightarrow \quad$ Range $\in [-4, 3]$

$\therefore \quad |a + b| = 1$

29. (0) $\because f(x) = \dfrac{\sin x}{\sqrt{1 + \tan^2 x}} + \dfrac{\cos x}{\sqrt{1 + \cot^2 x}}$

$= \sin x \cdot |\cos x| + \cos x |\sin x|$

Period of $f(x) = 2\pi$

$\therefore \quad f(x) = \begin{cases} \sin 2x, & x \in [0, \pi/2) \\ 0, & x \in [\pi/2, \pi) \\ -\sin 2x, & x \in [\pi, 3\pi/2) \\ 0, & x \in [3\pi/2, 2\pi) \end{cases}$

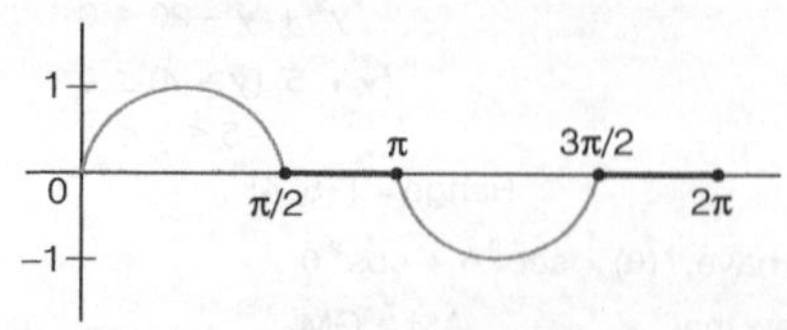

$\therefore$ Range $\in [-1, 1] \Rightarrow |a + b| = 0$

30. (2) We have, $f(x) = \log_{\sqrt{2}}[2 - \log_2(16\sin^2 x + 1)]$

its domain is $2 - \log_2(16\sin^2 x + 1) > 0$

$\Rightarrow \quad -\log_2(16\sin^2 x + 1) > -2$

$\Rightarrow \quad 16\sin^2 x + 1 < 4$

$\Rightarrow \quad \sin^2 x < \dfrac{3}{16}$

$\therefore \quad 0 \leq \sin^2 x < \dfrac{3}{16}$

Now, $\quad 1 \leq 16\sin^2 x + 1 < 4$

$\log_2 1 \leq \log_2(16\sin^2 x + 1) < \log_2 4$

$\Rightarrow \quad 0 \leq \log_2(16\sin^2 x + 1) < 2$

$\Rightarrow \quad -2 < -\log_2(16\sin^2 x + 1) \leq 0$

$\Rightarrow \quad 0 < 2 - \log_2(16\sin^2 x + 1) \leq 2$

$\log_{\sqrt{2}} 0 < \log_{\sqrt{2}}[2 - \log_2(16\sin^2 x + 1)] \leq \log_{\sqrt{2}} 2$

$-\infty < f(x) \leq 2$

But it is $(-\infty, a]$

By comparing, we get

$a = 2$

DPP-5 Identical, Even and Odd Functions

1. (b) For identical functions domain and range should be same.

(a) $f(x) = \sin^{-1} x + \cos^{-1} x = \frac{\pi}{2}$ for all $-1 \le x \le 1$.

$$g(x) = \pi/2 \text{ for all } x \in R.$$

$\therefore$ f and g are not identical.

(b) $f(x) = \tan^{-1} x + \cot^{-1} x = \frac{\pi}{2}$ for all $x \in R$

$$g(x) = \frac{\pi}{2} \text{ for all } x \in R.$$

$\therefore$ f and g are identical.

(c) $f(x) = \sec^{-1} x + \text{cosec}^{-1} x = \frac{\pi}{2}$ for all $|x| \ge 1$.

$$g(x) = \frac{\pi}{2} \text{ for all } x \in R.$$

$\therefore$ f and g are not identical.

2. (b,c,d)

(b) We have, $f(x) = \tan(\tan^{-1} x) = x, x \in R$

and $g(x) = \cot(\cot^{-1} x) = x, x \in R$

Thus, $f(x) = g(x)$

Hence, f and g are identical

(c) We have, $f(x) = \text{sgn}(x) = \begin{cases} 1, & \text{if } x > 0 \\ 0, & \text{if } x = 0 \\ -1, & \text{if } x < 0 \end{cases}$

and $g(x) = \text{sgn}[\text{sgn}(x)] = \begin{cases} 1, & \text{if } \text{sgn}(x) = 1 \\ 0, & \text{if } \text{sgn}(x) = 0 \\ -1, & \text{if } \text{sgn}(x) = -1 \end{cases}$

$$= \begin{cases} 1, & \text{if } x > 0 \\ 0, & \text{if } x = 0 \\ -1, & \text{if } x < 0 \end{cases}$$

Thus, $f(x) = g(x)$

Hence, f and g are identical.

(d) We have, $f(x) = \cot^2 x \cdot \cos^2 x = \frac{\cos^2 x}{\sin^2 x} \cdot \cos^2 x = \frac{\cos^4 x}{\sin^2 x}$

and $g(x) = \cot^2 x - \cos^2 x = \frac{\cos^2 x}{\sin^2 x} - \cos^2 x$

$$= \cos^2 x\left(\frac{1}{\sin^2 x} - 1\right) = \frac{\cos^2 x(1 - \sin^2 x)}{\sin^2 x}$$

$$= \frac{\cos^2 x \cdot \cos^2 x}{\sin^2 x} = \frac{\cos^4 x}{\sin^2 x}$$

Thus, $f(x) = g(x)$

Hence, f and g are identical.

3. (b)

(a) We have, $f(x) = \sqrt{x^2} = |x|, x \in R$

and $g(x) = (\sqrt{x})^2 = x, x \ge 0$

Clearly, $f(x) \ne g(x)$ for all x, as

(b) We have, $f(x) = \sec(\sec^{-1} x) = x, |x| \ge 1$

and $g(x) = \text{cosec}(\text{cosec}^{-1} x) = x, |x| \ge 1$

Clearly, $f(x) = g(x)$

Hence, f and g are identical.

(c) We have, $f(x) = \sqrt{\frac{1 + \cos 2x}{2}} = \sqrt{\frac{2\cos^2 x}{2}} = \sqrt{\cos^2 x} = |\cos x|$

and $g(x) = \cos x$

Clearly, $f(x) \ne g(x)$ for all x.

(d) We have, $f(x) = x$ and $g(x) = \log_e x = x, x > 0$

Then, $f(x) = g(x) = x$

Hence, f and g are not identical.

4. (d)

(a) We have, $f(x) = \sqrt{1 + \sin x}$

$$= \sqrt{\left(\cos\frac{x}{2} + \sin\frac{x}{2}\right)^2} = \left|\cos\frac{x}{2} + \sin\frac{x}{2}\right|$$

and $g(x) = \sin\frac{x}{2} + \cos\frac{x}{2}$

Clearly, $f(x) \ne g(x)$ for all x.

(b) We have,

$$f(x) = \sin^{-1}\left(\frac{2x}{1 + x^2}\right) = 2\tan^{-1} x, \text{ if } -1 \le x \le 1$$

and $g(x) = 2\tan^{-1} x$

Thus, $f(x) \ne g(x)$

Hence, f and g are not identical in their common domain.

(c) We have, $f(x) = \sqrt{x^2} = |x|, x \in R$

and $g(x) = (\sqrt{x})^2 = x, x \ge 0$

Clearly, $f(x) \ne g(x)$ for all x

(d) We have,

$$f(x) = \log_e x^3 + \log_e x^5$$

$$= \log_e(x^3 \cdot x^5) = \log_e x^8 = 8\log_e x, x > 0$$

and $g(x) = 8\log_e x, x > 0$

Hence, $f(x)$ and $g(x)$ are identical.

5. (i) Let $f(x) = \tan x$, then

$$f(-x) = \tan(-x) = -\tan x = -f(x)$$

$\therefore$ $f(x)$ is an odd function.

(ii) Let $f(x) = \cos x$, then

$$f(-x) = \cos(-x) = \cos x = f(x)$$

$\therefore$ $f(x)$ is an even function.

(iii) Let $f(x) = \sin(x^2 + 1)$, then

$$f(-x) = \sin[(-x)^2 + 1] = \sin(x^2 + 1) = f(x)$$

$\therefore$ $f(x)$ is an even function.

(iv) Let $f(x) = x + x^2$, then

$$f(-x) = (-x) + (-x)^2 = -x + x^2 \ne f(x) \text{ and } -f(x)$$

$\therefore$ $f(x)$ is neither even nor odd function.

(v) Let $f(x) = x - x^3$, then

$$f(-x) = (-x) - (x)^3 = -x + x^3$$

$$= -(x - x^3) = -f(x)$$

$\therefore$ $f(x)$ is an odd function.

(vi) We have, $f(x) = x \cdot \left(\frac{a^x - 1}{a^x + 1}\right)$, then

$$f(-x) = (-x)\left(\frac{a^{-x} - 1}{a^{-x} + 1}\right) = (-x)\left(\frac{\frac{1}{a^x} - 1}{\frac{1}{a^x} + 1}\right)$$

$$= (-x)\left(\frac{1 - a^x}{1 + a^x}\right) = x\left(\frac{a^x - 1}{a^x + 1}\right) = f(x)$$

$\therefore$ $f(x)$ is an even function.

(vii) We have, $f(x) = \log(x + \sqrt{x^2 + 1})$, then

$$f(-x) = \log(-x + \sqrt{(-x)^2 + 1})$$

$$= \log(-x + \sqrt{x^2 + 1}) = \log(\sqrt{x^2 + 1} - x)$$

$$= \log\left[\frac{(\sqrt{x^2 + 1} - x)(\sqrt{x^2 + 1} + x}{(\sqrt{x^2 + 1} + x)}\right]$$

$$= \log\left(\frac{x^2+1-x^2}{\sqrt{x^2+1}+x}\right) = \log\left(\frac{1}{\sqrt{x^2+1}+x}\right)$$

$$= -\log(\sqrt{x^2+1}+x) = -f(x)$$

$\therefore$ $f(x)$ is an odd function.

(viii) We have, $f(x) = \sin x + \cos x$, then

$$f(-x) = \sin(-x) + \cos(-x)$$
$$= -\sin x + \cos x \neq f(x) \text{ and } -f(x)$$

$\therefore$ $f(x)$ is neither an odd function nor even function.

(ix) We have, $f(x) = (x^2 - 1)|x|$, then

$$f(-x) = [(-x)^2 - 1)]|-x|$$
$$= (x^2 - 1)|x| = f(x)$$

$\therefore$ $f(x)$ is an even function.

(x) We have, $f(x) = \dfrac{(1+2^x)^7}{2^x}$, then

$$f(-x) = \frac{(1+2^{-x})^7}{2^{-x}} = \frac{(2^x+1)^7}{2^{7x}} \times \frac{1}{2^{-x}}$$
$$= \frac{(2^x+1)^7}{2^{6x}} = \frac{(1+2^x)^7}{2^{6x}}$$
$$\neq f(x) \text{ and } -f(x)$$

$\therefore$ $f(x)$ is neither an odd function nor even function.

(xi) We have, $f(x) = \dfrac{\sec x + x^2 - 9}{x\sin x}$, then

$$f(-x) = \frac{\sec(-x) + (-x)^2 - 9}{(-x)\sin(-x)}$$
$$= \frac{\sec x + x^2 - 9}{(-1)(-x)\sin x} \quad [\because \sec(-\theta) = \sec\theta]$$
$$= \frac{\sec x + x^2 - 9}{x\sin x} = f(x)$$

$\therefore$ $f(x)$ is an even function.

(xii) We have, $f(x) = \sqrt{1+x+x^2} - \sqrt{1-x+x^2}$, then

$$f(-x) = \sqrt{1+(-x)+(-x)^2} - \sqrt{1-(-x)+(-x)^2}$$
$$= \sqrt{1-x+x^2} - \sqrt{1+x+x^2}$$
$$= -(\sqrt{1+x+x^2} - \sqrt{1-x+x^2})$$
$$= -f(x)$$

$\Rightarrow$ $f(x)$ is an odd function.

(xiii) We have, $f(x) = \begin{cases} x|x|, & x \le -1 \\ [1+x]+[1-x], & -1 < x < 1, \text{ then} \\ -x|x|, & x \ge 1 \end{cases}$

$$f(-x) = \begin{cases} (-x)|x|, & -x \le -1 \\ [1-x]+[1+x], & -1 < (-x) < 1 \\ -(-x)|-x|, & -x \ge 1 \end{cases}$$

$$= \begin{cases} -x|x|, & x \ge 1 \\ [1-x]+[1+x], & -1 < x < 1 \\ x|x|, & x \le -1 \end{cases}$$

$$= \begin{cases} x|x|, & x \le -1 \\ [1+x]+[1-x], & -1 < x < 1 = f(x) \\ -x|x|, & x \ge 1 \end{cases}$$

$\therefore$ $f(x)$ is an even function.

(xiv) We have, $f(-x) = \dfrac{-2x(\sin 1(-x)) + \tan(-x)}{2\left[\dfrac{-x}{\pi}\right] + 4 - 3}$

$$= \frac{2x(\sin x + \tan x)}{1 + 2\left(-1 - \left[\dfrac{x}{\pi}\right]\right)}$$
$$= \frac{2x(\sin x + \tan x)}{-2\left[\dfrac{x}{\pi}\right] - 1} = -f(x)$$

$\therefore$ $f(x)$ is odd.

DPP-6 **Periodicity of Functions**

1. (c) We have, $f(x) = x - [x]$

$$f(x) = \begin{cases} x+2; & -2 \le x < -1 \\ x+1; & -1 \le x < 0 \\ x; & 0 \le x < 1 \\ x-1; & 1 \le x < 2 \\ x-2; & 2 \le x < 3 \end{cases}$$

On sketching the graph

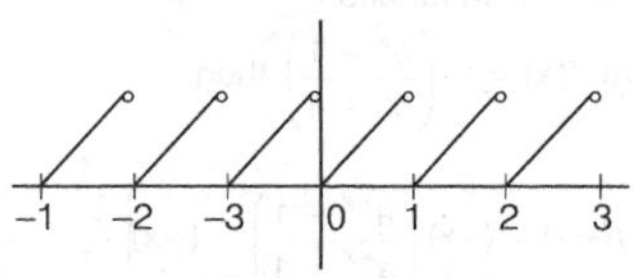

We see that function repeats same trend at constant interval [0, 1); [1, 2), ⋯ . So, breadth that interval is 1.

Hence, its period is 1.

Alternate Method

We know that, $f(x+T) = f(x)$

$$f(x+n) = x + n - [x+n];\ n \in I$$
$$f(x+n) = x + n - [x] - n = x - [x] = f(x)$$

Least positive value of n is 1.

Hence, its period is 1.

2. (b) Given, $f(x) = \sin^2 x$

$$f(x) = \frac{1-\cos 2x}{2}$$

$\because$ $\quad$ Period of $f(x) = \dfrac{2\pi}{2} = \pi$

[period of $f(ax) = \dfrac{T}{|a|}$, where T is the period of $f(x)$]

3. (b) We have, $f(x) = \sin^{-1}(\sin x)$

We know that, if $f(x)$ is not periodic and $g(x)$ is periodic, then $fog\ (x)$ is periodic with period of $g(x)$.

Hence, $f(x) = \sin^{-1}(\sin x)$ is periodic and its period $= 2\pi$.

4. (b) We have, $f(x) = |\cos x|$

$\Rightarrow$ $\quad f(x) = \sqrt{\cos^2 x} = \sqrt{\dfrac{1+\cos 2x}{2}}$

$\therefore$ $\quad$ Period $= \dfrac{2\pi}{2} = \pi$

5. (a) We have, $f(x) = |\cos 2x|$

We know that, period of $|\cos x|$ is π.

$\therefore$ Period of $|\cos 2x| = \dfrac{\pi}{2}$

6. (a) We have, $f(x) = \tan 4x$

We know that period of $\tan x$ is π.

$\therefore$ Period of $\tan 4x$ is $\frac{\pi}{4}$.

7. (c) We have, $f(x) = \tan(3x + 2)$

Its period is $\frac{\pi}{3}$.

8. (d) We have, $f(x) = \sin\left(\frac{x}{3}\right) + \cos\left(\frac{x}{2}\right)$

Let $f_1(x) = \sin\left(\frac{x}{3}\right)$ and $f_2(x) = \cos\left(\frac{x}{2}\right)$

$$T_1 = \text{Period of } f_1(x) = \frac{2\pi}{\frac{1}{3}} = 6\pi$$

$$T_2 = \text{Period of } f_2(x) = \frac{2\pi}{\frac{1}{2}} = 4\pi$$

$\Rightarrow$ Period of $f(x)$ = LCM of of T_1 and T_2

So, period $= 12\pi$

9. (a) We have, $f(x) = |\sin x| + |\cos x|$

Let T_1 = Period of $|\sin x| = \pi$

and T_2 = Period of $|\cos x| = \pi$

$\because f(x)$ is even function and both are complementary to each other.

$\therefore$ Its period is $\frac{1}{2}$ (LCM of T_1 and T_2)

$$= \frac{1}{2}\pi = \frac{\pi}{2}$$

10. (b) We have, $f(x) = \sin^4 x + \cos^4 x$

$$f(x) = (\sin^2 x + \cos^2 x)^2 - 2\sin^2 x \cos^2 x$$
$$= 1 - \frac{1}{2}(2\sin x \cdot \cos x)^2$$
$$= 1 - \frac{1}{2}\sin^2 2x$$
$$= 1 - \frac{1}{2}\left(\frac{1 - \cos 4x}{2}\right)$$
$$= 1 - \frac{1}{4}(1 - \cos 4x)$$

$\therefore$ Its period $= \frac{2\pi}{4} = \frac{\pi}{2}$

11. (c) We have, $f(x) = \sec(\sin x)$, $f(x + \pi) = \sec(\sin(x + \pi))$

$$= \sec(-\sin x)$$
$$= \sec(\sin x)$$
$$= f(x)$$

Hence its period is π.

12. (d) We have, $f(x) = \sin(\sqrt{[a]}x)$

Its period $= \frac{2\pi}{\sqrt{[a]}} = \pi$ (given)

$\Rightarrow \sqrt{[a]} = 2$

$\Rightarrow [a] = 4$

$\Rightarrow 4 \le a < 5$

$\therefore a \in [4, 5)$

13. (a) We have,

$$f(x) = (x + a) - [x + b] + \sin \pi x + \cos 2\pi x + \sin 3\pi x + \cos 4\pi x + \cdots + \sin(2n - 1)\pi x + \cos 2n\pi x$$

To period of $x + a - [x + b] + b - b$

$\Rightarrow x + b - [x + b] + a - b$

$\Rightarrow \{x + b\} + a - b$

Hence, we see that its period is 1.

Now, period of $\sin \pi x$ is $\frac{2\pi}{\pi} = 2$

Period of $\cos 2\pi x$ is $\frac{2\pi}{2\pi} = 1$

Similarly, period of $\cos 2n\pi x = \frac{2\pi}{2n\pi} = \frac{1}{n}$.

$\therefore$ Period of $f(x)$ is LCM of all above period, which is 2.

14. (c) We have, $f(x) = \sin\frac{\pi}{4}[x] + \cos\frac{\pi x}{2} + \cos\frac{\pi}{3}[x]$

Let T_1 = Period of $\sin\frac{\pi}{4}[x]$

$\Rightarrow \sin\frac{\pi}{4}[x + T_1] = \sin\frac{\pi}{4}[x]$

$\Rightarrow \frac{\pi}{4}[x + T_1] = 2n\pi + \frac{\pi}{4}[x]$

$\Rightarrow [x + T_1] = 8n + [x]$

$\therefore T_1 = 8n$

Its minimum value is 8.

Hence, $T_1 = 8$

$$T_2 = \text{Period of } \cos\frac{\pi x}{2} = \frac{2\pi}{\frac{\pi}{2}} = 4$$

$$T_3 = \text{Period of } \cos\frac{\pi}{3}[x] = 6$$

$\therefore$ Period of $[x]$ = LCM of T_1, T_2 and $T_3 = 24$

15. (a) We have,

$$f(x) = [x] + \left[x + \frac{1}{3}\right] + \left[x + \frac{2}{3}\right] - 3x + 15$$
$$= -x + [x] - \left(x + \frac{1}{3}\right) + \left[x + \frac{1}{3}\right] - \left(x + \frac{2}{3}\right) + \left[x + \frac{2}{3}\right] + 15 + \frac{1}{3} + \frac{2}{3}$$

$$\Rightarrow f(x) = -\{x\} - \left\{x + \frac{1}{3}\right\} - \left\{x + \frac{2}{3}\right\} + 16$$

$$\therefore f\left(x + \frac{1}{3}\right) = -\left\{x + \frac{1}{3}\right\} - \left\{x + \frac{2}{3}\right\} - \{x + 1\} + 16$$
$$= -\left\{x + \frac{1}{3}\right\} - \left\{x + \frac{2}{3}\right\} - \{x\} + 16 \text{ as } \{x + 1\} = \{x\}$$

$$\Rightarrow f\left(x + \frac{1}{3}\right) = f(x)$$

$\therefore$ Period of $f(x)$ is $\left(\frac{1}{3}\right)$.

16. (a) We have, $f(x) = \sin^4 3x + \cos^4 3x$

We know that period of $\sin^4 x + \cos^4 x$ is $\frac{\pi}{2}$.

$\therefore$ Period of $f(x) = \frac{1}{3}\left(\frac{\pi}{2}\right) = \frac{\pi}{6}$

17. (a,b,d) (a) $y = [x + 1]$ is not periodic because $x + 1$ is not periodic.

(b) $y = \sin(x)^2$ is not periodic because x^2 is not periodic.

(c) $y = \sin^2 x$ is periodic because $\sin x$ is periodic.

(d) $y = \sin^{-1} x$ is not periodic.

18. (c) We have, $f(x) = \sin\left(\frac{\pi}{2}[x]\right)$

To check $f(x)$ is even or odd.

$$f(-x) = \sin\left(\frac{\pi}{2}[-x]\right)$$

Clearly, we cannot say anything because x is not given integer or not integer.

Hence, $f(x)$ is not even or odd.

To check periodicity

$$f(x+T)=f(x)$$

$$f(x+T)=\sin\left(\frac{\pi}{2}[x+T]\right)=\sin\left(\frac{\pi}{2}[x]\right)$$

$\Rightarrow$ $\frac{\pi}{2}[x+T]=2n\pi+\frac{\pi}{2}[x]$

$\Rightarrow$ $[x+T]=4n+[x]$

$\Rightarrow$ $T+[x]=4n+[x]$

$\therefore$ $T=4n$

Hence, $f(x)$ is periodic with period 4.

19. (a,b,c)

(a) $f(x)=\cos(\cos^{-1}x)=x;\ -1\le x\le 1$, then

$$f(-x)=\cos\cos^{-1}(-x)=-x$$

$\therefore$ It is odd function.

Hence, it is incorrect.

(b) We have,

$$f(x)=\cos(\sin x)+\cos(\cos x)$$

Let it period be T.

$$f(x+T)=f(x)$$

$\cos(\sin(x+T))+\cos(\cos(x+T))=\cos(\sin x)+\cos(\cos x)$

Put $x=0$, we get

$$\cos\sin T+\cos\cos T=\cos 0+\cos 1$$

or $\cos\sin T+\cos\cos T=\cos\left(\cos\left(\frac{\pi}{2}\right)\right)+\cos\left(\sin\left(\frac{\pi}{2}\right)\right)$

By comparing, we get $T=\frac{\pi}{2}$

Hence, its period is $\frac{\pi}{2}$.

It is incorrect.

(c) We have, $f(x)=\cos(3\sin x)$

We know that, $-1\le\cos\theta\le 1$

$\Rightarrow$ $f(x)\in[-1,1]$

20. (a,b,c) We have, $f(x)=\dfrac{\sin\pi[x]}{\{x\}}$

Let T_1 be the period of $\sin\pi[x]$.

Then, $\sin\pi[T_1+x]=\sin\pi[x]$

$\Rightarrow$ $\pi[T_1+x]=2n\pi+\pi[x]$

$\Rightarrow$ $T_1+[x]=2n+[x]$

$\therefore$ $T_1=2n$

and minimum value is 2.

$$T_1=2$$

$T_2=$ Period of $\{x\}$ is 1.

Hence, period of $f(x)=$ LCM of (T_1 and T_2)

$=2$

To find range of $f(x)=\sin\pi[x]$ is always 0

Hence, range of $f(x)=0$, which is a singleton set.

Since, $f(x)$ is always 0, $\forall x\in R$.

$\therefore$ $f(x)$ is an even function.

21. (a)

(i) $f(x)=\sin\left(2\pi t+\frac{\pi}{3}\right)+2\sin\left(3\pi t+\frac{\pi}{4}\right)+3\sin(5\pi t)$

Let period of $\sin\left(2\pi t+\frac{\pi}{3}\right)=T_1=\frac{2\pi}{2\pi}=1$

Period of $\sin\left(3\pi t+\frac{\pi}{4}\right)=T_2=\frac{2\pi}{3\pi}=\frac{2}{3}$

Period of $3\sin 5\pi t=T_3=\frac{2\pi}{5\pi}=\frac{2}{5}$

Now, period of $f(x)=\dfrac{\text{LCM of numerator of } T_1, T_2 \text{ and } T_3}{\text{HCF of denominator of } T_1, T_2 \text{ and } T_3}$

$=\dfrac{\text{LCM of (1, 2 and 2)}}{\text{HCF of (1, 3 and 5)}}=2$

Thus, **(i) → (p)**

(ii) We have, $f(x)=\frac{1}{2}\left(\dfrac{\left|\sin\frac{\pi}{4}x\right|}{\cos\frac{\pi}{4}x}+\dfrac{\sin\frac{\pi}{4}x}{\left|\cos\frac{\pi}{4}x\right|}\right)$

We know that period of

$$\left|\sin\frac{\pi x}{4}\right| \text{ and } \left|\cos\frac{\pi x}{4}\right| \text{ is } \frac{\pi}{\frac{\pi}{4}}=4$$

and period of $\cos\frac{\pi x}{4}$ and $\sin\frac{\pi x}{4}$ is

$$=\frac{2\pi}{\frac{\pi}{4}}=8$$

Now, period of $\dfrac{\left|\sin\frac{\pi}{4}x\right|}{\cos\frac{\pi x}{4}}$ is 8.

and period of $\dfrac{\sin\frac{\pi x}{4}}{\left|\cos\frac{\pi x}{4}\right|}$ is 8.

Hence, period of $f(x)$ is 8.

Thus, **(ii) → (s)**

(iii) Let period of $f(x)$ is T.

Then, $\tan\left(\frac{\pi}{2}[x+T]\right)=\tan\frac{\pi}{2}[x]$

$\Rightarrow$ $\frac{\pi}{2}[x+T]=n\pi+\frac{\pi}{2}[x]$

$\Rightarrow$ $[x+T]=2n+[x]$

$\Rightarrow$ $T+[x]=2n+[x]$

$\therefore$ $T=2n$

and minimum +ve value of T is 2 and hence, its period is 2.

Thus, **(iii) → (p)**

(iv) By inspection, we check π can be period of $f(x)$ or not.

$$f(\pi+x)=\frac{\sin(\pi+x)+\sin 3(\pi+x)}{\cos(\pi+x)+\cos 3(\pi+x)}$$

$$=\frac{-\sin x-\sin 3x}{-\cos x-\cos 3x}$$

$$=\frac{\sin x+\sin 3x}{\cos x+\cos 3x}=f(x)$$

We see that, $f(\pi+x)=f(x)$.

Hence, π is the period of $f(x)$.

Thus, **(iv) → (q)**

Explanations for Q. Nos. (22-24)

We have,

$$f_1(x)=e^{\tan\left\{\frac{x}{4}\right\}}+\cos\pi\left(\frac{1-2[x]}{2}\right)+\sin\left(\frac{\pi[x]}{2}\right)$$

Let $T_1=$ Period of $e^{\tan\left\{\frac{x}{4}\right\}}=\dfrac{1}{\frac{1}{4}}=4$

$T_2=$ Period of $\cos\pi\left(\frac{1-2[x]}{2}\right)$

or $\cos\left(\frac{\pi}{2}-\pi[x]\right)=\sin\frac{\pi[x]}{2}$

$$\Rightarrow \quad \sin\frac{\pi}{2}[x+T_2]=\sin\frac{\pi}{2}[x]$$

$$\Rightarrow \quad \frac{\pi}{2}[x+T_2]=2n\pi+\frac{\pi}{2}[x]$$

$$\Rightarrow \quad [x+T_2]=4n+[x] \quad \Rightarrow \quad T_2=4n$$

$\because$ Its period $T_2 = 4$

So, period of $f_1(x)$ is $p = 4$.

We have, $y=\sqrt{2p+\frac{p}{2}[x]-[x]^2}$

We calculate $\quad p = 4$

$$\therefore \quad y=\sqrt{8+2[x]-[x]^2}$$

$$\text{Domain} = 8+2[x]-[x]^2 \ge 0$$

$$\Rightarrow \quad [x]^2-2[x]-8\le 0$$

$$\Rightarrow \quad ([x]-4)([x]+2)\le 0$$

$$\Rightarrow \quad -2\le [x]\le 4$$

$$\Rightarrow \quad -2\le x<5$$

$$[n_2\le [x]\le n_1 \quad \Rightarrow \quad n_2\le x<n_1+1, \text{ where } n, n_2\in I]$$

Since, we have domain of y is $[q, r)$.

Hence, by comparing, we get $q = -2$ and $r = 5$.

22. (c) $p = 4$, which is a composite number.

23. (a) The value of $r-q-1$

$$=5+2-1=6$$

24. (a) We have, $f_2(x)=\begin{cases}2+x; & x\ge 0\\ 2-x; & x<0\end{cases}$

$$f_2[f_2(x)]=\begin{cases}2+f_2(x); & f_2(x)\ge 0\\ 2-f_2(x); & f_2(x)<0\end{cases}$$

$$\Rightarrow \quad 2+f_2(x); f_2(x)\ge 2$$

$$f_2(f_2(x))=\begin{cases}2+2+x; & 2+x\ge 2\\ 2+2-x; & 2-x\ge 2\end{cases}$$

$$=\begin{cases}4+x; & x\ge 0\\ 4-x; & x\le 0\end{cases}=\begin{cases}4+x; & x\ge 0\\ 4-x; & x<0\end{cases}$$

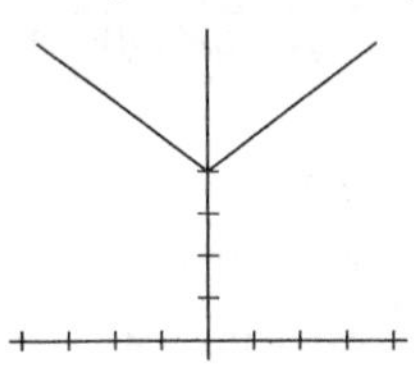

Hence, by graph the range of $f_2(x) = [4, \infty)$.

And we know that $p = 4$, hence range is $[p, \infty)$.

25. (4) Let period of $f(x)$ is T.

$$\Rightarrow \quad f(x+T)=f(x)$$

$$\Rightarrow \quad \sin\left(\frac{\pi}{2}[x+T]\right)=\sin\frac{\pi}{2}[x]$$

$$\Rightarrow \quad \frac{\pi}{2}[x+T]=2n\pi+\frac{\pi}{2}[x]$$

$$\Rightarrow \quad \pi[x+T]=4n\pi+[x]\pi$$

$$\Rightarrow \quad [x+T]=4n+[x]$$

$$\Rightarrow \quad T+[x]=4n+[x]$$

$$\Rightarrow \quad T=4n$$

So, minimum value is 4.

Hence, its period is 4.

26. (2) We have,

$$f(x)=e^{\cos^4\pi x+x-[x]+\cos\pi x}$$

Period of $\quad \cos^4\pi x=\frac{\pi}{\pi}=1=T_1 \quad$ [say]

Period of $\quad x-[x]=1=T_2 \quad$ [say]

Period of $\quad \cos\pi x=\frac{2\pi}{\pi}=2=T_3 \quad$ [say]

Now, period of $f(x)$ = LCM of T_1, T_2 and $T_3 = 2$.

Hence, its period is 2.

27. (8) We have,

$$f(x-1)+f(x+3)=f(x+1)+f(x+5) \quad \ldots\text{(i)}$$

Replacing x by $x+2$ in Eq. (i), we get

$$f(x+1)+f(x+5)=f(x+3)+f(x+7) \quad \ldots\text{(ii)}$$

On adding Eqs. (i) and (ii), we get

$$f(x-1)=f(x+7) \quad \ldots\text{(iii)}$$

Now, replacing x by $x+1$ in Eq. (iii), we get

$$f(x)=f(x+8)$$

Hence, period of $f(x)$ is 8.

28. (2) We have,

$$f(x+p)=1+\{1-3f(x)+3f^2(x)-f^3(x)\}^{1/3}$$

$$\Rightarrow \quad f(x+p)=1+[\{1-f(x)\}^3]^{1/3}$$

$$\Rightarrow \quad f(x+p)=1+1-f(x)$$

$$\Rightarrow \quad f(x+p)+f(x)=2 \quad \ldots\text{(i)}$$

Now, replacing x by $x+p$ in Eq. (i), we get

$$f(x+2p)+f(x+p)=2 \quad \ldots\text{(ii)}$$

On subtracting Eq. (i) from Eq. (ii), we get

$$f(x+2p)-f(x)=0$$

$$\therefore \quad f(x+2p)=f(x)$$

On comparing, we get $\lambda p = 2p$

$$\therefore \quad \lambda=2$$

29. (2) We have,

$$f(x+a)=\frac{1}{2}+\sqrt{f(x)-(f(x))^2} \quad \ldots\text{(i)}$$

Replacing x by $x+a$ in Eq. (i), we get

$$f(x+a+a)=\frac{1}{2}+\sqrt{f(x+a)-(f(x+a))^2} \quad \ldots\text{(ii)}$$

On putting the value of $f(x+a)$ from Eq. (i) in Eq. (ii), we get

$$f(x+2a)=\frac{1}{2}+\sqrt{\frac{1}{2}+\sqrt{f(x)-\{f(x)\}^2}-\left\{\frac{1}{2}+\sqrt{f(x)-\{f(x)\}^2}\right\}^2}$$

$$\Rightarrow f(x+2a)=\frac{1}{2}+\sqrt{\frac{1}{2}+\sqrt{f(x)-\{f(x)\}^2}-\frac{1}{4}-f(x)+(f(x))^2-\sqrt{f(x)-\{f(x)\}^2}}$$

$$\Rightarrow \quad f(x+2a)=\frac{1}{2}+\sqrt{\frac{1}{2}-\frac{1}{4}-f(x)+\{f(x)\}^2}$$

$$=\frac{1}{2}+\sqrt{\frac{1}{4}-f(x)+\{f(x)\}^2}$$

$$=\frac{1}{2}+\sqrt{\left(\frac{1}{2}-f(x)\right)^2}$$

$$=\frac{1}{2}+f(x)-\frac{1}{2}$$

$\therefore f(x+2a)=f(x)$

$\therefore$ Its period is $2a$ but it is given λa.

Hence, $\lambda = 2$.

DPP-7 Composition of Functions

1. Given, $f(x) = x^2 + x + 1$

and $g(x) = \sin x$

$\Rightarrow$ $fog(x) = f\{g(x)\} = \sin^2 x + \sin x + 1$

and $gof(x) = g\{f(x)\} = \sin(x^2 + x + 1)$

$\therefore$ $gof(x) \neq fog(x)$ **Hence proved.**

2. We have,

$f(x) = x^2, g(x) = \sin x$ and $h(x) = \sqrt{x}$

$\Rightarrow$ $(goh)(x) = g[h(x)] = g(\sqrt{x}) = \sin\sqrt{x}$

and $[fo(goh)](x) = f(\sin\sqrt{x}) = (\sin\sqrt{x})^2 = \sin^2\sqrt{x}$ ...(i)

Again $(fog)(x) = f(\sin x) = \sin^2 x$

and $[(fog)oh](x) = [(fog)(\sqrt{x})] = \sin^2\sqrt{x}$ ...(ii)

From Eqs. (i) and (ii)

$[fo(goh)](x) = [(fog)oh)]x)$ **Hence proved.**

3. (i) We have,

$f(x) = e^x$ and $g(x) = \log x$

$\therefore$ $fog(x) = f(\log x) = e^{\log x} = x;\ x > 0$

and $gof(x) = g(e^x) = \log e^x = x;\quad x \in R$

(ii) We have, $f(x) = |x|$ and $g(x) = \sin x$

$\therefore$ $fog(x) = f(\sin x) = |\sin x|$

and $gof(x) = g(|x|) = \sin|x|$

(iii) We have,

$f(x) = \sin^{-1} x$ and $g(x) = x^2$

$\therefore$ $fog(x) = f(x^2) = \sin^{-1} x^2$

and $gof(x) = g(\sin^{-1} x) = (\sin^{-1} x)^2$

(iv) We have,

$f(x) = x^2 + 2$ and $g(x) = 1 - \dfrac{1}{1-x}, x \neq 1,$

$\therefore$ $fog(x) = f\left(1 - \dfrac{1}{1-x}\right) = \left(1 - \dfrac{1}{1-x}\right)^2 + 2$

$= 1 + \left(\dfrac{1}{1-x}\right)^2 - 2\cdot 1\cdot \dfrac{1}{1-x} + 2$

$= 1 + \dfrac{1}{1 + x^2 - 2x} - \dfrac{2}{1-x} + 2$

$= \dfrac{1 + x^2 - 2x + 1}{1 + x^2 - 2x} - \dfrac{2(1-x)}{(1-x)^2} + 2$

$= \dfrac{2 + x^2 - 2x - 2 + 2x + 2(1 + x^2 - 2x)}{(1-x)^2}$

$= \dfrac{x^2 + 2 + 2x^2 - 4x}{(1-x)^2} = \dfrac{3x^2 - 4x + 2}{(1-x)^2}$

and $(gof)(x) = g(x^2 + 2) = 1 - \dfrac{1}{1 - (x^2 + 2)}$

$= \dfrac{1 - x^2 - 2 - 1}{1 - x^2 - 2} = \dfrac{-2 - x^2}{-1(1 + x^2)} = \dfrac{-(x^2 + 2)}{-(x^2 + 1)}$

$= \dfrac{x^2 + 2}{x^2 + 1}$

4. (d) We have,

$f(x) = e^{2x}$ and $g(x) = \log(\sqrt{x}), x > 0$

Then, $(fog)(x) = f(\log\sqrt{x})$

$= e^{2\log\sqrt{x}} = e^{\log x^{\frac{1}{2}\times 2}}$

$= x$

5. (a) We have, $f : [-6, 6] \to R$

Then, $f(x) = x^2 - 3$ for $x \in R$

$(fofof)(-1) + (fofof)(0) + (fofof)(1) = ?$

$\therefore$ $fof(x) = f(x^2 - 3)$

$= (x^2 - 3)^2 - 3 = x^4 + 9 - 6x^2 - 3$

$= x^4 + 6 - 6x^2$

and $(fofof)(x) = (x^4 + 6 - 6x^2)^2 - 3$

$\therefore$ $(fofof)(-1) = [(-1)^4 + 6 - 6(-1)^2]^2 - 3 = -2$

$(fofof)(0) = 36 - 3 = 33$

and $(fofof)(1) = (1 + 6 - 6)^2 - 3 = -2$

$\therefore$ $(fofof)(-1) + (fofof)(0) + (fofof)(1)$

$= -2 + 33 - 2 = 29$

Now, $f(y) = y^2 - 3 = 29$

$\Rightarrow$ $y^2 = 32$

$\Rightarrow$ $y = 4\sqrt{2}$

$\therefore$ $(fofof)(-1) + (fofof)(0) + (fofof)(1) = f(4\sqrt{2})$

6. (b) We have, $f(x) = \dfrac{x}{\sqrt{1 + x^2}}$

$\therefore$ $fof(x) = f\left(\dfrac{x}{\sqrt{1 + x^2}}\right) = \dfrac{x/\sqrt{1 + x^2}}{\sqrt{1 + \dfrac{x^2}{1 + x^2}}}$

$= \dfrac{x/\sqrt{1 + x^2}}{\sqrt{1 + 2x^2}/\sqrt{1 + x^2}} = \dfrac{x}{\sqrt{1 + 2x^2}}$

$\therefore$ $(fofof)(x) = \dfrac{x/\sqrt{1 + 2x^2}}{\sqrt{1 + \dfrac{x^2}{1 + 2x^2}}}$

$= \dfrac{x/\sqrt{1 + 2x^2}}{\sqrt{1 + 3x^2}/\sqrt{1 + 2x^2}} = \dfrac{x}{\sqrt{1 + 3x^2}}$

7. (a) We have,

$g(x) = x^2 + x - 2$

and $\dfrac{1}{2}(gof)(x) = 2x^2 - 5x + 2$

Now, $(gof)(x) = 4x^2 - 10x + 4$

$= (2x - 3)^2 - 5 + 2x$

and $g\{f(x)\} = (2x - 3)^2 + (2x - 3) - 2$

$\therefore$ $f(x) = 2x - 3$

8. (a) We have,

$f(x) = e^x$ and $g(x) = \log_e x$

$\therefore$ $f\{g(x)\} = f(\log_e x) = e^{\log_e x} = x;\ x > 0$

and $g\{f(x)\} = g(e^x) = \log_e e^x = x$

$\therefore$ $f\{g(x)\} \neq g\{f(x)\}$

9. (d) We have,

$f(x) = \dfrac{x}{x+1}$ and $g(x) = \dfrac{x}{1-x}$

$\therefore$ $(fog)(x) = f\left(\dfrac{x}{1-x}\right) = \dfrac{\dfrac{x}{1-x}}{\dfrac{x}{1-x} + 1}$

$= \dfrac{\dfrac{x}{1-x}}{\dfrac{x + 1 - x}{1-x}} = x$

10. (a) Given, $f(x) = (x+1)^2$ and $g(x) = x^2 + 1$

$\therefore \quad (fog)(x) = f(x^2+1)$

$= (x^2+1+1)^2 = (x^2+2)^2$

$\therefore \quad (fog)(-3) = [(-3)^2+2]^2 = 121$

11. (b) Given, $g(x) = 1+\sqrt{x}$

and $f\{g(x)\} = 3+2\sqrt{x}+x$

$= (1+\sqrt{x})^2 + 2$

$\therefore \quad f(x) = x^2+2$

12. (b) Given,

$$g(x) = 1+\{x\},\ f(x) = \begin{cases} -1, & x<0 \\ 0, & x=0, \\ 1, & x>0 \end{cases}$$

where $\{x\}$ represents the fractional part function.

$$\therefore \quad f\{g(x)\} = \begin{cases} -1, & 1+\{x\}<0 \\ 0, & 1+\{x\}=0 \\ 1, & 1+\{x\}>0 \end{cases}$$

$= 1, \forall x \in R, \qquad [\because 1+\{x\}>0]$

13. (d) Given, $f(x) = \dfrac{\alpha x}{x+1}, x \neq -1$

and $f\{f(x)\} = x \Rightarrow \dfrac{\alpha\left(\dfrac{\alpha x}{x+1}\right)}{\dfrac{\alpha x}{x+1}+1} = x$

$\Rightarrow \quad \dfrac{\alpha^2 x}{(\alpha+1)x+1} = x$

$\Rightarrow \quad (\alpha+1)x^2 + (1-\alpha^2)x = 0 \qquad \dots(i)$

$\therefore \quad \alpha+1=0$ and $1-\alpha^2=0$

[as it is true, $\forall x \neq -1$, Eq. (i) in an identity]

$\therefore \quad \alpha = -1$

14. (b) Given, $f(x) = \dfrac{2x+1}{3x-2}$

$$\therefore \quad (fof)(x) = \left[f\, f\left(\frac{2x+1}{3x-2}\right)\right] = \frac{2\cdot\left(\dfrac{2x+1}{3x-2}\right)+1}{3\cdot\left(\dfrac{2x+1}{3x-2}\right)-2}$$

$$= \frac{\dfrac{4x+2+3x-2}{3x-2}}{\dfrac{6x+3-6x+4}{3x-2}} = \frac{7x}{7} = x$$

$\therefore \quad fof(2) = 2$

15. (c) Given, $f(x) = (a-x^n)^{1/n}$

$\therefore \quad f[f(x)] = f(a-x^n)^{1/n}$

$= [a-[(a-x^n)^{1/n}]^n]^{1/n}$

$= [a-(a-x^n)]^{1/n}$

$= [a-a+x^n]^{1/n} = (x^n)^{1/n} = x$

16. (c) Given, domain of $f(x)$ is $[-5, 7]$

and $x \in [-5, 7]$

and $g(x) = |2x+5|$

$\therefore \quad fog(x) = f\{g(x)\} = f[|2x+5|]$

$\Rightarrow \quad -5 \le |2x+5| \le 7 \Rightarrow 0 \le |2x+5| \le 7$

$\Rightarrow \quad -7 \le 2x+5 \le 7$

$\Rightarrow \quad -12 \le 2x \le 2$

$\Rightarrow \quad -6 \le x \le 1$

$\therefore \quad fog(x) \in [-6, 1]$

17. (a) Given, $f(x) = \dfrac{ax+b}{cx+d}$

$$\therefore \quad (fof)(x) = f\left(\frac{ax+b}{cx+d}\right) = \frac{a\cdot\dfrac{ax+b}{cx+d}+b}{c\cdot\dfrac{ax+b}{cx+d}+d}$$

$$= \frac{a^2x+ab+bcx+bd}{cax+cb+cdx+d^2}$$

$$= \frac{a^2x+bcx+ab+bd}{d^2+cax+cdx+bc}$$

But $(fof)(x) = x$

$$\therefore \quad \frac{a^2x+bc\,x+ab+bd}{d^2+cax+cdx+bc} = x$$

$$fof(x) = \frac{(a^2+bc)x+(ab+bd)}{(ac+cd)x+(bc+d^2)} = x$$

On comparing the coefficient of both sides, we get

$(a^2+bc)x+(ab+bd) = (ac+cd)x^2+(bc+d^2)x$

$\Rightarrow \quad a^2+bc = bc+d^2 \Rightarrow a = d$ or $a = -d$

and $ab+bd = 0 \Rightarrow b = 0$ or $a = -d$

and $ac+cd = 0 \Rightarrow c = 0$ or $a = -d$

which can be simultaneously true for $a = -d$ i.e. $a+d = 0$.

18. (c) Given, $f(x) = \cot^{-1}(x) : R^+ \to \left(0, \dfrac{\pi}{2}\right)$

and $g(x) = 2x - x^2 : R \to R$

$\therefore \quad f\{g(x)\} = \cot^{-1} g(x)$

$= \cot^{-1}(2x-x^2)$

So, the range of $f\{g(x)\}$ is $\left[\dfrac{\pi}{4}, \dfrac{\pi}{2}\right)$.

19. (0) Given,

$$f(x) = \begin{cases} 1+x^2, & x \le 1 \\ x+1, & 1 < x \le 2 \end{cases}$$

and $g(x) = 1-x, -2 \le x \le 1$

$$\text{Now,}\quad fog(x) = f\{g(x)\} = \begin{cases} 1+\{g(x)\}^2, & g(x) \le 1 \\ g(x)+1, & 1 < g(x) \le 2 \end{cases}$$

$$= \begin{cases} 2-2x+x^2, & 0 < x < 1 \\ 2-x, & -1 \le x < 0 \end{cases}$$

On comparing, we get

$\therefore \quad a+b+c = 0+1+(-1) = 0$

20. (6) Given, $f(x) = \begin{cases} 1+x, & 0 \le x \le 2 \\ 3-x, & 2 < x \le 3 \end{cases}$

$$\text{Now,}\quad (fof)(x) = \begin{cases} 1+f(x), & 0 \le f(x) \le 2 \\ 3-f(x), & 2 < f(x) \le 3 \end{cases}$$

$$= \begin{cases} 2+x, & -1 \le x \le 1 \\ 2-x, & 1 < x \le 2 \\ 4-x, & 2 < x \le 4 \end{cases}$$

On comparing, we get

$a+b+c+d = -1+1+2+4 = 6$

DPP-8 Types of Functions

1. (c) We have, $f(x) = x^3 - 1$

$$f'(x) = 3x^2 \Rightarrow f(x) > 0, \forall x \in R$$

So, $f(x)$ is increasing,for all $x \in R$.
Now, $f(x)$ is one-one function.
So, the range of $f(x)$ is $(-\infty, \infty)$.
$\therefore f(x)$ is also onto function.
Since, $f(x)$ is one-one and onto.
Hence, $f(x)$ is a bijection.

2. (a) We have, gof is injective and surjective such that $gof : A \in C$

Let $\quad h(x) = gof(x)$

$\therefore \quad h(y) = gof(y)$

$h(x) = h(y) \Rightarrow gof(x) = gof(y)$

$\Rightarrow \quad g\{f(x)\} = g\{f(y)\} \Rightarrow f(x) = f(y)$ ($\because g$ is injective)

$\Rightarrow \quad x = y$ ($\because f$ is injective)

Thus, $\quad gof(x) = gof(y), \forall x, y \in R$

$\therefore$ gof is injective iff g and f both injective.

Let $z \in C$ such that $g(y) = z$ ($\because g$ is surjective)

and $\quad y \in B$ such that $f(x) = y$ ($\because f$ is surjective)

$\therefore \quad gof(x) = g(y) = z$

Every element of C is the some element of A.
So, gof is surjective iff g and f both are surjective.

3. (d) We have, $f(x) = \cos x$

$$f\left(\frac{\pi}{4}\right) = \cos\frac{\pi}{4} = \frac{1}{\sqrt{2}}$$

$$\Rightarrow \quad f\left(\frac{7\pi}{4}\right) = \cos\left(\frac{7\pi}{4}\right) = \cos\left(2\pi - \frac{\pi}{4}\right) = \cos\frac{\pi}{4} = \frac{1}{\sqrt{2}}$$

$$\Rightarrow \quad f\left(\frac{\pi}{4}\right) = f\left(\frac{7\pi}{4}\right) = \frac{1}{\sqrt{2}}$$

$\therefore \cos x$ is not one-one function and $\cos x$ lies between -1 to 1.
$\therefore \cos x$ is not onto function.
Hence, $f(x)$ is neither one-one nor onto function.

4. (b) We have,
$f(x) = (x - 1)(x - 2)(x - 3)$
The graph of $f(x)$ is

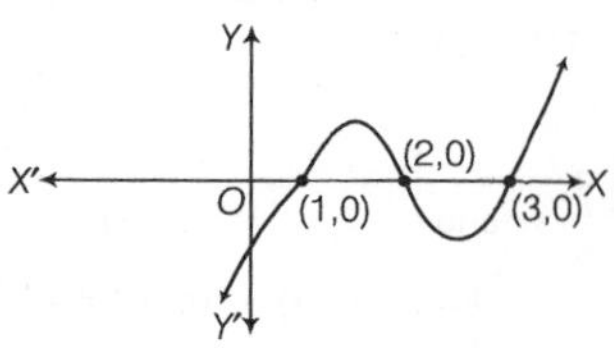

From the above graph, $f(x)$ is not one-one and range of $f(x)$ is $(-\infty, \infty)$.
Hence, $f(x)$ is onto function.

5. (d) We have, $f(z) = |z|$

Let $\quad z_1 = 1 + i, |z_1| = \sqrt{2}$

and $\quad z_2 = 1 - i \quad |z_2| = \sqrt{2}$

$\Rightarrow \quad f(z_1) = f(z_2)$, but $z_1 \neq z_2$

$\therefore f(z)$ is not one-one function.

Also, $\quad |z| \geq 0$

$\therefore f(z)$ is also not onto function.
Hence, $f(z)$ is neither one-one nor onto function.

6. (c) We have, $f(x) = x^3 + 5x + 1$

$$f'(x) = 3x^2 + 5 \Rightarrow f'(x) > 0, \quad \forall x \in R$$

$\therefore \quad f(x)$ is one-one function and range of $f(x)$ is $(-\infty, \infty)$.
$\therefore f(x)$ is also onto function.
Therefore , $f(x)$ is one-one function and onto, $\forall x \in R$.

7. (c) We have, set A has 3 elements and set B has 4 elements.
$\therefore$ Total number of injective from set A to set $B = {}^4P_3 = 4! = 24$

8. (b) We have, $f(x) = \dfrac{x - m}{x - n}$

$$f(x_1) = \frac{x_1 - m}{x_1 - n} \text{ and } f(x_2) = \frac{x_2 - m}{x_2 - n}$$

Let $\quad f(x_1) = f(x_2) \Rightarrow \dfrac{x_1 - m}{x_1 - n} = \dfrac{x_2 - m}{x_2 - m}$

$\Rightarrow \quad x_1 = x_2$, so $f(x)$ is one-one function

Let $\quad f(x) = y$

$\Rightarrow \quad y = \dfrac{x - m}{x - n} \Rightarrow x = \dfrac{m - ny}{1 - y}$

Clearly, $y \in R - \{1\}$, so f is not onto.
$\therefore f(x)$ is one-one and into function.

9. (a) Let $f(x) = 2x - 5$, then $f(x_1) = f(x_2)$

$\Rightarrow \quad 2x_1 - 5 = 2x_2 - 5 \Rightarrow x_1 = x_2$

and $\quad$ range of $f(x)$ is $(-\infty, \infty)$.

$\therefore \quad 2x - 5$ is bijective.

Let $\quad g(x) = |x|$

$\quad g(-1) = g(1) = 1$

$\Rightarrow \quad g(x)$ is not one-one function.

$\therefore \quad g(x)$ is not bijective.

Similarly, x^2 and $x^2 + 1$ is not one-one function.
Hence, $(2x - 5)$ is only bijective.

10. (c) We have, $f(x) = \dfrac{x^2 - 4}{x^2 + 4}$

$$f(x_1) = \frac{x_1^2 - 4}{x_1^2 + 4} \text{ and } f(x_2) = \frac{x_2^2 - 4}{x_2^2 + 4}$$

$$f(x_1) = f(x_2) \Rightarrow \frac{x_1^2 - 4}{x_1^2 + 4} = \frac{x_2^2 - 4}{x_2^2 + 4}$$

$$\Rightarrow \quad x_1^2x_2^2 + 4x_1^2 - 4x_2^2 - 16 = x_1^2x_2^2 - 4x_1^2 + 4x_2^2 - 16$$

$$\Rightarrow \quad 8x_1^2 - 8x_2^2 = 0$$

$$\Rightarrow \quad x_1^2 - x_2^2 = 0 \Rightarrow (x_1 + x_2)(x_1 - x_2) = 0$$

$$x_1 = -x_2 \text{ or } x_1 = x_2$$

$\therefore f(x)$ is many-one function.
Let $y \in R$ such that $f(x) = y$

$$\Rightarrow \quad y = \frac{x^2 - 4}{x^2 + 4} \Rightarrow x^2y + 4y = x^2 - 4$$

$$\Rightarrow \quad x^2(1 - y) = 4(y + 1)$$

$$\Rightarrow \quad x^2 = \frac{4(y + 1)}{1 - y}, x^2 \geq 4 \quad \Rightarrow \quad \frac{4(y + 1)}{1 - y} \geq 4$$

$$\Rightarrow \quad \frac{y + 1}{1 - y} \geq 1 \quad \Rightarrow \quad \frac{y + 1}{1 - y} - 1 \geq 0$$

$$\Rightarrow \quad \frac{y + 1 - 1 + y}{1 - y} \geq 0 \quad \Rightarrow \quad \frac{2y}{1 - y} \geq 0$$

$$\overset{-}{}\quad \overset{+}{0 \text{ ------ } 1}\quad \overset{-}{}$$

$$0 \leq y < 1$$

Hence, the range of $f(x)$ is $[0, 1)$ i.e. into

11. (a) We have, $f(x) = 2x + \sin x$

$$f'(x) = 2 + \cos x$$

$$f'(x) > 0, \forall x \in R$$

$\therefore f(x)$ is one-one function, so range of $f(x)$ is $(-\infty, \infty)$.
Hence, $f(x)$ is also onto function.

12. (c) We have, $f(n) = \begin{cases} \frac{n-1}{2}, & \text{if } n \text{ is odd} \\ \frac{n}{2}, & \text{if } n \text{ is even} \end{cases}$

$\therefore \quad f(3) = \frac{3-1}{2} = 1$ and $f(2) = \frac{2}{2} = 1$

Thus, $\quad f(3) = f(2)$, but $3 \neq 2$

So, f is many-one function.

Surjectively : Let n be any arbitrary element of N. If n is odd natural number, then $2m-1$ is also an odd natural number such that

$$f(2m-1) = \frac{2m-1-1}{2} = m-1$$

If n is even natural number, then $2m$ is also an even natural number such that

$$f(2m) = \frac{2m}{2} = m$$

Thus, for every $m \in N$ (whether even or odd).

So, f is an onto function.

$\therefore$ $f(x)$ is many-one and onto function.

13. (b) We have, $f(x) = \frac{x}{1+x}$

$\therefore \quad f(x_1) = f(x_2) \quad \Rightarrow \quad \frac{x_1}{1+x_1} = \frac{x_2}{1+x_2}$

$\Rightarrow \quad x_1 + x_1x_2 = x_2 + x_1x_2 \quad \Rightarrow \quad x_1 = x_2$

So, $f(x)$ is one-one function.

Let $y \in (0, \infty)$, such that $f(x) = y$

$\Rightarrow \quad y = \frac{x}{1+x} \quad \Rightarrow \quad y + xy = x$

$\Rightarrow \quad x = \frac{y}{1-y} \quad \Rightarrow \quad \frac{y}{1-y} \geq 0$ as $x \geq 0$

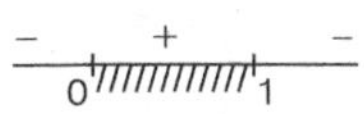

$\Rightarrow \quad 0 \leq y < 1$

$\therefore \quad f(x)$ is not onto function.

14. (d) We have, $f(x) = [x]$

When $0 \leq x < 1$, then $\quad f(x) = 0$

$\therefore$ f is not one-one function.

Range of $f(x)$ is an integer.

$\therefore$ f is neither one-one nor onto function.

15. (d) We have, $f(x) = x + \sqrt{x^2}$

$\Rightarrow \quad f(x) = x + |x| \qquad [\because \sqrt{x^2} = |x|]$

$\Rightarrow \quad f(x) = \begin{cases} 0, & x \leq 0 \\ 2x, & x > 0 \end{cases}$

$\therefore$ $f(x)$ is neither one-one nor onto function.

16. (d) We have, $f(n) = (n+5)^2$

$f(n_1) = f(n_2) \quad \Rightarrow \quad (n_1+5)^2 = (n_2+5)^2$

$\Rightarrow \quad n_1 = n_2$

$\therefore$ $f(n)$ is one-one function.

Let $y \in N$ is such that $f(n) = y \Rightarrow y = (n+5)^2$

$\Rightarrow \quad n \notin N$ for $y = 2$. So, f is not onto.

17. (b) We have, $f(x) = \tan^{-1}\left(\frac{2x}{1-x^2}\right), x \in (-1, 1)$

$\Rightarrow \quad f(x) = 2\tan^{-1} x, x \in (-1, 1)$

$\Rightarrow f(x)$ is bijective

iff $2\tan^{-1} x \in \left(-\frac{\pi}{2}, \frac{\pi}{2}\right)$

So, the interval of B is $\left(-\frac{\pi}{2}, \frac{\pi}{2}\right)$.

18. (a) We have, $f(x) = \frac{x^3}{3} + \frac{x^2}{2} + ax + b, x \in R$

$$f'(x) = x^2 + x + a$$

$\Rightarrow f(x)$ is injective function

iff $f'(x) \geq 0, \forall x \in R$.

For $f'(x) \geq 0, D \leq 0 \Rightarrow 1 - 4a \leq 0$

$\Rightarrow \quad 1 \leq 4a \quad \Rightarrow \quad a \geq \frac{1}{4}$

$\therefore$ Least value of $a = \frac{1}{4}$

19. (b) If $a = -1$, then $f'(x) = x^2 + x - 1$

$D = 1 + 4 > 0$

$f(x)$ is increasing and decreasing, $\forall x \in R$.

$\therefore$ $f(x)$ is not one-one function.

Range of $f(x)$ is $(-\infty, \infty)$ $\qquad [\because f(x)$ is cubic polynomial$]$

$\therefore$ $f(x)$ is many-one and onto function.

20. (a) We have, $f(x) = \frac{x^3}{3} + \frac{x^2}{2} + ax + b$

$f(x)$ is invertible iff $f(x)$ is bijective.

$f(x)$ is one-one iff $a \geq \frac{1}{4}$ $\qquad$ [from Q. No.18]

$\therefore \quad a \in \left[\frac{1}{4}, \infty\right), b \in R$

$f(x)$ is a cubic polynomial.

$\therefore$ $f(x)$ is also onto function.

21. (b) We have, $f(x) = x^2 - 2x - 1$

$\therefore \quad f'(x) = 2x - 2$

For injective, $\quad f'(x) \geq 0$

or $\quad f'(x) \leq 0, \quad \forall x \in (-\infty, a)$

For $f'(x) \leq 0 \quad \Rightarrow \quad 2x - 2 \leq 0$

$x \leq 1$

$\therefore$ Largest value of a is 1.

For surjective,

$f(x) = x^2 - 2x - 1$

$f(x) = (x-1)^2 - 2$

$\Rightarrow \quad f(x)$ is surjective, if $f(x) \in [-2, \infty)$

$\therefore \quad a = 1$ and $b = -2$

and $\quad a + b = 1 - 2 = -1$

22. (c) We have, $g(x) = f(x) + 3x - 1$

$\Rightarrow \quad g(x) = x^2 - 2x - 1 + 3x - 1 \quad [\because f(x) = x^2 - 2x - 1]$

$\Rightarrow \quad g(x) = x^2 + x - 2$

and $\quad y = g|x|$

$\Rightarrow \quad y = |x|^2 + |x| - 2$

$\therefore$ The least value of y is -2. $\qquad (\because |x| \geq 0)$

23. (a) We have, $f(x) = x^2 - 2x - 1$

Let $\quad f(x) = y \Rightarrow f^{-1}(y) = x$

$\Rightarrow \quad y = x^2 - 2x - 1$

$\Rightarrow \quad y = x^2 - 2x - 1 + 1 - 1$

$\Rightarrow \quad y = (x-1)^2 - 2$

$\Rightarrow \quad y + 2 = (x-1)^2$

$\Rightarrow \quad x - 1 = \pm\sqrt{y+2}$

$\Rightarrow \quad x = 1 \pm \sqrt{y+2}$

$\Rightarrow \quad x = 1 + \sqrt{y+2} \qquad [\because x \in [1, \infty)]$

$\therefore \quad f^{-1}(y) = 1 + \sqrt{y+2}$

$\Rightarrow \quad f^{-1}(x) = 1 + \sqrt{x+2}$

24. (a) We have, $f(x) = x^2 - 2x - 1$

$\Rightarrow \quad f(|x|) = |x|^2 - 2|x| - 1$

$\Rightarrow \quad k = |x|^2 - 2|x| - 1$

$\Rightarrow \quad k = |x|^2 - 2|x| - 1 + 1 - 1$

$k = (|x| - 1)^2 - 2$

$\therefore$ Range of k is $(-2, -1)$

25. (1) We have,

$f(x) = x^2 - 4x + 5$

$\Rightarrow \quad f(x) = (x - 2)^2 + 1$

$\therefore \quad f(x) \in [1, \infty]$

But $\quad f : [2, \infty) \rightarrow [1, \infty)$

Hence, the value of a is 1.

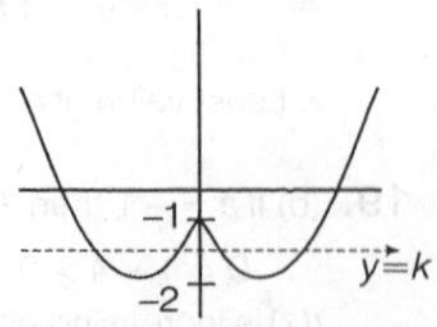

26. (6) We have, $f(x) = \left(\frac{4a - 7}{3}\right)x^3 + (a - 3)x^2 + x + 5$

$\therefore \quad f'(x) = (4a - 7)x^2 + 2(a - 3)x + 1$

For one-one function,

$f'(x) > 0$ or $f'(x) < 0$ for in domain

$\Rightarrow \quad f(x) > 0$ iff $D < 0$

$\Rightarrow \quad D = \{2(a - 3)\}^2 - 4(4a - 7)(1) < 0$

$\Rightarrow \quad a^2 - 6a + 9 - 4a + 7 < 0$

$\Rightarrow \quad a^2 - 10a + 16 < 0$

$\Rightarrow \quad (a^2 - 8)(a - 2) < 0$

$\Rightarrow \quad a \in (2, 8)$

But $a \in [\lambda, u] \Rightarrow \lambda = 2, \mu = 8$

$\therefore \quad |\lambda - \mu| = |2 - 8| = 6$

27. (1)

	Case I	Case II	Case III
$f(x) = 1$	T	F	F
$f(y) \neq 1$	F	T	F
$f(z) \neq 2$	F	F	T

Case I $f(y) = 1$, $f(z) = 2$, then $f(x) = 1$ (not true)

Case II $f(z) = 2$, $f(y) \neq 1$, $f(x) \neq 1$ (not possible)

Case III $f(y) = 1$, $f(z) = 3$, $f(x) = 2$

$\therefore \quad f(y) = 1$

28. (6) Number of different sets containing exactly 3 elements of $B = {}^5C_3 = 10$

Then, number of onto functions from A to the set containing 3 elements $= 10\,[3^4 - 3\{(2)^4 - 2\} - 3(1)^4]$

$= 10\,[81 - 3(16) + 3]$

$= 10 \times 36 = 360$

$\therefore \quad \frac{k}{60} = 6$

29. (3) Number of one-one function $= {}^5P_3 = 60$

Number of strictly monotonic functions $= 10 + 10 = 20$

$\therefore \quad \frac{\text{Number of one-one functions}}{\text{Number of strictly monotonic}} = \frac{60}{20} = 3$

30. (5) Here, $f(x) - 10x = (x - 1)(x - 2)(x - 3)(x - \alpha)$

$\therefore \quad f(12) = 120 + (11 \times 10 \times 9)(12 - \alpha)$

$f(-8) = -80 + (-9) \times (-10) \times (-11)(-8 - \alpha)$

$\Rightarrow \quad f(12) + f(-8) = 40 + (9 \times 10 \times 11) \times 20 = 19840 = (3968)\,5$

$\therefore \quad \lambda = 5$

DPP-9 **Inverse of Functions**

1. (b) Let $\quad y = \frac{e^x - e^{-x}}{e^x + e^{-x}} + 2 \quad \Rightarrow \quad y = \frac{e^{2x} - 1}{e^{2x} + 1} + 2$

$\Rightarrow \quad e^{2x} = \frac{1 - y}{y - 3} = \frac{y - 1}{3 - y} \quad \Rightarrow \quad x = \frac{1}{2}\log\left(\frac{y - 1}{3 - y}\right)$

$\Rightarrow \quad f^{-1}(y) = \log\left(\frac{y - 1}{3 - y}\right)^{1/2}$

$\therefore \quad f^{-1}(x) = \log\left(\frac{x - 1}{3 - x}\right)^{1/2}$

2. (c) Let $y = 2^{x(x-1)}$ or $x^2 - x = \log_2 y$

or $\quad x^2 - x - \log_2 y = 0$

or $\quad x = \frac{1 \pm \sqrt{1 + 4\log_2 y}}{2}$

$\therefore \quad f^{-1}(x) = \frac{1}{2}[1 + \sqrt{1 + 4\log_2 x}]$

As, $f^{-1} : [1, \infty) \rightarrow [1, \infty)$, neglecting $\frac{1}{2}[1 - \sqrt{1 + 4\log_2 x}]$

3. (c) Given, $f(x) = 4x + 3$

and $f : N \rightarrow Y$, where $Y = \{y \in N : y = 4x + 3\}$

Clearly, the given function is one-one and onto.

Let $y = 4x + 3 \quad \Rightarrow \quad y - 3 = 4x$

$\Rightarrow \quad x = \frac{y - 3}{4}$

$\therefore \quad f^{-1}(x) = \frac{x - 3}{4}$

4. (b) Given, $f(x) = 3x - 4$

Clearly, $f(x)$ is one-one and onto.

Let $\quad y = 3x - 4$

$\Rightarrow \quad x = \frac{y + 4}{3}$

$\therefore \quad f^{-1}(x) = \frac{x + 4}{3}$

5. (a) Given, $f(x) = \frac{1}{x - 1}, x \neq 1$

Let $f(x_1) = f(x_2) \quad \Rightarrow \quad \frac{1}{x_1 - 1} = \frac{1}{x_2 - 1}$

$\Rightarrow \quad x_1 - 1 = x_2 - 1$

$\Rightarrow \quad x_1 = x_2$

Hence, $f(x)$ is one-one.

Let $\quad y = \frac{1}{x - 1} \quad \Rightarrow x - 1 = \frac{1}{y}$

$\Rightarrow \quad x = \frac{1}{y} + 1,$

$\therefore$ Range $\in R - \{0\}$

f is onto, so f^{-1} exists.

(b) Given, $y = x^2 = f(x)$

Let $\quad f(x_1) = f(x_2)$

$\Rightarrow \quad x_1^2 = x_2^2$

$\Rightarrow \quad x_1 = \pm x_2$

f is not one-one, so f^{-1} does not exist.

(c) Given, $f(x) = x^2,\ x > 0$

Let $f(x_1) = f(x_2) \Rightarrow x_1^2 = x_2^2$

$\Rightarrow x_1 = \pm x_2$

$\Rightarrow x_1 = x_2\ (x > 0)$

$\therefore$ f is one-one.

Now, $y = x^2$

$x = \sqrt{y} \Rightarrow y > 0$

f is not onto, so f^{-1} does not exist.

(d) Given, $f(x) = x^2,\ x \le 0$

Let $f(x_1) = f(x_2)$

$x_1^2 = x_2^2 \Rightarrow x_1 = \pm x_2$

$x_1 = x_2$ (as $x \le 0$)

$\therefore$ f is one-one.

Range $\Rightarrow y \ge 0$

f is into, so f^{-1} does not exist.

6. (a) Given, $f(x) = \dfrac{10^x - 10^{-x}}{10^x + 10^{-x}} \Rightarrow f(x) = \dfrac{10^{2x} - 1}{10^{2x} + 1}$

Let $y = \dfrac{10^{2x} - 1}{10^{2x} + 1} \Rightarrow y(10^{2x} + 1) = 10^{2x} - 1$

$\Rightarrow 10^{2x}(y - 1) + (y + 1) = 0$

$\Rightarrow 10^{2x} = \dfrac{-(y+1)}{y-1} \Rightarrow 10^{2x} = \dfrac{y+1}{1-y}$

$\Rightarrow x = \dfrac{1}{2}\log_{10}\left(\dfrac{y+1}{1-y}\right)$

$\therefore f^{-1}(x) = \dfrac{1}{2}\log_{10}\left(\dfrac{x+1}{1-x}\right)$

7. (c) Given, $f : (2, 4) \to (1, 3)$

Defined by $f(x) = x - \left[\dfrac{x}{2}\right]$

Let $y = x - 1$ [as $x \in (2, 4)$]

$\Rightarrow x = y + 1$

$\therefore f^{-1}(x) = x + 1$

8. (b)Given $f(x) = \dfrac{1 - x}{1 + x^2}$

Let $y = \dfrac{1-x}{1+x^2} \Rightarrow y + yx^2 = 1 - x$

$\Rightarrow yx^2 + x + (y - 1) = 0$

$\Rightarrow x = \dfrac{-1 \pm \sqrt{1 - 4y(y-1)}}{2y} \Rightarrow x = \dfrac{-1 \pm \sqrt{1 + 4y(1-y)}}{2y}$

$\Rightarrow x = \dfrac{-1 \pm \sqrt{1 + 4y - 4y^2}}{2y}$

$\Rightarrow f^{-1}(x) = \dfrac{-1 \pm \sqrt{1 + 4x - 4x^2}}{2x}$

$\therefore f^{-1}(x) = \dfrac{-1 + \sqrt{1 + 4x - 4x^2}}{2x},\ x \ne 0$

and $f^{-1}(x) = 1$, when $x = 0$.

9. (d) Given, $f(x) = x + (-1)^{x-1}$

Now $fof\,(x) = f\{f(x)\} = f(x) + (-1)^{f(x)-1}$

$= x + (-1)^{x-1} + (-1)^{x + (-1)^{x-1} - 1}$

$= x + (-1)^{x-1} + (-1)^{x-1} \cdot (-1)^{(-1)^{x-1}}$

$= x - (-1)^x + (-1)^x\,(-1)^{-(-1)^x}$

$= x - (-1)^x + (-1)^x = x$

$\therefore f^{-1}(x) = x + (-1)^{x-1}$

10. (a) Given, $f(x) = \log_a x^2,\ a > 1$

Let $y = \log_a x^2$, then $a^y = x^2 \Rightarrow x = \sqrt{a^y}$

$\Rightarrow f^{-1}(x) = \sqrt{a^x} \Rightarrow f^{-1}(b + c) = \sqrt{a^{b+c}}$

$f^{-1}(b + c) = \sqrt{a^b \cdot a^c}$ ...(i)

Now, $\left.\begin{aligned} f^{-1}(b) &= \sqrt{a^b} \\ f^{-1}(c) &= \sqrt{a^c} \end{aligned}\right\}$ ...(ii)

Using Eq. (ii) in Eq. (i), we get

$f^{-1}(b + c) = f^{-1}(b) \cdot f^{-1}(c)$

11. (a) We have, $f(x) = \dfrac{x + a}{bx^2 + cx + 2}$, where $a, b, c \in R$, $f(-1) = 0$ and $y = 1$ is an asymptote of $y = f(x)$.

$\therefore a = 1, b = 0, c = 1 \Rightarrow f(x) = \dfrac{x+1}{x+2}$

12. (c) We have, $f^{-1}(x) = \dfrac{2x - 1}{1 - x}$

Now, $f^{-1}(x) : R - \{-1\} \to R$

and $y = \dfrac{2x-1}{1-x} \Rightarrow x = \dfrac{y+1}{y+2};\ y \ne -2$

$\therefore$ Range of $f^{-1}(x)$ is $R - \{-2\} \subset$ Codomain

Thus, f is one-one and into.

13. (b) Asymptote of curve $f(x) = \dfrac{x-1}{x+2}$ are $y = 1$ and $x = -2$, also asymptote of curve $f^{-1}(x) = \dfrac{2x-1}{1-x}$ are $y = -2$ and $x = 1$.

$\therefore$ Area bounded = Area of Rectangle = $3 \times 3 = 9$

14. (4) We have, $f(x) = 2\cos^2 x + \sqrt{3}\sin 2x + 1$

$\Rightarrow f(x) = 1 + \cos 2x + \sqrt{3}\sin 2x + 1$

$\Rightarrow f(x) = 2 + 2\left(\dfrac{1}{2}\cos 2x + \dfrac{\sqrt{3}}{2}\sin 2x\right)$

$\Rightarrow f(x) = 2 + 2\sin\left(2x + \dfrac{\pi}{6}\right)$

$\therefore$ Range of $f(x) = [0, 4]$

Hence, $f^{-1}(x)$ exists, if $B \in [0, 4]$, but $B \in [a, b]$.

Now, $(a + b) = (0 + 4) = 4$

15. (3) If $f(x) = mx + b$, then $f^{-1}(x) = \dfrac{x - b}{m}$ and their point of intersection can be found by setting $x = my + b$, since they intersect at $y = x$.

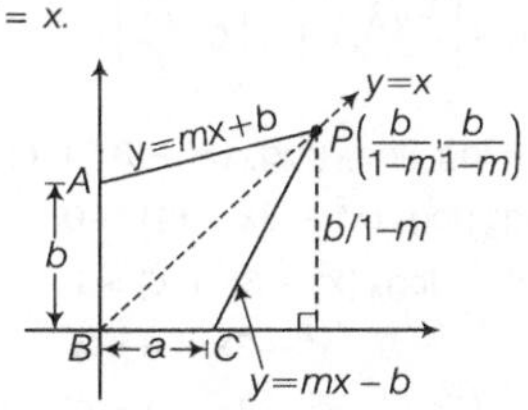

Thus, $x = \dfrac{b}{1-m}$ and the point of intersection is $\left(\dfrac{b}{1-m}, \dfrac{b}{1-m}\right)$.

Region R can be broken up into congruent Δ's PAB and ΔPCB, which both have a base of b and a height of $\left(\dfrac{b}{1-m}\right)$.

$\therefore$ Area of $R = 2 \cdot \left(\dfrac{b}{2}\right) \cdot \left(\dfrac{b}{1-m}\right) = \dfrac{b^2}{1-m} = 49$

For $m = \dfrac{9}{25}$, $b^2 = \dfrac{16}{25} \times 49 \Rightarrow b = \dfrac{28}{5} = \dfrac{\lambda}{\mu}$

$\therefore |\lambda - 5\mu| = |28 - 25| = 3$

Revisal Problems (JEE Main)

1. (d) We have, $f(x) = \sqrt{x - \sqrt{1 - x^2}}$

It is defined when

$$x - \sqrt{1 - x^2} \geq 0 \quad \text{and} \quad 1 - x^2 \geq 0$$

$$x \geq \sqrt{1 - x^2}$$

On squaring both sides, we get

$$x^2 \geq 1 - x^2$$

$$\Rightarrow \quad 2x^2 \geq 1$$

$$\Rightarrow \quad x^2 \geq \frac{1}{2}$$

$$x \leq -\frac{1}{2} \quad \text{or} \quad x \geq \frac{1}{\sqrt{2}}$$

But x is never be negative.

Now, $\quad D_1 \equiv x \geq \dfrac{1}{\sqrt{2}}$

and $\quad 1 - x^2 \geq 0$

$\Rightarrow \quad D_2 \equiv -1 < x \leq 1$

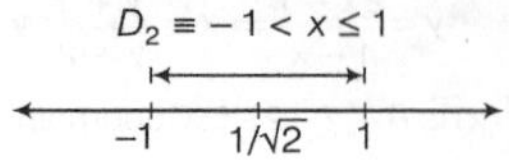

Combining D_1 and D_2, we get domain is $\left[\dfrac{1}{\sqrt{2}}, 1\right]$

2. (c) We have, $f(x)$ is defined, when $-5 \leq x \leq 7$ and $fog(x) = f[g(x)]$ is defined when

$$-5 \leq g(x) \leq 7 \text{ i.e. } -5 \leq |2x + 5| \leq 7$$

$$-7 \leq 2x + 5 \leq 7$$

$$-12 \leq 2x \leq 2$$

$$-6 \leq x \leq 1$$

Hence, the domain is $[-6, 1]$.

3. (c) We have, $f(x) = \sin^{-1}[2 - 4x^2]$

Its domain is $\quad -1 \leq [2 - 4x^2] \leq 1 \quad \Rightarrow \quad -1 \leq 2 - 4x^2 < 2$

$$\Rightarrow \quad -3 \leq -4x^2 < 0$$

$$\Rightarrow \quad 0 < 4x^2 \leq 3$$

$$\Rightarrow \quad 0 < x^2 \leq \frac{3}{4}$$

$$\Rightarrow \quad -\frac{\sqrt{3}}{2} \leq x < 0 \quad \text{or} \quad 0 < x \leq \frac{\sqrt{3}}{2}$$

Hence, domain is $\left[\dfrac{-\sqrt{3}}{2}, 0\right) \cup \left(0, \dfrac{\sqrt{3}}{2}\right]$

4. (c) Given, $f(x) = \log_2[\log_3\{\log_4(x^2 - 3x + 6)\}]$ its domain is

$$D_1 \equiv \log_3\{\log_4(x^2 - 3x + 6)\} > 0$$

$$\Rightarrow \quad \log_4(x^2 - 3x + 6) > 1$$

$$\Rightarrow \quad x^2 - 3x + 6 > 4$$

$$\Rightarrow \quad x^2 - 3x + 2 > 0$$

$$\Rightarrow \quad (x - 2)(x - 1) > 0$$

$$\Rightarrow \quad x < 1 \quad \text{or} \quad x > 2$$

$$D_2 \equiv \log_4(x^2 - 3x + 6) > 0$$

$$x^2 - 3x + 6 > 1$$

$$x^2 - 3x + 5 > 0$$

Its roots are imaginary, hence it is defined for all $x \in R$

$$D_3 \equiv x^2 - 3x + 6 > 0$$

It is also defined for all $x, x \in R$.

On combining D_1, D_2 and D_3, we get

$\therefore$ Domain of $f(x) = (-\infty, 1) \cup (2, \infty)$

5. (d) We have,

$$f(x) = \log_3|\log_e x|$$

Its domain is $|\log_e x| \neq 0$ and $x > 0$

$$\Rightarrow \quad x \neq 1 \quad \text{and} \quad x > 0$$

$\therefore$ Its domain is $(0, \infty) \sim \{1\}$

6. (a) Given, $f(x) = x^{\frac{1}{\log_{10} x}}$

It is defined, when $\log_{10} x \neq 0$ and $x > 0$

$$\Rightarrow \quad x \neq 1 \quad \text{and} \quad x > 0$$

On combining these above two equations, we get

$$(0,1) \cup (1, \infty)$$

7. (b) We have,

$$f(x) = \sqrt{e^{\cos^{-1}(\log_4 x^2)}}$$

It is defined, when

$$e^{\cos^{-1}(\log_4 x^2)} \geq 0$$

and $\quad -1 \leq \log_4 x^2 \leq 1$

and $\quad x \neq 0.$

We know that exponential functions are always positive. Hence, $e^{\cos^{-1}(\log_4 x^2)}$ is true for all x.

Now, $\quad -1 \leq \log_4 x^2 \leq 1$

$$4^{-1} \leq x^2 \leq 4$$

$$\frac{1}{4} \leq x^2 \leq 4$$

$$-2 \leq x \leq -\frac{1}{2} \quad \text{or} \quad \frac{1}{2} \leq x \leq 2$$

Hence, domain is $\left[-2, \dfrac{1}{2}\right] \cup \left[\dfrac{1}{2}, 2\right]$

8. (b) We have, $f(x) = \cot^{-1}\left\{\dfrac{x}{\sqrt{x^2 - [x]^2}}\right\}$

We know that, domain of $\cot^{-1} x$ is R.

$f(x)$ is defined, when $x^2 - [x]^2 \neq 0$

i.e. $\quad x^2 \neq [x]^2$

$\Rightarrow \quad x^2 \neq 0$ or integer

Hence, domain of$(x) = R - \{\pm\sqrt{n} : n \geq 0; n \in Z\}$

9. (a) We have, $f(x) = \sqrt{3 - 2^x - 2^{1-x}} + \sqrt{\sin^{-1} x}$

It is defined when $3 - 2^x - 2^{1-x} \geq 0$; $\sin^{-1} x \geq 0$

and $\quad -1 \leq x \leq 1$

Retain $\quad 3 - 2^x - 2^{1-x} \geq 0$

or $\quad 3 - 2^x - \dfrac{2}{2^x} \geq 0$

Putting $2^x = y \Rightarrow \dfrac{2}{y} + y - 3 \leq 0$

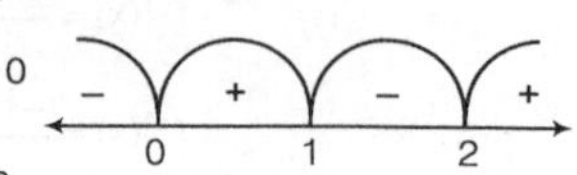

$$\frac{y^2 - 3y + 2}{y} \leq 0$$

$$\frac{(y - 2)(y - 1)}{y} \leq 0$$

$$y < 0 \quad \text{or} \quad 1 \leq y \leq 2$$

i.e. $\quad 2^x < 0$ or $1 \leq 2^x \leq 2$

2^x is always + ve, $0 \leq x \leq 1$ ...(i)

$\sin^{-1} \geq 0 \Rightarrow 0 \leq x \leq 1$ ...(ii)

and $\quad -1 < x \leq 1$ ...(iii)

On combining Eqs. (i), (ii) and (iii), we get

$$x \in [0, 1]$$

10. (a) We have, $f(x)$ is defined, when $0 < x < 1$.

The domain of $f(\sin x)$ is

$$0 < \sin x < 1$$
$$0 < x < \pi$$

and $\sin x$ is periodic with period 2π

$$2n\pi < x < (2n+1)\pi;\ n \in Z$$

11. (a) We have, $f(x) = \cos^{-1}\left(\frac{x^2}{x^2+1}\right)$

Its domain is $x \in R$.

To find range, put $y = \cos^{-1}\frac{x^2}{x^2+1}$

$$\cos y = \frac{x^2}{x^2+1}$$
$$x^2 = \frac{\cos y}{1-\cos y}$$

since $x^2 \geq 0$, hence $\frac{\cos y}{1-\cos y} \geq 0$

or $\frac{\cos y}{\cos y - 1} \leq 0$

$$0 \leq \cos y < 1$$

which is possible when

$$0 < y \leq \frac{\pi}{2}$$

$\therefore$ Range $= \left(0, \frac{\pi}{2}\right]$

12. (a) We have, $f(x) = x^2 + \frac{1}{x^2+1}$

We can rewrite $f(x)$ as

$$f(x) = x^2 + 1 + \frac{1}{x^2+1} - 1$$

We know that AM $\geq$ GM

$$\frac{x^2+1+\frac{1}{x^2+1}}{2} \geq (x^2+1)\cdot\frac{1}{x^2+1}$$
$$x^2+1+\frac{1}{x^2+1} \geq 2$$

$\therefore$ $x^2+1+\frac{1}{x^2+1} - 1 \geq 1$

Range of $f(x)$ is $[1, \infty)$.

13. (d) We have, $f(x) = \log_5(25 - x^2)$

$f(x)$ is defined when $25 - x^2 > 0$

$$-5 < x < 5$$
$$0 \leq x^2 < 25$$
$$-25 < -x^2 \leq 0$$
$$0 < 25 - x^2 \leq 25$$
$$\log_5 0 < \log_5(25 - x^2) \leq \log_5 25$$
$$-\infty < f(x) \leq 2$$

Hence, range is $(-\infty, 2]$.

14. (c) We have,

$$f(x) = x^3 - x^2 + 4x + 2\sin^{-1}x$$
$$f'(x) = 3x^2 - 2x + 4 + \frac{2}{\sqrt{1-x^2}}$$
$$f'(x) > 0, \forall x \in [0, 1)$$

So, $f(x)$ is strictly increasing function.

$\therefore$ Range $= [f(0), f(1)] = [0, 4 + \pi]$

15. (d) Let $y = \frac{x^2+x+2}{x^2+x+1}$

$\Rightarrow$ $xy^2 + xy + y = x^2 + x + 2$

$\Rightarrow$ $x^2(y-1) + x(y-1) + y - 2 = 0$

We have $x \in R, \Delta \geq 0$

$$(y-1)^2 - 4(y-1))(y-2) \geq 0$$
$$(y-1)[y-1-4y+8] \geq 0$$
$$(y-1)(-3y+7) \geq 0$$
$$(y-1)(3y-7) \leq 0$$

$\therefore$ $1 \leq y \leq \frac{7}{3}$, but $y \neq 1$

$\therefore$ Range of $f(x)$ is $\left(1, \frac{7}{3}\right]$.

16. (b) We have, $f(x) = [9^x - 3^x + 1]$

Take $9^x - 3^x + 1 = \left(3^x - \frac{1}{2}\right)^2 - \frac{1}{4} + 1$

$$= \left(3^x - \frac{1}{2}\right)^2 + \frac{3}{4} \geq \frac{3}{4} \text{ for all } x \in (-\infty, 1)$$

$\because$ $\frac{3}{4} \leq 9^x - 3^x + 1 < 7$

Now, $[9^x - 3^x + 1] = \{0, 1, 2, 3, 4, 5, 6\}$

Hence, range of $f(x) = \{0, 1, 2, 3, 4, 5, 6\}$.

17. (b) We have,

$$f(x) = \sin^{-1}\left(\frac{2-|x|}{4}\right) + \cos^{-1}\left(\frac{2-|x|}{4}\right) + \tan^{-1}\left(\frac{2-|x|}{4}\right),$$

then domain of $\sin^{-1}\left(\frac{2-|x|}{4}\right)$ and $\cos^{-1}\left(\frac{2-|x|}{4}\right)$ is identical and is

$$-1 \leq \frac{2-|x|}{4} \leq 1$$
$$-4 \leq 2 - |x| \leq 4, \Rightarrow -6 \leq -|x| \leq 2$$
$$-2 \leq |x| \leq 6, \Rightarrow -6 \leq x \leq 6$$

and domain of $\tan^{-1}\left(\frac{2-|x|}{4}\right)$ is $x \in R$.

Hence, domain of $f(x)$ is $[-6, 6]$.

18. (c) We have, $f(x) = \sqrt[4]{\log_3\left\{\frac{1}{|\cos x|}\right\}}$

It is defined when

(i) $\log_3\left\{\frac{1}{|\cos x|}\right\} > 0$; (ii) $\frac{1}{|\cos x|} > 0$; (iii) $|\cos x| \neq 0$

On solving

(i) $\frac{1}{|\cos x|} \geq 1 \Rightarrow |\cos x| \leq 1$

which is always true for $x \in R$.

(ii) $\frac{1}{|\cos x|} > 0$, it is also true for all $x \in R$.

(iii) $|\cos x| \neq 0$ i.e. $x \neq \frac{\pi}{2}, \frac{3\pi}{2}, \ldots$

Hence, domain of $f(x)$ is $R - \left\{(2n+1)\frac{\pi}{2}\right\}; n \in Z$

19. (a) We have, $f(x) = \cos^{-1}(x + [x])$

Its domain is $-1 \leq x + [x] \leq 1$

Case I If x is an integer.

Let $x = n$, where $n \in I$

$$x + [x] = n + n = 2n$$
$$-1 \leq 2n \leq 1 \Rightarrow -\frac{1}{2} \leq x \leq \frac{1}{2}$$

$\therefore$ $x = 0$ ...(i)

Case II If x is not an integer.

Put $x = n + k, n \in I$ and $0 < k < 1$

$$x + [x] = n + k + n = 2n + k$$

$$-1 \le 2n + k \le 1 \Rightarrow n = 0$$

$\therefore \quad x = k, 0 < k < 1 \quad \ldots$(ii)

From Eqs. (i) and (ii), we get

Domain of $f = [0, 1)$

20. (d) We have, $f(x) = \underbrace{\log_{10}\log_{10}\log_{10}\cdots\log_{10}}_{n \text{ times}} x$

It is defined $\quad \underbrace{\log_{10}\log_{10}\cdots\log_{10}}_{(n-1)\text{ times}} x > 0$

$$\underbrace{\log_{10}\log_{10}\cdots\log_{10}}_{(n-2)\text{ times}} x > 10$$

$$\log_{10}\underbrace{\log_{10}\cdots\log_{10}}_{(n-3)\text{ times}} x > 10$$

$$\underbrace{\log_{10}\log_{10}\cdots\log_{10}}_{} x > 10^{10}$$

Similarly, we solve all elements logarithms, we see that no option are matching.

21. (a) We have, $f(x) = \dfrac{1}{\sqrt{[x]^2 - [x] - 6}}$

It domain is $\quad [x]^2 - [x] - 6 > 0$

$$([x] - 3)([x] + 2) > 0$$

$\Rightarrow \quad [x] < -2 \text{ or } [x] > 3$

$$x < -2 \text{ or } x \ge 4$$

Hence, domain is $(-\infty, -2) \cup [4, \infty)$.

22. (c) We have, $f(x) = \dfrac{\log_2(x + 3)}{x^2 + 3x + 2}$

It is defined (i) $x + 3 > 0$ (ii) $x^2 + 3x + 2 \ne 0$

Solving (i) we get, $x > -3$

Solving (ii), $(x + 2)(x + 1) \ne 0 \Rightarrow x \ne -1, -2$

Hence, domain of $f(x)$ is

$$(-3, \infty) - \{-2, -1\}$$

23. (b) Since, $f : z \to z$

and $f(x) = ax^2 + bx + c$

$\therefore \quad f(1) = \text{Integer} \Rightarrow a + b + c = \text{Integer} \quad \ldots$(i)

$\quad f(0) = \text{Integer} \Rightarrow c = \text{Integer} \quad \ldots$(ii)

On subtracting Eqs. (i) and (ii) we get

$$a + b = \text{Integer}$$

24. (b) We have, $f(x) = \begin{cases} 2x + \cos x; & \cos x \ge 0 \\ 2x - \cos x; & \cos x < 0 \end{cases}$

$$f'(x) = \begin{cases} 2 - \sin x; & \cos x \ge 0 \Rightarrow 0 \le x \le \dfrac{\pi}{2} \\ 2 + \sin x; & \cos x < 0 \Rightarrow \dfrac{\pi}{2} < x < \dfrac{3\pi}{2} \end{cases}$$

Clearly, we see that for any value of x.

$f'(x)$ is always positive, hence it is always strictly increasing, therefore $f(x)$ is one-one.

And we see that $f(x)$ takes all real values, $\forall x \in R$.

Hence, it is onto.

25. (c) We have, $f(x) = ax + \sin x$ is bijective

To check one-one, $f'(x) = a + \cos x$

Clearly, it is strictly increasing or strictly decreasing, if $\cos x = 0$

and $\quad -1 \le \cos x \le 1$

$\therefore$ It is one-one when $\quad a \in R - [-1, 1]$

and the range of $f(x)$ is all real numbers.

Hence, $\quad a \in R - [-1, 1]$.

26. (b) We have, $f(x) = \tan^{-1}(x^2 + x + \lambda)$ is onto.

Hence, range of $f(x)$ must be A.

i.e. $\quad 0 \le \tan^{-1}(x^2 + x + \lambda) < \dfrac{\pi}{2}$

$$0 \le x^2 + x + \lambda < \infty$$

i.e. $\quad x^2 + x + \lambda \ge 0$

$\Rightarrow \quad D \le 0$

$\Rightarrow \quad 1 - 4\lambda \le 0$

$\Rightarrow \quad 4\lambda \ge 1$

$\therefore \quad \lambda \ge 1/4$

27. (c) We have, $f(x) = 2^x + 2^{|x|} = \begin{cases} 2^{x+1}, & x \ge 0 \\ 2^x + 2^{-x}, & x < 0 \end{cases}$

$$f'(x) = \begin{cases} 2^{x+1} \cdot \log 2; & x \ge 0 \\ (2^x - 2^{-x})\log 2; & x < 0 \end{cases}$$

We see that $f'(x) > 0, \quad \forall x \in R$.

Hence, it is strictly increasing.

$\therefore f(x)$ is one-one.

For onto, $\quad f(x) = 2^x + 2^{|x|}$

which is exponential function and it is always + ve and it is not equal to codomain, hence it is into.

28. (b) We have, $f(x) = 3x - 5$

It domain and range are R. Hence, it is bijective, hence inverse is defined.

Let $\quad y = 3x - 5$

$$x = \frac{y + 5}{3}$$

Hence, $\quad f^{-1}(x) = \dfrac{x + 5}{3}$

29. (b) We have, $f(x) = x(2 - x) = 2x - x^2$

$$f'(x) = 2 - 2x = 2(1 - x)$$

and $\quad f'(x) > 0, \quad \forall x < 1$

$\because \quad f(x)$ is one-one

Let $y = 2x - x^2 + 1 - 1 = 1 - (x^2 - 2x + 1) = 1 - (x - 1)^2$

Its range is $y \le 1$, therefore is onto.

We see that $f(x)$ is bijective i.e.

$$y = 1 - (x - 1)^2$$

$$(x - 1)^2 = 1 - y$$

$$x = 1 \pm \sqrt{1 - y}$$

Negative sign, gives $x \le 1$

$\therefore \quad f^{-1}(x) = 1 - \sqrt{1 - x}$

30. (c) We have, $f(x) = x - [x]$

which is a periodic function, hence never one-one.

Therefore, $f^{-1}(x)$ is never defined.

31. (a) We have, $f : A \to B$ and $f(x) = x^2 - x + 1$

Clearly, $f(x)$ is bijective in the given domain and codomain.

We have, $f(x) = f^{-1}(x) \Rightarrow \quad f(x) = x$

$\Rightarrow \quad x^2 - x + 1 = x$

$\Rightarrow \quad x^2 - 2x + 1 = 0$

$\Rightarrow \quad (x - 1)^2 = 0$

$\Rightarrow \quad x = 1$

32. (a) Number of onto from E to F

$$= \sum_{r=1}^{2} (-1)^{2-r} \cdot {}^2C_r \cdot (r)^4$$

$$= (-1)^{2-1} \cdot {}^2C_1 \cdot (1)^4 + (-1)^0 \cdot {}^2C_2 \cdot (2)^4$$

$$= -2 + 16 = 14$$

33. (b) We have, $g(x) = 1 + x - [x]$

i.e. $\quad g(x) = 1 + \{x\}$

and $\quad f(x) = \begin{cases} -1; & x < 0 \\ 0; & x = 0 = \text{sgn}(x) \\ 1; & x > 0 \end{cases}$

$\therefore \quad f[g(x)] = 1 + \{g(x)\} = 1 + \{\text{sgn}(x)\} = 1$

34. (a) We have, $f(x) = 2x + \sin x$.

$$f'(x) = 2 + \cos x > 0, \forall x \in R$$

Hence, $f(x)$ is strictly increasing function, therefore $f(x)$ is one-one.

Clearly, range of $f(x)$ is all real numbers, which is equal to codomain.

Hence, it is onto.

35. (d) We have, $f(x) = 5 - 3\sin x$

It is a periodic function.

$\therefore$ It is many-one.

and range of $f(x)$ is $[2, 8]$, which is not equal to codomain.

Hence, it is into and many one.

36. (b) We have, $f(x) = e^{x^3 - 3x + 2}$

$$f'(x) = e^{x^3 - 3x + 2}(3x^2 - 3)$$

$$f'(x) > 0, \text{ when } 3x^2 - 3 > 0$$

$$x^2 > 1$$

$$x < -1 \quad \text{or} \quad x > 1$$

So, in the interval $x < -1$. It is strictly increasing, therefore, $f(x)$ is one-one.

The range of $f(x)$ is $(f(-\infty), f(-1)]$

[because $f(x)$ is strictly increasing]

$(0, e^4]$, which is not equal to codomain.

Hence, $f(x)$ is one-one and into.

37. (b) If A having m elements and B having n elements, then total number of onto from A to B is

$$\sum_{r=1}^{n} (-1)^{n-r}\, {}^nC_r\, (r)^m$$

Hence, number of onto

$$\sum_{r=1}^{3} (-1)^{3-r}\, {}^3C_r\, (r)^4 = {}^3C_1 - {}^3C_2(2)^4 + (3)^4$$

$$= 3 - 3 \times 16 + 81 = 84 - 48 = 36$$

38. (b) We have, $f(x) = 1 + \alpha x$

It is bijective, $\forall x \in R$.

Hence, it is invertible.

Now, $\quad f(x) = f^{-1}(x) \quad$ or $\quad fof(x) = fof^{-1}(x) = x$

$\Rightarrow \quad 1 + \alpha f(x) = x$

$\Rightarrow \quad 1 + \alpha(1 + \alpha x) = x$

$\Rightarrow \quad 1 + \alpha + \alpha^2 x = x$

$\Rightarrow \quad x(1 - \alpha^2) = 1 + \alpha$

$\therefore$ It is satisfy only for $\alpha = -1$.

39. (c) We have, $f(x) = 4x - x^2$

$$f'(x) = 4 - 2x < 0$$

$\therefore \quad x > 2$

It is one-one.

We have, $f(x)$ is invertible.

Hence, we have to find range of $f(x)$ when $x \in [2, \infty)$

and $\quad f(-x) = -f(x) \;\Rightarrow\; f(0) = 0$

and $\quad f(x + 2) = f(x) \;\Rightarrow\; f(4) = f(2) = f(0) = 0$

$\therefore \quad \text{Range} = (f(\infty), f(2)] = (-\infty, 4]$

$\because \quad x = (-\infty, 4]$

40. (a) We have, $f(x) = \begin{cases} |x|; & x \le 1 \\ 2 - x; & x > 1 \end{cases}$

$$fof(x) = \begin{cases} f|x|; & x \le 1 \\ f(2 - x); & x > 1 \end{cases}$$

$$f|x| = \begin{cases} |x|; & |x| \le 1 \text{ and } x \le 1 \\ 2 - |x|; & |x| > 1 \text{ and } x \le 1 \end{cases}$$

$$= \begin{cases} |x|; & -1 \le x \le 1 \\ 2 - |x|; & x < -1 \end{cases}$$

$$f(2 - x) = \begin{cases} |2 - x| & 2 - x \le 1 \text{ and } x > 1 \\ 2 - (2 - x); & 2 - x \le \text{ and } x > 1 \end{cases}$$

it has no common part.

$$= \{ |2 - x| \,;\, x > 1$$

$$\therefore \quad f[f(x)] = \begin{cases} 2 - |x|; & x < -1 \\ |x|; & -1 \le x \le 1 \\ |2 - x|; & x > 1 \end{cases}$$

41. (b) We have, $f(x) = x^2 + 6x - 8$

$$f'(x) = 2x + 6$$

It is one-one either $x > -3$ or $x < -3$.

Range $\quad x^2 + 6x + 9 - 17$

$$(x + 3)^2 - 17$$

Range of $f(x)$ is $[-17, \infty)$.

$\therefore \quad A = [-3, \infty) \quad \text{and} \quad B = [-17, \infty)$

42. (a) We have,

$$f(x + y) = f(x) + f(y) - xy - 1 \text{ and } f(1) = 1$$

On putting $y = 1$, we get

$$f(x + 1) = f(x) + f(1) - x - 1$$

$\Rightarrow \quad f(n + 1) = f(n) - n < f(n)$

$\Rightarrow \quad f(n + 1) < f(n)$

So, $f(n) < f(n - 1) < f(n - 2) \cdots < f(3) < f(2) \cdots < f(1) < 1$

$\therefore \quad f(n) = n$ holds only for $n = 1$.

43. (a) We have, $f(x) = \sin(\sqrt{[\lambda]}x$

It is period is given by π.

$\therefore \quad \dfrac{2\pi}{\sqrt{[\lambda]}} = \pi \Rightarrow \sqrt{[\lambda]} = 2$

$\Rightarrow \quad [\lambda] = 4 \Rightarrow 4 \le \lambda < 5$

$\therefore \quad \lambda \in [4, 5)$

44. (d) We have, $f(x) = 2\cos 5x + 3\sin\sqrt{5}x$

Let period of $2\cos 5x$ is $T_1 = \dfrac{2\pi}{5}$

and period of $3\sin\sqrt{5}x$ is $T_2 = \dfrac{2\pi}{\sqrt{5}}$

Here, T_1 is rational and T_2 is irrational.

Hence, period of is not defined.

Therefore, $f(x)$ is not a periodic function.

45. (a) We have, $f(x) = \min\{x - [x], -x - [-x]\}$

or $\quad f(x) = \min\{\{x\}, \{-x\}\}$

where, $\{x\}$ is a fractional part of x sketching the graph of $f(x)$

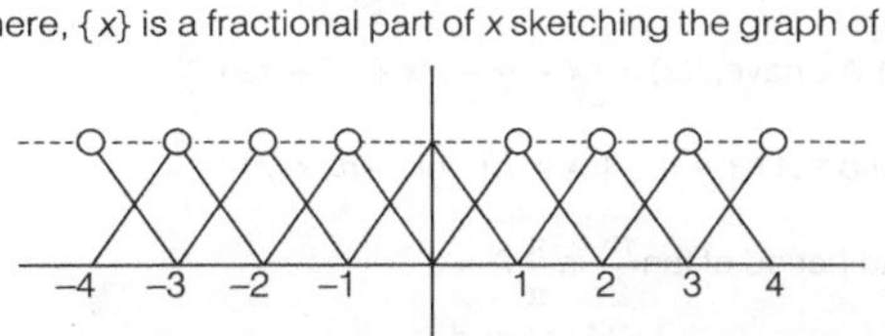

we see that the curve repeats triangle at unit interval.

Hence, its period is 1.

46. (d) $f(-x) = -f(x) \Rightarrow f(0) = 0$
and $f(x+2) = f(x) \Rightarrow f(4) = f(2) = f(0) = 0$

47. (c) We have, $f(x+y) = f(x) + f(y)$

Put $y = -x$, $f(0) = f(x) + f(-x)$ ...(i)

Put $x = y = 0$, $f(0) = 2f(0) \Rightarrow f(0) = 0$ ...(ii)

From Eqs. (i) and (ii), we get

$$f(-x) = -f(x)$$

$\therefore$ odd which is an odd function.

48. (b) Given, $f(x) = \frac{1}{3}\{\sin 3x + |\sin 3x| + [\sin 3x]\}$

Let period of $\sin 3x$ is $T_1 = \frac{2\pi}{3}$, period of $|\sin 3x|$ is $T_2 = \frac{\pi}{3}$

and period of $[\sin 3x]$ is $T_3 = \frac{2\pi}{3}$

(we know that 2π is the period of $[\sin x]$)

Hence, period of $f(x) = \frac{\text{LCM of } \{2\pi, \pi, 2\pi\}}{\text{HCF of } \{3, 3, 3\}} = \frac{2\pi}{3}$

49. (c) We have, $f(x) = \frac{\sin 2nx}{1 + \cos^2 nx} = \frac{\sin 2nx}{1 + \frac{1 + \cos 2nx}{2}}$

$\because$ $f(x) = \frac{2\sin 2nx}{3 + \cos 2nx}$

period of $2\sin 2nx$ is $\frac{2\pi}{2n} = \frac{\pi}{n}$

and period of $3 + \cos 2nx$ is $\frac{2\pi}{2n} = \frac{\pi}{n}$

$\therefore$ Period of $f(x)$ is $\frac{\pi}{n}$, which is given by $\frac{\pi}{6}$.

$\because$ $\frac{\pi}{n} = \frac{\pi}{6}$ $\therefore$ $n = 6$

50. (b) We know that

if $f(-x) = f(x)$, then $f(x)$ is symmetrical about line $x = 0$.

$\therefore$ If $f(x)$ is symmetrical about $x = 2$

$\therefore$ $f(2 + x) = f(2 - x)$

51. (b) We have, $f(x) = e^{\sin 3\pi\{x\} + \tan \pi[x]}$

The period of $\sin 3\pi\{x\}$ is 1

and let T be the period of $\tan \pi[x]$.

$\therefore$ $\tan \pi[x + T] = \tan \pi[x]$

$\pi[x + T] = n\pi + \pi[x]$

$T + [x] = n + [x]$

$T = n$

$\therefore$ Period of $\tan \pi[x]$ is 1.

Hence, period of $f(x)$ is 1.

52. (c) We have, $f(x) = \tan(\sqrt{[n]}\, x)$

Its period is $\frac{\pi}{\sqrt{[n]}} = \frac{\pi}{3}$

$\therefore$ $\sqrt{[n]} = 3$, $[n] = 9$

$9 \le n < 10$

$\therefore$ $n \in [9, 10)$

53. (b) We have, $f(x) = nx + n - [nx + n] + \tan\frac{\pi x}{2}$

Period of $nx + n - [nx + n]$ or $\{nx + n\}$ is $\frac{1}{n}$

and period of $\tan\frac{\pi x}{2}$ is $\frac{\pi}{2}\cdot 2 = 2$

Period of $f(x) = \frac{\text{LCM of 1 and 2}}{\text{HCF of } n \text{ and } 1}$

$\therefore$ Period of $f(x) = 2$.

54. (a) Statement I is correct.

Let $h(x) = f[g(x)]$

$h'(x) = f'[g(x)] \cdot g'(x)$

which is strictly increasing or strictly decreasing, according as $f(x)$ and $g(x)$ are strictly increasing or strictly decreasing.

Statement II is also correct and correct explanation of Statement I.

55. (d) Statement I is false because in $[\alpha, \beta]$ some points may be where the $f(x)$ is discontinuous or not defined, hence it is not a generalised form, so Statement II is correct.

56. (b) We have,

$$f(x) = x^2 + x - 1$$
$$f'(x) = 2x + 1$$

which is one-one, when $x > -\frac{1}{2}$.

In [0, 3], $f(x)$ is one-one and its range is $[f(0), f(3)] = [1, 13]$ which is codomain.

Hence, its invertible and its inverse is $\frac{-1 + \sqrt{4x - 3}}{2}$

$\therefore$ Statement I is correct.

and Statement II is also correct but not correct explanation of Statement I.

57. (d) Statement I is false , as the intersection points will coincide if one is inverse of other e.g. $f(x) = y = \frac{1}{x}$ and $f^{-1}(x)$ intersect at infinite points other than $y = x$.

But Statement II is correct because in inverse function domain and range are interchange.

58. (a) We know that,

$$a^3 + b^3 + c^3 = 3abc$$

$\Rightarrow$ $(a + b + c)(a^2 + b^2 + c^2 - ab - bc - ca) = 0$

$\Rightarrow$ $(a + b + c)\frac{1}{2}\{(a - b)^2 + (b - c)^2 + (c - a)^2\} = 0$

$\therefore$ $a + b + c = 0$

$[\because (a - b)^2 + (b - c)^2 + (c - a)^2 > 0 \text{ as } a > b > c]$

But the equation $ax^2 + bx + c = 0$

$\therefore$ $x = 1$ is a root. $(\because a + b + c = 0)$

If other root is α, then

$$\alpha \cdot 1 = \frac{c}{a} \Rightarrow \alpha = \frac{c}{a} < 1$$

which is positive or negative.

Hence, both roots are real and distinct.

59. (b) We have, $f(x) = ({}^{x+1}C_{2x-8})({}^{2x-8}C_{x+1})$

It is defined, when $x + 1 > 0 \Rightarrow x > -1$ ...(i)

$2x - 8 > 0 \Rightarrow x > 4$ ...(ii)

and $x + 1 \ge 2x - 8 \Rightarrow x \le 9$

also $2x - 8 \ge x + 1 \Rightarrow x \ge 9$

$\Rightarrow$ $x = 9$ ...(iii)

$\therefore$ Singleton set

and range of $f(x)$ is $f(9) = {}^{10}C_{10} \cdot {}^{10}C_{10=1}$

60. (d) $f(x) = ax^3 + bx + c$

Since, $\alpha + \beta + \gamma = 0$

$\alpha \ge 0, \beta \ge 0$ and $\gamma \ge 0$

$\therefore$ $\alpha = \beta = \gamma = 0$

$\Rightarrow$ $f(x) = ax^3$,

Thus, Statement I is incorrect.

Revisal Problems (JEE Advanced)

1. (b) We have, $f(x) + 2f\left(\frac{2002}{x}\right) = 3x$ ...(i)

Replacing x by $\frac{2002}{x}$ in Eq. (i), we get

$$f\left(\frac{2002}{x}\right) + 2f(x) = \frac{6006}{x} \quad ...(ii)$$

On multiplying by 2 in Eq. (ii) and then subtracting from Eq. (i), we get

$$-3f(x) = 3x - \frac{2 \times 6006}{x}$$

$$-f(x) = x - \frac{4004}{x}$$

$$f(x) = \frac{4004}{x} - x$$

$$\therefore \quad f(2) = \frac{4004}{2} - 2 = 2000$$

2. (c) We know that, AM $\geq$ GM

then $\frac{a^2 + b^2}{2} \geq \sqrt{a^2b^2}$

$\because$ $a^2 + b^2 \geq 2ab$ ...(i)

Similarly, $b^2 + c^2 \geq 2bc$ ...(ii)

and $c^2 + a^2 \geq 2ca$ ...(iii)

On adding Eq. (i), (ii) and (iii), we get

$$2(a^2 + b^2 + c^2) \geq 2(ab + bc + ca)$$

$\therefore$ $ab + bc + ca \leq 1$

also $a^2 + b^2 - c^2 < 2ab$

$b^2 + c^2 - a^2 < 2bc$

$c^2 + a^2 - b^2 < 2ac$

On adding $a^2 + b^2 + c^2 < 2(ab + bc + ca)$

$\frac{1}{2} < ab + bc + ca$

$\therefore$ $ab + bc + ca \in \left(\frac{1}{2}, 1\right]$

3. (b) We have,

$$f(x) = \sqrt{\frac{8}{1-x} + \frac{8}{1+x}} = \sqrt{\frac{16}{1-x^2}} = \frac{4}{\sqrt{1-x^2}}$$

$$\therefore \quad f(x) = \frac{4}{\sqrt{1-x^2}}$$

$$f(\sin x) = \frac{4}{|\cos x|} \text{ and } f(\cos x) = \frac{4}{|\sin x|}$$

Now, we have

$$g(x) = \frac{4}{f(\sin x)} + \frac{4}{f(\cos x)} = \frac{4|\cos x|}{4} + \frac{4|\sin x|}{4}$$

$$\therefore \quad g(x) = |\sin x| + |\cos x|$$

which is even function and both are complementary to each other.

Hence, period of $g(x) = \frac{\pi}{2}$.

4. (c) We have,

$$f(x, n) = \sum_{k=1}^{n} \log_x \frac{k}{x} = \sum_{k=1}^{n} \log_x k - \log_x x = \sum_{k=1}^{n} \frac{\log k}{\log x} - 1$$

$$= \frac{1}{\log x}[\log 1 + \log 2 + \log 3 + \dots + \log n] - \sum_{k=1}^{n} 1$$

$$= \frac{1}{\log x} \log 1 \cdot 2 \cdot 3 \cdot 4 \dots n - n$$

$$\therefore \quad f(x, n) = \frac{\log n!}{\log x} - n$$

Now, we have, $f(x, 10) = f(x, 11)$

$$\Rightarrow \quad \frac{\log 10!}{\log x} - 10 = \frac{\log 11!}{\log x} - 11$$

$$\Rightarrow \quad \log 10! = \log 11! - \log x$$

$$\Rightarrow \quad \log x = \log 11! - \log 10! = \log 11$$

$$\therefore \quad x = 11$$

5. (a) In this case, by choosing $x = 1$, we see that

$$f(y) = \frac{f(1)}{y} \text{ for all } y$$

put $y = 30$

$$\Rightarrow \quad f(1) = 30 \cdot f(30) \cdot = 30 \cdot (20) = 600$$

Now, $f(40) = \frac{f(1)}{40} = \frac{600}{40} = 15$

6. (c) We have,

$$f(x) = \sin^2 x + \cos^4 x + 2$$

Its period is T_1.

Now, $f(x) = \frac{1 - \cos 2x}{2} + \left(\frac{1 + \cos 2x}{2}\right)^2 + 2$

$$= \frac{1 - \cos 2x}{2} + \frac{1}{4}(1 + \cos^2 2x + 2\cos 2x) + 2$$

$$= \frac{1}{4}[2 - 2\cos 2x + 1 + \cos^2 2x + 2\cos 2x + 8]$$

$$= \frac{1}{4}[11 + \cos^2 2x] = \frac{1}{4}\left[11 + \frac{1 + \cos 4x}{2}\right]$$

$$= \frac{1}{8}[23 + \cos 4x]$$

$$\therefore \quad T_1 = \frac{2\pi}{4} = \frac{\pi}{2}$$

We have,

$$g(x) = \cos(\cos x) + \cos(\sin x)$$

Its period is T_2,

$$\because \quad g(x + T_2) = g(x)$$

$$\cos[\cos(x + T_2)] + \cos[\sin(x + T_2)] = \cos(\cos x) + \cos(\sin x)$$

Put $x = 0$

$$\cos(\cos T_2) + \cos(\sin T_2) = \cos 1 + \cos 0$$

$$\therefore \cos(\cos T_2) + \cos(\sin T_2) = \cos\left(\cos\frac{\pi}{2}\right) + \cos\left(\sin\frac{\pi}{2}\right)$$

On comparing both sides, we get

$$\therefore \quad T_2 = \frac{\pi}{2}$$

We see that $T_1 = T_2$

7. (b) $|x - p| = x - p$, since $x \geq p$

$|x - 15| = 15 - x$ $[\because x \leq 15]$

$|x - (p + 15)| = (p + 15) - x$ $[\because 15 + p > x]$

$\therefore$ Expression reduces to

$$E = (x - p) + (15 - x) + (p + 15 - x)$$

$$= 30 - x$$

$\therefore$ E_{min} occurs, when $x = 15$.

$$\Rightarrow \quad E_{min} = 15$$

8. (d) We have,

$$g\left(-1, -\frac{3}{2}\right) = -1 - \left(-\frac{3}{2}\right) = \frac{1}{2}$$

$$g(-4, -1.75) = -1.75 - (-4)$$

$$= 2.25$$

Now, $f\left(g\left(-1, -\frac{3}{2}\right), g(-4, -1.75)\right) = (2.25)^{\frac{1}{2}} = 1.5$

9. (d) We have to find solution set of

$$(f(x))^2 \ge f(x) + 6$$

$$(f(x))^2 - f(x) - 6 \ge 0$$

$$(f(x) - 3)(f(x) + 2) \ge 0$$

$$\Rightarrow \quad f(x) \le -2 \text{ or } f(x) \ge 3.$$

We have, $f(x) < 3$ solution set is $0 < x < \infty$.

$\therefore$ Solution set of $f(x) \ge 3$ is $-\infty < x \le 0$

and $f(x) > -2$ solution set $-\infty < x < 5$

$\therefore \quad f(x) \le -2$

It solution set $-5 \le x < \infty$

Hence, complete solution set is

$$(-\infty, 0] \cup [-5, \infty)$$

10. (d)

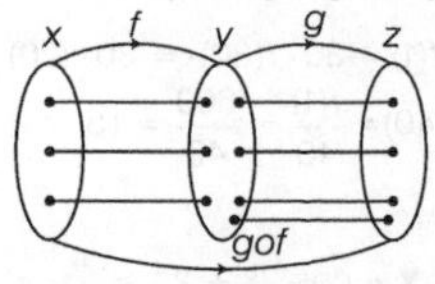

$\Rightarrow A$ is correct

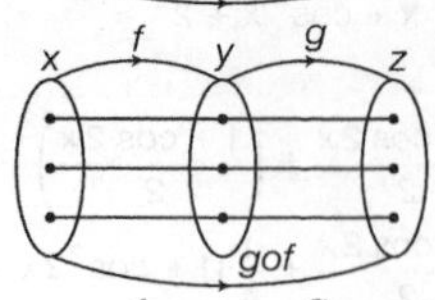

$\Rightarrow B$ is correct

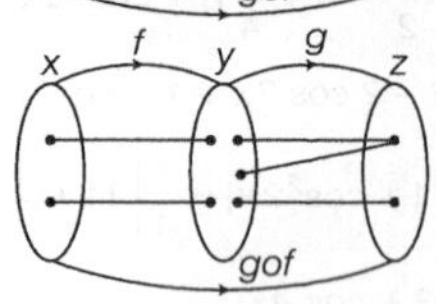

$\Rightarrow C$ is correct

11. (d)

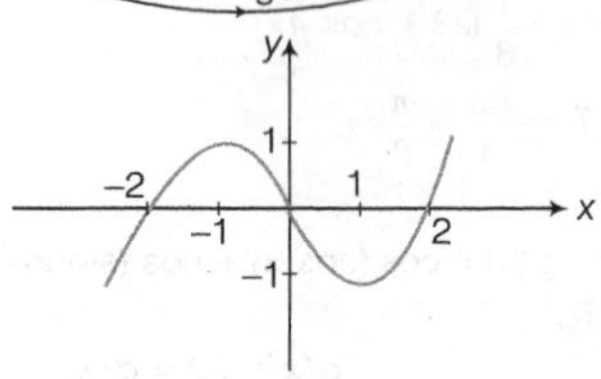

Graph for g(x)

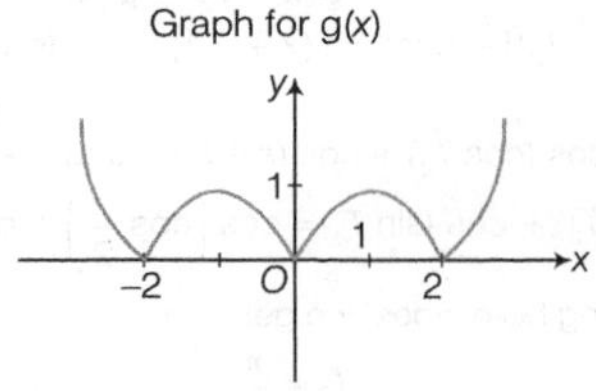

Graph for $|g(x)|$

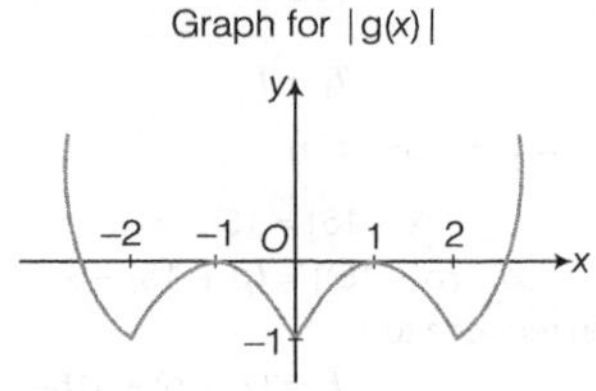

Graph for $|g(x)|-1$

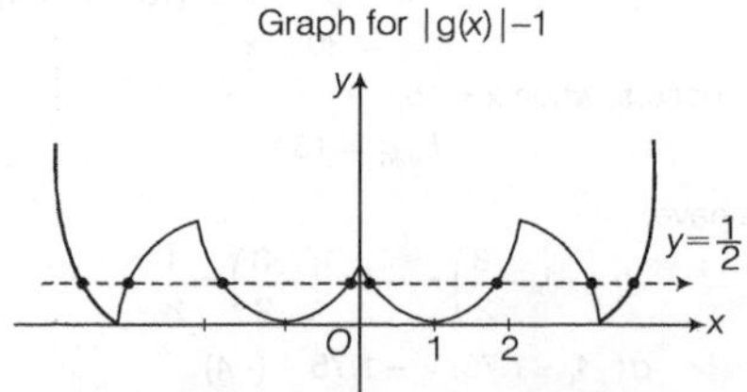

Graph for $|g(x)-1| \Rightarrow ||g(x)|-1| = \frac{1}{2}$

$\therefore$ 8 solutions

12. (d)

(I) If $f(x) = x$ and $g(x) = -x$

$\Rightarrow f(x) + g(x) = 0, \therefore$ many one.

(II) If $f(x) = x$ and $g(x) = x^3$

$\Rightarrow \quad f(x) \cdot g(x) = x^4$, i.e. many one.

(III) If $f(x) = \sin x$, which is odd but not one-one or $f(x) = x^2 \cdot \sin x$, which is odd but many one.

$\therefore$ (I), (II) and (III) all are incorrect.

13. (a) We have, $f(x)$ is one-one.

Hence, its inverse is defined.

We know that function and its inverse is symmetrical about the line $y = x$.

It means X-intercept becomes Y-intercept in at $f^{-1}(x)$ and *vice-versa*.

$\therefore$ The sum of all intercepts of $y = f^{-1}(x)$ is also 5.

14. (b) We have, $f(x) = x^2 + bx + c$

and $\quad f(2 + t) = f(2 - t)$

$$(2 + t)^2 + b(2 + t) + c = (2 - t)^2 + b(2 - t) + C$$

$$\Rightarrow \quad b = -4$$

$$\therefore \quad f(x) = x^2 - 4x + c$$

Now, $\quad f(1) = c - 3$

$$f(2) = c - 4$$

$$f(4) = c$$

We see that, $f(2) < f(1) < f(4)$

15. (d) We have, $[x]\{x\} = 1$

or $\quad \{x\} = \frac{1}{[x]} \qquad [0 \le \{x\} < 1]$

$$\therefore \quad 0 \le \frac{1}{[x]} < 1$$

$$\therefore \quad 1 < [x] < \infty \Rightarrow 2 \le x < \infty$$

But $\quad \{x\} = \frac{1}{[x]}, \{x\} = \frac{1}{m}$

$\therefore$ We see that, if $[x] = m$

Then, $\quad \{x\} = \frac{1}{m}$

$$\therefore \quad x = [x] + \{x\} = m + \frac{1}{m}, m \in N$$

But $m \ne 1$ i.e. $\left\{x = M + \frac{1}{m} : M \in N - \{1\}\right\}$

16. (c) Let $y = \frac{\{x\}}{1 + \{x\}}$

$$y + y\{x\} = \{x\}$$

$$\{x\}(y - 1) = -y$$

$$\{x\} = \frac{y}{1 - y}$$

We know that, $\quad 0 \le \{x\} < 1$

$$\therefore \quad 0 \le \frac{y}{1 - y} < 1$$

On solving $\quad \frac{y}{1 - y} \ge 0 \Rightarrow \frac{y}{y - 1} \le 0$

$$0 \le y < 1 \qquad \text{...(i)}$$

On solving $\quad \frac{y}{1 - y} - 1 < 0$

$$\frac{y - 1 + y}{1 - y} < 0, \frac{2y - 1}{y - 1} > 0$$

$$y < \frac{1}{2} \text{ or } y > 1 \qquad \text{...(ii)}$$

On combining Eqs. (i) and (ii), we get

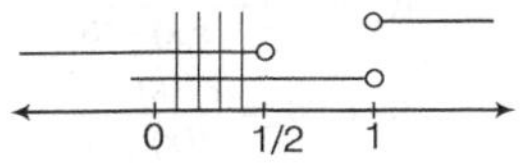

$$0 \le y < \frac{1}{2}$$

$\therefore$ Range $= \left[0, \frac{1}{2}\right)$

17. (b) We have, $fog\ (x) = gof\ (x)$

$$\therefore \quad \frac{g(x)}{1+g(x)} = \frac{r\,f(x)}{1-f(x)}$$

$$\Rightarrow \quad \frac{\frac{rx}{1-x}}{1+\frac{rx}{1-x}} = \frac{\frac{rx}{1+x}}{1-\frac{x}{1+x}}$$

$$\Rightarrow \quad \frac{rx}{1-x+rx} = \frac{rx}{1}$$

$$\Rightarrow \quad rx - rx\,(1 - x + rx) = 0$$

$$rx\,(1 - 1 + x - rx) = 0$$

$$r[x\,x(1-r)] = 0$$

$$\therefore \quad r(r-1) = 0$$

$$r = 0 \text{ and } 1$$

Number of elements in set $S = 2$.

18. (c) We have, $f(k) = \frac{k}{2009}$

$\therefore\ f(2009 - k) = \frac{2009-k}{2009} = 1 - \frac{k}{2009}$

We have,

$$g(k) = \frac{f^4(k)}{[1-f(k)]^4 + [f(k)]^4}$$

$$g(k) = \frac{\left(\frac{k}{2009}\right)^4}{\left(1-\frac{k}{2009}\right)^4 + \left(\frac{k}{2009}\right)^4}$$

$$g(k) = \frac{k^4}{(2009-k)^4 + k^4} \quad \ldots(i)$$

$$\Rightarrow \quad g(2009-k) = \frac{[f(2009-k)]^4}{[1-f(2009-k)]^4 + [f(2009-k)]^4}$$

$$= \frac{\left(1-\frac{k}{2009}\right)^4}{\left(\frac{k}{2009}\right)^4 + \left(1-\frac{k}{2009}\right)^4}$$

$$g(2009-k) = \frac{(2009-k)^4}{k^4 + (2009-k)^4} \quad \ldots(ii)$$

Now, adding Eqs. (i) and (ii), we get

$$g(k) + g(2009-k) = 1$$

We have to find

$$\sum_{K=0}^{2009} g(k) = [g(0) + g(2009)] + [g(1) + g(2008)] + \ldots$$

$$[g(1004) + g(1005)]$$

i.e. $\underbrace{1 + 1 + 1 \ldots + 1}_{1005 \text{ times}}$

$$= 1005$$

19. (c) Let 'α' be the root of $f(x) = 0$

$$\Rightarrow \quad f(\alpha) = f(2 - \alpha) = 0$$

$f(x)$ has 4 distinct and two equal roots.

$\therefore$ Sum of roots $= 6$

20. (a) Let $f(x)$ be n degree polynomial, then

$$n = (n-1) + (n-2) \Rightarrow n = 3.$$

$\therefore f(x)$ is a polynomial fo degree 3.

21. (c) We have, $f^2(x)\, f\left(\frac{1-x}{1+x}\right) = 64x \quad \ldots(i)$

Replace $x \to \frac{1-x}{1+x}$

then, $$f^2\left(\frac{1-x}{1+x}\right) f(x) = 64\left(\frac{1-x}{1+x}\right) \quad \ldots(ii)$$

$$\Rightarrow \quad \left(\frac{64x}{f^2(x)}\right)^2 f(x) = 64\left(\frac{1-x}{1+x}\right)$$

$$\therefore \quad f^3(x) = 64x^2\left(\frac{1+x}{1-x}\right)$$

$$\Rightarrow \quad f(x) = 4\left(x^2\left(\frac{1+x}{1-x}\right)\right)^{1/3}$$

22. (c) Suppose $f(x)$ has degree < 2

$$\Rightarrow \quad f(x) = bx + c$$

$$\Rightarrow \quad (bx + c) + 1 = bx + c,$$

which is not possible.

Suppose $f(x)$ has degree > 2

$$\Rightarrow \quad f(x) = ax^n + bx^{n-1} + \cdots + ,\ (a \ne 0),$$

On comparing coefficients of x^{n-2}, we get

$$c = \frac{an(n-1) + 2c}{2} \Rightarrow a = 0$$

$\therefore f(x)$ is of degree 2.

23. (d) Since, f is many one. Let $f(\alpha) = f(\beta)$, where $\alpha \ne \beta$.

Now, $f(\alpha + y) = g\ [f(\alpha), y] = g\ [f(\beta), y] = f(\beta + y)$

$\therefore g$ is periodic.

Also for $\quad y = t - \alpha.$

$$\Rightarrow \quad f(t) = f(t + \beta - \alpha),\ \forall t \in R$$

$\therefore f$ is periodic.

24. (a) Let $x = I + f,\ 0 \le f < 1$ and $I \in$ Integer.

$$\therefore \quad 2I^2 + 3I = 4I + 4f$$

$$\Rightarrow \quad f = \frac{2I^2 - I}{4} \Rightarrow 0 \le \frac{2I^2 - I}{4} < 1$$

$$\Rightarrow \quad 0 \le 2I^2 - I < 4$$

$$\Rightarrow \quad 2I^2 - I \ge 0 \text{ and } 2I^2 - I - 4 < 0$$

$$\therefore \quad \left(I \le 0 \text{ or } I \ge \frac{1}{2}\right)$$

and $$\left(\frac{1-\sqrt{33}}{4} < I < \frac{1+\sqrt{33}}{4}\right)$$

$$\Rightarrow \quad I = -1, 0, 1$$

when $\quad I = -1 \Rightarrow f = 3/4 \Rightarrow x = -1/4$

when $\quad I = 0 \Rightarrow f = 0 \Rightarrow x = 0$

when $\quad I = 1 \Rightarrow f = 1/4 \Rightarrow x = 5/4$

Hence, difference between the maximum and minimum value $= 5/4 - (-1/4) = 3/2$

25. (a) Since, $f(1) + f(2) = 4f(2)$

$$\Rightarrow \quad f(2) = \frac{f(1)}{3} = \frac{2005}{3} = \frac{2005}{1+2} \quad \text{[given]}$$

Similarly, $$f(3) = \frac{2005}{1+2+3}$$

$$\therefore \quad f(2004) = \frac{2005}{1+2+3+\cdots+2004}$$

$$= \frac{2005}{\frac{2004}{2}(1+2004)} = \frac{1}{1002}$$

26. (a,b,c,d) Let $f(x) = x$ and $g(x) = \frac{1}{x}$

$\Rightarrow \quad f(x)\,g(x) = 1$

$\therefore$ It is periodic.

(b) Let $f(x) = x$ and $g(x) = -x$

$\Rightarrow \quad f(x) + g(x) = 0$

$\therefore$ It is periodic

(c) Let $f(x) = x$ and $g(x) = x$

$\Rightarrow \quad f(x) - g(x) = 0$

$\therefore$ It is periodic

(d) Let $f(x) = \sin\sqrt{x}$ and $g(x) = x^2$

$\therefore f[g(x)] = \sin x$, periodic.

27. (a,b) Given, $f(x) + f\left(\frac{1}{1-x}\right) = \frac{2}{x} - \frac{2}{1-x}$ ...(i)

Replacing x by $\frac{1}{1-x}$ and by $\left(1 - \frac{1}{x}\right)$ in Eq. (i) one by one and on solving, we get

$$f(x) = \frac{x+1}{x-1}$$

$$\therefore \quad f(2) + f(3) = 3 + 2 = 5.$$

also, $f(x) = \frac{x+1}{x-1}$ crosses Y-axis at $x = 1$.

28. (a,b,c) Let the equation be

$$a_0x^n + a_1x^{n-1} + a_2x^{n-2} + \cdots + a_n$$

Now, $\Sigma\alpha_i = \pm 1, \Sigma\alpha_i\alpha_j = \pm 1$

$\Rightarrow \quad \Sigma\alpha_i^2 \le 1 + 2 = 3$

$\therefore \quad n \le 3$

29. (a,b) We have,

$$g(x) = \sin(\sin^{-1}\sqrt{\{x\}}) + \cos(\sin^{-1}\sqrt{\{x\}}) - 1$$

$$= \sqrt{\{x\}} + \sqrt{1 - \{x\}} - 1$$

Case I

If $x \in I$

$\Rightarrow \quad g(x) = 0 + 1 - 1 = 0$

$\Rightarrow \quad g(-x) = g(x)$

Case II

If $x \notin I$,

$$\Rightarrow \quad g(-x) = \sqrt{\{-x\}} + \sqrt{1 - \{-x\}} - 1$$

$$= \sqrt{1 - \{x\}} + \sqrt{1 - (1 - \{x\})} - 1$$

$$= \sqrt{\{x\}} + \sqrt{1 - \{x\}} - 1 = g(x)$$

$\therefore g$ is even function and period is 1.

30. (c) Clearly,

$$f(x + 2\pi) = \sin(x + 2\pi + 3\sin(x + 2\pi))$$

$$= \sin(2\pi + (x + 3\sin x))$$

$$= \sin(x + 3\sin x) = f(x)$$

$\therefore f(x)$ has period 2π.

Statement II is false as $(x + 3\sin x)$ is non-periodic but $\sin(x + 3\sin x)$ is periodic.

31. (a) Here, $$f(x+10) = \frac{f(x) - 5}{f(x) - 3}$$

$$\Rightarrow \quad f(x+20) = \frac{2f(x) - 5}{f(x) - 2}$$

$$f(x+30) = \frac{3f(x) - 5}{f(x) - 1}$$

$$f(x+40) = \frac{3f(x+10) - 5}{f(x+10) - 1} = f(x)$$

$\therefore f(x)$ is periodic with period 40 and $f(10) = f(50)$.

32. (c) ***Case I*** $x^2 - 2x + a = x - 1, x > 1$

$\Rightarrow \quad x^2 - 3x + (a+1) = 0$

$\Delta > 0$

$\Rightarrow \quad 9 > 4(a+1)$

$\Rightarrow \quad a < 5/4$

Product of roots > 1

$\Rightarrow \quad a + 1 > 1$ or $a > 0$

$\therefore \quad 0 < a < 5/4$ ...(i)

Case II

$$x^2 - 2x + a = 1 - x, x < 1$$

$\Rightarrow \quad x^2 - x + (a-1) = 0 \Rightarrow \Delta > 0$

$\Rightarrow \quad 1 > 4(a-1)$

$a < 5/4$

and $\quad f(1) > 0$

$\Rightarrow \quad a - 1 > 0$

$\Rightarrow \quad a > 1$ ...(ii)

From Eq. (i) and (ii), $a \in (1, 5/4)$.

33. (a) Three roots occurs when given set of equations has a common root.

i.e. when $a = 1$

34. (b) ***Case I*** $x - 1 > 0$

$$x^2 - 2x + a = x - 1$$

$\Rightarrow \quad x^2 - 3x + (a+1) = 0$

$\therefore \quad \Delta = 0$

$9 - 4(a+1) = 0$

$\Rightarrow \quad a = \frac{5}{4}$

Case II $x < 1$

$$x^2 - 2x + a + x - 1 = 0$$

$$x^2 - x + (a-1) = 0$$

$\therefore \quad \Delta = 0 \Rightarrow 1 - 4(a-1) = 0$

$\Rightarrow \quad a = 5/4$

35. (c) $f(x) = \log_{\{x\}}[x], 1 < x < 5$

$$\{x\} \ne 0 \Rightarrow x \ne 2, 3, 4.$$

$\therefore f(x)$ is not defined at 3 points.

36. (c) Now, domain of $f(x) = \log_{\{x\}}[x]$ is $x > 1$ – (positive integer)

Domain of $g(x) = \log_{\{x\}}[x]$ if $x > 2$ – positive integer.

$\therefore \quad (A - B)$ domain $\in (1, 2)$.

37. (c) Domain of $h(x) = \log_{\{x\}}[x]$

$\Rightarrow \quad \{x\} \ne 0$

$\therefore \quad x \in R$ – (Integer) or $R - I$

38. (d) Here, $f_1^{-1} = f_2, f_2^{-1} = f_1, f_3^{-1} = f_4, f_4^{-1} = f_3$

Now, $\quad f_1 of = f_4$

$\Rightarrow \quad F = f_1^{-1} of_4 = f_1 of_4$

$\Rightarrow \quad F = f_4$

39. (a) $Gof_3 = f_6 \Rightarrow G = f_6\ of = f_6 of_3$

$\therefore \quad G(x) = f_6(1-x) = \dfrac{1-x}{1-x-1} = 1 - \dfrac{1}{x} = f_5(x)$

40. (c) $J = f_3^{-1} of_4 of_2^{-1} = f_3 of_4 of_2$

$\therefore \quad J = f_3 o\, f_4\left(\dfrac{1}{x}\right) = f_3\left(\dfrac{1}{1-\dfrac{1}{x}}\right) = f_3\left(\dfrac{x}{x-1}\right)$

$= 1 - \dfrac{x}{x-1} = f_4(x)$

41. (1) Given, $\dfrac{(\sin x)^{2y}}{(\cos x)^{y^2/2}} + \dfrac{(\cos x)^{2y}}{(\sin x)^{y^2/2}} \geq 2\sqrt{\dfrac{(\sin x \cos x)^{2y}}{(\cos x \sin x)^{y^2/2}}}$ [AM ≥ GM]

$\geq 2\,(\sin x \cos x)^{y - \frac{y^2}{4}}$

$\Rightarrow \quad \sin 2x \geq 2(\sin x \cos x)^{y-\frac{y^2}{4}}$

$\therefore \quad y - \dfrac{y^2}{4} \geq 1$

$\Rightarrow \quad y^2 - 4y + 4 \leq 0$

$\Rightarrow \quad (y-2)^2 \leq 0$

$\therefore \quad y = 2$ and $\sin x = \cos x$

$\Rightarrow \quad x = \pi/4,\ y = 2$

∴ Number of pairs (x, y) is 1.

42. (1) As, $f([i]) = \dfrac{1}{1+[i]}$

$\Rightarrow \quad f(0) = 1, f(1) = \dfrac{1}{2}, f(2) = \dfrac{1}{3}, f(3) = \dfrac{1}{4}, f(4) = \dfrac{1}{5}$

$\Rightarrow \quad g(x) = (1+x) f(x) - 1,$

is a polynomial of degree 5 with roots 0, 1, 2, 3, 4.

$\Rightarrow \quad g(x) = Ax(x-1)(x-2)(x-3)(x-4)$

$\Rightarrow \quad A = \dfrac{1}{5!}$

$\Rightarrow \quad (1+x) f(x) - 1 = \dfrac{1}{5!} x(x-1)(x-2)(x-3)(x-4)$

$\therefore \quad 7 f(6) - 1 = \dfrac{6 \times 5 \times 4 \times 3 \times 2}{5!}$

$\Rightarrow \quad 7 f(6) - 1 = 6$

$\Rightarrow \quad f(6) = 1$

43. (9) $\therefore f(m) - f(n) = 3$

If $\quad m = 125, n = 124$

$\Rightarrow \quad (m-n)_{\min} = 1$

$m = 24, n = 5$

$\Rightarrow \quad f(m) - f(n) = 3$

$\Rightarrow \quad (m-n)_{\max} = 19$

$\therefore \quad \dfrac{(m-n)_{\max.} - (m-n)_{\min}}{2} = \dfrac{19-1}{2} = 9$

44. (2) Given, $f[xf(y)] = x^p y^4$.

Put $x = \dfrac{1}{f(y)}$

$\therefore \quad f(1) = \left(\dfrac{1}{f(y)}\right)^p y^4 = \dfrac{y^4}{[f(y)]^p},$

for $\quad y = 1, f(1) = \dfrac{1}{[f(1)]^p}$

$\Rightarrow \quad f(1) = 1$, So, $f(y) = y^{4/p}$

Hence, $\quad f(xy^{4/p}) = x^p y^4$

Put $y = z^{p/4}$

$\Rightarrow \quad f(xz) = x^p z^p$

$\Rightarrow \quad f(x) = x^p$

$\therefore \quad \dfrac{4}{p} = p$

$\Rightarrow \quad p = 2$

JEE Main & AIEEE Archive

1. (d) Given, $f_k(x) = \dfrac{1}{k}(\sin^k x + \cos^k x),$

where $x \in R$ and $k \geq 1$.

$f_4(x) - f_6(x) = \dfrac{1}{4}(\sin^4 x + \cos^4 x) - \dfrac{1}{6}(\sin^6 x + \cos^6 x)$

$= \dfrac{1}{4}(1 - 2\sin^2 x \cdot \cos^2 x) - \dfrac{1}{6}(1 - 3\sin^2 x \cdot \cos^2 x)$

$= \dfrac{1}{4} - \dfrac{1}{6} = \dfrac{1}{12}$

2. (d) Here, g is the inverse of $f(x)$.

$\Rightarrow fog\ (x) = x$

On differentiating w.r.t. x, we get

$f'\{g(x)\} \times g'(x) = 1$

$\Rightarrow \quad g'(x) = \dfrac{1}{f'\{g(x)\}} = 1 + \{g(x)\}^5 \quad \left[\because f'(x) = \dfrac{1}{1+x^5}\right]$

$\therefore \quad g'(x) = 1 + \{g(x)\}^5$

3. (c) **Statement I**

$A = \{(x, y) \in R \times R : y - x$ is an integer$\}$

(a) **Reflexive**

$xRx : (x - x)$ is an integer.

i.e. true

∴ A is reflexive.

(b) **Symmetric**

$xRy : (x - y)$ is an integer.

$\Rightarrow -(y - x)$ is an integer.

$\Rightarrow (y - x)$ is an integer.

$\Rightarrow yRx$

∴ A is symmetric.

(c) **Transitive**

xRy and yRz

$\Rightarrow (x - y)$ is an integer and $(y - z)$ is an integer.

$\Rightarrow (x - y) + (y - z)$ is an integer.

$\Rightarrow (x - z)$ is an integer.

$\Rightarrow xRz$

∴ A is transitive.

Hence, A is an equivalence relation.

Statement II

$B = \{(x, y) \in R \times R : x = \alpha y$ for some rational number $\alpha\}$

If $\alpha = \dfrac{1}{2}$, then for reflexive, we have

$xRx \Rightarrow x = \dfrac{1}{2}x$, which is not true, $\forall\ x \in R - \{0\}$.

∴ B is not reflexive on R.

Hence, B is not an equivalence relation on R.

Hence, Statement I is correct, Statement II is incorrect.

4. (b) Given, $y = \frac{1}{\sqrt{|x| - x}}$

For domain, $|x| - x > 0 \Rightarrow |x| > x$

ie. only possible, if $x < 0$.

$\therefore$ $x \in (-\infty, 0)$

5. (a) Given that $p(x) = f(x) - g(x)$ has only one root -1.

$\Rightarrow p(x) = (a - a_1)x^2 + (b - b_1)x + (c - c_1)$ has one root, -1 only.

$\Rightarrow p'(x)$ will also has root as -1.

$\Rightarrow p'(x) = 0$ at $x = -1$

$\Rightarrow 2(a - a_1)x + (b - b_1) = 0$ at $x = -1$.

$\Rightarrow -2(a - a_1) + (b - b_1) = 0$

$$\Rightarrow \frac{-(b - b_1)}{(a - a_1)} = -2 \quad \ldots\text{(i)}$$

Now, $p(x) = (a - a_1)x^2 + (b - b_1)x + (c - c_1)$

$$\Rightarrow \frac{p(x)}{a - a_1} = x^2 + \frac{b - b_1}{a - a_1}x + \frac{(c - c_1)}{a - a_1}$$

$\because$ $p(-1) = 0$

$$\therefore \quad 0 = (-1)^2 - \frac{(b - b_1)}{a - a_1} + \frac{(c - c_1)}{a - a_1}$$

$$\Rightarrow \quad 0 = 1 - 2 + \frac{(c - c_1)}{a - a_1} \quad \text{[using Eq. (i)]}$$

$$\Rightarrow \frac{c - c_1}{a - a_1} = 1 \quad \ldots\text{(ii)}$$

Also, given that $p(-2) = 2$

$\Rightarrow 4(a - a_1) - 2(b - b_1) + (c - c_1) = 2$ ...(iii)

From Eqs. (i), (ii) and (iii), we have

$4(a - a_1) - 4(a - a_1) + (a - a_1) = 2$

$\Rightarrow$ $a - a_1 = 2$

On substituting $a - a_1 = 2$ in Eq. (ii), we get

$c - c_1 = 2$

On substituting $a - a_1 = 2$ in Eq. (i), we get

$b - b_1 = 4$

Now, $p(2) = 4(a - a_1) + 2(b - b_1) + (c - c_1)$

$= 4 \times 2 + 2 \times 4 + 2 = 8 + 8 + 2 = 18$

6. (b) Given,

$R = \{(A, B) : A = P^{-1}BP$ for some invertible matrix $P\}$

For Statement I

(i) **Reflexive**

$ARA \Rightarrow A = P^{-1}AP$, which is true only, if $P = I$.

Since, $A = P^{-1}BP$ for some invertible matrix P.

$\therefore$ We can assume $P = I$

$\Rightarrow$ $ARA \Rightarrow A = I^{-1}AI$

$\Rightarrow$ $A = A$

$\Rightarrow R$ is Reflexive

Note *Here, due to some invertible matrix, P is used (reflexive) but if for all invertible matrix is used, then R is not reflexive.*

(ii) **Symmetric**

$ARB \Rightarrow A = P^{-1}BP$

$\Rightarrow$ $PAP^{-1} = P(P^{-1}BP)P^{-1}$

$\Rightarrow$ $PAP^{-1} = (PP^{-1})B(PP^{-1})$

$\therefore$ $B = PAP^{-1}$

Since, for some invertible matrix P, we can let

$Q = P^{-1}$

$\therefore$ $B = (P - 1)^{-1}AP^{-1}$

$\Rightarrow$ $B = Q^{-1}AQ$

$\Rightarrow BRA$

$\Rightarrow R$ is symmetric.

(iii) **Transitive**

ARB and BRC

$\Rightarrow$ $A = P^{-1}BP$

and $B = P^{-1}CP$

$\Rightarrow$ $A = P^{-1}(P^{-1}CP)P$

$= (P^{-1})^2C(P)^2$

So, ARC, for some $P^2 = P$

$\Rightarrow R$ is transitive

So, R is an equivalence relation.

For Statement II It is always true that

$(MN)^{-1} = N^{-1}M^{-1}$

Hence, both statements are true but Statement II is not the correct explanation of Statement I.

7. (a) $f(x) = (x - 1)^2 + 1$, as $x \geq 1$

When $f(x) = f^{-1}(x)$

$\Rightarrow$ $f(x) = x$

$\Rightarrow$ $(x - 1)^2 + 1 = x$

$\Rightarrow$ $(x - 1)^2 = x - 1$

$\Rightarrow$ $(x - 1)^2 - (x - 1) = 0$

$\Rightarrow$ $(x - 1)\{x - 1 - 1\} = 0$

$\Rightarrow$ $x = 1, 2$

$\therefore$ $\{x : f(x) = f^{-1}(x)\} = \{1, 2\}$

Also, let $f(x) = y$

$\therefore$ $y = (x - 1)^2 + 1$

$\Rightarrow$ $(x - 1) = \pm\sqrt{y - 1}$

$\therefore$ $x = 1 \pm \sqrt{y - 1}$

$\because$ $x \geq 1$

Neglecting $1 - \sqrt{y - 1}$, then

$x = 1 + \sqrt{y - 1}$

or $f^{-1}(y) = 1 + \sqrt{y - 1}$

$\therefore$ $f^{-1}(x) = 1 + \sqrt{x - 1}$

So, both statements are correct and Statement II is correct explanation of Statement I.

8. (c) Given, relation R is defined as

$R = \{(x, y) \mid x, y$ are real numbers and $x = wy$ for some rational number $w\}$

(i) **Reflexive** $xRx \Rightarrow x = wx$

$\therefore$ $w = 1 \in$ rational number

The relation R is reflexive.

(ii) **Symmetric** $xRy \neq yRx$ as $0R1$

But $1R0 \Rightarrow 1 = w \cdot (0)$

which is not true for any rational number.

The relation R is not symmetric.

Thus, R is not equivalence relation.

Now, for relation S which is defined as

$S = \left\{\left(\frac{m}{n}, \frac{p}{q}\right) \middle| m, n, p \text{ and } q \in \text{integers such that } n, q \neq 0\right.$ and $qm = pn\}$

(i) **Reflexive** $\frac{m}{n}R\frac{m}{n} \Rightarrow mn = mn$ [correct]

$\therefore$ The relation S is reflexive.

(ii) **Symmetric** $\frac{m}{n}R\frac{p}{q} \Rightarrow mq = np$

$\Rightarrow$ $np = mq$

$\Rightarrow$ $\frac{p}{q}R\frac{m}{n}$

$\therefore$ The relation S is symmetric.

(iii) **Transitive** $\frac{m}{n}R\frac{p}{q}$ and $\frac{p}{q}R\frac{r}{s}$

$\Rightarrow$ $mq = np$ and $ps = rq$

$\Rightarrow$ $mq \cdot ps = np \cdot rq \Rightarrow ms = nr$

$\Rightarrow$ $\frac{m}{n} = \frac{r}{s} \Rightarrow \frac{m}{n}R\frac{r}{s}$

The relation S is transitive.

Hence, the relation S is an equivalence relation.

9. (c) Given, $f(x) = x^3 + 5x + 1$

Now, $f'(x) = 3x^2 + 5 > 0, \ \forall \ x \in R$

Thus, $f(x)$ is strictly increasing function.

So, $f(x)$ is one-one function.

Clearly, $f(x)$ is a continuous function and also increasing on R.

$\therefore$ $\lim_{x \to -\infty} f(x) = -\infty$ and $\lim_{x \to \infty} = \infty$

Hence, $f(x)$ takes every value between $-\infty$ and ∞.

Thus, $f(x)$ is onto function.

10. (c) Given, $f(x) = (x+1)^2 - 1, x \ge -1$

$\Rightarrow$ $f'(x) = 2(x+1) \ge 0$, for $x \ge -1$

$\Rightarrow$ $f(x)$ is one-one.

Since, codomain of the given function is not given, hence it can be considered as R, the set of real and consequently f is not onto.

Hence, f is not bijective. Statement II is incorrect.

Also, $f(x) = (x+1)^2 - 1 \ge -1$ for $x \ge -1$

$\Rightarrow$ $R_f = [-1, \infty)$

Clearly, $f(x) = f^{-1}(x)$ at $x = 0$ and $x = -1$

$\therefore$ Statement I is correct.

11. (a) Since, $(1, 2) \in S$ but $(2, 1) \notin S$

So, S is not symmetric.

Hence, S is not an equivalence relation.

Given, $T = \{(x, y) : (x - y) \in I\}$

Now, $x - x = 0 \in I$, it is reflexive relation.

Again now, $(x - y) \in I$

$y - x \in I$, it is symmetric relation.

Let $x - y = I_1$

and $y - z = I_2$

Now, $x - z = (x - y) + (y - z) = I_1 + I_2 \in I$

So, T is also transitive.

Hence, T is an equivalence relation.

12. (a) The given function is $f : N \to Y$ defined as

$$f(x) = 4x + 3$$

where $Y = \{y \in N : y = 4x + 3 \text{ for some } x \in N\}$

Now, $f(x) = 4x + 3$

$\Rightarrow$ $f'(x) = 4 > 0$

$\Rightarrow f$ is a strictly increasing function.

$\Rightarrow f$ is one-one.

Also, here $y = \{7, 8, 9, \dots, \infty\}$

and range of $f = \{7, 8, 9, \dots, \infty\}$

$\because$ Range of $f = y =$ Codomain of f

$\Rightarrow f$ is onto.

$\Rightarrow f$ is invertible.

Now, let $y \in Y$ such that $f(x) = y$

$\Rightarrow$ $x = f^{-1}(y)$

$x \in N$

$\Rightarrow$ $y = 4x + 3$

$\Rightarrow$ $x = \frac{y-3}{4}$

$\therefore$ Inverse of f is $f^{-1}(y) = \frac{y-3}{4}$.

13. (d) Given that,

$$f(x) = 4^{-x^2} + \cos^{-1}\left(\frac{x}{2} - 1\right) + \log(\cos x)$$

Here, 4^{-x^2} is defined for $\left(-\frac{\pi}{2}, \frac{\pi}{2}\right)$,

$\cos^{-1}\left(\frac{x}{2} - 1\right)$ is defined, if $-1 \le \frac{x}{2} - 1 \le 1$.

$\Rightarrow$ $0 \le \frac{x}{2} \le 2$

$\Rightarrow$ $0 \le x \le 4$

And $\log(\cos x)$ is defined, if $\cos x > 0$.

$\Rightarrow$ $-\frac{\pi}{2} < x < \frac{\pi}{2}$

Hence, $f(x) = 4^{-x^2} + \cos^{-1}\left(\frac{x}{2} - 1\right) + \log(\cos x)$

is defined, if $x \in \left[0, \frac{\pi}{2}\right)$.

14. (a) Let $W = \{CAT, TOY, YOU, \dots\}$

Clearly, R is reflexive and symmetric but not transitive.

$[\because {}_{CAT}R_{TOY}, {}_{TOY}R_{YOU} \nRightarrow {}_{CAT}R_{YOU}]$

15. (d) Since, for every elements of A, there exists elements (3, 3), (6, 6), (9, 9), (12, 12) $\in R \Rightarrow R$ is reflexive relation.

Now, $(6, 12) \in R$ but $(12, 6) \notin R$, so it is not a symmetric relation.

Also, $(3, 6), (6, 12) \in R$

$\Rightarrow$ $(3, 12) \in R$

$\therefore$ R is transitive relation.

16. (a) Given, $x \in (-1, 1)$

$\Rightarrow$ $\tan^{-1} x \in \left(-\frac{\pi}{4}, \frac{\pi}{4}\right)$

$\Rightarrow$ $2\tan^{-1} x \in \left(-\frac{\pi}{2}, \frac{\pi}{2}\right)$

Given that,

$$f(x) = \tan^{-1}\left(\frac{2x}{1 - x^2}\right) = 2\tan^{-1} x \quad [\because x^2 < 1]$$

So, $f(x) \in \left(-\frac{\pi}{2}, \frac{\pi}{2}\right)$

Hence, function is one-one onto.

17. (d) Given,

$$f(x - y) = f(x)f(y) - f(a - x)f(a + y)$$

Let $x = 0 = y$

$\Rightarrow$ $f(0) = [f(0)]^2 - [f(a)]^2$

$\Rightarrow$ $1 = 1 - [f(a)]^2$ [given, $f(0) = 1$]

$\Rightarrow$ $f(a) = 0$

$\therefore$ $f(2a - x) = f\{a - (x - a)\} = f(a)f(x - a) - f(a - a)f(x)$

$= 0 - f(x) \cdot 1 = -f(x)$

18. (b) Given, graph is symmetrical about the line $x = 2$.

$\therefore$ $f(2 + x) = f(2 - x)$

19. (b) Given,

$f(x) = \frac{\sin^{-1}(x - 3)}{\sqrt{9 - x^2}}$ will be defined, if

$-1 \le (x - 3) \le 1$

$\Rightarrow$ $2 \le x \le 4$ ...(i)

and $9 - x^2 > 0$

$\Rightarrow$ $-3 < x < 3$ ...(ii)

From Eqs. (i) and (ii), we get

$2 \le x < 3$

Hence, domain of the given function is [2, 3).

20. (c) Given that, $f(n) = \begin{cases} \dfrac{n-1}{2}, & \text{when } n \text{ is odd} \\ -\dfrac{n}{2}, & \text{when } n \text{ is even} \end{cases}$

and $f : N \to I$, where N is the set of natural numbers and I is the set of integers.

Let $x, y \in N$ and both are even.

Then, $f(x) = f(y)$

$\Rightarrow \quad -\dfrac{x}{2} = -\dfrac{y}{2}$

$\Rightarrow \quad x = y$

Again, $x, y \in N$ and both are odd.

Then, $f(x) = f(y)$

$\Rightarrow \quad \dfrac{x-1}{2} = \dfrac{y-1}{2}$

$\Rightarrow \quad x = y$

So, mapping is one-one.

Since, each negative integer is an image of even natural number and positive integer is an image of odd natural number. So, mapping is onto. Hence, mapping is one-one and onto.

21. (d) We have, $\sum_{r=1}^{n} f(r) = f(1) + f(2) + f(3) + \ldots + f(n)$

$= f(1) + 2f(1) + 3f(1) + \ldots + nf(1)$ $\quad [\because f(x+y) = f(x) + f(y)]$

$= (1 + 2 + 3 + \ldots + n)f(1) = f(1)\Sigma n$

$= \dfrac{7n(n+1)}{2}$ $\quad [\because f(1) = 7, \text{ given}]$

22. (d) Given, $f(x) = \dfrac{3}{4 - x^2} + \log_{10}(x^3 - x)$

For domain of $f(x)$,

$x^3 - x > 0$

$\Rightarrow \quad x(x-1)(x+1) > 0$

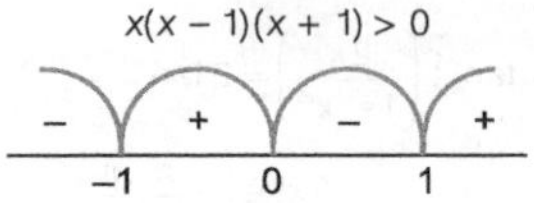

$\Rightarrow \quad x \in (-1, 0) \cup (1, \infty)$ and $4 - x^2 \neq 0$

$\Rightarrow \quad x \neq \pm 2$

$\Rightarrow \quad x \in (-\infty, -2) \cup (-2, 2) \cup (2, \infty)$

So, common region is $(-1, 0) \cup (1, 2) \cup (2, \infty)$.

23. (b) Given that, $f(x) = \log(x + \sqrt{x^2 + 1})$

Now, $f(-x) = \log(-x + \sqrt{x^2 + 1})$

$\therefore \quad f(x) + f(-x) = \log(x + \sqrt{x^2+1}) + \log(-x + \sqrt{x^2+1})$

$= \log(1) = 0$

Hence, $f(x)$ is an odd function.

24. (b) We know that,

$\sin^2\theta = \dfrac{1 - \cos 2\theta}{2} = \dfrac{1}{2} - \dfrac{1}{2}\cos 2\theta$

$\therefore$ Period of $\sin^2\theta = \dfrac{2\pi}{2} = \pi$

25. (a) Since, domain of $\sin^{-1} x$ is $[-1, 1]$.

$\therefore \quad -1 \leq \log_3\left(\dfrac{x}{3}\right) \leq 1$

$\Rightarrow \quad 3^{-1} \leq \dfrac{x}{3} \leq 3$

$\Rightarrow \quad 1 \leq x \leq 9$

Hence, domain of $\sin^{-1}\left[\log_3\left(\dfrac{x}{3}\right)\right]$ is $[1, 9]$.

26. (b) Given, $f(x) = \sin^4 x + \cos^4 x$

$\therefore \quad f(x) = (\sin^2 x + \cos^2 x)^2 - 2\sin^2 x \cos^2 x$

$= 1 - \dfrac{1}{2}(2\sin x \cos x)^2 = 1 - \dfrac{1}{2}(\sin 2x)^2$

$= 1 - \dfrac{1}{2}\left(\dfrac{1 - \cos 4x}{2}\right) = \dfrac{3}{4} + \dfrac{1}{4}\cos 4x$

$\therefore$ The period of $f(x) = \dfrac{2\pi}{4} = \dfrac{\pi}{2}$

$[\because \cos x \text{ is periodic with period } 2\pi]$

27. (a) Given, $f(x) = \sqrt{\log_{10}\left(\dfrac{5x - x^2}{4}\right)}$

For domain of $f(x)$,

$\log_{10}\left(\dfrac{5x - x^2}{4}\right) \geq 0$

$\Rightarrow \quad \dfrac{5x - x^2}{4} \geq 1$

$\Rightarrow \quad x^2 - 5x + 4 \leq 0$

$\Rightarrow \quad (x-1)(x-4) \leq 0$

$\therefore \quad x \in [1, 4]$

JEE Advanced & IIT JEE Archive

1. (a,b,c)

(a) $f(x) = \sin\left(\dfrac{\pi}{6}\sin\left(\dfrac{\pi}{2}\sin x\right)\right), x \in R.$

$= \sin\left(\dfrac{\pi}{6}\sin\theta\right), -\dfrac{\pi}{2} \leq \theta \leq \dfrac{\pi}{2} = \sin\alpha, -\dfrac{\pi}{6} \leq \alpha \leq \dfrac{\pi}{6}$

$\therefore$ Range of $f(x) \in \left[-\dfrac{1}{2}, \dfrac{1}{2}\right]$

(b) $(fog)(x) = f[g(x)] = f(t), -\dfrac{\pi}{2} \leq t \leq \dfrac{\pi}{2}$

$= \sin\left[\dfrac{\pi}{6}\cdot\sin\left(\dfrac{\pi}{2}\sin t\right)\right], -\dfrac{\pi}{2} \leq t \leq \dfrac{\pi}{2}$

$\therefore \quad f[g(x)] = f(t)$ has same range as of $f(x)$.

$\therefore \quad$ Range of $(fog)(x) \in \left[-\dfrac{1}{2}, \dfrac{1}{2}\right]$

(c) $\lim_{x \to 0} \dfrac{\sin\left(\dfrac{\pi}{6}\sin\left(\dfrac{\pi}{2}\sin x\right)\right)}{\dfrac{\pi}{2}\sin x}$

$\Rightarrow \quad \lim_{x \to 0} \dfrac{\sin\left(\dfrac{\pi}{6}\sin\left(\dfrac{\pi}{2}\cdot\sin x\right)\right)}{\dfrac{\pi}{6}\sin\left(\dfrac{\pi}{2}\cdot\sin x\right)} \cdot \dfrac{\dfrac{\pi}{6}\cdot\sin\left(\dfrac{\pi}{2}\cdot\sin x\right)}{\left(\dfrac{\pi}{2}\cdot\sin x\right)}$

$\Rightarrow \quad 1 \times \dfrac{\pi}{6} \times 1 = \dfrac{\pi}{6}$

(d) $(gof)(x) = g[f(x)]$

$= \dfrac{\pi}{2}\sin[f(x)]$

$\therefore \quad (gof)(x) = 1$

2. (a,b,c) **Plan**

(i) If $f'(x) > 0, \forall x \in (a,b)$, then $f(x)$ is an increasing function in (a,b) and thus, $f(x)$ is one-one function in (a,b).

(ii) If range of $f(x)$ = codomain of $f(x)$, then $f(x)$ is an onto function.

(iii) A function $f(x)$ is said to be odd function, if $f(-x) = -f(x), \forall x \in R$ i.e., $f(-x) + f(x) = 0, \forall x \in R$

Given, $f(x) = [\ln(\sec x + \tan x)]^3$

$$f'(x) = \frac{3[\ln(\sec x + \tan x)]^2(\sec x \tan x + \sec^2 x)}{(\sec x + \tan x)}$$

$$f'(x) = 3\sec x[\ln(\sec x + \tan x)]^2 > 0, \forall x \in \left(\frac{-\pi}{2}, \frac{\pi}{2}\right)$$

$f(x)$ is an increasing function.

$\therefore$ $f(x)$ is an one-one function.

$(\sec x + \tan x) = \tan\left(\frac{\pi}{4} + \frac{x}{2}\right)$, as $x \in \left(-\frac{\pi}{2}, \frac{\pi}{2}\right)$, then

$0 < \tan\left(\frac{\pi}{4} + \frac{x}{2}\right) < \infty$

$$0 < \sec x + \tan x < \infty$$

$\Rightarrow$ $$-\infty < \ln(\sec x + \tan x) < \infty$$

$$-\infty < [\ln(\sec x + \tan x)]^3 < \infty$$

$\Rightarrow$ $$-\infty < f(x) < \infty$$

$\therefore$ Range of $f(x)$ is R and thus $f(x)$ is an onto function.

$$f(-x) = [\ln(\sec x - \tan x)]^3$$

$$= \left[\ln\left(\frac{1}{\sec x + \tan x}\right)\right]^3$$

$$f(-x) = -[\ln(\sec x + \tan x)]^3$$

Now, $f(x) + f(-x) = 0$

$\Rightarrow f(x)$ is an odd function.

3. (a,d) **Plan** If a continuous function has values of opposite sign inside an interval, then it has a root in that interval.

$$f, g : [0,1] \to R$$

We take two cases.

Let f and g attain their common maximum value at p.

$\Rightarrow$ $$f(p) = g(p),$$

where $p \in [0,1]$.

Let f and g attain their common maximum value at different points.

$\Rightarrow$ $$f(a) = M \text{ and } g(b) = M$$

$\Rightarrow$ $$f(a) - g(a) > 0$$

and $$f(b) - g(b) < 0$$

$\Rightarrow f(c) - g(c) = 0$ for some $c \in [0,1]$ as f and g are continuous functions.

$\Rightarrow$ $f(c) - g(c) = 0$ for some $c \in [0,1]$ for all cases. ...(i)

Option (a) $\Rightarrow f^2(c) - g^2(c) + 3[f(c) - g(c)] = 0$

which is true from Eq. (i).

Option (d) $\Rightarrow f^2(c) - g^2(c) = 0$, which is true from Eq. (i).

Now, if we take $f(x) = 1$ and $g(x) = 1, \forall x \in [0,1]$

Option (b) and (c) does not hold.

4. (d) (i) For such questions, we need to properly define the functions and then we draw their graphs.

(ii) From the graphs, we can examine the function for continuity, differentiability, one-one and onto.

Given, $$f_1(x) = \begin{cases} -x, & x < 0 \\ e^x, & x \ge 0 \end{cases}$$

$$f_2(x) = x^2, x \ge 0$$

$$f_3(x) = \begin{cases} \sin x, & x < 0 \\ x, & x \ge 0 \end{cases}$$

$$f_4(x) = \begin{cases} f_2[f_1(x)], & x < 0 \\ f_2[f_1(x)] - 1, & x \ge 0 \end{cases}$$

Now, $$f_2[f_1(x)] = \begin{cases} x^2, & x < 0 \\ e^{2x}, & x \ge 0 \end{cases}$$

$\Rightarrow$ $$f_4 = \begin{cases} x^2, & x < 0 \\ e^{2x} - 1, & x \ge 0 \end{cases}$$

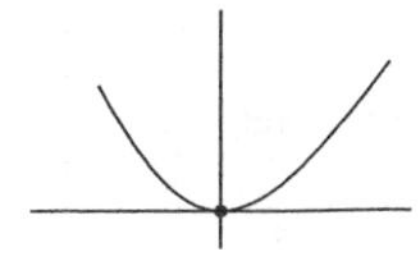

As $f_4(x)$ is continuous $f'_4(x) = \begin{cases} 2x, & x < 0 \\ 2e^{2x}, & x > 0 \end{cases}$

$f_4'(0)$ is not defined. Its range is $[0, \infty)$.

Thus, range = codomain = $[0, \infty)$ thus, f_4 is onto.

Also, horizontal line (drawn parallel to X-axis) meets the curve more than once, thus function is not one-one.

Hence, (i)$\to$(p), (ii)$\to$(r), (iii)$\to$(q), (iv)$\to$(s)

5. (b,d) $f(2n) = a_n$, $f(2n^+) = a_n$

$$f(2n^-) = b_n + 1$$

$\Rightarrow$ $$a_n - b_n = 1$$ [option (b)]

$$f(2n+1) = a_n$$

$$f((2n+1)^-) = a_n$$

$$f((2n+1)^+) = b_{n+1} - 1$$

$\Rightarrow$ $a_n = b_{n+1} - 1$ or $a_n - b_{n+1} = -1$

or $$a_{n-1} - b_n = -1$$ [option (d)]

6. (b) **Plan** To check nature of function.

(i) One-One To check one-one, we must check whether $f'(x) > 0$ or $f'(x) < 0$ in given domain.

(ii) Onto To check onto, we must check

Range = Co-domain

Description of Situation To find range in given domain $[a,b]$ put $f'(x) = 0$ and find $x = \alpha_1, \alpha_2, \ldots, \alpha_n \in [a,b]$

Now find, $\{f(a), f(\alpha_1), f(\alpha_2), \ldots, f(\alpha_n), f(b)\}$

its the greatest and least value gives you range.

Given, $f : [0,3] \to [1,29]$

$$f(x) = 2x^3 - 15x^2 + 36x + 1$$

$\therefore$ $$f'(x) = 6x^2 - 30x + 36 = 6(x^2 - 5x + 6)$$

$$= 6(x-2)(x-3)$$

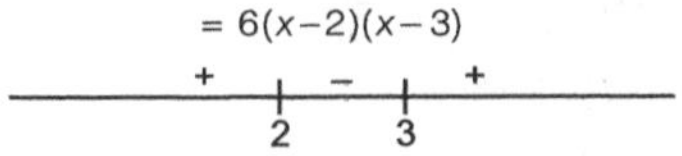

For given domain [0, 3], $f(x)$ is increasing as well as decreasing.

$\Rightarrow f$ is many-one.

Now, put $$f'(x) = 0$$

$\Rightarrow$ $$x = 2, 3$$

Thus, for Range $f(0) = 1, f(2) = 29, f(3) = 28$

$\Rightarrow$ $$\text{Range} \in [1, 29]$$

$\therefore$ f is onto.

7. (a) Here, $$f(x) = \frac{b - x}{1 - bx}$$

where $0 < b < 1, 0 < x < 1$

For function to be invertible it should be one-one onto.

$\therefore$ Check Range

Let $$f(x) = y$$

$\Rightarrow$ $$y = \frac{b - x}{1 - bx}$$

$\Rightarrow$ $$y - bxy = b - x$$

$\Rightarrow$ $$x(1 - by) = b - y$$

$\Rightarrow$ $$x = \frac{b - y}{1 - by}$$

where $$0 < x < 1$$

$\therefore$ $$0 < \frac{b - y}{1 - by} < 1$$

$$\frac{b-y}{1-by} > 0$$

and
$$\frac{b-y}{1-by} < 1$$

$$\Rightarrow \quad y < b \text{ or } y > \frac{1}{b} \quad \ldots\text{(i)}$$

$$\frac{(b-1)(y+1)}{1-by} < 0 - 1 < y < \frac{1}{b} \quad \ldots\text{(ii)}$$

From Eqs. (i) and (ii), we get

$$y \in \left(-1, \frac{1}{b}\right) \subset \text{codomain}$$

Thus, $f(x)$ is not invertible.

8. (b) Given, $f(x) = x^2, g(x) = \sin x$

$$(gof)(x) = \sin x^2$$

Now, $go(gof)(x) = \sin(\sin x^2)$

$\Rightarrow (fogogof)(x) = [\sin(\sin x^2)]^2 \quad \ldots\text{(i)}$

Again, $(gof)(x) = \sin x^2$

$(gogof)(x) = \sin(\sin x^2) \quad \ldots\text{(ii)}$

Given, $(fogogof)(x) = (gogof)(x)$

$\Rightarrow (\sin(\sin x^2))^2 = \sin(\sin x^2)$

$\Rightarrow \sin(\sin x^2)\{\sin(\sin x^2) - 1\} = 0$

$\Rightarrow \sin(\sin x^2) = 0$ or $\sin(\sin x^2) = 1$

$\Rightarrow \sin x^2 = 0$

or $\sin x^2 = \frac{\pi}{2}$

$\therefore x^2 = n\pi$ (i.e. not possible as $-1 \le \sin\theta \le 1$)

$\Rightarrow x = \pm\sqrt{n\pi}$

9. (d) Let $A \cap B = \phi, A, B \subset S$

$\Rightarrow 3^4 = \frac{81+1}{2} = 41$

$\therefore \frac{3^4+1}{2} = 41$

10. (b) Given, $f(x) = \frac{(x-1)(x-5)}{(x-2)(x-3)}$

The graph of $f(x)$ is shown

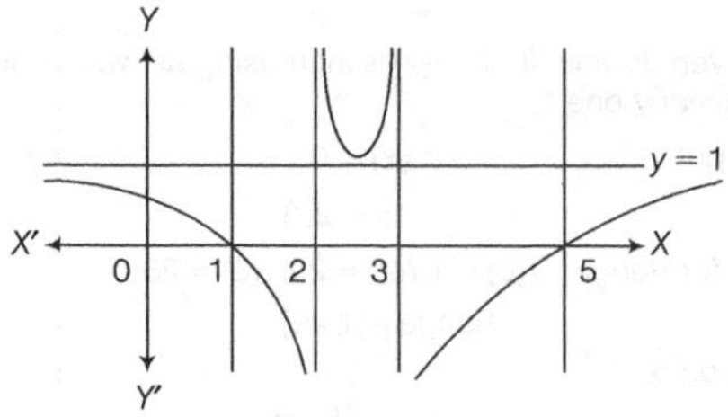

(i) If $-1 < x < 1 \Rightarrow 0 < f(x) < 1$

(ii) If $1 < x < 2 \Rightarrow f(x) < 0$

(iii) If $3 < x < 5 \Rightarrow f(x) < 0$

(iv) If $x > 5 \Rightarrow 0 < f(x) < 1$

Hence, (i)→(p), (ii)→(q), (iii)→(q), (iv)→(p)

11. (d) Given, $F(x) = \int \sin^2 x\,dx = \int \frac{1-\cos 2x}{2}dx$

$\Rightarrow F(x) = \frac{1}{4}(2x - \sin 2x) + C$

Since, $F(x+\pi) \ne F(x)$

Hence, Statement I is incorrect.
But Statement II is correct as $\sin^2 x$ is periodic with period π.

12. (b) Given, $f(x) = 2 + \cos x, \forall x \in R$

Statement I There exists a point $c \in [t, t+\pi]$, where $f'(c) = 0$
Hence, Statement I is correct.

Statement II $f(t) = f(t + 2\pi)$ is true. But Statement II is not a correct explanation for Statement I.

13. (a) Given, $y = -\frac{x^2}{2} + x + 1$

$\Rightarrow y - \frac{3}{2} = -\frac{1}{2}(x-1)^2$

$\Rightarrow$ It is symmetric about

$$x = 1$$

14. (d) Let $\phi(x) = f(x) - g(x) = \begin{cases} x, & x \in Q \\ -x, & x \notin Q \end{cases}$

Now, to check one-one
Take any straight line parallel to X-axis which will intersect $\phi(x)$ only at one point.
$\Rightarrow \phi(x)$ is one-one.
To check onto

As, $f(x) = \begin{cases} x, & x \in Q \\ -x, & x \notin Q \end{cases}$, which shows

$y = x$ and $y = -x$ for rational and irrational values.

$\Rightarrow y \in$ real numbers

$\therefore$ Range = Codomain $\Rightarrow$ onto

Thus, $f - g$ is one-one and onto.

15. (c) Since, only option (c) satisfy given definition

i.e. $f\{f^{-1}(B)\} = B$

Only, if $B \subseteq f(x)$

16. Given, $2\sin t = \frac{1-2x+5x^2}{3x^2-2x-1}, t \in \left[-\frac{\pi}{2}, \frac{\pi}{2}\right]$

Put, $2\sin t = y \Rightarrow -2 \le y \le 2$

$\therefore y = \frac{1-2x+5x^2}{3x^2-2x-1}$

$\Rightarrow (3y-5)x^2 - 2x(y-1) - (y+1) = 0$

Since, $x \in R - \{1, -1/3\}$ $\left(\text{as, } 3x^2 - 2x - 1 \ne 0 \Rightarrow (x-1)(x+1/3) \ne 0\right)$

$\therefore D \ge 0$

$\Rightarrow 4(y-1)^2 + 4(3y-5)(y+1) \ge 0$

$\Rightarrow y^2 - y - 1 \ge 0$

$\Rightarrow \left(y - \frac{1}{2}\right)^2 - \frac{5}{4} \ge 0$

$\Rightarrow \left(y - \frac{1}{2} - \frac{\sqrt{5}}{2}\right)\left(y - \frac{1}{2} + \frac{\sqrt{5}}{2}\right) \ge 0$

$\Rightarrow y \le \frac{1-\sqrt{5}}{2}$ or $y \ge \frac{1+\sqrt{5}}{2}$

$\Rightarrow 2\sin t \le \frac{1-\sqrt{5}}{2}$

or $2\sin t \ge \frac{1+\sqrt{5}}{2}$

$\Rightarrow \sin t \le \sin\left(-\frac{\pi}{10}\right)$

or $\sin t \ge \sin\left(\frac{3\pi}{10}\right)$

$\Rightarrow t \le -\frac{\pi}{10}$ or $t \ge \frac{3\pi}{10}$

Hence, range of t is $\left[-\frac{\pi}{2}, -\frac{\pi}{10}\right] \cup \left[\frac{3\pi}{10}, \frac{\pi}{2}\right]$.

3. Matrices and Determinants

DPP-1 Types of Matrices, Addition and Subtraction of Matrices

1. (c) The number of matrices can be

$3\times 4,\quad 4\times 3,$
$6\times 2,\quad 2\times 6,$
12×1
and $1\times 12,$
i.e 6

2. (b) Given, $a_{ij} = 2\hat{i} - \hat{j}$

$$[A]_{3\times 2} = \begin{bmatrix} a_{11} & a_{12} \\ a_{21} & a_{22} \\ a_{31} & a_{32} \end{bmatrix} = \begin{bmatrix} 1 & 0 \\ 3 & 2 \\ 5 & 4 \end{bmatrix}$$

3. (a) Given, $a_{ij} = \frac{1}{2}(3i - 2j)$

$$[A]_{2\times 2} = \begin{bmatrix} a_{11} & a_{12} \\ a_{21} & a_{22} \end{bmatrix}$$

$$= \begin{bmatrix} \frac{1}{2}(3-2) & \frac{1}{2}(3-4) \\ \frac{1}{2}(6-2) & \frac{1}{2}(6-4) \end{bmatrix}$$

$$= \begin{bmatrix} \frac{1}{2} & \frac{-1}{2} \\ \frac{4}{2} & \frac{2}{2} \end{bmatrix}$$

$$= \begin{bmatrix} \frac{1}{2} & \frac{-1}{2} \\ 2 & 1 \end{bmatrix}$$

4. (c) Given, matrix is a lower triangular matrix because all elements of right of the diagonal are zero.

5. (a) Additive commutativity exists in all square matrices.

6. (a) $\begin{bmatrix} 1 & 0 & 0 \\ 0 & 1 & 0 \\ 0 & 0 & 1 \end{bmatrix}, \begin{bmatrix} 1 & 0 \\ 0 & 1 \end{bmatrix}$

These are the identity matrices and also the scalar matrices as well.

7. (a)

The minimum number of zeros in upper triangular matrix	Order
1	2×2
$1+2=3$	3×3
$1+2+3=6$	4×4
$1+2+3+4=10$	5×5
$1+2+\cdots+(n-1)=\frac{n(n-1)}{2}$	$n\times n$

8. (a) Trace of $A = 1 + 7 + 9 = 17$

9. (d) Order of $A = q \times p$

$$A = \begin{bmatrix} a_{11} & a_{22} & a_{1p} \\ a_{21} & a_{22} & \vdots \\ a_{21} & \cdots & a_{qp} \end{bmatrix}_{q\times p}$$

10. (c) $\begin{bmatrix} 2 & -3 \\ 4 & 0 \end{bmatrix} - \begin{bmatrix} a & c \\ b & d \end{bmatrix} = \begin{bmatrix} 1 & 4 \\ 2 & -5 \end{bmatrix}$

$\Rightarrow \begin{bmatrix} 2-a & -3-c \\ 4-b & -d \end{bmatrix} = \begin{bmatrix} 1 & 4 \\ 2 & -5 \end{bmatrix}$

$\Rightarrow 2 - a = 1$
$\Rightarrow a = 1$
$-3 - c = 4$
$\Rightarrow c = -7$
$4 - b = 2$
$\Rightarrow b = 2$
and $-d = -5$
$\Rightarrow d = 5$

11. (b) $2A + 3B - 2(A + 2B)$

$$= \begin{bmatrix} 2 & -1 & 4 \\ 3 & 2 & 5 \end{bmatrix} - 2\begin{bmatrix} 5 & 0 & 3 \\ 1 & 6 & 2 \end{bmatrix}$$

$$= \begin{bmatrix} 2-10 & -1 & 4-6 \\ 3-2 & 2-12 & 5-4 \end{bmatrix}$$

$$= \begin{bmatrix} -8 & -1 & -2 \\ 1 & -10 & 1 \end{bmatrix}$$

So, $-B = \begin{bmatrix} -8 & -1 & -2 \\ 1 & -10 & 1 \end{bmatrix}$

$\Rightarrow B = \begin{bmatrix} 8 & 1 & 2 \\ -1 & 10 & -1 \end{bmatrix}$

12. (b) $5A - 3B - 2C$

$$= 5\begin{bmatrix} 1 & -2 \\ 3 & 0 \end{bmatrix} - 3\begin{bmatrix} -1 & 4 \\ 2 & 3 \end{bmatrix} - 2\begin{bmatrix} 0 & -1 \\ 1 & 0 \end{bmatrix}$$

$$= \begin{bmatrix} 5+3 & -10-12+2 \\ 15-6-2 & -9 \end{bmatrix}$$

$$= \begin{bmatrix} 8 & -20 \\ 7 & -9 \end{bmatrix}$$

13. (b) $\begin{bmatrix} x-2 & 1 \\ 3+1 & 4+y \end{bmatrix} = \begin{bmatrix} 1 & 1 \\ 4 & 2 \end{bmatrix}$

$\Rightarrow x - 2 = 1$
$\Rightarrow x = 3$
and $4 + y = 2$
$\Rightarrow y = 2 - 4 \Rightarrow y = -2$
$\therefore x = 3, y = -2$

14. (a) $\because x + y = 4$
and $x - y = 0$
On solving above equations, we get
$x = 2$ and $y = 2$
$\because 2x + z = 7$
$\Rightarrow 4 + z = 7$
$\Rightarrow z = 3$
and $2z + w = 10$
$\Rightarrow 6 + w = 10$
$\therefore w = 4$

15. (c) $a_1\, a_2\, a_3\, a_4 \cdots a_{25} = b_1 \cdot b_2 \cdots b_{25}$

DPP-2 **Multiplication and Transpose of a Matrix**

1. (d) Order of matrix A is 3×1.
Order of matrix B is 3×3.
$\because$ Number of columns of matrix $A \neq$ Number of rows of matrix B
Hence, AB does not exist.

2. (b) $(A+B)^2 = (A+B)(A+B)$
$$= A \cdot A + B \cdot B + A \cdot B + B \cdot A$$
$$= A^2 + B^2 + AB + BA$$

3. (a) Given, $A = \begin{bmatrix} 3 & -4 \\ 1 & -1 \end{bmatrix}$

Now, $A^2 = A \cdot A$
$$= \begin{bmatrix} 3 & -4 \\ 1 & -1 \end{bmatrix}\begin{bmatrix} 3 & -4 \\ 1 & -1 \end{bmatrix} = \begin{bmatrix} 9-4 & -12+4 \\ 3-1 & -4+1 \end{bmatrix} = \begin{bmatrix} 5 & -8 \\ 2 & -3 \end{bmatrix}$$

and $A^3 = A^2 \cdot A$
$$= \begin{bmatrix} 5 & -8 \\ 2 & -3 \end{bmatrix}\begin{bmatrix} 3 & -4 \\ 1 & -1 \end{bmatrix} = \begin{bmatrix} 15-8 & -20+8 \\ 6-3 & -8+3 \end{bmatrix} = \begin{bmatrix} 7 & -12 \\ 3 & -5 \end{bmatrix}$$

4. (a) Given $A = \begin{bmatrix} 0 & -\tan\frac{\alpha}{2} \\ \tan\frac{\alpha}{2} & 0 \end{bmatrix}$ $\quad \left[\because I = \begin{bmatrix} 1 & 0 \\ 0 & 1 \end{bmatrix}\right]$

$$\Rightarrow \quad (I-A)\begin{bmatrix} \cos\alpha & -\sin\alpha \\ \sin\alpha & \cos\alpha \end{bmatrix}$$
$$= \begin{bmatrix} 1 & \tan\frac{\alpha}{2} \\ -\tan\frac{\alpha}{2} & 1 \end{bmatrix}\begin{bmatrix} \cos\alpha & -\sin\alpha \\ \sin\alpha & \cos\alpha \end{bmatrix}$$
$$= \begin{bmatrix} \cos\alpha + \sin\alpha\tan\frac{\alpha}{2} & -\sin\alpha + \cos\alpha\tan\frac{\alpha}{2} \\ -\cos\alpha\tan\frac{\alpha}{2} + \sin\alpha & \sin\alpha\tan\frac{\alpha}{2} + \cos\alpha \end{bmatrix}$$
$$= \begin{bmatrix} \dfrac{1-\tan^2\frac{\alpha}{2} + 2\tan^2\frac{\alpha}{2}}{1+\tan^2\frac{\alpha}{2}} & \dfrac{-2\tan\frac{\alpha}{2} + \left(\tan\frac{\alpha}{2} - \tan^3\frac{\alpha}{2}\right)}{1+\tan^2\frac{\alpha}{2}} \\ \dfrac{-\left(\tan\frac{\alpha}{2} - \tan^3\frac{\alpha}{2}\right) + 2\tan\frac{\alpha}{2}}{1+\tan^2\frac{\alpha}{2}} & \dfrac{1-\tan^2\frac{\alpha}{2} + 2\tan^2\frac{\alpha}{2}}{1+\tan^2\frac{\alpha}{2}} \end{bmatrix}$$
$$= \begin{bmatrix} \dfrac{1+\tan^2\frac{\alpha}{2}}{1+\tan^2\frac{\alpha}{2}} & \dfrac{-\tan\frac{\alpha}{2}\left(1+\tan^2\frac{\alpha}{2}\right)}{1+\tan^2\frac{\alpha}{2}} \\ \dfrac{\tan\frac{\alpha}{2}\left(1+\tan^2\frac{\alpha}{2}\right)}{1+\tan^2\frac{\alpha}{2}} & \dfrac{1+\tan^2\frac{\alpha}{2}}{1+\tan^2\frac{\alpha}{2}} \end{bmatrix}$$
$$= \begin{bmatrix} 1 & -\tan\frac{\alpha}{2} \\ \tan\frac{\alpha}{2} & 1 \end{bmatrix}$$
$$= I + A$$

5. (b) $A = \text{diag}\{2, -1, 3\}$
and $\quad B = \text{diag}\{-1, 3, 2\}$
$$\Rightarrow \quad A = \begin{bmatrix} 2 & 0 & 0 \\ 0 & -1 & 0 \\ 0 & 0 & 3 \end{bmatrix}$$
$$\Rightarrow \quad B = \begin{bmatrix} -1 & 0 & 0 \\ 0 & 3 & 0 \\ 0 & 0 & 2 \end{bmatrix}$$
Then, $$A^2 = \begin{bmatrix} 4 & 0 & 0 \\ 0 & 1 & 0 \\ 0 & 0 & 9 \end{bmatrix}$$
$$\therefore \quad A^2B = \begin{bmatrix} 4 & 0 & 0 \\ 0 & 1 & 0 \\ 0 & 0 & 9 \end{bmatrix}\begin{bmatrix} -1 & 0 & 0 \\ 0 & 3 & 0 \\ 0 & 0 & 2 \end{bmatrix} = \begin{bmatrix} -4 & 0 & 0 \\ 0 & 3 & 0 \\ 0 & 0 & 18 \end{bmatrix}$$
$$= \text{diag}\{-4, 3, 18\}$$

6. (b) $\because B^T A^T = (AB)^T$
$$= \left(\begin{bmatrix} 2 & -1 \\ -7 & 4 \end{bmatrix}\begin{bmatrix} 4 & 1 \\ 7 & 2 \end{bmatrix}\right)^T = \begin{bmatrix} 8-7 & 2-2 \\ -28+28 & -7+8 \end{bmatrix}^T$$
$$= \begin{bmatrix} 1 & 0 \\ 0 & 1 \end{bmatrix}^T = \begin{bmatrix} 1 & 0 \\ 0 & 1 \end{bmatrix} = I$$
Hence, $B^T A^T$ is an identity matrix.

7. (c) If $AB = 0$.
Then, three cases may be possible:
Case I Both $A = 0$ and $B = 0$
Case II Either $A = 0$ or $B = 0$
Case III Neither $A = 0$ nor $B = 0$
Hence, it is not necessary that either $A = 0$ or $B = 0$.

8. (d) $A = \begin{bmatrix} 2 & -1 \\ -1 & 2 \end{bmatrix}$ and $B = \begin{bmatrix} 1 & 4 \\ -1 & 1 \end{bmatrix}$
$$AB = \begin{bmatrix} 2+1 & 8-1 \\ -1-2 & -4+2 \end{bmatrix} = \begin{bmatrix} 3 & 7 \\ -3 & -2 \end{bmatrix}$$
$$BA = \begin{bmatrix} -2 & 7 \\ -4 & 3 \end{bmatrix}$$
So, $\quad AB \neq BA$
Now, $$A^2 = \begin{bmatrix} 2 & -1 \\ -1 & 2 \end{bmatrix}\begin{bmatrix} 2 & -1 \\ -1 & 2 \end{bmatrix} = \begin{bmatrix} 4+1 & -2-2 \\ -2-2 & 1+4 \end{bmatrix} = \begin{bmatrix} 5 & -4 \\ -4 & 5 \end{bmatrix}$$
$$B^2 = \begin{bmatrix} 1 & 4 \\ -1 & 1 \end{bmatrix}\begin{bmatrix} 1 & 4 \\ -1 & 1 \end{bmatrix} = \begin{bmatrix} 1-4 & 4+4 \\ -1-1 & -4+1 \end{bmatrix} = \begin{bmatrix} -3 & 8 \\ -2 & -3 \end{bmatrix}$$
$\therefore \quad A^2 + 2AB + B^2$
$$= \begin{bmatrix} 5 & -4 \\ -4 & 5 \end{bmatrix} + \begin{bmatrix} 6 & 14 \\ -6 & -4 \end{bmatrix} + \begin{bmatrix} -3 & 8 \\ -2 & -3 \end{bmatrix} = \begin{bmatrix} 8 & 18 \\ -12 & -2 \end{bmatrix}$$

and $A^2 - 2AB + B^2$

$$= \begin{bmatrix} 5 & -4 \\ -4 & 5 \end{bmatrix} + \begin{bmatrix} -6 & -14 \\ 6 & 4 \end{bmatrix} + \begin{bmatrix} -3 & 8 \\ -2 & -3 \end{bmatrix}$$

$$= \begin{bmatrix} -4 & -10 \\ 0 & 6 \end{bmatrix}$$

Now, $(A + B)^2$

$$= \begin{bmatrix} 3 & 3 \\ -2 & 3 \end{bmatrix}\begin{bmatrix} 3 & 3 \\ -2 & 3 \end{bmatrix} = \begin{bmatrix} 9-6 & 9+9 \\ -6-6 & -6+9 \end{bmatrix}$$

$$= \begin{bmatrix} 3 & 18 \\ -12 & 3 \end{bmatrix}$$

$$\neq A^2 + 2AB + B^2$$

and $(A - B)^2$ $= \begin{bmatrix} 1 & -5 \\ 0 & 1 \end{bmatrix}\begin{bmatrix} 1 & -5 \\ 0 & 1 \end{bmatrix} = \begin{bmatrix} 1 & -5-5 \\ 0 & 1 \end{bmatrix}$

$$= \begin{bmatrix} 1 & -10 \\ 0 & 1 \end{bmatrix}$$

$$\neq A^2 - 2AB + B^2$$

9. (a) If $AB = 0$.

$$\Rightarrow \begin{bmatrix} \cos^2\theta & \cos\theta\sin\theta \\ \cos\theta\sin\theta & \sin^2\theta \end{bmatrix}\begin{bmatrix} \cos^2\phi & \cos\phi\sin\phi \\ \cos\phi\sin\phi & \sin^2\phi \end{bmatrix} = 0$$

$$\Rightarrow \begin{bmatrix} \cos^2\theta\cos^2\phi + \sin\theta\sin\phi\cos\theta\cos\phi & \cos^2\theta\cos\phi\sin\phi + \cos\theta\sin\theta\sin^2\phi \\ \cos\theta\cos^2\phi + \sin^2\theta\cos\phi\sin\phi & \cos\theta\sin\theta\cos\phi\sin\phi + \sin^2\theta\sin^2\phi \end{bmatrix} = 0$$

$$\Rightarrow \begin{bmatrix} \cos\theta\cos\phi\cos(\theta-\phi) & \cos\theta\sin\phi\cos(\theta-\phi) \\ \sin\theta\cos\phi\cos(\theta-\phi) & \sin\theta\sin\phi\cos(\theta-\phi) \end{bmatrix} = \begin{bmatrix} 0 & 0 \\ 0 & 0 \end{bmatrix}$$

Hence, $\theta - \phi$ is an odd multiple of $\frac{\pi}{2}$.

10. (d) We have, $X = \begin{bmatrix} 3 & -4 \\ 1 & -1 \end{bmatrix}$

$$X^2 = \begin{bmatrix} 3 & -4 \\ 1 & -1 \end{bmatrix}\begin{bmatrix} 3 & -4 \\ 1 & -1 \end{bmatrix} = \begin{bmatrix} 9-4 & -12+4 \\ 3-1 & -4+1 \end{bmatrix}$$

$$= \begin{bmatrix} 5 & -8 \\ 2 & -3 \end{bmatrix} \neq \begin{bmatrix} 3\cdot 2 & -4\cdot 2 \\ 2 & 2 \end{bmatrix}$$

Also, $\begin{bmatrix} 5 & -8 \\ 2 & -3 \end{bmatrix} \neq \begin{bmatrix} 2+2 & 5-2 \\ 2 & -2 \end{bmatrix}$

and $\begin{bmatrix} 5 & -8 \\ 2 & -3 \end{bmatrix} \neq \begin{bmatrix} 3^2 & (-4)^2 \\ 1 & (-1)^2 \end{bmatrix}$

11. (a) $\because AB = A$ and $BA = B$

$\therefore$ $B^2 = (BA)(BA)$

$= B(AB)A = B \cdot A \cdot A$

$= (BA) \cdot A = BA = B$

12. (a) $\because A^T = A$ and $B^T = B$

Now, $(A\,B\,A)^T$

$A^T\,(AB)^T = A^T\,(B^T A^T) = A\,(BA) = ABA$

13. (a) Given, $A^T = -A$

Now, $(A^n)^T = (A^T)^n = (-A)^n$

$= (-1)^n \cdot A^n = A^n$

14. (d) $\because |A| \neq 0$

Also, $|A| = |A^T|$

$|A\,A^T| = |A|\,|A^T| = |A|\,|A| = |A|^2$

and $|A^T A| = |A|^2 = |A^T|^2$

Hence, $|A| + |A^T| = |A| + |A| = 2|A| \neq 0$

15. (a)

(i) $(A^T)^T = A$

(ii) $(AB)^n = A^n B^n$, when $AB = BA$

(iii) $(A - I)(I + A) = 0$

$\Rightarrow$ $A^2 + A \cdot I - I \cdot A - I^2 = 0$

$\Rightarrow$ $A^2 - I^2 = 0$ $[\because A \cdot I = I \cdot A = A]$

$\Rightarrow$ $A^2 = I^2$

Similarly, $(A - I)(I + A) = 0$

(iv) $(A + B)(A - B) = A^2 - AB + BA - B^2$

$\neq A^2 - B^2$ $(\because AB \neq BA)$

16. (d) We have, $A = \begin{bmatrix} a & b \\ c & d \end{bmatrix}$

$$\Rightarrow A^2 = \begin{bmatrix} a & b \\ c & d \end{bmatrix}\begin{bmatrix} a & b \\ c & d \end{bmatrix}$$

$$= \begin{bmatrix} a^2 + bc & ab + bd \\ ac + cd & bc + d^2 \end{bmatrix}$$

$$(a + d) \cdot A = \begin{bmatrix} (a+d) \cdot a & (a+d) \cdot b \\ (a+d) \cdot c & (a+d) \cdot d \end{bmatrix}$$

$$= \begin{bmatrix} a^2 + ad & ab + bd \\ ac + cd & ad + d^2 \end{bmatrix}$$

$\because$ $x^2 - (a + d)x + k = 0$

$$\Rightarrow \begin{bmatrix} a^2 + bc & ab + bd \\ ac + cd & bc + d^2 \end{bmatrix} - \begin{bmatrix} a^2 + ad & ab + bd \\ ac + cd & ad + d^2 \end{bmatrix} + \begin{bmatrix} k & 0 \\ 0 & k \end{bmatrix} = \begin{bmatrix} 0 & 0 \\ 0 & 0 \end{bmatrix}$$

$$\Rightarrow \begin{bmatrix} bc - ad + k & 0 \\ 0 & bc - ad + k \end{bmatrix} = \begin{bmatrix} 0 & 0 \\ 0 & 0 \end{bmatrix}$$

$\Rightarrow$ $k = ad - bc$

17. (c) $\because \begin{bmatrix} 0 & 0 \\ 1 & 0 \end{bmatrix}\begin{bmatrix} 0 & 0 \\ 1 & 0 \end{bmatrix} = \begin{bmatrix} 0+0 & 0+0 \\ 0+0 & 0+0 \end{bmatrix} = \begin{bmatrix} 0 & 0 \\ 0 & 0 \end{bmatrix}$

Hence, $\begin{bmatrix} 0 & 0 \\ 1 & 0 \end{bmatrix}$ is a nilpotent matrix.

18. (a) $Q^2 = Q \cdot Q = PAP^T \cdot PAP^T$

$= PA(P^T \cdot P)AP^T = PA \cdot I \cdot AP^T = PA^2P^T$

Similarly, $Q^3 = PA^3P^T$

$Q^{2005} = PA^{2005}P^T$

Now, $X = P^T Q^{2005} P$

$= P^T\,(PA^{2005}P^T)P$

$= (P^T P)\,A^{2005}(P^T P)$

$= I \cdot A^{2005} \cdot I = A^{2005} = \begin{bmatrix} 1 & 2005 \\ 0 & 1 \end{bmatrix}$

19. (a) A and B matrices are of equal order.

So, $A + B$, $A - B$, $B - A$ are defined.

Also, $A + B = B + A$

20. (d) $\because$ $A^2 = B$

$$\Rightarrow \begin{bmatrix} \alpha & 0 \\ 1 & 1 \end{bmatrix}\begin{bmatrix} \alpha & 0 \\ 1 & 1 \end{bmatrix} = \begin{bmatrix} 1 & 0 \\ 5 & 1 \end{bmatrix}$$

$$\Rightarrow \begin{bmatrix} \alpha^2 & 0 \\ \alpha + 1 & 1 \end{bmatrix} = \begin{bmatrix} 1 & 0 \\ 5 & 1 \end{bmatrix}$$

$\therefore$ $\alpha^2 = 1$ and $\alpha + 1 = 5$

Hence, no real value of α exist.

21. (d) Given, A is a square matrix of order 2×2.

$$A = \begin{bmatrix} a_{11} & a_{12} \\ a_{21} & a_{22} \end{bmatrix} = \begin{bmatrix} 1-1 & 1-4 \\ 4-1 & 4-4 \end{bmatrix} = \begin{bmatrix} 0 & -3 \\ 3 & 0 \end{bmatrix}$$

$$A^T = \begin{bmatrix} 0 & 3 \\ -3 & 0 \end{bmatrix} = -A$$

Hence, A is a skew-symmetric matrix.

22. (c) $$A^2 = \begin{bmatrix} \frac{1}{\sqrt{2}} & \frac{1}{\sqrt{2}} \\ \frac{-1}{\sqrt{2}} & \frac{-1}{\sqrt{2}} \end{bmatrix} \begin{bmatrix} \frac{1}{\sqrt{2}} & \frac{1}{\sqrt{2}} \\ \frac{-1}{\sqrt{2}} & \frac{-1}{\sqrt{2}} \end{bmatrix} = \begin{bmatrix} \frac{1}{2} - \frac{1}{2} & \frac{1}{2} - \frac{1}{2} \\ \frac{-1}{2} + \frac{1}{2} & \frac{-1}{2} + \frac{1}{2} \end{bmatrix} = \begin{bmatrix} 0 & 0 \\ 0 & 0 \end{bmatrix}$$

Hence, A is nilpotent matrix.

23. (a) $\because$ $$A^2 = \begin{bmatrix} 0 & 0 & -1 \\ 0 & -1 & 0 \\ -1 & 0 & 0 \end{bmatrix} \begin{bmatrix} 0 & 0 & -1 \\ 0 & -1 & 0 \\ -1 & 0 & 0 \end{bmatrix} = \begin{bmatrix} 1 & 0 & 0 \\ 0 & 1 & 0 \\ 0 & 0 & 1 \end{bmatrix} = I$$

24. (b) $$A^2 = \begin{bmatrix} 1 & 1 & 0 \\ 1 & 2 & 1 \\ 2 & 1 & 0 \end{bmatrix} \begin{bmatrix} 1 & 1 & 0 \\ 1 & 2 & 1 \\ 2 & 1 & 0 \end{bmatrix} = \begin{bmatrix} 2 & 3 & 1 \\ 5 & 6 & 2 \\ 3 & 4 & 1 \end{bmatrix}$$

Now, $$A^3 = A \cdot A^2 = \begin{bmatrix} 1 & 1 & 0 \\ 1 & 2 & 1 \\ 2 & 1 & 0 \end{bmatrix} \begin{bmatrix} 2 & 3 & 1 \\ 5 & 6 & 2 \\ 3 & 4 & 1 \end{bmatrix} = \begin{bmatrix} 7 & 9 & 3 \\ 15 & 19 & 6 \\ 9 & 12 & 4 \end{bmatrix}$$

$$= A^3 - 3A^2 = \begin{bmatrix} 7 & 9 & 3 \\ 15 & 19 & 6 \\ 9 & 12 & 4 \end{bmatrix} - \begin{bmatrix} 6 & 9 & 3 \\ 15 & 18 & 6 \\ 9 & 12 & 3 \end{bmatrix} = \begin{bmatrix} 1 & 0 & 0 \\ 0 & 1 & 0 \\ 0 & 0 & 1 \end{bmatrix} = I$$

Hence, $A^3 - 3A^2 - I = 0.$

25. (b) $$\begin{bmatrix} \cos\alpha & \sin\alpha \\ -\sin\alpha & \cos\alpha \end{bmatrix} \begin{bmatrix} \cos\alpha & \sin\alpha \\ -\sin\alpha & \cos\alpha \end{bmatrix}^T = \begin{bmatrix} \cos\alpha & \sin\alpha \\ -\sin\alpha & \cos\alpha \end{bmatrix} \begin{bmatrix} \cos\alpha & -\sin\alpha \\ \sin\alpha & \cos\alpha \end{bmatrix}$$

$$= \begin{bmatrix} \cos^2\alpha + \sin^2\alpha & -\cos\alpha\sin\alpha + \sin\alpha\cos\alpha \\ -\sin\alpha\cos\alpha + \cos\alpha\sin\alpha & \sin^2\alpha + \cos^2\alpha \end{bmatrix} = \begin{bmatrix} 1 & 0 \\ 0 & 1 \end{bmatrix} = I$$

$\because$ If A is an orthogonal matrix, then $AA^T = I$.

$\therefore$ $\begin{bmatrix} \cos\alpha & \sin\alpha \\ -\sin\alpha & \cos\alpha \end{bmatrix}$ is an orthogonal matrix.

26. (a,b,c,d) If $AB = BA$

Then, $(AB)^2 = (AB)(AB) = A(BA)B = A(AB)B = A^2B^2$

Similarly, $(AB)^3 = A^3B^3$

By induction method, we can say that

$$(AB)^n = A^nB^n$$

Now, $A^2B = A \cdot (A \cdot B) = A(BA) = (AB)A = (BA)A = BA^2$

Similarly, $A^3B = BA^3$

By induction method, we can say that

$$A^nB = BA^n$$

Now, $A^2 - B^2 = A^2 + AB - AB - B^2 = A^2 + AB - BA - B^2 = (A - B)(A + B)$

$\therefore$ $A^4 - B^4 = A^4 + A^2B^2 - A^2B^2 - B^4 = A^4 + A^2B^2 - B^2A^2 - B^4$

$\left[\because AB = BA, \text{ so } (AB)^n = A^nB^n \Rightarrow (AB)^2 = (BA)^2 \Rightarrow A^2B^2 = B^2A^2\right]$

$$= (A^2 + B^2)(A^2 - B^2)$$

By induction method, we can say that

$$A^{2n} - B^{2n} = (A^n - B^n)(A^n + B^n)$$

By Binomial theorem,

$$(A + B)^n = A^n + {}^nC_1A^{n-1}B + {}^nC_2A^{n-2}B^2 + \cdots + B^n$$

27. (b) The elements on the main diagonal of a skew-symmetric matrix are also zero.

Also, $$\left(\frac{1}{2}(A + A^T)\right)^T = \frac{1}{2}(A + A^T)^T = \frac{1}{2}(A^T + (A^T)^T) = \frac{1}{2}(A^T + A)$$

and $$\left(\frac{1}{2}(A - A^T)\right)^T = \frac{1}{2}(A - A^T)^T = \frac{1}{2}(A^T - (A^T)^T) = \frac{1}{2}(A^T - A) = \frac{-1}{2}(A - A^T)$$

28. (a,c,d) $\because$ B is an idempotent matrix.

$\therefore$ $B^2 = B$

Given, $A = I - B$

Now, $AB = (I - B)B = I \cdot B - B^2 = B - B = 0$

Now, $BA = B(I - B) = B \cdot I - B^2 = B - B = 0$

Now, $A^2 = (I - B)(I - B) = I - B - B + B^2 = I - 2B + B^2 = I - 2B + B = I - B = A$

29. (b,c) $$A + 4I = \begin{bmatrix} 1 & -1 \\ 2 & 3 \end{bmatrix} + \begin{bmatrix} 4 & 0 \\ 0 & 4 \end{bmatrix} = \begin{bmatrix} 5 & -1 \\ 2 & 7 \end{bmatrix}$$

$A + 4I$ is not a symmetric matrix.

$$A - 4I = \begin{bmatrix} 1 & -1 \\ 2 & 3 \end{bmatrix} - \begin{bmatrix} 4 & 0 \\ 0 & 4 \end{bmatrix} = \begin{bmatrix} -3 & -1 \\ 2 & -1 \end{bmatrix}$$

$A - 4I$ is not a skew-symmetric matrix.

If $$B = \begin{bmatrix} \alpha & -1 \\ 2 & 5 \end{bmatrix}$$

Then, $$A - B = \begin{bmatrix} 1 & -1 \\ 2 & 3 \end{bmatrix} - \begin{bmatrix} \alpha & -1 \\ 2 & 5 \end{bmatrix} = \begin{bmatrix} 1-\alpha & 0 \\ 0 & -2 \end{bmatrix}$$

Hence, $A - B$ is not a diagonal matrix for any value of α.

Now, $$A^2 = \begin{bmatrix} 1 & -1 \\ 2 & 3 \end{bmatrix} \begin{bmatrix} 1 & -1 \\ 2 & 3 \end{bmatrix} = \begin{bmatrix} 1-2 & -1-3 \\ 2+6 & -2+9 \end{bmatrix} = \begin{bmatrix} -1 & -4 \\ 8 & 7 \end{bmatrix}$$

$$A^2 - 4A + 5I_2 = \begin{bmatrix} -1 & -4 \\ 8 & 7 \end{bmatrix} - \begin{bmatrix} 4 & -4 \\ 8 & 12 \end{bmatrix} + \begin{bmatrix} 5 & 0 \\ 0 & 5 \end{bmatrix} = \begin{bmatrix} 0 & 0 \\ 0 & 0 \end{bmatrix}$$

30. (b) $\begin{bmatrix}1&1\\0&1\end{bmatrix}\begin{bmatrix}1&2\\0&1\end{bmatrix}\begin{bmatrix}1&3\\0&1\end{bmatrix}\cdots\begin{bmatrix}1&n\\0&1\end{bmatrix}=\begin{bmatrix}1&378\\0&1\end{bmatrix}$

$$\Rightarrow \begin{bmatrix}1&3\\0&1\end{bmatrix}\begin{bmatrix}1&3\\0&1\end{bmatrix}\begin{bmatrix}1&4\\0&1\end{bmatrix}\cdots\begin{bmatrix}1&n\\0&1\end{bmatrix}=\begin{bmatrix}1&378\\0&1\end{bmatrix}$$

$$\Rightarrow \begin{bmatrix}1&6\\0&1\end{bmatrix}\begin{bmatrix}1&4\\0&1\end{bmatrix}\cdots\begin{bmatrix}1&n\\0&1\end{bmatrix}=\begin{bmatrix}1&378\\0&1\end{bmatrix}$$

and so on,

$$\Rightarrow \begin{bmatrix}1&1+2+3+\cdots+n\\0&1\end{bmatrix}=\begin{bmatrix}1&378\\0&1\end{bmatrix}$$

$$\therefore \quad 1+2+3+\cdots+n=378$$

$$\Rightarrow \quad \frac{n(n+1)}{2}=378$$

$$\Rightarrow \quad n(n+1)=27\times 28$$

Hence, $\quad n=27$

31. (4) $\because \quad A^T A = I$

$$\begin{bmatrix}a&b&c\\b&c&a\\c&a&b\end{bmatrix}\begin{bmatrix}a&b&c\\b&c&a\\c&a&b\end{bmatrix}=\begin{bmatrix}1&0&0\\0&1&0\\0&0&1\end{bmatrix}$$

$$=\begin{bmatrix}a^2+b^2+c^2 & ab+bc+ca & ac+ab+bc\\ ab+bc+ac & b^2+c^2+a^2 & bc+ca+ab\\ ca+ab+bc & bc+ac+ab & c^2+a^2+b^2\end{bmatrix}$$

$$=\begin{bmatrix}1&0&0\\0&1&0\\0&0&1\end{bmatrix}$$

$\because \quad a^3+b^3+c^3-3abc$

$= (a+b+c)(a^2+b^2+c^2-ab-bc-ca) \quad \ldots(i)$

and $\quad (a+b+c)^2 = a^2+b^2+c^2+2(ab+bc+ca)$

$= 1+2\times 0 = 1$

$\Rightarrow \quad a+b+c=1 \quad \ldots(ii)$

[$\because$ a, b, c are positive real numbers]

From Eqs. (i) and (ii), we get

$a^3+b^3+c^3-3\times 1 = 1\times(1-0) \quad [\because abc=1]$

$\Rightarrow \quad a^3+b^3+c^3=4$

32. (4) $\quad M=\begin{bmatrix}1&2\\2&3\end{bmatrix}$

$$M^2=\begin{bmatrix}1&2\\2&3\end{bmatrix}\begin{bmatrix}1&2\\2&3\end{bmatrix}=\begin{bmatrix}1+4&2+6\\2+6&4+9\end{bmatrix}=\begin{bmatrix}5&8\\8&13\end{bmatrix}$$

$\because \quad M^2-\lambda M-I_2=0$

$$\begin{bmatrix}5&8\\8&13\end{bmatrix}-\begin{bmatrix}\lambda&2\lambda\\2\lambda&3\lambda\end{bmatrix}-\begin{bmatrix}1&0\\0&1\end{bmatrix}=\begin{bmatrix}0&0\\0&0\end{bmatrix}$$

$$\Rightarrow \begin{bmatrix}5-\lambda-1 & 8-2\lambda\\ 8-2\lambda & 13-3\lambda-1\end{bmatrix}=\begin{bmatrix}0&0\\0&0\end{bmatrix}$$

$$\Rightarrow \quad 2\lambda=8$$

$$\therefore \quad \lambda=4$$

33. (2) $A^2=\begin{bmatrix}2&2\\2&2\end{bmatrix}$

$$\Rightarrow \quad A^3=\begin{bmatrix}2&2\\2&2\end{bmatrix}\begin{bmatrix}1&1\\1&1\end{bmatrix}=\begin{bmatrix}4&4\\4&4\end{bmatrix}$$

$$\Rightarrow \quad A^4=\begin{bmatrix}8&8\\8&8\end{bmatrix}$$

Hence, $\quad A^{100}=\begin{bmatrix}2^{99}&2^{99}\\2^{99}&2^{99}\end{bmatrix}$

$\because \quad A^{100}=\lambda^{99}A$

$$\Rightarrow \begin{bmatrix}2^{99}&2^{99}\\2^{99}&2^{99}\end{bmatrix}=\lambda^{99}\begin{bmatrix}1&1\\1&1\end{bmatrix}$$

$$\Rightarrow \quad 2^{99}\begin{bmatrix}1&1\\1&1\end{bmatrix}=\lambda^{99}\begin{bmatrix}1&1\\1&1\end{bmatrix}$$

$$\therefore \quad \lambda=2$$

34. (1) $\quad A=\begin{bmatrix}0&-1\\1&0\end{bmatrix}$

$$A^2=\begin{bmatrix}0&-1\\1&0\end{bmatrix}\begin{bmatrix}0&-1\\1&0\end{bmatrix}=\begin{bmatrix}-1&0\\0&-1\end{bmatrix}$$

$$\Rightarrow \quad A^4=\begin{bmatrix}-1&0\\0&-1\end{bmatrix}\begin{bmatrix}-1&0\\0&-1\end{bmatrix}=\begin{bmatrix}1&0\\0&1\end{bmatrix}$$

$\Rightarrow \quad A^8 = I\cdot I = I$

$\Rightarrow \quad A^{16} = I\cdot I = I$

$\because \quad A^{16}=\lambda I$

$\Rightarrow \quad I=\lambda I$

$\therefore \quad \lambda=1$

35. (5) $\quad \begin{bmatrix}a&b\\c&d\end{bmatrix}\begin{bmatrix}1\\-1\end{bmatrix}=\begin{bmatrix}-1\\2\end{bmatrix}$

$$\begin{bmatrix}a-b\\c-d\end{bmatrix}=\begin{bmatrix}-1\\2\end{bmatrix}$$

and $\quad A^2\begin{bmatrix}1\\-1\end{bmatrix}=\begin{bmatrix}1\\0\end{bmatrix}$

$$\Rightarrow \begin{bmatrix}a&b\\c&d\end{bmatrix}\begin{bmatrix}a&b\\c&d\end{bmatrix}\begin{bmatrix}1\\-1\end{bmatrix}=\begin{bmatrix}1\\0\end{bmatrix}$$

$$\begin{bmatrix}a&b\\c&d\end{bmatrix}\begin{bmatrix}a-b\\c-d\end{bmatrix}=\begin{bmatrix}1\\0\end{bmatrix}$$

$$\begin{bmatrix}a&b\\c&d\end{bmatrix}\begin{bmatrix}-1\\2\end{bmatrix}=\begin{bmatrix}1\\0\end{bmatrix}$$

$$\begin{bmatrix}-a+2b\\-c+2d\end{bmatrix}=\begin{bmatrix}1\\0\end{bmatrix}$$

Now, $\quad a-b=-1 \quad \ldots(i)$

$c-d=2 \quad \ldots(ii)$

$-a+2b=1 \quad \ldots(iii)$

$-c+2d=0 \quad \ldots(iv)$

On adding Eqs. (i), (ii), (iii) and (iv), we get

$b+d=2$

On adding Eqs. (i) and (ii), we get

$(a+c)-(b+d)=1$

$\Rightarrow \quad a+c-2=1$

$\Rightarrow \quad a+c=3$

Hence, $\quad a+b+c+d=2+3=5$

DPP-3 **Properties of Determinant**

1. (c) ∵ $$\begin{vmatrix} 4 & 2 \\ 1 & 1 \end{vmatrix}^2 = \begin{vmatrix} 3 & 2 \\ 1 & x \end{vmatrix} - \begin{vmatrix} x & 3 \\ -2 & 1 \end{vmatrix}$$

$\Rightarrow \quad (2)^2 = 3x - 2 - (x + 6)$

$\Rightarrow \quad 4 = 2x - 8 \quad \Rightarrow \quad 2x - 12 = 0$

$\therefore \quad x = 6$

2. (c) The sum of the products of the elements of any row of a determinant A with the cofactor of same row is always equal to determinant of A.

3. (b) $(AB)^T = B^T A^T$

4. (b) $$\begin{vmatrix} a-b & b-c & c-a \\ x-y & y-z & z-x \\ p-q & q-r & r-p \end{vmatrix}$$

On applying $C_1 \to C_1 + C_2 + C_3$

$$= \begin{vmatrix} a-b+b-c+c-a & b-c & c-a \\ x-y+y-z+z-x & y-z & z-x \\ p-q+q-r+r-p & q-r & r-p \end{vmatrix} = \begin{vmatrix} 0 & b-c & c-a \\ 0 & y-z & z-x \\ 0 & q-r & r-p \end{vmatrix}$$

$= 0$

5. (b) $$\begin{vmatrix} 1 & 4 & 20 \\ 1 & -2 & 5 \\ 1 & 2x & 5x^2 \end{vmatrix} = 0$$

On applying $R_2 \to R_2 - R_1$ and $R_3 \to R_3 - R_1$

$$\begin{vmatrix} 1 & 4 & 20 \\ 0 & -6 & -15 \\ 0 & 2x-4 & 5x^2-20 \end{vmatrix} = 0$$

$\Rightarrow \quad -6(5x^2 - 20) + 15(2x - 4) = 0$

$\Rightarrow \quad -30(x^2 - 4) + 15(2x - 4) = 0$

$\Rightarrow \quad -2(x^2 - 4) + (2x - 4) = 0$

$\Rightarrow \quad -2x^2 + 8 + 2x - 4 = 0$

$\Rightarrow \quad 2x^2 - 8 - 2x + 4 = 0$

$\Rightarrow \quad x^2 - x - 2 = 0$

$\therefore \quad x = -1, 2$

6. (d) $$\begin{vmatrix} x+1 & 3 & 5 \\ 2 & x+2 & 5 \\ 2 & 3 & x+4 \end{vmatrix} = 0$$

$\Rightarrow \quad (x + 1)(x^2 + 6x + 8 - 15) + 3(10 - 2x - 8) + 5(6 - 2x - 4) = 0$

$\Rightarrow \quad (x + 1)(x^2 + 6x - 7) + 3(2 - 2x) + 5(2 - 2x) = 0$

$\Rightarrow \quad x^3 + 6x^2 - 7x + x^2 + 6x - 7 + 6 - 6x + 10 - 10x = 0$

$\Rightarrow \quad x^3 + 7x^2 - 17x + 9 = 0$

$\Rightarrow \quad (x - 1)^2 (x + 9) = 0$

$\therefore \quad x = 1, 1, -9$

7. (d) $$\begin{vmatrix} a+b & a+2b & a+3b \\ a+2b & a+3b & a+4b \\ a+4b & a+5b & a+6b \end{vmatrix}$$

On applying $C_3 \to C_3 - C_2$

$$\begin{vmatrix} a+b & a+2b & b \\ a+2b & a+3b & b \\ a+4b & a+5b & b \end{vmatrix}$$

On applying $C_2 \to C_2 - C_1$

$$= \begin{vmatrix} a+b & b & b \\ a+2b & b & b \\ a+4b & b & b \end{vmatrix}$$

$= 0 \quad$ [∵ C_2 and C_3 are same]

8. (c) We have, $$\begin{vmatrix} 1 & 1 & 1 \\ a & b & c \\ a^3 & b^3 & c^3 \end{vmatrix}$$

On applying $C_2 \to C_2 - C_1$

and $C_3 \to C_3 - C_1$

$$= \begin{vmatrix} 1 & 0 & 0 \\ a & b-a & c-a \\ a^3 & b^3-a^3 & c^3-a^3 \end{vmatrix}$$

$= (b - a)(c^3 - a^3) - (b^3 - a^3)(c - a)$

$= (b - a)(c - a)[c^2 + a^2 + ac - b^2 - a^2 - ab]$

$= (b - a)(c - a)(c^2 - b^2 + ac - ab)$

$= (b - a)(c - a)[(c + b)(c - b) + a(c - b)]$

$= (b - a)(c - a)(c - b)(a + b + c)$

$= (a + b + c)(a - b)(b - c)(c - a)$

9. (b) $$f(1) = \begin{vmatrix} -2 & -16 & -78 \\ -4 & -48 & -496 \\ 1 & 2 & 3 \end{vmatrix}$$

$$f(3) = \begin{vmatrix} 0 & 0 & 0 \\ -2 & -32 & -392 \\ 1 & 2 & 3 \end{vmatrix} = 0$$

$\therefore \quad f(1)\, f(3) + f(3)\, f(5)\, f(5)\, f(1) = f(1)\, f(3)\, [1 + \{f(5)\}^2]$

$= 0 \times [1 + \{f(5)\}^2]$

$= 0$

10. (b) We have, $$\begin{vmatrix} a & b & c \\ b & c & a \\ c & a & b \end{vmatrix}$$

On applying $C_1 \to C_1 + C_2 + C_3$

$$= \begin{vmatrix} a+b+c & b & c \\ a+b+c & c & a \\ a+b+c & a & b \end{vmatrix}$$

$$= (a + b + c)\begin{vmatrix} 1 & b & c \\ 1 & c & a \\ 1 & a & b \end{vmatrix}$$

On applying $R_2 \to R_2 - R_1$ and $R_3 \to R_3 - R_1$

$$= (a + b + c)\begin{vmatrix} 1 & b & c \\ 0 & c-b & a-c \\ 0 & a-b & b-c \end{vmatrix}$$

$= (a + b + c)[(c - b)(b - c) - (a - b)(a - c)]$

$= (a + b + c)[-b^2 - c^2 + 2bc - a^2 + ac + ab - bc]$

$= -(a + b + c)(a^2 + b^2 + c^2 + ab + bc + ca)$

$= -[a^3 + b^3 + c^3 - 3abc]$

$= 3abc - a^3 - b^3 - c^3$

11. (b) $$\begin{vmatrix} a & b & a\alpha + b \\ b & c & b\alpha + c \\ a\alpha + b & b\alpha + c & 0 \end{vmatrix} = 0$$

On applying $R_3 \to R_3 - \alpha R_1$

$$\Rightarrow \quad \begin{vmatrix} a & b & a\alpha + b \\ b & c & b\alpha + c \\ b & c & -a\alpha^2 - b\alpha \end{vmatrix} = 0$$

On applying $R_3 \to R_3 - R_2$

$$\Rightarrow \quad \begin{vmatrix} a & b & a\alpha + b \\ b & c & b\alpha + c \\ 0 & 0 & -a\alpha^2 - 2b\alpha - c \end{vmatrix} = 0$$

On applying $C_3 \to C_3 - (\alpha C_1 + C_2)$

$$\Rightarrow \begin{vmatrix} a & b & 0 \\ b & c & 0 \\ 0 & 0 & -a\alpha^2 - 2b\alpha - c \end{vmatrix} = 0$$

$$\Rightarrow \begin{vmatrix} 0 & 0 & -a\alpha^2 - 2b\alpha - c \\ b & c & 0 \\ a & b & 0 \end{vmatrix} = 0$$

$$\Rightarrow (a\alpha^2 + 2b\alpha + c)(b^2 - ac) = 0$$

$$\Rightarrow b^2 - ac = 0$$

$\therefore$ a, b, c are in GP.

12. (a) We have, $\begin{vmatrix} 1+x & 1 & 1 \\ 1 & 1+y & 1 \\ 1 & 1 & 1+z \end{vmatrix}$

On applying $C_1 \to \frac{C_1}{x}, C_2 \to \frac{C_2}{y}, C_3 \to \frac{C_3}{z}$

$$= xyz \begin{vmatrix} \frac{1}{x}+1 & \frac{1}{y} & \frac{1}{z} \\ \frac{1}{x} & 1+\frac{1}{y} & \frac{1}{z} \\ \frac{1}{x} & \frac{1}{y} & 1+\frac{1}{z} \end{vmatrix}$$

On applying $C_1 \to C_1 + C_2 + C_3$

$$= xyz \begin{vmatrix} 1+\frac{1}{x}+\frac{1}{y}+\frac{1}{z} & \frac{1}{y} & \frac{1}{z} \\ 1+\frac{1}{x}+\frac{1}{y}+\frac{1}{z} & 1+\frac{1}{y} & \frac{1}{z} \\ 1+\frac{1}{x}+\frac{1}{y}+\frac{1}{z} & \frac{1}{y} & 1+\frac{1}{z} \end{vmatrix}$$

$$= xyz\left(1+\frac{1}{x}+\frac{1}{y}+\frac{1}{z}\right) \begin{vmatrix} 1 & \frac{1}{y} & \frac{1}{z} \\ 1 & 1+\frac{1}{y} & \frac{1}{z} \\ 1 & \frac{1}{y} & 1+\frac{1}{z} \end{vmatrix}$$

On applying $R_2 \to R_2 - R_1$ and $R_3 \to R_3 - R_1$

$$= xyz\left(1+\frac{1}{x}+\frac{1}{y}+\frac{1}{z}\right) \begin{vmatrix} 1 & \frac{1}{y} & \frac{1}{z} \\ 0 & 1 & 0 \\ 0 & 0 & 1 \end{vmatrix}$$

$$= xyz\left(1+\frac{1}{x}+\frac{1}{y}+\frac{1}{z}\right)$$

13. (a) $\begin{vmatrix} a & a^2 & a^3+1 \\ b & b^2 & b^3+1 \\ c & c^2 & c^3+1 \end{vmatrix} = 0$

$$\Rightarrow \begin{vmatrix} a & a^2 & a^3 \\ b & b^2 & b^3 \\ c & c^2 & c^3 \end{vmatrix} + \begin{vmatrix} a & a^2 & 1 \\ b & b^2 & 1 \\ c & c^2 & 1 \end{vmatrix} = 0$$

$$\Rightarrow abc \begin{vmatrix} 1 & a & a^2 \\ 1 & b & b^2 \\ 1 & c & c^2 \end{vmatrix} + \begin{vmatrix} 1 & a & a^2 \\ 1 & b & b^2 \\ 1 & c & c^2 \end{vmatrix} = 0$$

$$\Rightarrow \begin{vmatrix} 1 & a & a^2 \\ 1 & b & b^2 \\ 1 & c & c^2 \end{vmatrix}(1+abc) = 0$$

$$\Rightarrow (a-b)(b-c)(c-a)(1+abc) = 0$$

$$\therefore \quad 1 + abc = 0 \qquad [\because a \neq b \neq c]$$

14. (c) $\begin{vmatrix} 1+\sin^2\theta & \sin^2\theta & \sin^2\theta \\ \cos^2\theta & 1+\cos^2\theta & \cos^2\theta \\ 4\sin 4\theta & 4\sin 4\theta & 1+4\sin 4\theta \end{vmatrix} = 0$

On applying $R_1 \to R_1 + R_2 + R_3$

$$\Rightarrow \begin{vmatrix} 2+4\sin 4\theta & 2+4\sin 4\theta & 2+4\sin 4\theta \\ \cos^2\theta & 1+\cos^2\theta & \cos^2\theta \\ 4\sin 4\theta & 4\sin 4\theta & 1+4\sin 4\theta \end{vmatrix} = 0$$

$$[\because \sin^2\theta + \cos^2\theta = 1]$$

$$\Rightarrow (2+4\sin\theta)\begin{vmatrix} 1 & 1 & 1 \\ \cos^2\theta & 1+\cos^2\theta & \cos^2\theta \\ 4\sin 4\theta & 4\sin 4\theta & 1+4\sin 4\theta \end{vmatrix} = 0$$

On applying $C_2 \to C_2 - C_1$ and $C_3 \to C_3 - C_1$

$$\Rightarrow (2+4\sin\theta)\begin{vmatrix} 1 & 0 & 0 \\ \cos^2\theta & 1 & 0 \\ 4\sin 4\theta & 0 & 1 \end{vmatrix} = 0$$

$$\Rightarrow 2(1+2\sin 4\theta) = 0 \Rightarrow \sin 4\theta = \frac{-1}{2}$$

15. (a) $\begin{vmatrix} x+1 & x+2 & x+3 \\ x+2 & x+3 & x+4 \\ x+a & x+b & x+c \end{vmatrix} = 0$

On applying $C_2 \to C_2 - C_1$ and $C_3 \to C_3 - C_2$

$$\Rightarrow \begin{vmatrix} x+1 & 1 & 1 \\ x+2 & 1 & 1 \\ x+a & b-a & c-b \end{vmatrix} = 0 \Rightarrow \begin{vmatrix} x+1 & x+2 & x+a \\ 1 & 1 & b-a \\ 1 & 1 & c-b \end{vmatrix} = 0$$

On applying $R_3 \to R_3 - R_2$

$$\Rightarrow \begin{vmatrix} x+1 & x+2 & x+a \\ 1 & 1 & b-a \\ 0 & 0 & c+a-2b \end{vmatrix} = 0$$

$$\Rightarrow \begin{vmatrix} x & x & x \\ 1 & 1 & b-a \\ 0 & 0 & c+a-2b \end{vmatrix} + \begin{vmatrix} 1 & 2 & a \\ 1 & 1 & b-a \\ 0 & 0 & c+a-2b \end{vmatrix} = 0$$

$$\Rightarrow x\begin{vmatrix} 1 & 1 & 1 \\ 1 & 1 & b-a \\ 0 & 0 & c+a-2b \end{vmatrix} + \begin{vmatrix} 1 & 2 & a \\ 1 & 1 & b-a \\ 0 & 0 & c+a-2b \end{vmatrix} = 0$$

On applying $R_2 \to R_2 - R_1$

$$\begin{vmatrix} 1 & 2 & a \\ 0 & -1 & b-2a \\ 0 & 0 & c+a-2b \end{vmatrix} = 0 \Rightarrow -c-a+2b = 0$$

$$\Rightarrow b = \frac{a+c}{2}$$

So, a, b, c are in an AP.

16. (b) $\begin{vmatrix} 1 & 1 & 1 \\ \cos(nx) & \cos(n+1)x & \cos(n+2)x \\ \sin(nx) & \sin(n+1)x & \sin(n+2)x \end{vmatrix}$

$$= \begin{vmatrix} 1 & 0 & 0 \\ \cos(nx) & \cos(n+1)x - \cos(nx) & \cos(n+2)x - \cos(nx) \\ \sin(nx) & \sin(n+1)x - \sin(nx) & \sin(n+2)x - \sin(nx) \end{vmatrix}$$

$$= [\cos(n+1)x - \cos(nx)][\sin(n+2)x - \sin(nx)] - [\cos(n+2)x - \cos(nx)][\sin(n+1)x - \sin(nx)]$$

$$= \left(-2\sin\frac{(2n+1)x}{2}\sin\frac{x}{2}\right)\left(2\cos\frac{(2n+2)x}{2}\cos x\right) - \left(-2\sin\frac{(2n+2)x}{2}\sin x\right)\left(2\cos\frac{(2n+1)x}{2}\cos\frac{x}{2}\right)$$

$$= 4\left[-\sin\frac{x}{2}\cos x \sin\frac{(2n+1)x}{2}\cos(n+1)x + \sin(n+1)x \sin x \cos\frac{(2n+1)x}{2}\cos\frac{x}{2}\right]$$

17. (c) $\begin{vmatrix} b^2-ab & b-c & bc-ac \\ ab-a^2 & a-b & b^2-ab \\ bc-ac & c-a & ab-a^2 \end{vmatrix}$

$$= \begin{vmatrix} b(b-a) & b-c & c(b-a) \\ a(b-a) & a-b & b(b-a) \\ c(b-a) & c-a & a(b-a) \end{vmatrix}$$

$$= (b-a)^2 \begin{vmatrix} b & b-c & c \\ a & a-b & b \\ c & c-a & a \end{vmatrix}$$

On applying $C_2 \to C_2 + C_3$

$$= (b-a)^2 \begin{vmatrix} b & b & c \\ a & a & b \\ c & c & a \end{vmatrix} = 0 \quad [\because C_1 \text{ and } C_2 \text{ are same}]$$

18. (c) We have, $\begin{vmatrix} b^2+c^2 & a^2 & a^2 \\ b^2 & c^2+a^2 & b^2 \\ c^2 & c^2 & a^2+b^2 \end{vmatrix}$

$$= \begin{vmatrix} b^2+c^2 & b^2 & c^2 \\ a^2 & c^2+a^2 & c^2 \\ a^2 & b^2 & a^2+b^2 \end{vmatrix}$$

On applying $C_1 \to C_1 - C_2 - C_3$

$$= \begin{vmatrix} 0 & b^2 & c^2 \\ -2c^2 & c^2+a^2 & c^2 \\ -2b^2 & b^2 & a^2+b^2 \end{vmatrix}$$

$$= -2 \begin{vmatrix} 0 & b^2 & c^2 \\ c^2 & c^2+a^2 & c^2 \\ b^2 & b^2 & a^2+b^2 \end{vmatrix}$$

On applying $C_2 \to C_2 - C_1$ and $C_3 \to C_3 - C_1$

$$= -2 \begin{vmatrix} 0 & b^2 & c^2 \\ c^2 & a^2 & 0 \\ b^2 & 0 & a^2 \end{vmatrix}$$

$$= -2[-b^2a^2c^2 - a^2b^2c^2] = 4a^2b^2c^2$$

19. (b) $\begin{vmatrix} a & b & a+b \\ b & c & b+c \\ a+b & b+c & 0 \end{vmatrix} = 0$

On applying $R_3 \to R_3 - R_1$

$$\Rightarrow \begin{vmatrix} a & b & a+b \\ b & c & b+c \\ b & c & -a-b \end{vmatrix} = 0$$

On applying $R_3 \to R_3 - R_2$

$$\Rightarrow \begin{vmatrix} a & b & a+b \\ b & c & b+c \\ 0 & 0 & -a-2b-c \end{vmatrix} = 0$$

On applying $C_3 \to C_3 - (C_1 + C_2)$

$$\Rightarrow \begin{vmatrix} a & b & 0 \\ b & c & 0 \\ 0 & 0 & -a-2b-c \end{vmatrix} = 0$$

$\Rightarrow \quad a[-ac - 2bc - c^2] + b[ab + 2b^2 + bc] = 0$

$\Rightarrow \quad -a^2c - 2abc - ac^2 + ab^2 + 2b^3 + b^2c = 0$

$\Rightarrow \quad 2b(b^2 - ac) + a(b^2 - ac) + c(b^2 - ac) = 0$

$\Rightarrow \quad (2b + a + c)(b^2 - ac) = 0$

$\Rightarrow \quad b^2 - ac = 0$

$\therefore$ a, b, c are in GP.

20. (a) $\begin{vmatrix} (a^x+a^{-x})^2 & (a^x-a^{-x})^2 & 1 \\ (b^x+b^{-x})^2 & (b^x-b^{-x})^2 & 1 \\ (c^x+c^{-x})^2 & (c^x-c^{-x})^2 & 1 \end{vmatrix}$

$$= \begin{vmatrix} a^{2x}+a^{-2x}+2 & a^{2x}+a^{-2x}-2 & 1 \\ b^{2x}+b^{-2x}+2 & b^{2x}+b^{-2x}-2 & 1 \\ c^{2x}+c^{-2x}+2 & c^{2x}+c^{-2x}-2 & 1 \end{vmatrix}$$

On applying $C_2 \to C_2 - C_1$

$$= \begin{vmatrix} a^{2x}+a^{-2x}+2 & -4 & 1 \\ b^{2x}+b^{-2x}+2 & -4 & 1 \\ c^{2x}+c^{-2x}+2 & -4 & 1 \end{vmatrix}$$

$$= -4 \begin{vmatrix} a^{2x}+a^{-2x}+2 & 1 & 1 \\ b^{2x}+b^{-2x}+2 & 1 & 1 \\ c^{2x}+c^{-2x}+2 & 1 & 1 \end{vmatrix}$$

$$= 0 \quad [\because C_2 \text{ and } C_3 \text{ are same}]$$

21. (b) $\begin{vmatrix} x^n & x^{n+2} & x^{n+3} \\ y^n & y^{n+2} & y^{n+3} \\ z^n & z^{n+2} & z^{n+3} \end{vmatrix} = \begin{vmatrix} x^n & x^2 \cdot x^n & x^3 \cdot x^n \\ y^n & y^2 \cdot y^n & y^3 \cdot y^n \\ z^n & z^2 \cdot z^n & z^3 \cdot z^n \end{vmatrix}$

$$= x^n y^n z^n \begin{vmatrix} 1 & x^2 & x^3 \\ 1 & y^2 & y^3 \\ 1 & z^2 & z^3 \end{vmatrix}$$

$$= x^n y^n z^n (x-y)(y-z)(z-x)(xy+yz+zx)$$

$\because$ $x^n y^n z^n (x-y)(y-z)(z-x)(xy+yz+zx)$

$$= (x-y)(y-z)(z-x)\left(\frac{1}{x} + \frac{1}{y} + \frac{1}{z}\right)$$

$\therefore \quad n = -1$

22. (c) $\because a = p$th term $= m + (p-1)d$

$b = q$th term $= m + (q-1)d$

$c = r$th term $= m + (r-1)d$

where m is the first term and d is the common difference of an AP.

So, $\begin{vmatrix} a & p & 1 \\ b & q & 1 \\ c & r & 1 \end{vmatrix} = \begin{vmatrix} m+(p-1)d & p & 1 \\ m+(q-1)d & q & 1 \\ m+(r-1)d & r & 1 \end{vmatrix}$

On applying $C_1 \to C_1 + dC_2 - mC_3$

$$= \begin{vmatrix} -d & p & 1 \\ -d & q & 1 \\ -d & r & 1 \end{vmatrix} = -d \begin{vmatrix} 1 & p & 1 \\ 1 & q & 1 \\ 1 & r & 1 \end{vmatrix} = 0$$

$[\because C_1 \text{ and } C_3 \text{ are same}]$

23. (b) $\begin{vmatrix} x! & (x+1)! & (x+2)! \\ (x+1)! & (x+2)! & (x+3)! \\ (x+2)! & (x+3)! & (x+4)! \end{vmatrix}$

By taking out common $x!, (x+1)!, (x+2)!$ from R_1, R_2, R_3 respectively

$$= x!(x+1)!(x+2)! \begin{vmatrix} 1 & (x+1) & (x+2)(x+1) \\ 1 & (x+2) & (x+3)(x+2) \\ 1 & (x+3) & (x+4)(x+3) \end{vmatrix}$$

On applying $C_3 \to C_3 - C_2$

$$= x!(x+1)!(x+2)! \begin{vmatrix} 1 & (x+1) & (x+1)^2 \\ 1 & (x+2) & (x+2)^2 \\ 1 & (x+3) & (x+3)^2 \end{vmatrix}$$

$$= x!\,(x+1)!\,(x+2)!\,[(-1)(-1)(2)]$$

$$= 2x!(x+1)!(x+2)!$$

24. (a) $\begin{vmatrix} 1 & 1 & 1 \\ 1 & \omega^2 & \omega \\ 1 & \omega & \omega^2 \end{vmatrix}$

On applying $R_2 \to R_2 - R_1$ and $R_3 \to R_3 - R_1$

$$= \begin{vmatrix} 1 & 1 & 1 \\ 0 & \omega^2 - 1 & \omega - 1 \\ 0 & \omega - 1 & \omega^2 - 1 \end{vmatrix} = (\omega^2 - 1)^2 - (\omega - 1)^2$$

$$= (\omega - 1)^2 [(\omega + 1)^2 - 1] = (\omega - 1)^2 [\omega^2 + 1 + 2\omega - 1]$$

$$= (\omega - 1)^2 (\omega^2 + 2\omega)$$

$$= (\omega^2 + 1 - 2\omega)(\omega^2 + 2\omega)$$

$$= \omega^4 + \omega^2 - 2\omega^3 + 2\omega^3 + 2\omega - 4\omega^2$$

$$= \omega - 3\omega^2 + 2\omega = 3\omega - 3\omega^2 = 3(\omega - \omega^2)$$

$$= 3\left[\frac{-1 + i\sqrt{3} - (-1 - i\sqrt{3})}{2}\right]$$

$$= 3 \times \frac{2i\sqrt{3}}{2} = 3\sqrt{3}i$$

25. (a) $U_n = \begin{vmatrix} n & 1 & 5 \\ n^2 & 2N + 1 & 2N + 1 \\ n^3 & 3N^2 & 3N \end{vmatrix}$

$$\sum_{n=1}^{N} U_n = \begin{vmatrix} \frac{N(N+1)}{2} & 1 & 5 \\ \frac{N(N+1)(2N+1)}{6} & 2N + 1 & 2N + 1 \\ \frac{N^2(N+1)^2}{4} & 3N^2 & 3N \end{vmatrix}$$

$$= \frac{N(N+1)}{2} \begin{vmatrix} 1 & 1 & 5 \\ \frac{2N+1}{3} & 2N + 1 & 2N + 1 \\ \frac{N(N+1)}{2} & 3N^2 & 3N \end{vmatrix}$$

$$= \frac{N(N+1)(2N+1)}{6} \begin{vmatrix} 1 & 1 & 5 \\ 1 & 3 & 3 \\ \frac{N(N+1)}{6} & 3N^2 & 3N \end{vmatrix}$$

$$= \frac{N^2(N+1)(2N+1)}{6} \begin{vmatrix} 1 & 1 & 5 \\ 1 & 3 & 3 \\ N+1 & 6N & 6 \end{vmatrix}$$

$$= \frac{N^2(N+1)(2N+1)}{6} [(18 - 18N) + (3N + 3 - 6) + 5(6N - 3N - 3)]$$

$$= \frac{N^2(N+1)(2N+1)}{6} [18 - 18N + 3N - 3 + 15N - 15]$$

$$= \frac{N^2(N+1)(2N+1)}{6} \times 0 = 0$$

26. (c) $D_r = \begin{vmatrix} 2^{r-1} & 2 \cdot 3^{r-1} & 4 \cdot 5^{r-1} \\ x & y & z \\ 2^n - 1 & 3^n - 1 & 5^n - 1 \end{vmatrix}$

$$\sum_{r=1}^{n} D_r = \begin{vmatrix} \frac{2^{n-1}}{2-1} & \frac{2(3^n - 1)}{3-1} & \frac{4(5^n - 1)}{5-1} \\ x & y & z \\ 2^n - 1 & 3^n - 1 & 5^n - 1 \end{vmatrix}$$

$$= \begin{vmatrix} 2^n - 1 & 3^n - 1 & 5^n - 1 \\ x & y & z \\ 2^n - 1 & 3^n - 1 & 5^n - 1 \end{vmatrix}$$

$= 0$ [$\because R_1$ and R_3 are same]

27. (b) $|A| = -1$ and $|B| = 3$

Now, $|3AB| = 27|AB|$ [$\because$ order of AB is 3×3]

$= 27\,|A|\,|B|$

$= 27 \times (-1) \times 3$

$= -81$

28. (a,d) $\Delta = \begin{vmatrix} x & 2y - z & -z \\ y & 2x - z & -z \\ z & 2y - z & 2x - 2y + z \end{vmatrix}$

On applying $R_1 \to R_1 - R_2$

$$= \begin{vmatrix} x - y & 2(y - x) & 0 \\ y & 2x - z & -z \\ z & 2y - z & 2x - 2y + z \end{vmatrix}$$

$$= (x - y) \begin{vmatrix} 1 & -2 & 0 \\ y & 2x - z & -z \\ z & 2y - z & 2x - 2y + z \end{vmatrix}$$

On applying $C_2 \to C_2 + 2C_1$

$$= (x - y) \begin{vmatrix} 1 & 0 & 0 \\ y & 2x + 2y - z & -z \\ z & 2y + z & 2x - 2y + z \end{vmatrix}$$

$$= (x - y)[(2x + 2y - z)(2x - 2y + z) + z(2y + z)]$$

$$= (x - y)[4x^2 - (2y - z)^2 + 2yz + z^2]$$

$$= (x - y)[4x^2 - 4y^2 + 4yz + 2yz]$$

$$= (x - y)(4x^2 - 4y^2 + 6yz)$$

29. (a,b,c) $\Delta = \begin{vmatrix} -x & a & b \\ b & -x & a \\ a & b & -x \end{vmatrix}$

On applying $R_1 \to R_1 + R_2 + R_3$

$$\begin{vmatrix} a + b - x & a + b - x & a + b - x \\ b & -x & a \\ a & b & -x \end{vmatrix} = (a + b - x) \begin{vmatrix} 1 & 1 & 1 \\ b & -x & a \\ a & b & -x \end{vmatrix}$$

On applying $C_2 \to C_2 - C_1$ and $C_3 \to C_3 - C_1$

$$= (a + b - x) \begin{vmatrix} 1 & 0 & 0 \\ b & -x - b & a - b \\ a & b - a & -x - a \end{vmatrix}$$

$$= (a + b - x)[(-x - b)(-x - a) - (a - b)(b - a)]$$

$$= (a + b - x)[x^2 + (a + b)x + ab + a^2 + b^2 - 2ab]$$

$$= (a + b - x)(x^2 + (a + b)x + a^2 + b^2 - ab)$$

30. (b,d) $\begin{vmatrix} b & c & b\alpha + c \\ c & d & c\alpha + d \\ b\alpha + c & c\alpha + d & a\alpha^3 - c\alpha \end{vmatrix} = 0$

On applying $R_3 \to R_3 - \alpha R_1$

$$\begin{vmatrix} b & c & b\alpha + c \\ c & d & c\alpha + d \\ c & d & a\alpha^3 - (b + c)\alpha - \alpha c \end{vmatrix} = 0$$

On applying $R_3 \to R_3 - R_2$

$$\Rightarrow \begin{vmatrix} b & c & b\alpha + c \\ c & d & c\alpha + d \\ 0 & 0 & a\alpha^3 - (b + c)\alpha - 2c\alpha - d \end{vmatrix} = 0$$

$$\Rightarrow \begin{vmatrix} 0 & 0 & a\alpha^3 - (b + c)\alpha - 2c\alpha - d \\ c & d & c\alpha + d \\ b & c & b\alpha + c \end{vmatrix} = 0$$

$$\Rightarrow [a\alpha^3 - (b + c)\alpha - 2c\alpha - d](c^2 - bd) = 0$$

$\therefore b, c, d$ are in GP

and α is a root of $ax^3 - bx^2 - 3cx - d = 0$.

31. (a,b) $\Delta = \begin{vmatrix} a & a^2 & 0 \\ 1 & 2a+b & (a+b)^2 \\ 0 & 1 & 2a+3b \end{vmatrix}$

On applying $R_2 \to aR_2 - R_1$

$$= \frac{1}{a}\begin{vmatrix} a & a^2 & 0 \\ 0 & a^2+ab & a(a+b)^2 \\ 0 & 1 & 2a+3b \end{vmatrix}$$

$$= \{(a^2+ab)(2a+3b) - a(a+b)^2\}$$

$$= a\{(a+b)(2a+3b) - (a+b)^2\}$$

$$= a(a+b)\{2a+3b-a-b\}$$

$$\therefore \quad = a(a+b)(a+2b)$$

32. (c,d) $A = \begin{bmatrix} \alpha & 2 \\ 2 & \alpha \end{bmatrix}$

and $\quad |A^3| = 125$

$$A^2 = \begin{bmatrix} \alpha & 2 \\ 2 & \alpha \end{bmatrix}\begin{bmatrix} \alpha & 2 \\ 2 & \alpha \end{bmatrix} = \begin{bmatrix} \alpha^2+4 & 4\alpha \\ 4\alpha & 4+\alpha^2 \end{bmatrix}$$

$$\therefore \quad A^3 = A^2 \cdot A$$

$$= \begin{bmatrix} \alpha^2+4 & 4\alpha \\ 4\alpha & \alpha^2+4 \end{bmatrix}\begin{bmatrix} \alpha & 2 \\ 2 & \alpha \end{bmatrix}$$

$$= \begin{bmatrix} \alpha^3+12\alpha & 6\alpha^2+8 \\ 6\alpha^2+8 & \alpha^3+12\alpha \end{bmatrix}$$

$$\Rightarrow \quad |A|^3 = (\alpha^3+12\alpha)^2 - (6\alpha^2+8)^2$$

If $\alpha = 3$, then

$$= (27+36)^2 - (54+8)^2$$

$$= (63)^2 - (62)^2$$

$$= (63+62) \times 1 = 125$$

If $\alpha = -3$

Then, $\quad |A|^3 = (-27-36)^2 - (54+8)^2 = 125$

Explanations for Q. Nos. (33-34)

$$t_1 = \frac{a_{11}+a_{12}+a_{13}}{3}$$

$$t_2 = \frac{a_{21}+a_{22}+a_{23}}{3}$$

$$t_3 = \frac{a_{31}+a_{32}+a_{33}}{3}$$

Now, $\quad \Sigma a_{ij} = a_{11}+a_{12}+a_{13}+a_{21}+a_{22}+a_{23} + a_{31}+a_{32}+a_{33}$

$$\Rightarrow \quad \Sigma a_{ij} = (a_{11}+a_{12}+a_{21}) + (a_{22}+a_{23}+a_{32}) + (a_{13}+a_{33}+a_{31})$$

$$= 0+0+0 = 0$$

Also, $\quad t_1+t_2+t_3 = \frac{1}{3}(\Sigma a_{ij}) = \frac{1}{3} \times 0 = 0$

$$\because \quad A = \begin{bmatrix} a_{11} & a_{12} & a_{13} \\ a_{21} & a_{22} & a_{23} \\ a_{31} & a_{32} & a_{33} \end{bmatrix}$$

According to the given condition,

$$a_{11}+a_{12}+a_{21} = 0$$

$$a_{22}+a_{13}+a_{31} = 0$$

and $\quad a_{23}+a_{32}+a_{33} = 0$

If $i = j = k$,

$$a_{ij} = a_{ii} = a_{ii} = 0$$

$$\Rightarrow \quad 3a_{ii} = 0$$

$$\Rightarrow \quad a_{ii} = 0$$

$\therefore \; a_{11} = a_{22} = a_{33} = 0$ and if $k = i$

$$\Rightarrow \quad a_{ij} + a_{ji} = a_{ii} = 0$$

$$\Rightarrow \quad a_{ij} = -a_{ji}$$

Then, $\quad A = \begin{bmatrix} 0 & a_{12} & a_{13} \\ -a_{12} & 0 & a_{23} \\ -a_{13} & -a_{23} & 0 \end{bmatrix}$

So, A is a skew-symmetric matrix.

Then, $|A| = 0 \;\Rightarrow\; |A|^2 = 0$

33. (d) **34.** (c)

35. (2) $A = \begin{bmatrix} 1 & 1 \\ 1 & 1 \end{bmatrix}$

Suppose $n = 1$.

$$|A - I| = 1 - \lambda$$

$$\Rightarrow \quad \begin{vmatrix} 0 & 1 \\ 1 & 0 \end{vmatrix} = 1 - \lambda$$

$$\Rightarrow \quad -1 = 1 - \lambda$$

$$\therefore \quad \lambda = 2$$

36. (2) $|2A^9B^{-1}| = 2^2|A^9B^{-1}|$

[$\because A^9B^{-1}$ is a matrix of order 2×2]

$$= 4|A^9||B^{-1}|$$

$$= \frac{4|A|^9}{|B|} = \frac{4 \times (-1)^9}{2}$$

$$= -2$$

Modulus of $|2A^9B^{-1}|$ is 2.

37. (1) $A = \begin{bmatrix} 1 & \tan x \\ -\tan x & 1 \end{bmatrix}$

$$A^T = \begin{bmatrix} 1 & -\tan x \\ \tan x & 1 \end{bmatrix}$$

$$A^{-1} = \frac{1}{1+\tan^2 x}\begin{bmatrix} 1 & -\tan x \\ \tan x & 1 \end{bmatrix}$$

$$\therefore \quad A^TA^{-1} = \frac{1}{\sec^2 x}\begin{bmatrix} 1 & -\tan x \\ \tan x & 1 \end{bmatrix}\begin{bmatrix} 1 & -\tan x \\ \tan x & 1 \end{bmatrix}$$

$$= \frac{1}{\sec^2 x}\begin{bmatrix} 1-\tan^2 x & -2\tan x \\ 2\tan x & 1-\tan^2 x \end{bmatrix}$$

$$= \begin{bmatrix} \cos 2x & -\sin 2x \\ \sin 2x & \cos 2x \end{bmatrix}$$

Now, $|A^TA^{-1}| = \cos^2 2x + \sin^2 2x = 1$

That means, $f(x) = 1$

Hence, $f(f(f(f \cdots f(x))))$ is always 1.

Alternate Method

$|A^TA^{-1}| = |A^T||A^{-1}| = |A| \cdot |A^{-1}| = 1$

$$\therefore \quad f(x) = 1$$

Hence, $f(f(f \cdots f(x))) = 1$

38. (6) $f(x) = \begin{vmatrix} 1+\sin^2 x & \cos^2 x & 4\sin 2x \\ \sin^2 x & 1+\cos^2 x & 4\sin 2x \\ \sin^2 x & \cos^2 x & 1+4\sin 2x \end{vmatrix}$

On applying $C_1 \to C_1 + C_2 + C_3$

$$= \begin{vmatrix} 2+4\sin 2x & \cos^2 x & 4\sin 2x \\ 2+4\sin 2x & 1+\cos^2 x & 4\sin 2x \\ 2+4\sin 2x & \cos^2 x & 1+4\sin 2x \end{vmatrix}$$

[$\because \sin^2 x + \cos^2 x = 1$]

$$= (2+4\sin 2x)\begin{vmatrix} 1 & \cos^2 x & 4\sin 2x \\ 1 & 1+\cos^2 x & 4\sin 2x \\ 1 & \cos^2 x & 1+4\sin 2x \end{vmatrix}$$

On applying $R_2 \to R_2 - R_1$ and $R_3 \to R_3 - R_1$

$$= (2 + 4\sin 2x)\begin{vmatrix} 1 & \cos^2 x & 4\sin 2x \\ 0 & 1 & 0 \\ 0 & 0 & 1 \end{vmatrix}$$

$$= (2 + 4\sin 2x)$$

Hence, maximum value of $f(x)$ is 6.

39. (2) $f(x) = \begin{vmatrix} 1 + a^2x & (1 + b^2)x & (1 + c^2)x \\ (1 + a^2)x & 1 + b^2x & (1 + c^2)x \\ (1 + a^2)x & (1 + b^2)x & 1 + c^2x \end{vmatrix}$

On applying $C_1 \to C_1 + C_2 + C_3$

$$= \begin{vmatrix} 1 + 2x + (a^2 + b^2 + c^2)x & (1 + b^2)x & (1 + c^2)x \\ 1 + 2x + (a^2 + b^2 + c^2)x & 1 + b^2x & (1 + c^2)x \\ 1 + 2x + (a^2 + b^2 + c^2)x & (1 + b^2)x & 1 + c^2x \end{vmatrix}$$

$$= \begin{vmatrix} 1 + 2x - 2x & (1 + b^2)x & (1 + c^2)x \\ 1 + 2x - 2x & 1 + b^2x & (1 + c^2)x \\ 1 + 2x - 2x & (1 + b^2)x & 1 + c^2x \end{vmatrix}$$

$$= \begin{vmatrix} 1 & (1 + b^2)x & (1 + c^2)x \\ 1 & 1 + b^2x & (1 + c^2)x \\ 1 & (1 + b^2)x & 1 + c^2x \end{vmatrix}$$

On applying $R_2 \to R_2 - R_1$ and $R_3 \to R_3 - R_1$

$$= \begin{vmatrix} 1 & (1 + b^2)x & (1 + c^2)x \\ 0 & 1 - x & 0 \\ 0 & 0 & 1 - x \end{vmatrix}$$

$= (1 - x)^2$ {Polynomial of degree 2}

40. (4) $\because \frac{5}{3}|A^2BC^{-1}| = \frac{5}{3}|A \cdot A \cdot B \cdot C^{-1}|$

$$= \frac{5}{3}|A|\,|A|\,|B|\,|C^{-1}|$$

$$= \frac{5}{3}|A|^2|B| \times \frac{1}{|C|}$$

$$= \frac{5}{3} \times 4 \times 3 \times \frac{1}{5}$$

$$= 4$$

DPP-4 **Adjoint and Inverse of a Matrix**

1. (c) $(\text{adj } A^T) - (\text{adj } A)^T = 0$ $\quad [\because \text{adj}(A^T) = (\text{adj } A)^T]$

2. (b) $\text{adj}(AB) = (\text{adj} B) \cdot (\text{adj } A)$

3. (b) $\because A = \begin{bmatrix} 4 & 2 \\ 3 & 3 \end{bmatrix}$

$\Rightarrow |A| = 12 - 6 = 6$

$$\text{adj}(\text{adj } A) = |A|^{n-2}A = |A|^{2-2}A \quad [\because n = 2]$$

$$= 6^0\begin{bmatrix} 4 & 2 \\ 3 & 3 \end{bmatrix} \quad [\because 6^0 = 1]$$

$$= \begin{bmatrix} 4 & 2 \\ 3 & 3 \end{bmatrix}$$

4. (a) $|\text{adj } A| = |A|^{n-1}$

5. (a) $\text{adj}(A^T) = B$ $\quad \ldots$(i)

$(\text{adj } (A^T))^T = B^T$

$\Rightarrow (\text{adj } (A))^T = B^T$ $\quad [\because A^T = A]$

$\Rightarrow \text{adj}(A^T) = B^T$ $\quad \ldots$(ii)

From Eqs. (i) and (ii), we get

$\therefore B = B^T$

6. (a) Adjoint of the matrix, $\begin{bmatrix} 1 & -2 & -3 \\ 0 & 2 & -1 \\ -4 & 5 & 2 \end{bmatrix}$ is

$$= \begin{bmatrix} (4 + 5) & -(-4 - 15) & (2 - 6) \\ -(-4) & (2 + 12) & -(-1) \\ (8) & -(5 - 8) & 2 \end{bmatrix} = \begin{bmatrix} 9 & 19 & -4 \\ 4 & 14 & 1 \\ 8 & 3 & 2 \end{bmatrix}$$

7. (d) $(A^T)^{-1} = (A^{-1})^T$

$$(A^3)^{-1} = (A^{-1})^3$$

$$(\text{adj } A)^{-1} = \text{adj}(A^{-1})$$

$\therefore$ All are correct.

8. (a) Let $A\begin{bmatrix} 4 & 7 \\ 1 & 2 \end{bmatrix}$

$\therefore A^{-1} = \frac{1}{|A|}\text{adj}(A) = \frac{1}{8 - 7}\begin{bmatrix} 2 & -7 \\ -1 & 4 \end{bmatrix} = \begin{bmatrix} 2 & -7 \\ -1 & 4 \end{bmatrix}$

9. (a) Given, $A = \begin{bmatrix} 1 & 2 \\ -4 & -1 \end{bmatrix}$

$\Rightarrow A^{-1} = \frac{1}{|A|}\text{adj}(A) = \frac{1}{(-1 + 8)}\begin{bmatrix} 1 & 2 \\ -4 & -1 \end{bmatrix} = \frac{1}{7}\begin{bmatrix} 1 & 2 \\ -4 & -1 \end{bmatrix}$

10. (b) $(AB)^{-1} = B^{-1}A^{-1}$

11. (b) $\because A = \begin{bmatrix} 2 & 0 & 0 \\ 0 & 2 & 0 \\ 0 & 0 & 2 \end{bmatrix}$

$\therefore A^{-1} = \frac{1}{|A|}\text{adj}(A) = \frac{1}{8}\begin{bmatrix} 4 & 0 & 0 \\ 0 & 4 & 0 \\ 0 & 0 & 4 \end{bmatrix} = \begin{bmatrix} 1/2 & 0 & 0 \\ 0 & 1/2 & 0 \\ 0 & 0 & 1/2 \end{bmatrix}$

12. (c) Let $A = \begin{bmatrix} 3 & -2 & -1 \\ -4 & 1 & -1 \\ 2 & 0 & 1 \end{bmatrix}$

$\therefore A^{-1} = \frac{1}{|A|}\text{adj}(A)$

Now, $|A| = 3 \times 1 + (-2)[-2 + 4] - 1 \times (-2) = 3 - 4 + 2 = 1$

Then, $\text{adj}(A) = \begin{bmatrix} 1 & -(-2) & (2 + 1) \\ -(-4 + 2) & (3 + 2) & -(-3 - 4) \\ -2 & -(4) & (3 - 8) \end{bmatrix}$

$$= \begin{bmatrix} 1 & 2 & 3 \\ 2 & 5 & 7 \\ -2 & -4 & -5 \end{bmatrix}$$

$\therefore A^{-1} = \begin{bmatrix} 1 & 2 & 3 \\ 2 & 5 & 7 \\ -2 & -4 & -5 \end{bmatrix}$

13. (d) $A = \begin{bmatrix} 1 & 0 & -k \\ 2 & 1 & 3 \\ k & 0 & 1 \end{bmatrix}$

A is invertible, when $|A| \neq 0$.

$\Rightarrow 1\,(1) - k\,(-k) \neq 0$

$\Rightarrow 1 + k^2 \neq 0 \Rightarrow k^2 \neq -1$

i.e. A is invertible for all real values of k.

14. (a) Given, $N = \begin{bmatrix} -4 & -3 & -3 \\ 1 & 0 & 1 \\ 4 & 4 & 3 \end{bmatrix}$

$$\text{adj}(N) = \begin{bmatrix} (-4) & -(-9+12) & -3 \\ -(3-4) & (-12+12) & -(-4+3) \\ 4 & -(-16+12) & 3 \end{bmatrix}$$

$$= \begin{bmatrix} -4 & -3 & -3 \\ 1 & 0 & 1 \\ 4 & 4 & 3 \end{bmatrix}$$

$$= N$$

15. (d) Given, $B = |A|\, A^{-1} \Rightarrow B = -5A^{-1}$

$$\Rightarrow \quad |B| = |-5A^{-1}|$$

$$\Rightarrow \quad |B| = (-5)^3\, |A^{-1}|$$

[$\because A^{-1}$ is a square matrix of order 3]

$$\Rightarrow \quad |B| = \frac{-125}{|A|}$$

$$\Rightarrow \quad |B| = \frac{-125}{-5} = 25$$

16. (b) $\because A^3 + 2A^2 + 3A + 5I = 0$

17. (b) $KI = \begin{bmatrix} K & 0 & 0 \\ 0 & K & 0 \\ 0 & 0 & K \end{bmatrix}$

$$\text{adj}(KI) = \begin{bmatrix} K^2 & 0 & 0 \\ 0 & K^2 & 0 \\ 0 & 0 & K^2 \end{bmatrix}$$

$$= K^2 \begin{bmatrix} 1 & 0 & 0 \\ 0 & 1 & 0 \\ 0 & 0 & 1 \end{bmatrix} = K^2 I$$

18. (c) $\because \quad (A+B)(A-B) = A^2 - B^2$

$$\Rightarrow \quad A^2 - AB + BA - B^2 = A^2 - B^2$$

$$\Rightarrow \quad -AB + BA = 0$$

$$\Rightarrow \quad AB = BA$$

Now, $\quad (ABA^{-1})^2 = (BAA^{-1})^2 = (B \cdot I)^2 = (B)^2$

19. (d) Adjoint of a symmetric matrix is symmetric matrix. Adjoint of a unit matrix is a unit matrix.

$$A(\text{adj}\, A) = (\text{adj}\, A)\, A = |A|\, I$$

Adjoint of a diagonal matrix is a diagonal matrix.

20. (c) Non-singular matrix has a unique inverse. Determinant of a non-singular matrix is non-zero.

If $A^T = A$. Then, A is a square matrix.

If $|A| \neq 0$

Then, $\quad |\text{adj}\, A| = |A|^{n-1}$

21. (a) Suppose

$A = \begin{bmatrix} 0 & -3 & 1 \\ 3 & 0 & 0 \\ -1 & 0 & 0 \end{bmatrix}$ which is a skew-symmetric matrix.

Then, $\quad |A| = (0) - 3(-2) + (-6)$

$$= 0 + 6 - 6 = 0$$

Hence, A is a singular matrix.

22. (a) $\because |A| \neq 0$

and $\quad |B| \neq 0$

So, $\quad |AB| = |A|\,|B| \neq 0$

23. (d) $A = \begin{bmatrix} 1 & -1 & 1 \\ 2 & 1 & -3 \\ 1 & 1 & 1 \end{bmatrix}$

$$A^{-1} = \frac{1}{|A|}\text{adj}(A)$$

$$= \frac{1}{4+5+1}\begin{bmatrix} 4 & -(-2) & 2 \\ -5 & 0 & -(-5) \\ 1 & -2 & 3 \end{bmatrix}$$

$$= \frac{1}{10}\begin{bmatrix} 4 & 2 & 2 \\ -5 & 0 & 5 \\ 1 & -2 & 3 \end{bmatrix}$$

$$\because \quad 10B = \begin{bmatrix} 4 & 2 & 2 \\ -5 & 0 & \alpha \\ 1 & -2 & 3 \end{bmatrix}$$

$$\Rightarrow \quad \frac{10}{10}\begin{bmatrix} 4 & 2 & 2 \\ -5 & 0 & 5 \\ 1 & -2 & 3 \end{bmatrix} = \begin{bmatrix} 4 & 2 & 2 \\ -5 & 0 & \alpha \\ 1 & -2 & 3 \end{bmatrix} \quad [\because B = A^{-1}]$$

$$\therefore \quad \alpha = 5$$

24. (c) $A = \begin{bmatrix} 1 & 0 & 0 \\ 0 & 1 & 1 \\ 0 & -2 & 4 \end{bmatrix}$

$$A^{-1} \frac{1}{|A|}\text{adj}(A)$$

$$= \frac{1}{6}\begin{bmatrix} 6 & -(0) & 0 \\ -(0) & 4 & -1 \\ 0 & -(-2) & 1 \end{bmatrix}$$

$$\Rightarrow \quad 6A^{-1} = \begin{bmatrix} 6 & 0 & 0 \\ 0 & 4 & -1 \\ 0 & 2 & 1 \end{bmatrix} \quad \ldots\text{(i)}$$

Now, $\quad A^2 = \begin{bmatrix} 1 & 0 & 0 \\ 0 & 1 & 1 \\ 0 & -2 & 4 \end{bmatrix}\begin{bmatrix} 1 & 0 & 0 \\ 0 & 1 & 1 \\ 0 & -2 & 4 \end{bmatrix}$

$$= \begin{bmatrix} 1 & 0 & 0 \\ 0 & -1 & 5 \\ 0 & -10 & 14 \end{bmatrix} \quad \ldots\text{(ii)}$$

$$\because \quad A^{-1} = \frac{1}{6}(A^2 + cA + dI) \quad \text{[given]}$$

$$\Rightarrow \quad 6A^{-1} = A^2 + cA + dI$$

From Eqs. (i) and (ii), we get

$$\Rightarrow \begin{bmatrix} 6 & 0 & 0 \\ 0 & 4 & -1 \\ 0 & 2 & 1 \end{bmatrix} = \begin{bmatrix} 1 & 0 & 0 \\ 0 & -1 & 5 \\ 0 & -10 & 14 \end{bmatrix} + \begin{bmatrix} c & 0 & 0 \\ 0 & c & c \\ 0 & -2c & 4c \end{bmatrix} + \begin{bmatrix} d & 0 & 0 \\ 0 & d & 0 \\ 0 & 0 & d \end{bmatrix}$$

$$\Rightarrow \begin{bmatrix} 6 & 0 & 0 \\ 0 & 4 & -1 \\ 0 & 2 & 1 \end{bmatrix} = \begin{bmatrix} 1+c+d & 0 & 0 \\ 0 & -1+c+d & 5+c \\ 0 & -10-2c & 14+4c+d \end{bmatrix}$$

$$\Rightarrow \quad c + d = 5$$

and $\quad 5 + c = -1$

$$\Rightarrow \quad c = -6$$

and $\quad d = 11$

25. (a) $\because A^2 - A + I = 0$

$$A^{-1} \cdot A \cdot A - A^{-1} \cdot A + A^{-1} \cdot I = 0$$

$$\Rightarrow \quad I \cdot A - I + A^{-1} = 0$$

$$\therefore \quad A^{-1} = I - A$$

26. (a,d) $\text{adj} P = \begin{bmatrix} 1 & 4 & 4 \\ 2 & 1 & 7 \\ 1 & 1 & 3 \end{bmatrix}$

$$|\text{adj} P| = 1(3-7) + 4(7-6) + 4(2-1)$$
$$= -4 + 4 + 4 = 4$$

$\because \quad |\text{adj } P| = |P|^2$

$\Rightarrow \quad |P|^2 = 4$

$\therefore \quad |P| = 2 \quad \text{or} \quad |P| = -2$

27. (c) Given, $A = \begin{bmatrix} a & b \\ c & d \end{bmatrix}$

$$A^2 = \begin{bmatrix} a & b \\ c & d \end{bmatrix}\begin{bmatrix} a & b \\ c & d \end{bmatrix}$$
$$= \begin{bmatrix} a^2 + bc & ab + bd \\ ac + cd & bc + d^2 \end{bmatrix}$$

$\because \quad A^2 = I$

$\Rightarrow \quad \begin{bmatrix} a^2 + bc & b(a+d) \\ c(a+d) & bc + d^2 \end{bmatrix} = \begin{bmatrix} 1 & 0 \\ 0 & 1 \end{bmatrix}$

$\Rightarrow \quad a + d = 0 \quad \begin{bmatrix} \text{if } b \neq 0 \text{ and } c \neq 0 \\ \text{i.e. } A \neq I \text{ and } A \neq -I \end{bmatrix}$

$\Rightarrow \quad d = -a$

That means, $A = \begin{bmatrix} a & b \\ c & -a \end{bmatrix}$

$$tr(A) = 0$$

Now, $|A| = -a^2 - bc = -a^2 - (1 - a^2) \quad [\because a^2 + bc = 1]$

$$= -a^2 - 1 + a^2 = -1$$

28. (b) $\because \text{adj}(\text{adj } A) = |A|^{n-2} A = |A|^{2-2} A \quad [\because n = 2]$

$$= |A|^0 A = A$$

and $|\text{adj } A| = |A|^{n-1} = |A|^{2-1} \quad [\because n = 2]$

$$= |A|$$

Both statements are correct but Statement II is not the correct explanation of Statement I.

29. (c) Determinant of a skew-symmetric matrix of odd order is zero and of even order is a non-zero perfect square.

For any matrix A,

$$|A^T| = A$$

and $\quad |-A| = |A| \quad$ [when A is of even order]

and $\quad |-A| = -|A| \quad$ [when A is of odd order]

$\therefore$ Statement II is incorrect.

30. (d) $A = \begin{bmatrix} \alpha & 0 \\ 0 & \beta \end{bmatrix}$ and $B = \begin{bmatrix} 0 & \gamma \\ \gamma & 0 \end{bmatrix}$

$$AB = \begin{bmatrix} \alpha & 0 \\ 0 & \beta \end{bmatrix}\begin{bmatrix} 0 & \gamma \\ \gamma & 0 \end{bmatrix} = \begin{bmatrix} 0 & \alpha\gamma \\ \beta\gamma & 0 \end{bmatrix}$$
$$BA = \begin{bmatrix} 0 & \gamma \\ \gamma & 0 \end{bmatrix}\begin{bmatrix} \alpha & 0 \\ 0 & \beta \end{bmatrix} = \begin{bmatrix} 0 & \gamma\beta \\ \gamma\alpha & 0 \end{bmatrix}$$
$$AB - BA = \begin{bmatrix} 0 & \alpha\gamma \\ \beta\gamma & 0 \end{bmatrix} - \begin{bmatrix} 0 & \gamma\beta \\ \gamma\alpha & 0 \end{bmatrix}$$
$$= \begin{bmatrix} 0 & \gamma(\alpha - \beta) \\ \gamma(\beta - \alpha) & 0 \end{bmatrix} = \gamma(\alpha - \beta)\begin{bmatrix} 0 & 1 \\ -1 & 0 \end{bmatrix}$$

$$|AB - BA| = \gamma(\alpha - \beta)$$

Also, $AB - BA$ is never be an identity matrix.

Case I If $\alpha = \beta$, then $AB - BA$ is a singular matrix.

Case II If $\alpha \neq \beta$ and α, β, γ are non-zero.

Then, $AB - BA$ is a non-singular matrix.

Statement I is incorrect and Statement II is correct.

31. (a) Here, $B = \begin{bmatrix} x & c & -b \\ -c & x & a \\ b & -a & x \end{bmatrix}$

$$\text{adj}(B) = \begin{bmatrix} a^2 + x^2 & ab - cx & ac + bc \\ ab + cx & b^2 + x^2 & bc - ax \\ ac - bx & bc + ax & c^2 + x^2 \end{bmatrix}^T$$

$\therefore \quad A^T = \text{adj } B$

$\Rightarrow \quad |A^T| = |A| = |\text{adj } B| = |B^2|$

$\therefore$ Both Statements are correct and Statement II is correct explanation of Statement I.

32. (a) $\text{adj}(\text{adj } A)$

$$= |A|^{n-2} A = |A|^{3-2} A = |A| A$$

Given, $A = \begin{bmatrix} 1 & 2 & 3 \\ 2 & 3 & 4 \\ 5 & 6 & 8 \end{bmatrix}$

$\therefore \quad |A| = (24 - 24) + 2(20 - 16) + 3(12 - 15)$

$$= 0 + 2(4) + 3(-3) = 8 - 9 = -1$$

So, $\text{adj}(\text{adj } A) = (-1)A = -A$

33. (b) $|\text{adj}(\text{adj } A)| = |A|^{(n-1)^2} = |A|^{(3-1)^2} = |A|^4$

$$= (-1)^4 = 1$$

34. (b) $B = \begin{bmatrix} 3 & 2 & 5 \\ 2 & 3 & 8 \\ 7 & 2 & 9 \end{bmatrix}$

$$|B| = 3(27 - 16) + 2(56 - 18) + 5(4 - 21)$$
$$= 3(11) + 2(38) + 5(-17)$$
$$= 33 + 76 - 85 = 33 - 9 = 24$$

Now, $|\text{adj} B| = |B|^{n-1} = |B|^{3-1} = |B|^2 = (24)^2 = 576$

35. (b) $\because A = \begin{bmatrix} 1 & 2 & 3 \\ 2 & 3 & 4 \\ 5 & 6 & 8 \end{bmatrix}$

$$B = \begin{bmatrix} 3 & 2 & 5 \\ 2 & 3 & 8 \\ 7 & 2 & 9 \end{bmatrix}$$

$\therefore \quad AB = \begin{bmatrix} 1 & 2 & 3 \\ 2 & 3 & 4 \\ 5 & 6 & 8 \end{bmatrix}\begin{bmatrix} 3 & 2 & 5 \\ 2 & 3 & 8 \\ 7 & 2 & 9 \end{bmatrix}$

$$= \begin{bmatrix} 3 + 4 + 21 & 2 + 6 + 6 & 5 + 16 + 27 \\ 6 + 6 + 28 & 4 + 9 + 8 & 10 + 24 + 36 \\ 15 + 12 + 56 & 10 + 18 + 16 & 25 + 48 + 72 \end{bmatrix}$$
$$= \begin{bmatrix} 28 & 14 & 48 \\ 40 & 21 & 70 \\ 83 & 44 & 145 \end{bmatrix}$$

Now, $|AB| = 28[3045 - 3080] + 14[5810 - 5800] + 48[1760 - 1743]$

$$= 28 \times (-35) + 14 \times 10 + 48 \times 17$$
$$= -980 + 140 + 816$$
$$= -24$$

$\therefore \quad |\text{adj}(AB)| = |AB|^{n-1} = |AB|^{3-1}$

$$= (-24)^2 = (24)^2 = 576$$

36. (b) $\because |A - \lambda I| = 0$

If $\quad \lambda = 0$

Then, $\quad |A| = 0$

37. (d) Given, $AA' = I$

and $|A| = 1$

Suppose $A\begin{bmatrix}1 & 0\\0 & 1\end{bmatrix} = I$

$\Rightarrow \quad |A - I| = 0$

If $A = \begin{bmatrix}-1 & 0\\0 & -1\end{bmatrix}$

$\Rightarrow \quad A - I = \begin{bmatrix}-2 & 0\\0 & -2\end{bmatrix}$

$\therefore \quad |A - I| = 4$

38. (9) Given, $A^2 = A$

and $(I + A)^{10} = I + kA$, where $k = 2\cdot 2^{\lambda} - 1$

Now, $(I + A)^{10} = I + A + \dfrac{A^2}{2!} + \dfrac{A^3}{3!} + \cdots + \dfrac{A^{10}}{10!}$

$= I + A + \dfrac{A}{2!} + \dfrac{A}{3!} + \cdots + \dfrac{A}{10!}$

$= I + A\left(1 + \dfrac{1}{2!} + \dfrac{1}{3!} + \cdots + \dfrac{1}{10!}\right)$

$= I + A\,(1 + 1)^{10}$

$= I + A \times (2)^{10}$

$= I + kA$ (given)

$k = 2^{\lambda + 1} - 1$

$\Rightarrow \quad k = 2^{10}$

$\Rightarrow \quad 2^{\lambda + 1} - 1 = 2^{10}$

$\Rightarrow \quad 2^{\lambda + 1 - 10} = 1$

$\Rightarrow \quad \lambda + 1 - 10 = 0$

$\Rightarrow \quad \lambda = 9$

39. (10) Given, $A(\text{adj}\,A) = \begin{bmatrix}10 & 0\\0 & 10\end{bmatrix}$

$\Rightarrow \quad |A(\text{adj}\,A)| = 100$

$\Rightarrow \quad |A||\text{adj}\,A| = 100$

$\Rightarrow \quad |A|\,|A|^{2-1} = 100$

$\Rightarrow \quad |A|\cdot|A| = 100 \quad \Rightarrow \quad |A|^2 = 100$

$\therefore \quad |A| = 10$

40. (4) Given, $A = \begin{bmatrix}2k-1 & 2\sqrt{k} & 2\sqrt{k}\\ 2\sqrt{k} & 1 & -2k\\ -2\sqrt{k} & 2k & -1\end{bmatrix}$

$\Rightarrow |A| = (2K - 1)\,[-1 + 4k^2] + 2\sqrt{k}\,[4K\sqrt{k} + 2\sqrt{k}] + 2\sqrt{k}\,[4k\sqrt{k} + 2\sqrt{k}]$

$= (2k - 1)(-1 + 4k^2) + 4k(2k + 1) + 4k\,[2k + 1]$

$= -2k + 8k^3 + 1 - 4k^2 + 16k^2 + 8k$

$= 8k^3 + 12k^2 + 6k + 1 = (2k + 1)^3$

and $B = \begin{bmatrix}0 & 2k-1 & \sqrt{k}\\ 1-2k & 0 & 2\sqrt{k}\\ -\sqrt{k} & -2\sqrt{k} & 0\end{bmatrix}$

$\Rightarrow |B| = 0$

[$\because B$ is a skew-symmetric matrix of odd order]

Now, $|\text{adj}\,A| + |\text{adj}\,B| = |A|^{3-1} + |B|^{3-1}$

$= |A|^2 + |B|^2 = (2k + 1)^6$

Given, $|\text{adj}\,A| + |\text{adj}\,B| = 10^6$

$\Rightarrow \quad (2k + 1)^6 = 10^6$

$\Rightarrow \quad 2k + 1 = 10$

$\Rightarrow \quad 2k = 9 \Rightarrow k = 4\cdot 5$

$[k] = 4$

DPP-5 Solving of Equations

1. (a) Given, $x + 2y + 3z = 1, \quad 2x + y + 3z = 2$

$5x + 5y + 9z = 4$

Now, $\begin{vmatrix}1 & 2 & 3\\2 & 1 & 3\\5 & 5 & 9\end{vmatrix}$

$= 1(9 - 15) + 2(15 - 18) + 3(10 - 5)$

$= -6 + 2(-3) + 3(5) = -6 - 6 + 15$

$= (-12) + 15 = -3 \neq 0$

Hence, the system of equations has unique solution.

2. (a) Given, $x_2 - x_3 = 1$

$-x_1 + 2x_3 = -2$

$x_1 - 2x_2 = 3$

Now, $[A : B] = \left|\begin{array}{ccc:c}0 & 1 & -1 & 1\\-1 & 0 & 2 & -2\\1 & -2 & 0 & 3\end{array}\right|$

On applying $R_3 \to R_3 + R_2$

$= \left|\begin{array}{ccc:c}0 & 1 & -1 & 1\\-1 & 0 & 2 & -2\\0 & -2 & 2 & 1\end{array}\right|$

On applying $R_3 \to R_3 + 2R_1$

$= \left|\begin{array}{ccc:c}0 & 1 & -1 & 1\\-1 & 0 & 2 & -2\\0 & 0 & 0 & 3\end{array}\right|$

Then, Rank $A = 2$ and Rank $AB = 3$

Hence, Rank $A \neq$ Rank AB

Therefore, the system of equations has no solution.

3. (d) Given, $x - y + z = 2, 2x + y - z = 5$

$4x - y + z = 10$

$|A| = \begin{vmatrix}1 & -1 & 1\\2 & 1 & -1\\4 & -1 & 1\end{vmatrix}$

$(1)(1 - 1) - 1(-4 - 2) + 1(-2 - 4) = 6 - 6 = 0$

Now, $[A : B] = \left|\begin{array}{ccc:c}1 & -1 & 1 & 2\\2 & 1 & -1 & 5\\4 & -1 & 1 & 10\end{array}\right|$

On applying $R_3 \to R_3 - 4R_1$ and $R_2 \to R_2 - 2R_1$

$= \left|\begin{array}{ccc:c}1 & -1 & 1 & 2\\0 & 3 & -3 & 1\\0 & 3 & -3 & 2\end{array}\right|$

On applying $R_3 \to R_3 - R_2$

$= \left|\begin{array}{ccc:c}1 & -1 & 1 & 2\\0 & 3 & -3 & 1\\0 & 0 & 0 & 1\end{array}\right|$

That means, Rank $A \neq$ Rank AB.

Hence, the system of equations has no solution.

4. (d) Given, $3x - 2y + z = 0$, $\lambda x - 14y + 15z = 0$,

$x + 2y - 3z = 0$

The system of equations has a solution other than $x = y = z = 0$, when

$$\begin{vmatrix} 3 & -2 & 1 \\ \lambda & -14 & 15 \\ 1 & 2 & -3 \end{vmatrix} = 0$$

$$\Rightarrow \quad 3(42 - 30) - 2(15 + 3\lambda) + (2\lambda + 14) = 0$$
$$\Rightarrow \quad 3 \times 12 - 30 - 6\lambda + 2\lambda + 14 = 0$$
$$\Rightarrow \quad -4\lambda + 20 = 0$$
$$\therefore \quad \lambda = 5$$

5. (a) $2kx - 2y + 3z = 0$

$x + ky + 2z = 0$

$2x + kz = 0$

The system of equations has a non-trivial solution, when

$$\begin{vmatrix} 2k & -2 & 3 \\ 1 & k & 2 \\ 2 & 0 & k \end{vmatrix} = 0$$

$$\Rightarrow \quad 2k(k^2) - 2(4 - k) + 3(-2k) = 0$$
$$\Rightarrow \quad 2k^3 - 8 + 2k - 6k = 0$$
$$\Rightarrow \quad 2k^3 - 4k - 8 = 0$$
$$\Rightarrow \quad k^3 - 2k - 4 = 0$$
$$\Rightarrow \quad (k - 2)(k^2 + 2k + 2) = 0$$
$$\therefore \quad k = 2$$

6. (c) Given, $x + ky - z = 0$

$3x - ky - z = 0$

$x - 3y + z = 0$

The system of equations has a non-zero solution, when

$$\begin{vmatrix} 1 & k & -1 \\ 3 & -k & -1 \\ 1 & -3 & 1 \end{vmatrix} = 0$$

$$\Rightarrow \quad 1(-k - 3) + k(-1 - 3) - 1(-9 + k) = 0$$
$$\Rightarrow \quad -k - 3 - 4k + 9 - k = 0$$
$$\Rightarrow \quad -6k + 6 = 0$$
$$\therefore \quad k = 1$$

7. (c) Given, $\alpha x + y + z = \alpha - 1$, $x + \alpha y + z = \alpha - 1$

$x + y + \alpha z = \alpha - 1$

Now, $$[A : B] = \left[\begin{array}{ccc:c} \alpha & 1 & 1 & \alpha - 1 \\ 1 & \alpha & 1 & \alpha - 1 \\ 1 & 1 & \alpha & \alpha - 1 \end{array}\right]$$

On applying $R_2 \to \alpha R_2 - R_1$ and $R_3 \to \alpha R_3 - R_1$

$$= \left[\begin{array}{ccc:c} \alpha & 1 & 1 & \alpha - 1 \\ 0 & \alpha^2 - 1 & \alpha - 1 & \alpha(\alpha - 1) - (\alpha - 1) \\ 0 & \alpha - 1 & \alpha^2 - 1 & \alpha(\alpha - 1) - (\alpha - 1) \end{array}\right]$$

On applying $R_3 \to (\alpha + 1)R_3 - R_2$

$$= \left[\begin{array}{ccc:c} \alpha & 1 & 1 & \alpha - 1 \\ 0 & \alpha^2 - 1 & \alpha - 1 & (\alpha - 1)^2 \\ 0 & 0 & (\alpha + 1)(\alpha^2 - 1) - (\alpha - 1) & (\alpha + 1) - (\alpha - 1)^2 \end{array}\right]$$

$$= \left[\begin{array}{ccc:c} \alpha & 1 & 1 & \alpha - 1 \\ 0 & \alpha^2 - 1 & \alpha - 1 & (\alpha - 1)^2 \\ 0 & 0 & (\alpha - 1)(\alpha^2 + 2\alpha) & (\alpha + 1) - (\alpha - 1)^2 \end{array}\right]$$

Rank $(A) \neq$ Rank (AB), when $\alpha = -2$.

Hence, the system of equations has no solution, when $\alpha = -2$.

8. (b) We have,

$x + 4y - z = 0$

$3x - 4y - z = 0$

$x - 3y + z = 0$

Now, $$\Delta = \begin{vmatrix} 1 & 4 & -1 \\ 3 & -4 & -1 \\ 1 & -3 & 1 \end{vmatrix}$$

$$= 1(-4 - 3) + 4(-1 - 3) - 1(-9 + 4)$$
$$= -7 - 16 + 5$$
$$= -23 + 5$$
$$= -18 \neq 0$$

$\because \Delta$ is not equal to zero.

So, the system of equations has unique solution.

9. (d) The given system of equation is homogeneous.

According to given condition, the system has infinitely many solutions.

10. (d) We have,

$x + ky + 3z = 0$

$3x + ky - 2z = 0$

$2x + 3y - 4z = 0$

As, the given system of equations has non-trivial solution.

$$\begin{vmatrix} 1 & k & 3 \\ 3 & k & -2 \\ 2 & 3 & -4 \end{vmatrix} = 0$$

$$\Rightarrow \quad 1(-4k + 6) + k(-4 + 12) + 3(9 - 2k) = 0$$
$$\Rightarrow \quad -4k + 6 + 8k + 27 - 6k = 0$$
$$\Rightarrow \quad -2k + 33 = 0 \Rightarrow k = \frac{33}{2}$$

11. (b) We have, $a + b - 2c = 0$, $2a - 3b + c = 0$

$a - 5b + 4c = \alpha$

Now, $$[A : B] = \left[\begin{array}{ccc:c} 1 & 1 & -2 & 0 \\ 2 & -3 & 1 & 0 \\ 1 & -5 & 4 & \alpha \end{array}\right]$$

On applying $R_3 \to R_3 - R_1$ and $R_2 \to R_2 - 2R$

$$= \left[\begin{array}{ccc:c} 1 & 1 & -2 & 0 \\ 0 & -5 & 5 & 0 \\ 0 & -6 & 6 & \alpha \end{array}\right]$$

On applying $R_3 \to \frac{R_3}{6} - \frac{R_2}{5}$

$$= \left[\begin{array}{ccc:c} 1 & 1 & -2 & 0 \\ 0 & -5 & 5 & 0 \\ 0 & 0 & 0 & \frac{\alpha}{6} \end{array}\right]$$

$\therefore$ Rank (A) = Rank (AB), when $\alpha = 0$

Hence, the system of equations is consistent, when $\alpha = 0$.

12. (a) Given, $x + y + z = 2$

$2x + y - z = 3$

$3x + 2y + kz = 4$

The given system of equations has a unique solution, when

$$\begin{vmatrix} 1 & 1 & 1 \\ 2 & 1 & -1 \\ 3 & 2 & k \end{vmatrix} \neq 0$$

$$\Rightarrow \quad 1(k + 2) + (-3 - 2k) + (4 - 3) \neq 0$$
$$\Rightarrow \quad k + 2 - 3 - 2k + 1 \neq 0$$
$$\Rightarrow \quad -k \neq 0$$
$$\therefore \quad k \neq 0$$

13. (b) $(k+1)x + 8y = 4k$

$kx + (k+3)y = 3k - 1$

Now, $[A:B] = \begin{vmatrix} k+1 & 8 & : & 4k \\ k & k+3 & : & 3k-1 \end{vmatrix}$

On applying $R_2 \to (k+1)R_2 - kR_1$

$$= \begin{vmatrix} k+1 & 8 & : & 4k \\ 0 & (k+1)(k+3) - 8k & : & (k+1)(3k-1) - 4k^2 \end{vmatrix}$$

$$= \begin{vmatrix} k+1 & 2 & : & 4k \\ 0 & k^2 - 4k + 3 & : & -k^2 + 2k - 1 \end{vmatrix}$$

$$= \begin{vmatrix} k+1 & 2 & : & 4k \\ 0 & (k-3)(k-1) & : & -(k-1)^2 \end{vmatrix}$$

When $k = 1$

Then, Rank (A) = Rank $(AB) < 2$

Hence, for $k = 1$, the given system of equations has infinitely many solutions.

14. (a) We have, $x + y + z = \lambda$

$5x - y + \mu z = 10$

$2x + 3y - z = 6$

The given system of equations has unique solution when

$$\begin{vmatrix} 1 & 1 & 1 \\ 5 & -1 & \mu \\ 2 & 3 & -1 \end{vmatrix} = 0$$

That means, it depends on μ only.

15. (d) We have, $2x - y - z = 12$

$x - 2y + z = 4$

$x + y + \lambda z = 4$

Now, $[A:B] = \begin{vmatrix} 2 & -1 & -1 & : & 12 \\ 1 & -2 & 1 & : & 4 \\ 1 & 1 & \lambda & : & 4 \end{vmatrix}$

On applying $R_3 \to 2R_3 - R_1$ and $R_2 \to 2R_2 - R_1$

$$= \begin{vmatrix} 2 & -1 & -1 & : & 12 \\ 0 & -3 & 3 & : & -4 \\ 0 & 3 & 2\lambda + 1 & : & -4 \end{vmatrix}$$

On applying $R_3 \to R_3 + R_2$

$$= \begin{vmatrix} 2 & -1 & -1 & : & 12 \\ 0 & -3 & 3 & : & -4 \\ 0 & 0 & 2\lambda + 4 & : & -8 \end{vmatrix}$$

For $\lambda = -2$,

Rank $A \neq$ Rank (AB)

i.e. for $\lambda = -2$, the given system of equations has no solution.

16. (c) $M^2N^2(M^TN)^{-1}(MN^{-1})^T$

$= M^2N^2N^{-1}(M^T)^{-1} \cdot (N^{-1})^T M^T$

$= M^2 \cdot N \cdot (-M)^{-1} \cdot (-N)^{-1}(-M) = -M^2 \cdot N \cdot (NM)^{-1} \cdot M$

$= -M^2 \cdot N \cdot (MN)^{-1} \cdot M = -M^2 \cdot N \cdot N^{-1} \cdot M^{-1} \cdot M = -M^2$

17. (d) $Q = \begin{bmatrix} 2^2a_{11} & 2^3a_{12} & 2^4a_{13} \\ 2^3a_{21} & 2^4a_{22} & 2^5a_{23} \\ 2^4a_{31} & 2^5a_{32} & 2^6a_{33} \end{bmatrix}$

$$|Q| = 2^2 \cdot 2^3 \cdot 2^4 \begin{vmatrix} a_{11} & 2a_{12} & 2^2a_{13} \\ a_{21} & 2a_{22} & 2^2a_{23} \\ a_{31} & 2a_{32} & 2^2a_{33} \end{vmatrix}$$

$$= 2^9 \cdot 2 \cdot 2^2 \begin{vmatrix} a_{11} & a_{12} & a_{13} \\ a_{21} & a_{22} & a_{23} \\ a_{31} & a_{32} & a_{33} \end{vmatrix}$$

$= 2^{12} \cdot 2 = 2^{13}$

18. (d) Given, $P^T = 2P + I \Rightarrow P = 2P^T + I$

$\Rightarrow \quad P = 2(2P + I) + I$

$\Rightarrow \quad P = 4P + 3I$

$\Rightarrow \quad 3(P + I) = 0$

$\Rightarrow \quad P + I = 0$

$\Rightarrow \quad PX + X = 0$

$\therefore \quad PX = -X$

19. (c) For no solution

$$\frac{k+1}{k} = \frac{8}{k+3} \neq \frac{4k}{3k-1}$$

$\Rightarrow \quad (k+1)(k+3) = 8k$

or $\quad k^2 - 4k + 3 = 0$

$\therefore \quad k = 1, 3$

For $k = 1$, $\quad \dfrac{k+1}{k} = \dfrac{8}{k+3} = \dfrac{4k}{3k-1}$

For $k = 3$, $\quad \dfrac{k+1}{k} = \dfrac{8}{k+3} \neq \dfrac{4k}{3k-1}$

$\therefore k = 3$, neglecting $k = 1$.

20. (b,c) If $\alpha = -2$

Then, $\quad -x - 1 = 0$

and $\quad x + y - 1 = 0$

$\Rightarrow \quad x = -1, y = 2$

If $\quad \alpha = 1$

Then, $\quad 8x + 27y - 64 = 0$

$2x + 3y - 4 = 0$

$x + y - 1 = 0$

$$[A:B] = \begin{vmatrix} 1 & 1 & 0 & : & 1 \\ 2 & 3 & 0 & : & 4 \\ 8 & 27 & 0 & : & 64 \end{vmatrix}$$

On applying $R_3 \to R_3 - 8R_1$ and $R_2 \to R_2 - 2R_1$

$$= \begin{vmatrix} 1 & 1 & 0 & : & 1 \\ 0 & 1 & 0 & : & 2 \\ 0 & 19 & 0 & : & 56 \end{vmatrix}$$

On applying $R_3 \to R_3 - 19R_2$

$$= \begin{vmatrix} 1 & 1 & 0 & : & 1 \\ 0 & 1 & 0 & : & 2 \\ 0 & 0 & 0 & : & 16 \end{vmatrix}$$

So, Rank $(A) \neq$ Rank (AB)

So, the system of equations is inconsistent.

Similarly, for $\alpha = 0$ and $\alpha = -3$.

We can prove that the given system of equations is inconsistent.

Explanations for Q. Nos. (21-23)

Given, $\quad A = \begin{bmatrix} 1 & 0 & 0 \\ 2 & 1 & 0 \\ 3 & 2 & 1 \end{bmatrix}$

$$A^{-1} = \frac{1}{1}\begin{bmatrix} 1 & 0 & 0 \\ -2 & 1 & 0 \\ 1 & -2 & 1 \end{bmatrix} = \begin{bmatrix} 1 & 0 & 0 \\ -2 & 1 & 0 \\ 1 & -2 & 1 \end{bmatrix}$$

$\because \quad AU_1 = \begin{bmatrix} 1 \\ 0 \\ 0 \end{bmatrix}$

$$\Rightarrow \quad U_1 = A^{-1}\begin{bmatrix} 1 \\ 0 \\ 0 \end{bmatrix} = \begin{bmatrix} 1 & 0 & 0 \\ -2 & 1 & 0 \\ 1 & -2 & 1 \end{bmatrix}\begin{bmatrix} 1 \\ 0 \\ 0 \end{bmatrix} = \begin{bmatrix} 1 \\ -2 \\ 1 \end{bmatrix}$$

$\because \quad AU_2 = \begin{bmatrix} 2 \\ 3 \\ 0 \end{bmatrix}$

$$\Rightarrow \quad U_2 = A^{-1}\begin{bmatrix}2\\3\\0\end{bmatrix} = \begin{bmatrix}1&0&0\\-2&1&0\\1&-2&1\end{bmatrix}\begin{bmatrix}2\\3\\0\end{bmatrix} = \begin{bmatrix}2\\-4+3\\2-6\end{bmatrix} = \begin{bmatrix}2\\-1\\-4\end{bmatrix}$$

$$\because \quad AU_3 = \begin{bmatrix}2\\3\\1\end{bmatrix}$$

$$\Rightarrow \quad U_3 = A^{-1}\begin{bmatrix}2\\3\\1\end{bmatrix} = \begin{bmatrix}1&0&0\\-2&1&0\\1&-2&1\end{bmatrix}\begin{bmatrix}2\\3\\1\end{bmatrix} = \begin{bmatrix}2\\-4+3\\2-6+1\end{bmatrix} = \begin{bmatrix}2\\-1\\-3\end{bmatrix}$$

So, $$U = \begin{bmatrix}1&2&2\\-2&-1&-1\\1&-4&-3\end{bmatrix}$$

$$|U| = 1(3-4) + 2(-1-6) + 2(8+1)$$
$$= -1 - 14 + 18 = -15 + 18 = 3$$

Now, $$U^{-1} = \frac{1}{3}\begin{bmatrix}-1&-2&0\\-7&-5&-3\\9&6&3\end{bmatrix}$$

Sum of elements of

$$U^{-1} = \frac{1}{3}(-1-2-7-5-3+9+6+3)$$
$$= \frac{1}{3}(-18+18) = 0$$

$$[3\ \ 2\ \ 0]\begin{bmatrix}1&2&2\\-2&-1&-1\\1&-4&-3\end{bmatrix}\begin{bmatrix}3\\2\\0\end{bmatrix} = [3\ \ 2\ \ 0]\begin{bmatrix}7\\-8\\-5\end{bmatrix}$$
$$= [21-16] = [5]$$

21. (a) **22.** (b) **23.** (a)

Explanation for Q. Nos. (24-25)

$$A^2 = \begin{bmatrix}1&2&1\\0&1&-1\\3&-1&1\end{bmatrix}\begin{bmatrix}1&2&1\\0&1&-1\\3&-1&1\end{bmatrix} = \begin{bmatrix}4&3&0\\-3&2&-2\\6&4&5\end{bmatrix}$$

$$A^3 = \begin{bmatrix}4&3&0\\-3&2&-2\\6&4&5\end{bmatrix}\begin{bmatrix}1&2&1\\0&1&-1\\3&-1&1\end{bmatrix} = \begin{bmatrix}4&11&1\\-9&-2&-7\\21&11&7\end{bmatrix}$$

$$A^3 + kA^2 + lA + mI = 0$$

$$\begin{bmatrix}4+4k+l+m & 11+3k+2l & 1+l\\ -9-3k & -2+2k+l+m & -7-2k-l\\ 21+6k+3l & 11+4k & 7+5k+l+m\end{bmatrix} = 0$$

$\Rightarrow \quad l = -1, k = -3$

and $\quad m = 9$

So, $\quad (k+m+l) = -3+9-1 = 5$

$\because \quad A^3 - 3A^2 - A + 9I = 0$

$\Rightarrow \quad A^2 - 3A - I + 9A^{-1} = 0$

$\Rightarrow \quad -9A^{-1} = A^2 - 3A - I$

$\therefore \quad A^{-1} = \frac{-1}{9}(A^2 - 3A - I)$

24. (c) **25.** (c)

26. (d) We have,

$x + y - z = b_1, \ 2x + z = b_2$

and $\quad x - y + az = b_3$

$$\because \quad A = \begin{bmatrix}1&1&-1\\2&0&1\\1&-1&a\end{bmatrix}$$

For $a = 1$, $$A = \begin{bmatrix}1&1&-1\\2&0&1\\1&-1&1\end{bmatrix}$$

$\Rightarrow |A| = 1(1) + 1(1-2) - 1(-2) = 1 - 1 + 2 = 2 \neq 0$

$$A^{-1} = \frac{1}{2}\begin{bmatrix}1&0&1\\-1&2&-3\\-2&2&-2\end{bmatrix} = \begin{bmatrix}1/2&0&1/2\\-1/2&1&-3/2\\-1&1&-1\end{bmatrix}$$

27. (a) $AX = 0$ has non-trivial solution,

when $|A| = 0$.

$$\Rightarrow \quad \begin{vmatrix}1&1&-1\\2&0&1\\1&-1&a\end{vmatrix} = 0$$

$\Rightarrow \quad 1(1) + 1(1-2a) - 1(-2) = 0$

$\Rightarrow \quad 1 + 1 - 2a + 2 = 0$

$\Rightarrow \quad 4a = 2a$

$\therefore \quad a = 2$

28. (a) A can be written as

$$\begin{bmatrix}0&0&1\\0&0&1\\1&1&1\end{bmatrix}\begin{bmatrix}0&1&0\\1&0&1\\0&1&1\end{bmatrix}$$

$$\begin{bmatrix}0&1&1\\1&0&0\\1&0&1\end{bmatrix}\begin{bmatrix}1&0&0\\0&1&1\\0&1&1\end{bmatrix}\begin{bmatrix}1&0&1\\0&1&0\\1&0&1\end{bmatrix}$$

$$\begin{bmatrix}1&1&0\\1&1&0\\0&0&1\end{bmatrix}\begin{bmatrix}1&0&1\\0&0&1\\1&1&0\end{bmatrix}$$

$$\begin{bmatrix}1&1&0\\1&0&1\\0&1&0\end{bmatrix}\begin{bmatrix}1&1&1\\1&0&0\\1&0&0\end{bmatrix}$$

$$\begin{bmatrix}0&0&1\\0&1&1\\1&1&0\end{bmatrix}\begin{bmatrix}0&1&0\\1&1&1\\0&1&0\end{bmatrix}$$

$$\begin{bmatrix}0&1&1\\1&1&0\\1&0&0\end{bmatrix}$$

Alternate Method

Case I All three diagonal elements are 1.

Number of matrices $= {}^3C_1 = 3$.

Case II Two diagonal elements are zero and one element is one.

Number of matrices $= {}^3C_1 \cdot {}^3C_1 = 9$

$\therefore$ Total matrices $= 3 + 9 = 12$

29. (b) Given, $A\begin{bmatrix}x\\y\\z\end{bmatrix} = \begin{bmatrix}1\\0\\0\end{bmatrix}$

For unique solution, $\det(A) \neq 0$.

Case I $$\det(A) = \begin{vmatrix}1&a&b\\a&1&c\\b&c&1\end{vmatrix} = 2abc \neq 0$$

Here, a, b, c is selected from 1, 0, 0. [no case is possible]

Case II

(i) $$\det(A) = \begin{vmatrix}1&a&b\\a&0&c\\b&c&0\end{vmatrix} = 2abc - c^2 \neq 0$$

Here, a, b, c are selected from 1, 1, 0. [2 cases are possible]

(ii) $$\det(A) = \begin{vmatrix}0&a&b\\a&1&c\\b&c&0\end{vmatrix} = 2abc - b^2 \neq 0$$

Here, a, b, c are selected from 1, 1, 0. [2 cases are possible]

(iii) $\det(A) = \begin{vmatrix} 0 & a & b \\ a & 0 & c \\ b & c & 1 \end{vmatrix}$

$$2abc - a^2 \neq 0$$

Here, a, b, c are selected from 1, 1, 0. [2 cases are possible]
Hence, there are exactly 6 matrices for unique solution.

30. (b) Given, $A \begin{bmatrix} x \\ y \\ z \end{bmatrix} = \begin{bmatrix} 1 \\ 0 \\ 0 \end{bmatrix}$

***Case* I** $\begin{bmatrix} 1 & a & b \\ a & 1 & c \\ b & c & 1 \end{bmatrix}\begin{bmatrix} x \\ y \\ z \end{bmatrix} = \begin{bmatrix} 1 \\ 0 \\ 0 \end{bmatrix}$

a, b, c are selected from 1, 0, 0.

$\Rightarrow$
$$x + ay + bz = 1$$
$$ax + y + cz = 0$$
$$bx + cy + z = 0$$

(i) If $a = 1, b = c = 0$, then
$$x + y = 1$$
Inconsistent system of equations
$$x + y = 0$$

(ii) If $a = 0 = c, b = 1$, then $x + z = 1, y = 0$.
Inconsistent system of equations, $x + z = 0$.

(iii) If $c = 1, a = b = 0$, then
$$x = 1, z = 0, y = 0$$

***Case* II**

(i) $\begin{bmatrix} 1 & a & b \\ a & 0 & c \\ b & c & 0 \end{bmatrix}\begin{bmatrix} x \\ y \\ z \end{bmatrix} = \begin{bmatrix} 1 \\ 0 \\ 0 \end{bmatrix}$

a, b, c are selected from 1, 1, 0.

$\Rightarrow$
$$x + ay + bz = 1$$
$$ax + cz = 0$$
$$bx + cy = 0$$

Clearly, in all three cases, solutions are possible, so system is consistent.

(ii) $\begin{bmatrix} 0 & a & b \\ a & 1 & c \\ b & c & 0 \end{bmatrix}\begin{bmatrix} x \\ y \\ z \end{bmatrix} = \begin{bmatrix} 1 \\ 0 \\ 0 \end{bmatrix}$

$\Rightarrow$
$$ay + bz = 1$$
$$ax + y + cz = 0$$
$$bx + cy = 0$$

Clearly, $b = 0, a = c = 1$
$$y = 1$$
$$x + y + z = 0$$

Inconsistent system, $y = 0$
More than 2 matrices are possible.

DPP-6 Rank of Matrix, Differentiation and Product of Determinants

1. We have, $\begin{vmatrix} (a-x)^2 & (a-y)^2 & (a-z)^2 \\ (b-x)^2 & (b-y)^2 & (b-z)^2 \\ (c-x)^2 & (c-y)^2 & (c-z)^2 \end{vmatrix}$

$$= \begin{vmatrix} a^2 - 2ax + x^2 & a^2 - 2ay + y^2 & a^2 - 2az + z^2 \\ b^2 - 2bx + x^2 & b^2 - 2by + y^2 & b^2 - 2bz + z^2 \\ c^2 - 2cx + x^2 & c^2 - 2cy + y^2 & c^2 - 2cz + z^2 \end{vmatrix}$$

$$= \begin{vmatrix} a^2 - 2ax + x^2 & b^2 - 2bx + x^2 & c^2 - 2cx + x^2 \\ a^2 - 2ay + y^2 & b^2 - 2by + y^2 & c^2 - 2cy + y^2 \\ a^2 - 2az + z^2 & b^2 - 2bz + z^2 & c^2 - 2cz + z^2 \end{vmatrix}$$

$$= \begin{vmatrix} 1 & x & x^2 \\ 1 & y & y^2 \\ 1 & z & z^2 \end{vmatrix} \times \begin{vmatrix} a^2 & -2a & 1 \\ b^2 & -2b & 1 \\ c^2 & -2c & 1 \end{vmatrix} = \begin{vmatrix} 1 & x & x^2 \\ 1 & y & y^2 \\ 1 & z & z^2 \end{vmatrix}\begin{vmatrix} a^2 & 2a & 1 \\ b^2 & 2b & 1 \\ c^2 & 2c & 1 \end{vmatrix}$$

$$= 2\begin{vmatrix} 1 & x & x^2 \\ 1 & y & y^2 \\ 1 & z & z^2 \end{vmatrix}\begin{vmatrix} 1 & a & a^2 \\ 1 & b & b^2 \\ 1 & c & c^2 \end{vmatrix}$$

2. $\Delta = \begin{vmatrix} 1 & \cos(\beta - \alpha) & \cos(\gamma - \alpha) \\ \cos(\alpha - \beta) & 1 & \cos(\gamma - \beta) \\ \cos(\alpha - \gamma) & \cos(\beta - \gamma) & 1 \end{vmatrix}$

$$= \begin{vmatrix} 1 & \cos\beta\cos\alpha + \sin\beta\sin\alpha & \cos\alpha\cos\gamma + \sin\alpha\sin\gamma \\ \cos\alpha\cos\beta + \sin\alpha\sin\beta & 1 & \cos\gamma\cos\beta + \sin\gamma\sin\beta \\ \cos\alpha\cos\gamma + \sin\alpha\sin\gamma & \cos\beta\cos\gamma + \sin\beta\sin\gamma & 1 \end{vmatrix}$$

$$= \begin{vmatrix} 0 & \cos\alpha & \sin\alpha \\ 0 & \cos\beta & \sin\beta \\ 0 & \cos\gamma & \sin\gamma \end{vmatrix}\begin{vmatrix} 0 & \cos\alpha & \sin\alpha \\ 0 & \cos\beta & \sin\beta \\ 0 & \cos\gamma & \sin\gamma \end{vmatrix} = 0 \times 0 = 0$$

3. $S_r = \alpha^r + \beta^r + \gamma^r$

$$\begin{vmatrix} S_0 & S_1 & S_2 \\ S_1 & S_2 & S_3 \\ S_2 & S_3 & S_4 \end{vmatrix} = \begin{vmatrix} 3 & \alpha + \beta + \gamma & \alpha^2 + \beta^2 + \gamma^2 \\ \alpha + \beta + \gamma & \alpha^2 + \beta^2 + \gamma^2 & \alpha^3 + \beta^3 + \gamma^3 \\ \alpha^2 + \beta^2 + \gamma^2 & \alpha^3 + \beta^3 + \gamma^3 & \alpha^4 + \beta^4 + \gamma^4 \end{vmatrix}$$

$$= \begin{vmatrix} 1 & 1 & 1 \\ \alpha & \beta & \gamma \\ \alpha^2 & \beta^2 & \gamma^2 \end{vmatrix}\begin{vmatrix} 1 & 1 & 1 \\ \alpha & \beta & \gamma \\ \alpha^2 & \beta^2 & \gamma^2 \end{vmatrix}$$

$$= (\alpha - \beta)((\beta - \gamma)(\gamma - \alpha)(\alpha - \beta)(\beta - \gamma)(\gamma - \alpha)$$
$$= (\alpha - \beta)^2(\beta - \gamma)^2(\gamma - \alpha)^2$$

4. $\begin{vmatrix} 3 & 1 + S_1 & 1 + S_2 \\ 1 + S_1 & 1 + S_2 & 1 + S_3 \\ 1 + S_2 & 1 + S_3 & 1 + S_4 \end{vmatrix}$

$$= \begin{vmatrix} 1 + 1 + 1 & 1 + \alpha + \beta & 1 + \alpha^2 + \beta^2 \\ 1 + \alpha + \beta & 1 + \alpha^2 + \beta^2 & 1 + \alpha^3 + \beta^3 \\ 1 + \alpha^2 + \beta^2 & 1 + \alpha^3 + \beta^3 & 1 + \alpha^4 + \beta^4 \end{vmatrix}$$

$$= \begin{vmatrix} 1 & 1 & 1 \\ 1 & \alpha & \beta \\ 1 & \alpha^2 & \beta^2 \end{vmatrix}\begin{vmatrix} 1 & 1 & 1 \\ 1 & \alpha & \beta \\ 1 & \alpha^2 & \beta^2 \end{vmatrix} = [(1 - \alpha)(\alpha - \beta)(\beta - 1)]^2$$

$$= [(\beta - \alpha)(\alpha - 1)(\beta - 1)]^2$$
$$= [(\beta - \alpha)\{\alpha\beta - (\alpha + \beta) + 1\}]^2$$
$$= \left\{(\beta - \alpha)\left(\frac{c}{a} + \frac{b}{a} + 1\right)\right\}^2$$
$$= \frac{(\beta - \alpha)^2}{a^2} \times (a + b + c)^2$$
$$= \frac{(b^2 - 4ac)}{a^2} \times \frac{(a + b + c)^2}{a^2}$$
$$= \frac{(b^2 - 4ac)(a + b + c)^2}{a^4}$$

5. $$\begin{vmatrix} a^2+b^2+c^2 & bc+ca+ab & bc+ca+ab \\ bc+ca+ab & a^2+b^2+c^2 & bc+ca+ab \\ bc+ca+ab & bc+ca+ab & a^2+b^2+c^2 \end{vmatrix}$$

$$= \begin{vmatrix} a & b & c \\ b & c & a \\ c & a & b \end{vmatrix} \times \begin{vmatrix} a & b & c \\ b & c & a \\ c & a & b \end{vmatrix} = \begin{vmatrix} a & b & c \\ b & c & a \\ c & a & b \end{vmatrix}^2 = K$$

where K is positive number.

6. $$\because \ \Delta = \begin{vmatrix} f & g & h \\ (xf)' & (xg)' & (xh)' \\ (x^2f)'' & (x^2g)'' & (x^2h)'' \end{vmatrix}$$

$$= \begin{vmatrix} f & g & h \\ xf'+f & xg'+g & xh'+h \\ x^2f''+4xf'+2f & x^2g''+4xg'+2g & x^2h''+4xh'+2h \end{vmatrix}$$

On applying $R_2 \to R_2 - R_1$ and $R_3 \to R_3 - 4R_2 - 2R_1$

$$\Delta = \begin{vmatrix} f & g & h \\ xf' & xg' & xh' \\ x^2f'' & x^2g'' & x^2h'' \end{vmatrix} = x \begin{vmatrix} f & g & h \\ f' & g' & h' \\ x^2f'' & x^2g'' & x^2h'' \end{vmatrix}$$

$$= \begin{vmatrix} f & g & h \\ f' & g' & h' \\ x^3f'' & x^3g'' & x^3h'' \end{vmatrix}$$

$$\text{Now, } \Delta' = \begin{vmatrix} f' & g' & h' \\ f' & g' & h' \\ x^3f'' & x^3g'' & x^3h'' \end{vmatrix} + \begin{vmatrix} f & g & h \\ f'' & g'' & h'' \\ (x^3f'') & (x^3g'') & (x^3h'') \end{vmatrix}$$

$$+ \begin{vmatrix} f & g & h \\ f' & g' & h' \\ (x^3f'')' & (x^3g'')' & (x^3h'')' \end{vmatrix}$$

$$= 0 + 0 + \begin{vmatrix} f & g & h \\ f' & g' & h' \\ (x^3f'')' & (x^3g'')' & (x^3h'')' \end{vmatrix}$$

$$= \begin{vmatrix} f & g & h \\ f' & g' & h' \\ (x^3f'')' & (x^3g'')' & (x^3h'')' \end{vmatrix}$$

7. $$\Delta(x) = \begin{vmatrix} \sin(x+\alpha) & \cos(x+\alpha) & a + x\sin\alpha \\ \sin(x+\beta) & \cos(x+\beta) & b + x\sin\beta \\ \sin(x+\gamma) & \cos(x+\gamma) & c + x\sin\gamma \end{vmatrix}$$

$$= \begin{vmatrix} \sin x\cos\alpha + \cos x\sin\alpha & \cos x\cos\alpha - \sin x\sin\alpha & a + x\sin\alpha \\ \sin x\cos\beta + \cos x\sin\beta & \cos x\cos\beta - \sin x\sin\beta & b + x\sin\beta \\ \sin x\cos\gamma + \cos x\sin\gamma & \cos x\cos\gamma - \sin x\sin\gamma & c + x\sin\gamma \end{vmatrix}$$

$$= \begin{vmatrix} \sin x\cos\alpha + \cos x\sin\alpha & \sin x\cos\beta + \cos x\sin\beta & \sin x\cos\gamma + \cos x\sin\gamma \\ \cos x\cos\gamma - \sin x\sin\alpha & \cos x\cos\beta - \sin x\sin\beta & \cos x\cos\gamma - \sin x\sin\alpha \\ a + x\sin\alpha & b + x\sin\beta & c + x\sin\gamma \end{vmatrix}$$

$$= \begin{vmatrix} \sin x & \cos x & 0 \\ \cos x & -\sin x & 0 \\ 0 & x & 1 \end{vmatrix} \begin{vmatrix} \cos\alpha & \sin\alpha & a \\ \cos\beta & \sin\beta & b \\ \cos\gamma & \sin\gamma & c \end{vmatrix}$$

$$= - \begin{vmatrix} \cos\alpha & \sin\alpha & a \\ \cos\beta & \sin\beta & b \\ \cos\gamma & \sin\gamma & c \end{vmatrix}$$

[which is independent of x]

8. Given, $$f(x) = \begin{vmatrix} x^n & \sin x & \cos x \\ n! & \sin\frac{n\pi}{2} & \cos\frac{n\pi}{2} \\ a & a^2 & a^3 \end{vmatrix}$$

$$f'(x) = \begin{vmatrix} nx^{n-1} & \cos x & -\sin x \\ n! & \sin\frac{n\pi}{2} & \cos\frac{n\pi}{2} \\ a & a^2 & a^3 \end{vmatrix}$$

$$f''(x) = \begin{vmatrix} n(n-1)x^{n-2} & -\sin x & -\cos x \\ n! & \sin\frac{n\pi}{2} & \cos\frac{n\pi}{2} \\ a & a^2 & a^3 \end{vmatrix}$$

$$f'''(x) = \begin{vmatrix} n(n-1)(n-2)x^{n-3} & -\cos x & \sin x \\ n! & \sin\frac{n\pi}{2} & \cos\frac{n\pi}{2} \\ a & a^2 & a^3 \end{vmatrix}$$

$$\text{Hence, } f^n(x) = \begin{vmatrix} n! & \sin\left(\frac{n\pi}{2} + x\right) & \cos\left(\frac{n\pi}{2} + x\right) \\ n! & \sin\left(\frac{n\pi}{2}\right) & \cos\left(\frac{n\pi}{2}\right) \\ a & a^2 & a^3 \end{vmatrix}$$

$$\text{Now, } \quad f^n(0) = \begin{vmatrix} n! & \sin\left(\frac{n\pi}{2}\right) & \cos\left(\frac{n\pi}{2}\right) \\ n! & \sin\left(\frac{n\pi}{2}\right) & \cos\left(\frac{n\pi}{2}\right) \\ a & a^2 & a^3 \end{vmatrix}$$

$= 0$ [$\because R_1$ and R_2 are identical]

9. (d) Given, $$A = \begin{bmatrix} y+a & b & c \\ a & y+b & c \\ a & b & y+c \end{bmatrix}$$

$$|A| = \begin{vmatrix} y+a & b & c \\ a & y+b & c \\ a & b & y+c \end{vmatrix}$$

On applying $R_1 \to R_1 + R_2 + R_3$

$$= \begin{vmatrix} y+a+b+c & b & c \\ y+a+b+c & y+b & c \\ y+a+b+c & b & y+c \end{vmatrix}$$

$$= (y+a+b+c) \begin{vmatrix} 1 & b & c \\ 1 & y+b & c \\ 1 & b & y+c \end{vmatrix}$$

$\therefore$ On applying $R_2 \to R_2 - R_1$ and $R_3 \to R_3 - R_1$

$$= (y+a+b+c) \begin{vmatrix} 1 & b & c \\ 0 & y & 0 \\ 0 & 0 & y \end{vmatrix}$$

A is of rank 3, if $y \neq -(a+b+c)$ and $y \neq 0$.

10. (c) Order of A is $m \times 1$ and order of B is $1 \times n$.

Order of AB is $m \times n$.

Since, A and B are non-zero matrices.

$\therefore$ AB will be a non-zero matrix.

The matrix AB will have atleast one non-zero element obtained by multiplying corresponding non-zero elements of A and B.

All the two -rowed minors of AB clearly vanish.

$\therefore$ Rank of $AB = 1$

Revisal Problems (JEE Main)

1. (a) Given, $A = \begin{bmatrix} 3 & 4 \\ 2 & 4 \end{bmatrix}$ and $B = \begin{bmatrix} -2 & -2 \\ 0 & -1 \end{bmatrix}$

Now, $(A + B) = \begin{bmatrix} 1 & 2 \\ 2 & 3 \end{bmatrix}$

$\therefore$ $(A + B)^{-1} = \frac{1}{3-4}\begin{bmatrix} 3 & -2 \\ -2 & 1 \end{bmatrix} = \begin{bmatrix} -3 & 2 \\ 2 & -1 \end{bmatrix}$

2. (b) $\because AB = A$ and $BA = B$

Now, $B = BA = B(AB) = (BA)B = B \cdot B = B^2$

and $A = AB = A(BA) = (AB)A = A \cdot A = A^2$

3. (c) $\because AB = B$ and $BA = A$

Then, $A^2 + B^2 = A \cdot A + B \cdot B = A \cdot (BA) + B(AB) = (AB)A + (BA)B = BA + AB = A + B$

4. (b) Given, $|A| = 0$

Now, $|\text{adj } A| = |A|^{n-1} = 0$

5. (c) Given, $A = \begin{bmatrix} a & 0 & 0 \\ 0 & a & 0 \\ 0 & 0 & a \end{bmatrix}$

Then, $|A| = a^3$

$\because$ $|\text{adj } A| = |A|^{n-1} = |A|^2 \quad [\because n = 3]$

$= (a^3)^2 = a^6$

6. (c) Given, $A \cdot (\text{adj } A) = \begin{bmatrix} 10 & 0 \\ 0 & 10 \end{bmatrix}$

$\Rightarrow$ $|A(\text{adj } A)| = 100$

$\Rightarrow$ $||A|\, I| = 100$

$\Rightarrow$ $\begin{vmatrix} |A| & 0 \\ 0 & |A| \end{vmatrix} = 100$

$\Rightarrow$ $|A|^2 = 100$

$\therefore$ $|A| = 10$

7. (a) Given, $A = \begin{bmatrix} 1 & a \\ 0 & 1 \end{bmatrix}$

Now, $A^2 = \begin{bmatrix} 1 & a \\ 0 & 1 \end{bmatrix}\begin{bmatrix} 1 & a \\ 0 & 1 \end{bmatrix} = \begin{bmatrix} 1 & 2a \\ 0 & 1 \end{bmatrix}$

$\therefore$ $A^3 = A^2 \cdot A = \begin{bmatrix} 1 & 2a \\ 0 & 1 \end{bmatrix}\begin{bmatrix} 1 & a \\ 0 & 1 \end{bmatrix} = \begin{bmatrix} 1 & 3a \\ 0 & 1 \end{bmatrix}$

Hence, $A^n = \begin{bmatrix} 1 & na \\ 0 & 1 \end{bmatrix}$

8. (d) $(I - A)^{-1} = I + A + A^2 + A^3 + A^4 + A^5 + \dots \infty$

[here, $A^5 = A^6 = \dots = 0$]

$\therefore$ $(I - A)^{-1} = I + A + A^2 + A^3 + A^4$.

9. (c) Given, $A = \begin{bmatrix} \cos\theta & -\sin\theta \\ \sin\theta & \cos\theta \end{bmatrix}$

$\because$ $A^T + A = I_2$

$\Rightarrow$ $\begin{bmatrix} \cos\theta & \sin\theta \\ -\sin\theta & \cos\theta \end{bmatrix} + \begin{bmatrix} \cos\theta & -\sin\theta \\ \sin\theta & \cos\theta \end{bmatrix} = \begin{bmatrix} 1 & 0 \\ 0 & 1 \end{bmatrix}$

$\Rightarrow$ $\begin{bmatrix} 2\cos\theta & 0 \\ 0 & 2\cos\theta \end{bmatrix} = \begin{bmatrix} 1 & 0 \\ 0 & 1 \end{bmatrix}$

$\Rightarrow$ $\cos\theta = \frac{1}{2}$

$\therefore$ $\theta = 2n\pi + \frac{\pi}{3}, n \in Z$

10. (a) Given, $A = \begin{bmatrix} 1 & a \\ 0 & 1 \end{bmatrix}$

Now, $A^n = \begin{bmatrix} 1 & na \\ 0 & 1 \end{bmatrix}$

$\frac{1}{n}A^n = \frac{1}{n}\begin{bmatrix} 1 & na \\ 0 & 1 \end{bmatrix} = \begin{bmatrix} \frac{1}{n} & a \\ 0 & \frac{1}{n} \end{bmatrix}$

Hence, $\lim_{n \to \infty} \frac{1}{n}A^n = \begin{bmatrix} \lim_{n \to \infty} \frac{1}{n} & a \\ 0 & \lim_{n \to \infty} \frac{1}{n} \end{bmatrix} = \begin{bmatrix} 0 & a \\ 0 & 0 \end{bmatrix} \quad \left[\because \frac{1}{\alpha} = 0\right]$

11. (b) Given, $(A + B)^2 = A^2 + B^2 \Rightarrow AB = -BA$

$\Rightarrow$ $\begin{bmatrix} 1 & -1 \\ 2 & -1 \end{bmatrix}\begin{bmatrix} a & 1 \\ b & -1 \end{bmatrix} = -\begin{bmatrix} a & 1 \\ b & -1 \end{bmatrix}\begin{bmatrix} 1 & -1 \\ 2 & -1 \end{bmatrix}$

$\Rightarrow$ $\begin{bmatrix} a-b & 1+1 \\ 2a-b & 2+1 \end{bmatrix} = -\begin{bmatrix} a+2 & -a-1 \\ b-2 & -b+1 \end{bmatrix}$

$\Rightarrow$ $\begin{bmatrix} a-b & 2 \\ 2a-b & 3 \end{bmatrix} = \begin{bmatrix} -a-2 & a+1 \\ -b+2 & b-1 \end{bmatrix}$

$\therefore$ $a + 1 = 2$, i.e. $a = 1$

and $b - 1 = 3$, i.e. $b = 4$

12. (a) $A = \begin{bmatrix} 1 & -3 & -4 \\ -1 & 3 & 4 \\ 1 & -3 & -4 \end{bmatrix}$

Now, $A^2 = \begin{bmatrix} 1 & -3 & -4 \\ -1 & 3 & 4 \\ 1 & -3 & -4 \end{bmatrix}\begin{bmatrix} 1 & -3 & -4 \\ -1 & 3 & 4 \\ 1 & -3 & -4 \end{bmatrix} = \begin{bmatrix} 0 & 0 & 0 \\ 0 & 0 & 0 \\ 0 & 0 & 0 \end{bmatrix}$

Hence, $A^2 = 0$

$\therefore$ A is nilpotent of index 2.

13. (d) Given, $x + y + z = 6$, $x + 2y + 3z = 10$

$x + 2y + \lambda z = 12$

$[A : B] = \left|\begin{matrix} 1 & 1 & 1 & : & 6 \\ 1 & 2 & 3 & : & 10 \\ 1 & 2 & \lambda & : & 12 \end{matrix}\right|$

On applying $R_2 \to R_2 - R_1$ and $R_3 \to R_3 - R_1$

$= \left|\begin{matrix} 1 & 1 & 1 & : & 6 \\ 0 & 1 & 2 & : & 4 \\ 0 & 1 & \lambda - 1 & : & 6 \end{matrix}\right|$

On applying $R_3 \to R_3 - R_2$

$= \left|\begin{matrix} 1 & 1 & 1 & : & 6 \\ 0 & 1 & 2 & : & 4 \\ 0 & 0 & \lambda - 3 & : & 2 \end{matrix}\right|$

So, Rank$(A) \neq$ Rank(AB), when $\lambda = 3$, i.e. the system of equations becomes inconsistent for $\lambda = 3$.

14. (b) Given,
$$x+y+z=6$$
$$x+2y+3z=10$$
$$x+2y+\lambda z=\mu$$

Now, $[A:B]=\begin{vmatrix}1&1&1&:&6\\1&2&3&:&10\\1&2&\lambda&:&\mu\end{vmatrix}$

On applying $R_2\to R_2-R_1$ and $R_3\to R_3-R_1$
$$=\begin{vmatrix}1&1&1&:&6\\0&1&2&:&4\\0&1&\lambda-1&:&\mu-6\end{vmatrix}$$

On applying $R_3\to R_3-R_2$
$$=\begin{vmatrix}1&1&1&:&6\\0&1&2&:&4\\0&0&\lambda-3&:&\mu-10\end{vmatrix}$$

$\text{Rank}(A)=2,\ \text{ when } \lambda=3$

and $\text{Rank}(AB)=3,\ \text{ when } \mu\neq 10$

So, we can say that Rank $(A)\neq$ Rank (AB), when
$\lambda=3$ and $\mu\neq 10$

15. (b) We have, $A^2-A+2I=0$

Multiply both sides by A^{-1}

$\Rightarrow \quad A^{-1}\cdot A\cdot A-A^{-1}\cdot A+2A^{-1}\cdot I=A^{-1}\cdot 0$

$\Rightarrow \quad I\cdot A-I+2A^{-1}=0 \quad [\because A^{-1}A=I]$

$\Rightarrow \quad 2A^{-1}=I-A$

$\therefore \quad A^{-1}=\frac{1}{2}(I-A)$

16. (b) We have $A^3=0 \Rightarrow -A^3=0$ [adding I on both sides]

$\Rightarrow \quad -A^3+I=I$

$\Rightarrow \quad -A^3+A^2-A^2+A-A+I=I$

$\Rightarrow \quad I+A+A^2-A-A^2-A^3=I$

$\Rightarrow \quad (I+A+A^2)-A(I+A+A^2)=I$

$\Rightarrow \quad (I-A)(I+A+A^2)=I$

$\Rightarrow \quad (I-A)^{-1}(I-A)(I+A+A^2)=(I-A)^{-1}\cdot I$

$\therefore \quad (I-A)^{-1}=I\cdot(I+A+A^2)=(I+A+A^2)$

17. (b) We have, $|AB|\,|AB|=|AB|\,|B|=|A|\times 0=0$

[$\because$ The determinant of a skew-symmetric matrix of odd order is zero]

18. (c) Given, $\begin{bmatrix}a&b^3\\2&0\end{bmatrix}=\begin{bmatrix}1&8\\2&0\end{bmatrix}$

$\Rightarrow \quad a=1$ and $b^3=8$

$\therefore \quad b=2$

Now, $\begin{bmatrix}a&b\\2&0\end{bmatrix}^{-1}=\frac{-1}{2b}\begin{bmatrix}0&-b\\-2&a\end{bmatrix}=\frac{-1}{4}\begin{bmatrix}0&-2\\-2&1\end{bmatrix}=\begin{bmatrix}0&\frac{1}{2}\\\frac{1}{2}&\frac{-1}{4}\end{bmatrix}$

19. (a) We have, $A=\begin{bmatrix}0&1&-1\\2&1&3\\3&2&1\end{bmatrix}$

$\Rightarrow \quad |A|=0(1-6)+1(9-2)-1(4-3)$
$=0+7-1=6$

Now, $(A(\text{adj }A)\cdot A^{-1})A=(|A|\cdot I\cdot A^{-1})A$
$=(\text{adj }A)\,A=|A|\,I$
$=6\begin{bmatrix}1&0&0\\0&1&0\\0&0&1\end{bmatrix}$

20. (b) We have, $\begin{vmatrix}a&1&1\\1&b&1\\1&1&c\end{vmatrix}$

$=a(bc-1)+(1-c)+(1-b)$
$=abc-a+1-c+1-b$
$=2-(a+b+c)+abc>0$

When, $(a+b+c)<abc+2$

$\because \quad \text{AM}>\text{GM}$

$\therefore \quad (a+b+c)>3(abc)^{1/3}$

Hence, $abc+2>3(abc)^{1/3}$

$\Rightarrow \quad x^3+2>3x \quad$ [where, $(abc)^{1/3}=x$]

$\Rightarrow \quad x^3-3x+2>0$

$\Rightarrow \quad (x-1)(x^2+x-2)>0 \quad \Rightarrow \quad (x-1)^2(x+2)>0$

$\Rightarrow \quad x+2>0 \quad \Rightarrow \quad x>-2$

$\Rightarrow \quad (abc)^{1/3}>-2$

$\therefore \quad abc>-8$

21. (b) We have, $\begin{vmatrix}b+c&a-b&a\\c+a&b-c&b\\a+b&c-a&c\end{vmatrix}$

On applying $R_1\to R_2+R_2+R_3$
$$=\begin{vmatrix}2(a+b+c)&0&a+b+c\\c+a&b-c&b\\a+b&c-a&c\end{vmatrix}$$
$$=(a+b+c)\begin{vmatrix}2&0&1\\c+a&b-c&b\\a+b&c-a&c\end{vmatrix}$$
$=(a+b+c)\,[2(bc-c^2)-2bc+2ab+c^2-a^2$
$-(a+b)(b-c)]$
$=(a+b+c)\,[2bc-2c^2-2bc+2ab+c^2-a^2$
$-ab+ac-b^2+bc]$
$=(a+b+c)\,[-a^2-b^2-c^2+ab+bc+ac]$
$=-(a^3+b^3+c^3-3abc)=3abc-a^3-b^3-c^3$

22. (a) We have, $\Delta=\begin{vmatrix}\cos\alpha&-\sin\alpha&1\\\sin\alpha&\cos\alpha&1\\\cos(\alpha+\beta)&-\sin(\alpha+\beta)&1\end{vmatrix}$

$$=\begin{vmatrix}\cos\alpha&-\sin\alpha&1\\\sin\alpha&\cos\alpha&1\\\cos\alpha\cos\beta-\sin\alpha\sin\beta&-\sin\alpha\cos\beta-\cos\alpha\sin\beta&1\end{vmatrix}$$

On applying $R_3\to-\cos\beta\,R_1+\sin\beta\,R_2+R_3$
$$=\begin{vmatrix}\cos\alpha&-\sin\alpha&1\\\sin\alpha&\cos\alpha&1\\0&0&-\cos\beta+\sin\beta+1\end{vmatrix}$$
$=\cos\alpha\,[-\cos\alpha\cos\beta+\cos\alpha\sin\beta+\cos\alpha]$
$-\sin\alpha\,[\sin\alpha\cos\beta-\sin\alpha\sin\beta-\sin\alpha]$
$=-\cos^2\alpha\cos\beta+\cos^2\alpha\sin\beta+\cos^2\alpha$
$-\sin^2\alpha\cos\beta+\sin^2\alpha\sin\beta+\sin^2\alpha$
$=-\cos\beta+\sin\beta+1$

$\Rightarrow -\sqrt{(-1)^2+(1)^2}+1\leq-\cos\beta+\sin\beta+1\leq\sqrt{(-1)^2+(1)^2}+1$

$\Rightarrow \quad 1-\sqrt{2}\leq-\cos\beta+\sin\beta+1\leq 1+\sqrt{2}$

Hence, $\Delta\in[1-\sqrt{2},1+\sqrt{2}]$

23. (c) We have, $\begin{vmatrix}1&1&0\\0&1&1\\1&0&1\end{vmatrix}=1(1)+1(1)+0=2$

The largest value of a third order determinant whose elements are 0 and 1 is 2.

24. (d) $\because \alpha, \beta$ and γ are the roots of

$$x^3 + px + q = 0$$

$$\therefore \quad \alpha + \beta + \gamma = -p$$

$$\alpha\beta + \beta\gamma + \gamma\alpha = 0 \text{ and } \alpha\beta\gamma = -q$$

Now, $\begin{vmatrix} \alpha & \beta & \gamma \\ \beta & \gamma & \alpha \\ \gamma & \alpha & \beta \end{vmatrix}$

On applying $C_1 \to C_1 + C_2 + C_3$

$$= \begin{vmatrix} \alpha+\beta+\gamma & \beta & \gamma \\ \alpha+\beta+\gamma & \gamma & \alpha \\ \alpha+\beta+\gamma & \alpha & \beta \end{vmatrix}$$

$$= (\alpha+\beta+\gamma) \begin{vmatrix} 1 & \beta & \gamma \\ 1 & \gamma & \alpha \\ 1 & \alpha & \beta \end{vmatrix}$$

On applying $R_2 \to R_2 - R_1$ and $R_3 \to R_3 - R_1$

$$= (\alpha+\beta+\gamma) \begin{vmatrix} 1 & \beta & \gamma \\ 0 & \gamma-\beta & \alpha-\gamma \\ 0 & \alpha-\beta & \beta-\gamma \end{vmatrix}$$

$$= (\alpha+\beta+\gamma)[(\gamma-\beta)(\beta-\gamma) - (\alpha-\beta)(\alpha-\gamma)]$$

$$= -p[\beta\gamma - \gamma^2 - \beta^2 + \beta\gamma - \alpha^2 + \alpha\gamma + \beta\alpha - \beta\gamma]$$

$$= p[\alpha^2 + \beta^2 + \gamma^2 + \alpha\beta + \beta\gamma + \gamma\alpha]$$

$$= p[(\alpha+\beta+\gamma)^2 - (\alpha\beta + \beta\gamma + \gamma\alpha)]$$

$$= p[p^2 - 0] = p^3$$

25. (b) We have, $\begin{vmatrix} x^n & x^{n+2} & x^{n+3} \\ y^n & y^{n+2} & y^{n+3} \\ z^n & z^{n+2} & z^{n+3} \end{vmatrix}$

$$= x^n y^n z^n \begin{vmatrix} 1 & x^2 & x^3 \\ 1 & y^2 & y^3 \\ 1 & z^2 & z^3 \end{vmatrix}$$

$$= x^n y^n z^n (x-y)(y-z)(z-x)(xy + yz + zx) \quad \text{...(i)}$$

$$\because \begin{vmatrix} x^n & x^{n+2} & x^{n+3} \\ y^n & y^{n+2} & y^{n+3} \\ z^n & z^{n+2} & z^{n+3} \end{vmatrix}$$

$$= (x-y)(y-z)(z-x)\left(\frac{1}{x} + \frac{1}{y} + \frac{1}{z}\right) \quad \text{...(ii)}$$

From Eqs. (i) and (ii),

$$x^n y^n z^n (x-y)(y-z)(z-x)(xy + yz + zx) = (x-y)(y-z)(z-x)\left(\frac{1}{x} + \frac{1}{y} + \frac{1}{z}\right)$$

$$\Rightarrow \quad x^n y^n z^n = x^{-1} y^{-1} z^{-1}$$

$$\Rightarrow \quad (xyz)^n = (xyz)^{-1}$$

$$\Rightarrow \quad n = -1$$

26. (a) Given, $f(x) = \begin{vmatrix} 1 & x & x+1 \\ 2x & x(x-1) & (x+1)x \\ 3x(x-1) & x(x-1)(x-2) & x(x+1)(x-1) \end{vmatrix}$

$$= x^2(x-1) \begin{vmatrix} 1 & x & x+1 \\ 2 & x-1 & x+1 \\ 3 & x-2 & x+1 \end{vmatrix}$$

On applying $C_3 \to C_3 - C_2$

$$= x^2(x-1) \begin{vmatrix} 1 & x & 1 \\ 2 & x-1 & 2 \\ 3 & x-2 & 3 \end{vmatrix} = 0$$

Thus, $f(100) = 0$

27. (b) Since, D_i is formed with a distinct element.

$$\therefore \quad \sum_{i=1}^{n} D_i = 0$$

As the number of determinants having positive value = Number of determinants having negative value.

28. (c) We have, $f(x) = \begin{vmatrix} x & (1+\sin x)^3 & \cos x \\ 1 & \log(1+x) & 2 \\ x^2 & (1+x)^2 & 0 \end{vmatrix}$

$$= x[-2(1+x)^2] + (1+\sin x)^3 [2x^2] + [(1+x^2) - x^2 \log(1+x)]$$

$$= -2x(1 + x^2 + 2x) + (1+\sin x)^3 \times 2x^2 + \cos x[1 + x^2 + 2x - x^2 \log(1+x)]$$

Hence, the coefficient of x is -2.

29. (a) Given, $D_k = \begin{vmatrix} 1 & n & n \\ 2k & n^2+n+2 & n^2+n \\ 2k-1 & n^2 & n^2+n+2 \end{vmatrix}$

$$\sum_{k=1}^{n} D_k = \begin{vmatrix} n & n & n \\ n(n+1) & n^2+n+2 & n^2+n \\ n^2 & n^2 & n^2+n+2 \end{vmatrix}$$

On applying $C_2 \to C_2 - C_1$

$$= \begin{vmatrix} n & 0 & n \\ n(n+1) & 2 & n^2+n \\ n^2 & 0 & n^2+n+2 \end{vmatrix}$$

$$= 2n(n^2+n+2) - 2n^3 = 2n^2 + 4n$$

$$= 2n(n+2) = 48$$

$$\therefore \quad n = 4$$

30. (d) We have, $\begin{vmatrix} x^2+3x & x-1 & x+3 \\ x+1 & -2x & x-4 \\ x-3 & x+4 & 3x \end{vmatrix}$

$$= x^2 + 3x[-6x - (x^2 - 16)] + (x-1)[x^2 - 7x + 12 - 3x^2 - 3x] + (x+3)[x^2 + 5x + 4 + 2x^2 - 6x]$$

$$= (x^2 + 3x)(-x^2 - 6x + 16) + (x-1)[-2x^2 - 10x + 12] + (x+3)[3x^2 - x + 4]$$

$$= -x^4 - 3x^3 - 6x^3 - 18x^2 + 16x^2 + 48x - 2x^3 + 2x^2 - 10x^2 + 10x + 12x - 12 + 3x^4 + 9x^2 - x^2 - 3x + 4x + 12$$

$$= 2x^4 - 11x^3 - 12x^2 + 71x = ax^4 + bx^3 + cx^2 + dx + e$$

Hence, $\quad e = 0$

31. (a) Given, $I_n = \begin{vmatrix} 1 & k & k \\ 2n & k^2+k+1 & k^2+k \\ 2n-1 & k^2 & k^2+k+1 \end{vmatrix}$

$$\sum_{n=1}^{k} I_n = \begin{vmatrix} k & k & k \\ k(k+1) & k^2+k+1 & k^2+k \\ k(k+1)-k & k^2 & k^2+k+1 \end{vmatrix}$$

On applying $C_2 \to C_2 - C_1$

$$= \begin{vmatrix} k & 0 & k \\ k^2+k & 1 & k^2+k \\ k^2 & 0 & k^2+k+1 \end{vmatrix} = k(k^2+k+1) - k^3$$

$$= k(k+1) = 72$$

$$\therefore \quad k = 8$$

32. (b) Since, $\alpha + \beta + \gamma = \pi$

and $\quad e^{i\pi} = \cos\pi + i\sin\pi = -1$

Now, $\quad e^{i(\beta+\gamma)} = -e^{i(\pi-\alpha)}$

By taking $e^{i\alpha}$, $e^{i\beta}$ and $e^{i\gamma}$ common from C_1, C_2 and C_3 respectively.

We have, $\Delta = -\begin{vmatrix} e^{i\alpha} & e^{-i(\alpha+\gamma)} & e^{-i(\alpha+\beta)} \\ e^{-i(\beta+\gamma)} & e^{i\beta} & e^{-i(\alpha+\beta)} \\ e^{-i(\beta+\gamma)} & e^{-i(\alpha+\gamma)} & e^{i\gamma} \end{vmatrix}$

$$= \begin{vmatrix} e^{i\alpha} & -e^{i\beta} & -e^{i\gamma} \\ -e^{i\alpha} & e^{i\beta} & -e^{i\gamma} \\ -e^{i\alpha} & -e^{i\beta} & e^{i\gamma} \end{vmatrix}$$

$$= -e^{i(\alpha+\beta+\gamma)} \begin{vmatrix} 1 & -1 & 1 \\ -1 & 1 & -1 \\ -1 & -1 & 1 \end{vmatrix} = 1 \begin{vmatrix} 1 & -1 & -1 \\ -1 & 1 & -1 \\ -1 & -1 & 1 \end{vmatrix}$$

$= -4$

33. (b) We have, $\Delta = \begin{vmatrix} a_1 + b_1\omega & a_1\omega^2 + b_1 & a_1 + b_1\omega + c_1\omega^2 \\ a_2 + b_2\omega & a_2\omega^2 + b_2 & a_2 + b_2\omega + c_2\omega^2 \\ a_3 + b_3\omega & a_3\omega^2 + b_3 & a_3 + b_3\omega + c_3\omega^2 \end{vmatrix}$

$[\because 1 = \omega^3]$

$$= \begin{vmatrix} a_1\omega^3 + b_1\omega & a_1\omega^2 + b_1 & a_1 + b_1\omega + c_1\omega^2 \\ a_2\omega^3 + b_2\omega & a_2\omega^2 + b_2 & a_2 + b_2\omega + c_2\omega^2 \\ a_3\omega^3 + b_3\omega & a_3\omega^2 + b_3 & a_3 + b_3\omega + c_3\omega^2 \end{vmatrix}$$

$$= \omega \begin{vmatrix} a_1\omega^2 + b_1 & a_1\omega^2 + b_1 & a_1 + b_1\omega + c_1\omega^2 \\ a_2\omega^2 + b_2 & a_2\omega^2 + b_2 & a_2 + b_2\omega + c_2\omega^2 \\ a_3\omega^2 + b_3 & a_3\omega^2 + b_3 & a_3 + b_3\omega + c_3\omega^2 \end{vmatrix}$$

$= \omega \times 0 = 0$ $[\because C_1$ and C_2 are same$]$

34. (a) Given, $\Delta_r = \begin{vmatrix} 2^{r-1} & \dfrac{(r+1)!}{1+\dfrac{1}{r}} & 2r \\ a & b & c \\ 2^n - 1 & (n+1)! - 1 & n(n+1) \end{vmatrix}$

$$\sum_{r=1}^{n} \Delta_r = \begin{vmatrix} \Sigma 2^{r-1} & \Sigma \dfrac{(r+1)!}{1+\dfrac{1}{r}} & \Sigma 2r \\ a & b & c \\ 2^n - 1 & (n+1)! - 1 & n(n+1) \end{vmatrix}$$

$$= \begin{vmatrix} 2^n - 1 & (n+1)! - 1 & n(n+1) \\ a & b & c \\ 2^n - 1 & (n+1)! - 1 & n(n+1) \end{vmatrix}$$

$= 0$ $(\because R_1$ and R_3 are same$)$

35. (a) Given, if $ax + by + cz = 0$, $bx + cy + az = 0$ and $cx + ay + bz = 0$ has non-trivial solution,

When $\begin{vmatrix} a & b & c \\ b & c & a \\ c & a & b \end{vmatrix} = 0$

$\Rightarrow$ $(a+b+c)\begin{vmatrix} 1 & b & c \\ 1 & c & a \\ 1 & a & b \end{vmatrix} = 0$

On applying $R_3 \to R_3 - R_1$ and $R_2 \to R_2 - R_1$

$\Rightarrow$ $(a+b+c)\begin{vmatrix} 1 & b & c \\ 0 & c-b & a-c \\ 0 & a-b & b-c \end{vmatrix} = 0$

$\Rightarrow$ $(a+b+c)[-(c-b)^2 - (a-b)(a-c)] = 0$

$\Rightarrow$ $(a+b+c)[(b-c)^2 + (a-b)(a-c)] = 0$

$\Rightarrow (a+b+c)[b^2 + c^2 - 2bc + a^2 - ac - ab + bc] = 0$

$\Rightarrow$ $\dfrac{1}{2}(a+b+c)[(a-b)^2 + (b-c)^2 + (c-a)^2] = 0$

$\therefore$ $(a+b+c) = 0$

Either one or two of a, b, c are negative.

Hence, $ax^2 + bx + c$ has atleast one positive root.

36. (c) Since, $|A| = \pm 1$

and $A^{-1} = \dfrac{\text{adj}(A)}{|A|} = \pm\,\text{adj}(A)$ ...(i)

Also, all entries of A are integers.

$\therefore$ In adj A. all entries are integer ...(ii)

From Eqs. (i) and Eqs. (ii), we get

A^{-1} exists and all entries are integers.

37. (d) Let $A = \begin{bmatrix} \dfrac{1}{3} & 2/3 & 2/3 \\ 2/3 & \dfrac{1}{3} & -2/3 \\ 2/3 & -2/3 & \dfrac{1}{3} \end{bmatrix}$

where $AA^T = I$

Here, product of any two different rows is zero.

$\therefore$ Statement II is correct and Statement I is incorrect.

38. (b) Given, $A^2 - 3A + 2I = 0$

Then, A can be I or $2I$.

If $A = \begin{bmatrix} 3 & -2 \\ 1 & 0 \end{bmatrix}$

Then, $A^2 = \begin{bmatrix} 3 & -2 \\ 1 & 0 \end{bmatrix}\begin{bmatrix} 3 & -2 \\ 1 & 0 \end{bmatrix} = \begin{bmatrix} 7 & -6 \\ 3 & -2 \end{bmatrix}$

$\therefore$ $A^2 - 3A + 2I = \begin{bmatrix} 7 & -6 \\ 3 & -2 \end{bmatrix} - 3\begin{bmatrix} 3 & -2 \\ 1 & 0 \end{bmatrix} + 2\begin{bmatrix} 1 & 0 \\ 0 & 0 \end{bmatrix} = \begin{bmatrix} 0 & 0 \\ 0 & 0 \end{bmatrix}$

If $A = \begin{bmatrix} 3 & 1 \\ -2 & 0 \end{bmatrix}$

Then, $A^2 = \begin{bmatrix} 3 & 1 \\ -2 & 0 \end{bmatrix}\begin{bmatrix} 3 & 1 \\ -2 & 0 \end{bmatrix} = \begin{bmatrix} 7 & 3 \\ -6 & -2 \end{bmatrix}$

$\therefore$ $A^2 - 3A + 2I = \begin{bmatrix} 7 & 3 \\ -6 & -2 \end{bmatrix} - 3\begin{bmatrix} 3 & 1 \\ -2 & 0 \end{bmatrix} + 2\begin{bmatrix} 1 & 0 \\ 0 & 1 \end{bmatrix}$

$= \begin{bmatrix} 0 & 0 \\ 0 & 0 \end{bmatrix}$

If $|A - xI| = ax^2 + bx + c$

Hence, $aA^2 + bA + cI = 0$

Both Statements are correct but Statement II is not correct explanation of Statement I.

39. (c) Given, $A = \begin{bmatrix} 2 & 1+2i \\ 1-2i & 7 \end{bmatrix}$

$\therefore$ $|A| = 14 - (1-2i)(1+2i)$

$= 14 - [1+4] = 14 - 5 = 9$

If $A = \begin{bmatrix} a_{11} & a_{12} \\ a_{21} & a_{22} \end{bmatrix}$

a_{ij} being complex numbers, then $|A|$ may be real or complex.

Hence, Statement I is correct and Statement II is not correct.

40. (b) Let, $x = A(BA)$

$\therefore$ $x^T = (A(BA))^T = (BA)^T \cdot A^T = A^T B^T A^T = A(BA)$

$\Rightarrow$ $X^T = X$, X is symmetric.

Also, AB is symmetric.

$\Rightarrow$ $(AB)^T = AB$

or $B^T A^T = AB$ or $BA = AB$

i.e. A and B are commutative.

Revisal Problems (JEE Advanced)

1. (a) $\because A^2 = I \quad \Rightarrow \quad A \cdot A = I$

Also, $\quad A \cdot A^{-1} = I$

That means $\quad A \cdot A = A \cdot A^{-1} \qquad \ldots(\text{i})$

As $\quad |A| = (12 - 16) - 1(-12 + 9)$

$= -4 + 3$

$= -1 \neq 0$

Hence, A^{-1} exists.

On multiplying Eq. (i) by A^{-1}, we get

$$A^{-1} \cdot A \cdot A = A^{-1} \cdot A \cdot A^{-1}$$

$\Rightarrow \quad (A^{-1} \cdot A) \cdot A = (A^{-1} \cdot A) \cdot A^{-1}$

$\Rightarrow \quad A = A^{-1}$

Now, $\quad \left(\dfrac{A}{2}\right)^{-1} = \dfrac{A^{-1}}{2^{-1}}$

$= 2A^{-1}$

$= 2A \qquad (\because A^{-1} = A)$

2. (d) Let A be a 2×2 matrix.

Suppose $\quad A = \begin{bmatrix} a & b \\ c & d \end{bmatrix}$

If A is skew-symmetric.

Then, $\quad A = \begin{bmatrix} 0 & b \\ -b & 0 \end{bmatrix}$

If A is also idempotent,

then $\quad A^2 = A$

$\Rightarrow \quad \begin{bmatrix} 0 & b \\ -b & 0 \end{bmatrix}\begin{bmatrix} 0 & b \\ -b & 0 \end{bmatrix} = \begin{bmatrix} 0 & b \\ -b & 0 \end{bmatrix}$

$\Rightarrow \quad \begin{bmatrix} -b^2 & 0 \\ 0 & -b^2 \end{bmatrix} = \begin{bmatrix} 0 & b \\ -b & 0 \end{bmatrix}$

$\Rightarrow \quad b = 0$

Therefore, A is the null matrix.

3. (c) $\because \quad A(\text{adj } A) = (\text{adj } A)\, A = |A|\, I$

$\Rightarrow \quad (\text{adj } A) = |A|\, A^{-1}$

$\det(A^{-1}) = [\det(A)]^{-1}$

and $\quad (AB)^{-1} = B^{-1}A^{-1}$

But $\quad (A + B)^{-1} \neq B^{-1} + A^{-1}$

4. (d) $Tr(A + B) = Tr(A) + Tr(B)$

$Tr(I_3) = 3$

$Tr(KA) = K\,Tr(A)$

But $\quad Tr(A^2) \neq [Tr(A)]^2$

5. (d) We have, $A^3 = B^3$ and $A^2B = B^2A$

$\Rightarrow \quad A^3 - A^2B = B^3 - B^2A$

$\Rightarrow \quad A^2(A - B) + B^2(A - B) = 0$

$\Rightarrow \quad (A^2 + B^2)(A - B) = 0$

$\Rightarrow \quad |(A^2 + B^2)(A - B)| = 0$

or $\quad |A^2 + B^2|\,|A - B| = 0$

$\therefore$ Atleast one of $|A^2 + B^2|$ or $|A - B|$ must be zero.

6. (c) If $A^{-1} = A$ and $|A| \neq 0$

That means A^{-1} may exists.

We get, $\quad A \cdot A^{-1} = A \cdot A$

$\Rightarrow \quad I = A^2$

That means, A is an involuntary matrix.

7. (b) $\Delta_1 = \Delta_0^{2^1}$, as $|\text{Cofactors of } A_{n \times n}| = |A|^{n-1}$

Similarly, $\Delta_2 = |\text{Cofactors of } \Delta_1| = \Delta_1^2 = \Delta_0^4 = \Delta_0^{2^2}$

$\therefore \quad \Delta_n = |\text{Cofactors of } \Delta_{n-1}| = \Delta_{n-1}^2 = \Delta_0^{2^n}$

8. (a) $\begin{vmatrix} 1 + a + x & a + y & a + z \\ b + x & 1 + b + y & b + z \\ c + x & c + y & 1 + c + z \end{vmatrix}$

$= \begin{vmatrix} 1 + a + x & b + x & c + x \\ a + y & 1 + b + y & c + y \\ a + z & b + z & 1 + c + z \end{vmatrix}$

$= \begin{vmatrix} 1 + a + b + c + 3x & b + x & c + x \\ 1 + a + b + c + 3y & 1 + b + y & c + y \\ 1 + a + b + c + 3z & b + z & 1 + c + z \end{vmatrix}$

$= (1 + a + b + c)\begin{vmatrix} 1 & b + x & c + x \\ 1 & 1 + b + y & c + y \\ 1 & b + z & 1 + c + z \end{vmatrix} + 3\begin{vmatrix} x & b + x & c + x \\ y & 1 + b + y & c + y \\ z & b + z & 1 + c + z \end{vmatrix}$

$= (1 + a + b + c)\begin{vmatrix} 1 & b + x & c + x \\ 0 & y + 1 - x & y - x \\ 0 & z - x & 1 + z - x \end{vmatrix} + 3\begin{vmatrix} x & b & c \\ y & 1 + b & c \\ z & b & 1 + c \end{vmatrix}$

$= (1 + a + b + c)\,[(y + 1 - x)(1 + z - x) - (z - x)(y - x)]$
$\quad + 3\,[x\{(1 + b)(1 + c) - bc\} + b\,\{zc - y(1 + c)\}$
$\quad + c\,\{by - z(1 + b)\}]$

$= (1 + a + b + c)\,[y + yz - yx + 1 + z - x - x - zx + x^2$
$\quad - yz + zx + xy - x^2] + 3[x\,(1 + b + c) + b\,\{zc - y - yc\}$
$\quad + c\{by - z - zb\}]$

$= (1 + a + b + c)\,[1 + y + z - zx]$
$\quad + 3[x + bx + cx + bcz - by - bcy + bcy - cz - bcz]$

$= (1 + a + b + c)\,[1 + y + z - 2x]$
$\quad + 3\,(x + bx + cx - by - bcy + bcy - cz)$

$= (1 + a + b + c)[1 + y + z + x] - 3x\,(1 + a + b + c)$
$\quad + 3\,(x + bx + cx - by - cz)$

$= (1 + a + b + c)\,[1 + x + y + z] - 3x - 3ax - 3bx - 3cx$
$\quad + 3x + 3bx + 3cx - 3by - 3cz$

$= (1 + a + b + c)\,(1 + x + y + z) - 3\,(ax + by + cz)$

9. (c) $\quad A + 2B = \begin{bmatrix} 1 & 2 & 0 \\ 6 & -3 & 3 \\ -5 & 3 & 1 \end{bmatrix}$

$\Rightarrow \quad 2(A + 2B) = \begin{bmatrix} 2 & 4 & 0 \\ 12 & -6 & 6 \\ -10 & 6 & 2 \end{bmatrix}$

$\Rightarrow \quad 2A - B = \begin{bmatrix} 2 & -1 & 5 \\ 2 & -1 & 6 \\ 0 & 1 & 2 \end{bmatrix}$

$\Rightarrow \quad 5B = \begin{bmatrix} 0 & 5 & -5 \\ 10 & -5 & 0 \\ -10 & 5 & 0 \end{bmatrix}$

$\Rightarrow \quad B = \begin{bmatrix} 0 & 1 & -1 \\ 2 & -1 & 0 \\ -2 & 1 & 0 \end{bmatrix}$

Now, $A = \begin{bmatrix} 1 & 2 & 0 \\ 6 & -3 & 3 \\ -5 & 3 & 1 \end{bmatrix} - 2\begin{bmatrix} 0 & 1 & -1 \\ 2 & -1 & 0 \\ -2 & 1 & 0 \end{bmatrix}$

$= \begin{bmatrix} 1 & 0 & 2 \\ 2 & -1 & 3 \\ -1 & 1 & 1 \end{bmatrix}$

Now, $Tr(A) - Tr(B) = 1 - (-1) = 2$

10. (d) $X^2 = I$

$\Rightarrow$ X is an idempotent matrix.

$\therefore$ Number of solutions are more than 2.

11. (c) $\because S^3 = S \cdot S \cdot S$

$= (ABCD)(ABCD)(ABCD)$

$= (ABC)(DAB)(CDA)(BCD)$

$= D^T C^T B^T A^T = (BCD)^T A^T = (A^T)^T A^T$

$= AA^T = ABCD = S$

$\therefore$ $S^3 = S$ and $S^4 = S^2$

12. (b) As $x = 1$, $y = 3$ is one of the solution.

$\therefore$ $a - 3b = 2a - b$

and $(c+1)\cdot 1 + 3c = 10 - a + 3b$

$\Rightarrow$ $a = -2b$

and $4c = 9 - a + 3b$ ... (i)

Also, $\dfrac{a}{c+1} = \dfrac{-b}{c} = \dfrac{2a - b}{10 - a + 3b}$

As $a = -2b$

$\Rightarrow$ $\dfrac{a}{c+1} = \dfrac{-b}{c}$ or $\dfrac{-2b}{c+1} = \dfrac{-b}{c}$

$\Rightarrow$ $-2bc = -b(c+1)$

$\Rightarrow$ $b(1-c) = 0$

$\therefore$ Either $b = 0$ or $c = 1$

If $b = 0 \Rightarrow a = 0, c = \dfrac{9}{4}$

If $c = 1 \Rightarrow b = -1, a = 2$

$\therefore$ Number of triplets of a, b, c are exactly two.

13. (d) Given,

$$\begin{bmatrix} 2A+1 & -5 \\ -4 & A \end{bmatrix}^{-1} \begin{bmatrix} A-5 & B \\ 2A-2 & C \end{bmatrix} = \begin{bmatrix} 14 & D \\ E & F \end{bmatrix}$$

$\Rightarrow$ $\dfrac{1}{2A^2 + A - 20}\begin{bmatrix} A & 5 \\ 4 & 2A+1 \end{bmatrix}\begin{bmatrix} A-5 & B \\ 2A-2 & C \end{bmatrix} = \begin{bmatrix} 14 & D \\ E & F \end{bmatrix}$

$\Rightarrow$ $A(A-5) + 5(2A-2) = 14(2A^2 + A - 20)$

$\Rightarrow$ $A^2 - 5A + 10A - 10 = 28A^2 + 14A - 280$

$\Rightarrow$ $27A^2 + 9A - 270 = 0$

If α, β are the roots of above equation,

then $\alpha + \beta = \dfrac{1}{3}$ and $\alpha\beta = -10$

Now, $(\alpha - \beta)^2 = (\alpha + \beta)^2 - 4\alpha\beta$

$= \dfrac{1}{9} + 4 \times 10 = \dfrac{361}{9}$

$\therefore$ $(\alpha - \beta) = \dfrac{19}{3}$

14. (c) We have, $B = KA$

$|B| = |KA| = K^n |A|$

$\dfrac{1}{|B|^{-1}} = \dfrac{K^n}{|A|^{-1}}$

$\Rightarrow$ $|A|^{-1} = K^n |B|^{-1}$

$\therefore$ $|A^{-1}| = K^n |B^{-1}|$

15. (c) $\because A^2 = 2A - I$

Now, $A^3 = A \cdot A^2$

$= A(2A - I)$

$= 2A^2 - A$

$= 2(2A - I) - A$

$= 4A - A - 2I$

$= 3A - 2I$

Now, $A^4 = A^2 \cdot A^2$

$= (2A - I)(2A - I)$

$= 4A^2 - 2A - 2A + I$

$= 4A^2 - 4A + I$

$= 4(2A - I) - 4A + I$

$= 8A - 4I - 4A + I$

$= 4A - 3I$

Hence, $A^n = nA - (n-1)I$

16. (a) Given, $A = \begin{bmatrix} 2 & 1 \\ 4 & 1 \end{bmatrix}$

$B = \begin{bmatrix} 3 & 4 \\ 2 & 3 \end{bmatrix}$

$C = \begin{bmatrix} 3 & -4 \\ -2 & 3 \end{bmatrix}$

$BC = \begin{bmatrix} 3 & 4 \\ 2 & 3 \end{bmatrix}\begin{bmatrix} 3 & -4 \\ -2 & 3 \end{bmatrix}$

$= \begin{bmatrix} 9-8 & -12+12 \\ 6-6 & -8+9 \end{bmatrix}$

$= \begin{bmatrix} 1 & 0 \\ 0 & 1 \end{bmatrix} = I$

Now, $Tr(A) + Tr\left(\dfrac{ABC}{2}\right) + Tr\left(\dfrac{A(BC)^2}{4}\right) + Tr\left(\dfrac{A(BC)^3}{8}\right) + \cdots$

$= 3 + \dfrac{3}{2} + \dfrac{3}{4} + \dfrac{3}{8} + \cdots$

$= 3\left[1 + \dfrac{1}{2} + \dfrac{1}{4} + \dfrac{1}{8} + \cdots\right]$

$= \dfrac{3}{1 - \dfrac{1}{2}} = 6$

17. (d) $\begin{vmatrix} (1+x)^2 & (1-x)^2 & -(2+x^2) \\ 2x+1 & 3x & 1-5x \\ x+1 & 2x & 2-3x \end{vmatrix} + \begin{vmatrix} (1+x)^2 & 2x+1 & x+1 \\ (1-x)^2 & 3x & 2x \\ 1-2x & 3x-2 & 2x-3 \end{vmatrix}^T = 0$

$\Rightarrow$ $\begin{vmatrix} (1+x)^2 & (1-x)^2 & -(2+x^2) \\ 2x+1 & 3x & 1-5x \\ x+1 & 2x & 2-3x \end{vmatrix} + \begin{vmatrix} (1+x)^2 & (1-x)^2 & (1-2x) \\ 2x+1 & 3x & (3x-2) \\ x+1 & 2x & (2x-3) \end{vmatrix} = 0$

$\Rightarrow$ $\begin{vmatrix} (1+x)^2 & (1-x)^2 & -(1+x)^2 \\ (2x+1) & 3x & -(1+2x) \\ (x+1) & 2x & -(x+1) \end{vmatrix} = 0$

which is always true as C_1 is proportional to C_3.

18. (d) Given, $A = \begin{bmatrix} a+b & ab & 0 \\ 1 & a+b & ab \\ 0 & 1 & a+b \end{bmatrix}$

$$\therefore \quad |A| = (a+b)[a^2+b^2+2ab-ab] + ab(-a-b)$$
$$= (a+b)(a^2+b^2+ab) - a^2b - ab^2$$
$$= (a^2+b^2)(a+b) + a^2b + ab^2 - a^2b - ab^2$$
$$= (a^2+b^2)(a+b)$$

19. (b) We have, $|D_2| = \begin{vmatrix} 0 & 1 \\ 2 & 0 \end{vmatrix} = -2$

and $|D_3| = \begin{vmatrix} 0 & 3 & 1 \\ 3 & 0 & 3 \\ 3 & 3 & 0 \end{vmatrix}$

$$= 0 - 3(-9) + 1(9) = 36$$

$\therefore |D_2| + |D_3| = 34$

20. (d) $A = \begin{bmatrix} 1 & 2r-1 \\ 0 & 1 \end{bmatrix}$

$$\prod_{r=1}^{50} \begin{bmatrix} 1 & 2r-1 \\ 0 & 1 \end{bmatrix} = \begin{bmatrix} 1 & \prod_{r=1}^{50} (2r-1) \\ 0 & 1 \end{bmatrix}$$
$$= \begin{bmatrix} 1 & 2500 \\ 0 & 1 \end{bmatrix}$$

21. (b) On applying $R_1 \to R_1 - zR_2 + yR_3$

$$\Rightarrow \begin{vmatrix} 1+x^2+y^2+z^2 & z(1+x^2+y^2+z^2) & -y(1+x^2+y^2+z^2) \\ 2(xy-z) & 1+y^2-z^2-x^2 & 2(yz+x) \\ 2(xz+y) & 2(yz-x) & 1+z^2-x^2-y^2 \end{vmatrix}$$

$$\Rightarrow (1+x^2+y^2+z^2) \begin{vmatrix} 1 & z & -y \\ 2(xy-z) & 1+y^2-z^2-x^2 & 2(yz+x) \\ 2(xz+y) & 2(yz-x) & 1+z^2-x^2-y^2 \end{vmatrix}$$

On applying $C_2 \to C_2 - zC_1 - xC_3$

$$\Rightarrow (1+x^2+y^2+z^2) \begin{vmatrix} 1 & 0 & -y \\ 2(xy-z) & 1+x^2+y^2+z^2 & 2(yz+x) \\ 2(xz+y) & -x(1+x^2+y^2+z^2) & 1+z^2-x^2-y^2 \end{vmatrix}$$

$$\Rightarrow \quad (1+x^2+y^2+z^2)^2 \begin{vmatrix} 1 & 0 & -y \\ 2(xy-z) & 1 & 2(yz+x) \\ 2(xy+y) & -x & 1+z^2-x^2-y^2 \end{vmatrix}$$

On expanding, we get

$\Rightarrow (1+x^2+y^2+z^2)^3$

Alternate Method

Objective approach : put $y = z = 0$.

$\therefore$ Determinant comes out to be $(1+x^2)^3$

and option (b) comes out to be $(1+x^2)^3$.

22. (c) Since, $(2n+1)^2 - (2m+1)^2 = 4(m+n+1)(n-m)$

$\therefore$ Multiple of 8.

23. (b) Here, $(AB)(AB) = A(BA)B = A(A^2B)B = A^3B^2$

Now, $(AB)(AB)(AB) = (A^3B^2)AB$

$$= A^3B^2AB = A^3B(BA)B$$
$$= A^3B(A^2B)B = A^3(BA)\cdot AB^2$$
$$= A^3(A^2B)\cdot AB$$
$$= A^5BAB^2 = A^5\cdot A^2B\cdot B^2 = A^7\cdot B^3$$

So, $(AB)^n = A^{2^n-1}\cdot B^n$

$\therefore (AB)^{10} = A^{2^{10}-1}\cdot B^{10} \Rightarrow k = 2^{10}-1 = 1023$

24. (d) For non-trivial solution,

$$\begin{vmatrix} \alpha & \beta & \gamma \\ \beta & \gamma & \alpha \\ \gamma & \alpha & \beta \end{vmatrix} = 0 \Rightarrow \alpha^3+\beta^3+\gamma^3 - 3\alpha\beta\gamma = 0$$

$\Rightarrow (\alpha+\beta+\gamma)(\alpha^2+\beta^2+\gamma^2-\alpha\beta-\beta\gamma-\alpha\gamma) = 0$

$\Rightarrow (-a)[(\alpha+\beta+\gamma)^2 - 3(\alpha\beta+\beta\gamma+\gamma\alpha)] = 0$

$\Rightarrow (-a)[a^2-3b] = 0$

$\Rightarrow a^3 - 3ab = 0$

or $a^3 = 3ab$

$\therefore a^3 + 6c = 3ab + 6c$

25. (a) We have, $(A(A+B)^{-1}B)^{-1}(AB)$

$$= B^{-1}\cdot(A+B)\cdot A^{-1}(AB)$$
$$= B^{-1}\cdot(A+B)\cdot(A^{-1}A)\cdot B$$
$$= B^{-1}\cdot(A+B)\cdot I\cdot B$$
$$= (B^{-1}A + B^{-1}B)B$$
$$= (B^{-1}A + I)B$$
$$= B^{-1}AB + I\cdot B$$
$$= B^{-1}(BA) + B = (B^{-1}B)A + B$$
$$= IA + B$$
$$= A + B$$

26. (b) We have, $(A+B)^2$

$$= A^2 + AB + BA + B^2$$

$$\begin{bmatrix} \because A^2 = BA\cdot BA = B\cdot B\cdot A = B\cdot A = A \\ \text{and } B^2 = ABAB = A\cdot AB = AB = B \end{bmatrix}$$

$$= A + B + A + B$$
$$= 2(A+B)$$

Also, $(A+B)^3 = (A+B)^2\cdot(A+B)$

$$= 2(A+B)\cdot(A+B)$$
$$= 2(A+B)^2 = 2\times 2(A+B)$$
$$= 4(A+B)$$

Hence, $(A+B)^n = 2^{n-1}(A+B)$

27. (d) Given,

$$P = \begin{vmatrix} c-a & c-b & ab-c^2 \\ a-b & a-c & bc-a^2 \\ b-c & b-a & ca-b^2 \end{vmatrix} = \begin{vmatrix} a & b & 1 \\ b & c & 1 \\ c & d & 1 \end{vmatrix}^2 = (2010)^2$$

$$= (2\times 3\times 5\times 67)^2 = 2^2\times 3^2\times 5^2\times 67^2$$

$\therefore$ Number of divisors $= (2+1)(2+1)(2+1)(2+1) = 81$

28. (a) Here, $A^2 = \begin{bmatrix} a & b \\ c & d \end{bmatrix}\begin{bmatrix} a & b \\ c & d \end{bmatrix}$

$$= \begin{bmatrix} a^2+bc & ab+bd \\ ac+cd & bc+d^2 \end{bmatrix}$$
$$= \begin{bmatrix} a^2+ad-1 & b(a+d) \\ c(a+d) & d^2+ad-1 \end{bmatrix}$$
$$= \begin{bmatrix} a(a+d)-1 & b \\ c & d(a+d)-1 \end{bmatrix}$$
$$= \begin{bmatrix} a-1 & b \\ c & d-1 \end{bmatrix} = \begin{bmatrix} a & b \\ c & d \end{bmatrix} - \begin{bmatrix} 1 & 0 \\ 0 & 1 \end{bmatrix}$$

$\therefore \quad A^2 = A - I$

$\Rightarrow \quad A^3 = A^2 - A = (A-I) - A = -I$

$\therefore \quad A^3 = -I$

$\Rightarrow \quad A^{2010} = (A^3)^{670} = (-I)^{670} = I_2$

29. (d) $\left.\begin{array}{l}AB = A\\BA = B\end{array}\right\} \Rightarrow \left.\begin{array}{l}A^2 = A\\B^2 = B\end{array}\right\} \Rightarrow \left.\begin{array}{l}A^n = A\\B^n = B\end{array}\right\}$

and $(A + B)^2 = A^2 + B^2 + AB + BA$

$= A + B + A + B = 2(A + B)$

$\therefore\ (A^{2010} + B^{2010})^{2011} = (A + B)^{2011} = 2^{2011}(A + B)$

30. (c) Let $A = (a_{ij})_{3\times 3}$

and $X = \begin{bmatrix} x \\ y \\ z \end{bmatrix}$

Now, $X^T AX = 0$

$\Rightarrow\ [x\ \ y\ \ z]\begin{bmatrix} a_{11} & a_{12} & a_{13} \\ a_{21} & a_{22} & a_{23} \\ a_{31} & a_{32} & a_{33} \end{bmatrix}\begin{bmatrix} x \\ y \\ z \end{bmatrix} = 0$

$\Rightarrow a_{11}x^2 + a_{22}y^2 + a_{33}z^2 + (a_{12} + a_{21})_{xy} + (a_{13} + a_{31})xz + (a_{23} + a_{32})yz = 0$

$\Rightarrow\ a_{12} = -a_{21},\ a_{13} = -a_{31}$

$a_{23} = -a_{32}$ and $a_{11} = a_{22} = a_{33} = 0$

Hence, $a_{21} + a_{31} + a_{32} = -(a_{12} + a_{13} + a_{23})$

$= -(2008 + 2010 - 2012) = -2006$

31. (a,b,c)

Here, $A^2 = A$ and $B^2 = B$, also $AB \pm BA = 0$

$\therefore\ A^2 - B^2 = A - B$

$\Rightarrow\ (A - B)[(A + B) - I] = 0$

Either $A - B = 0$ or $A + B = I$ ... (i)

Now, $(A - B)^2 = A^2 + B^2 + AB - BA$

$= A^2 + B^2 + 0$

$\Rightarrow\ (A - B)^2 = A + B$

$\therefore\ |A - B|^2 = |A + B|$, [From Eq. (i), $|A + B| = 1$]

$\Rightarrow\ |A - B|^2 = 1$ or $|A - B| = \pm 1$ ... (ii)

Again, from Eq. (i), $A - B = 0$

$\Rightarrow\ |A - B| = 0$ ... (iii)

$\therefore\ |A - B| = 0, -1, 1$

32. (a,b) $AA^T = A^T A = I$

$\Rightarrow\ (A^T)^T A^T = A^T (A^T)^T = I$

Hence, A^T is orthogonal.

$\because\ AA^T = I$

$\Rightarrow\ (AA^T)^{-1} = I^{-1}$

$\Rightarrow\ (A^T)^{-1}A^{-1} = I \Rightarrow (A^{-1})^T A^{-1} = I$

Similarly, $A^{-1}(A^{-1})^T = I$

Hence, A^{-1} is orthogonal.

33. (a,b) The transpose of a diagonal matrix is itself.

Also, the multiplication of A and D will be always commutative, where A and D are matrices of order 3×3 and D is a diagonal matrix.

34. (a,b,c,d) If $|A| \neq 0$

$\Rightarrow\ A^{-1}$ exists.

$AB = 0$

$\Rightarrow\ A^{-1}(AB) = 0$

$\Rightarrow\ B = 0$

again if $|B| \neq 0$

$\Rightarrow\ B^T$ exists.

$AB = 0$

$\Rightarrow\ (AB)B^{-1} = 0$

$\Rightarrow\ A = 0$

$\because\ AB = 0$

$\Rightarrow\ |AB| = 0$

$\Rightarrow\ |A||B| = 0$

If $|A| \neq 0$

Then, $|B| = 0$

and $|B| \neq 0$

Then, $|A| = 0$

35. (a,b,c) Here, $B = A^{2^n} = A^{2^{n-1}\cdot 2} = (A^2)^{2^{n-1}}$

$= (A^{-1})^{2^{n-1}}$ [using $A^2 = A^{-1}$]

$= (A^{-1})^{2^{n-2}\cdot 2^1}$

$= ((A^{-1})^2)^{2^{n-2}}$

$= ((A^2)^{-1})^{2^{n-2}}$

$= ((A^{-1})^{-1})^{2^{n-2}}$

$= A^{2^{n-2}}$

$= C$

$\therefore\ B = C$ Option (c) is correct.

$\Rightarrow |B - C| = 0$, option (a) is correct.

$\Rightarrow\ (B + C)(B - C) = 0$, option (b) is correct.

36. (a,c) Given, $f(x) = \begin{vmatrix} p + \sin x & p + d + \sin x & -2d + \sin x \\ p + d + \sin x & p + 2d + \sin x & -1 + \sin x \\ p + 2d + \sin x & p + 3d + \sin x & 2d + \sin x \end{vmatrix}$

On applying $C_1 \to C_1 - C_2$

$= \begin{vmatrix} -d & p + d + \sin x & -2d + \sin x \\ -d & p + 2d + \sin x & -1 + \sin x \\ -d & p + 3d + \sin x & 2d + \sin x \end{vmatrix}$

$= -d\begin{vmatrix} 1 & p + d + \sin x & -2d + \sin x \\ 1 & p + 2d + \sin x & -1 + \sin x \\ 1 & p + 3d + \sin x & 2d + \sin x \end{vmatrix}$

On applying $R_2 \to R_2 - R_1$ and $R_3 \to R_3 - R_1$

$= -d\begin{vmatrix} 1 & p + d + \sin x & -2d + \sin x \\ 0 & d & -1 + 2d \\ 0 & 2d & 4d \end{vmatrix}$

$= -d[4d^2 + 2d - 4d^2] = -d \times 2d = -2d^2$

$\because\ \int_0^2 f(x)dx = -4$

$\Rightarrow\ -2d^2\int_0^2 dx = -4$

$\Rightarrow\ 4d^2 = -4$

$\Rightarrow\ d^2 = 1$

$\therefore\ d = \pm 1$

37. (c) Here, $B_0^2 = \begin{bmatrix} -4 & -3 & -3 \\ 1 & 0 & 1 \\ 4 & 4 & 3 \end{bmatrix}\begin{bmatrix} -4 & -3 & -3 \\ 1 & 0 & 1 \\ 4 & 4 & 3 \end{bmatrix}$

$= \begin{bmatrix} 1 & 0 & 0 \\ 0 & 1 & 0 \\ 0 & 0 & 1 \end{bmatrix} = I$

$\therefore$ det $[A_0 + A_0^2 B_0^2 + A_0^3 + A_0^4 B_0^4 + \ldots + 10 \text{ terms}]$

$\Rightarrow$ det $[A_0 + A_0^2 + A_0^3 + \ldots + 10 \text{ terms}]$ ... (i)

Also,

$A_0^2 = \begin{bmatrix} 2 & -2 & -4 \\ -1 & 3 & 4 \\ 1 & -2 & -3 \end{bmatrix}\begin{bmatrix} 2 & -2 & -4 \\ -1 & 3 & 4 \\ 1 & -2 & -3 \end{bmatrix} = \begin{bmatrix} 2 & -2 & -4 \\ -1 & 3 & 4 \\ 1 & -2 & -3 \end{bmatrix} = A_0$

$\therefore$ det $[A_0 + A_0^2 + A_0^3 + \ldots + 10 \text{ terms}]$

$\Rightarrow 10|A_0| = 0$, as $|A_0| = 0$

38. (c) Here, $B_n = \text{adj}(B_{n-1})$

$$\therefore \quad B_1 = \text{adj}(B_0) = \begin{bmatrix} -4 & -3 & -3 \\ 1 & 0 & 1 \\ 4 & 4 & 3 \end{bmatrix} = B_0$$

$$B_2 = \text{adj}(B_1) = \text{adj}(B_0) = B_0$$

$$\Rightarrow B_1 + B_2 + \dots + B_{49} = B_0 + B_0 + \dots + 49 \text{ times}$$
$$= 49\, B_0$$

39. (d) $A_0 X = B_0$

Here, $|A_0| = 0$

and $(\text{adj}\, A_0)\, B_0 \neq 0$

$\therefore$ No solution.

40. (a) We have, $C^T(A+B)C$

$$\Rightarrow \quad C^T(A+B)(A+B)^{-1}(A-B)$$
$$= C^T(A-B)$$
$$= [(A+B)^{-1}(A-B)]^T(A-B)$$
$$= (A-B)^T((A+B)^T)^{-1}(A-B)$$
$$= (A+B)(A-B)^{-1}(A-B)$$
$$= A+B$$

41. (b) We have,

$$C^T(A-B)C$$
$$= [(A+B)^{-1}(A-B)]^T(A-B)C$$
$$= (A-B)^T[(A+B)^T]^{-1}(A-B)C$$
$$= (A+B)(A-B)^{-1}(A-B)C$$
$$= (A+B)C$$
$$= (A+B)(A+B)^{-1}(A-B)$$
$$= A-B$$

42. (c) $C^T(2A)C = 2A$

or $C^TAC = A$

43. (c) Let A = Skew-symmetric matrix

and B = Scalar matrix

$$\therefore n(A \cup B) = n(A) + n(B) - n(A \cap B)$$
$$= 7^3 + 7 \cdot (1) - 1 = 7^3 + 6$$

44. (c) Let A = Lower diagonal matrix

B = Upper diagonal matrix

$$\therefore n(A \cup B) = n(A) + n(B) - n(A \cap B)$$
$$= 7^6 + 7^6 - 7^3 = 2 \cdot 7^6 - 7^3 = 7^3(2 \times 7^3 - 1)$$

45. (c) If (a, b, c) are trace elements,

Trace	Trace elements
9	3, 3, 3
8	3, 3, 2
7	3, 2, 2 or 3, 3, 1

$$\therefore \text{Number of matrices} = 1(7^6) + 3(7^6) + 6(7^6)$$
$$= 10(7^6)$$

46. (6) $\because$ $P^2 = I - P$

Now,
$$P^3 = P^2 \cdot P$$
$$= (I-P)P$$
$$= P - P^2$$
$$= P - (I-P)$$
$$= P - I + P$$
$$= 2P - I$$

Now,
$$P^4 = P^2 \cdot P^2$$
$$= (I-P)(I-P) = I^2 - 2P + P^2$$
$$= I - 2P + I - P$$
$$= 2I - 3P$$
$$\Rightarrow \quad P^5 = P^4 \cdot P$$
$$= (2I - 3P)P = 2P - 3P^2$$
$$= 2P - 3(I - P)$$
$$= 2P + 3P - 3I = 5P - 3I$$

Now,
$$P^6 = P^5 \cdot P$$
$$= (5P - 3I)P$$
$$= 5P^2 - 3P$$
$$= 5(I - P) - 3P$$
$$= 5I - 8P$$

Hence, $n = 6$

47. (2) We have, $(I - 0.4A)^{-1} = I - \lambda A$

$$\Rightarrow \quad I = (I - 0.4A)(I - \lambda A)$$
$$\Rightarrow \quad I = I - (\lambda + 0.4A) + 0.4\lambda A^2$$
$$\Rightarrow \quad 0 = -\lambda - 0.4A + 0.4\lambda A$$
$$\Rightarrow \quad 0 = -0.6\lambda A - 0.4A$$
$$\Rightarrow \quad \lambda = \frac{-0.4}{0.6} = \frac{-2}{3}$$

Now, $|3\lambda| = \left|3 \times \left(\frac{-2}{3}\right)\right| = 2$

48. (2) Let
$$A = \begin{bmatrix} a & b \\ c & d \end{bmatrix}$$
$$A^T = \begin{bmatrix} a & c \\ b & d \end{bmatrix}$$
$$\because \quad A + A^T = I$$
$$\Rightarrow \quad \begin{bmatrix} 2a & b+c \\ b+c & 2d \end{bmatrix} = \begin{bmatrix} 1 & 0 \\ 0 & 1 \end{bmatrix}$$
$$\Rightarrow \quad a = d = \frac{1}{2} \text{ and } b = -c$$

Hence,
$$A = \begin{bmatrix} \frac{1}{2} & -c \\ c & \frac{1}{2} \end{bmatrix}$$
$$\because \quad AA^T = I$$
$$\Rightarrow \quad \begin{bmatrix} \frac{1}{2} & -c \\ c & 1/2 \end{bmatrix} \begin{bmatrix} \frac{1}{2} & c \\ -c & \frac{1}{2} \end{bmatrix} = \begin{bmatrix} 1 & 0 \\ 0 & 1 \end{bmatrix}$$
$$\Rightarrow \quad \begin{bmatrix} \frac{1}{4} + c^2 & 0 \\ 0 & c^2 + \frac{1}{4} \end{bmatrix} = \begin{bmatrix} 1 & 0 \\ 0 & 1 \end{bmatrix}$$
$$\Rightarrow \quad \frac{1}{4} + c^2 = 1$$
$$\Rightarrow \quad c^2 = \frac{3}{4}$$
$$\Rightarrow \quad c = \frac{\pm\sqrt{3}}{2}$$

Hence,
$$A = \begin{bmatrix} \frac{1}{2} & \frac{\sqrt{3}}{2} \\ \frac{-\sqrt{3}}{2} & \frac{1}{2} \end{bmatrix}$$

and
$$A^T = \begin{bmatrix} \frac{1}{2} & \frac{-\sqrt{3}}{2} \\ \frac{\sqrt{3}}{2} & \frac{1}{2} \end{bmatrix}$$

49. (2) $\therefore \quad AA^T = I$

$$\Rightarrow \begin{bmatrix} a & b \\ c & d \end{bmatrix}\begin{bmatrix} a & c \\ b & d \end{bmatrix} = I$$

$$\Rightarrow \begin{bmatrix} a^2+b^2 & ac+bd \\ ac+bd & c^2+d^2 \end{bmatrix} = \begin{bmatrix} 1 & 0 \\ 0 & 1 \end{bmatrix}$$

$$\Rightarrow a^2 = 1, d^2 = 1$$

$$b = 0 \text{ and } c = 0$$

or $$b^2 = 1, c^2 = 1$$

$$a = 0 \text{ and } d = 0$$

Hence, $$A = \begin{bmatrix} 1 & 0 \\ 0 & 1 \end{bmatrix} \text{ or } A = \begin{bmatrix} 0 & 1 \\ 1 & 0 \end{bmatrix}$$

50. (7) We have, $$A = \begin{bmatrix} 3 & 11 \\ 2 & 8 \end{bmatrix}$$

$$\Rightarrow |A| = 24 - 22 = 2$$

Now, $$|A^{2011} - 5A^{2010}| = |A^{2010}(A-5)| = |A^{2010}||A-5| = (2)^{2010} \times (-28) = -7 \times (2)^{2012}$$

Hence, $\lambda = 7$

JEE Main & AIEEE Archive

1. (c) Let $\Delta = \begin{vmatrix} 3 & 1+f(1) & 1+f(2) \\ 1+f(1) & 1+f(2) & 1+f(3) \\ 1+f(2) & 1+f(3) & 1+f(4) \end{vmatrix}$

$$\Rightarrow \Delta = \begin{vmatrix} 3 & 1+\alpha+\beta & 1+\alpha^2+\beta^2 \\ 1+\alpha+\beta & 1+\alpha^2+\beta^2 & 1+\alpha^3+\beta^3 \\ 1+\alpha^2+\beta^2 & 1+\alpha^3+\beta^3 & 1+\alpha^4+\beta^4 \end{vmatrix}$$

$$= \begin{vmatrix} 1\cdot1+1\cdot1+1\cdot1 & 1\cdot1+1\cdot\alpha+1\cdot\beta & 1\cdot1+1\cdot\alpha^2+1\cdot\beta^2 \\ 1\cdot1+\alpha\cdot1+\beta\cdot1 & 1\cdot1+\alpha\cdot\alpha+\alpha\cdot\beta & 1\cdot1+\alpha\cdot\alpha^2+\beta\cdot\beta^2 \\ 1\cdot1+1\cdot\alpha^2+1\cdot\beta^2 & 1\cdot1+\alpha^2\cdot\alpha+\beta^2\cdot\beta & 1\cdot1+\alpha^2\cdot\alpha^2+\beta^2\cdot\beta^2 \end{vmatrix}$$

$$= \begin{vmatrix} 1 & 1 & 1 \\ 1 & \alpha & \beta \\ 1 & \alpha^2 & \beta^2 \end{vmatrix}\begin{vmatrix} 1 & 1 & 1 \\ 1 & \alpha & \beta \\ 1 & \alpha^2 & \beta^2 \end{vmatrix} = \begin{vmatrix} 1 & 1 & 1 \\ 1 & \alpha & \beta \\ 1 & \alpha^2 & \beta^2 \end{vmatrix}^2$$

On expanding, we get

$$\Delta = (1-\alpha)^2(1-\beta)^2(\alpha-\beta^2)$$

Hence,

$$K(1-\alpha)^2(1-\beta)^2(\alpha-\beta)^2 = (1-\alpha)^2(1-\beta)^2(\alpha-\beta)^2$$

$$\therefore \quad K = 1$$

2. (b) If A is a non-singular matrix, then $|A| \neq 0$.

$AA^T = A^TA$ and $B = A^{-1}A^T$

$BB' = (A^{-1}A^T)(A^{-1}A^T)^T$

$= A^{-1}A^T A(A^{-1})^T$ $\quad [\because (A')' = A]$

$= A^{-1}AA^T(A^{-1})^T$ $\quad [\because AA^T = A^TA]$

$= IA^T(A^{-1})^T$

$= A^T(A^{-1})^T$

$= (A^{-1}A)^T$ $\quad [\because (AB)^T = B^TA^T]$

$= I^T = I$

3. (b) Given equations can be written in matrix form as

$$AX = B$$

where, $A = \begin{bmatrix} k+1 & 8 \\ k & k+3 \end{bmatrix}, X = \begin{bmatrix} x \\ y \end{bmatrix}$

and $B = \begin{bmatrix} 4k \\ 3k-1 \end{bmatrix}$

For no solution, $|A| = 0$ and (adj A) $B \neq 0$

Now, $$|A| = \begin{vmatrix} k+1 & 8 \\ k & k+3 \end{vmatrix} = 0$$

$$\Rightarrow (k+1)(k+3) - 8k = 0$$

$$\Rightarrow k^2 + 4k + 3 - 8k = 0$$

$$\Rightarrow k^2 - 4k + 3 = 0$$

$$\Rightarrow (k-1)(k-3) = 0$$

$$\Rightarrow k = 1 \text{ or } = 3$$

Now, $\quad$ adj $A = \begin{bmatrix} k+3 & -8 \\ -k & k+1 \end{bmatrix}$

Now, (adj A) $B = \begin{bmatrix} k+3 & -8 \\ -k & k+1 \end{bmatrix}\begin{bmatrix} 4k \\ 3k-1 \end{bmatrix}$

$$= \begin{bmatrix} (k+3)(4k) - 8(3k-1) \\ -4k^2 + (k+1)(3k-1) \end{bmatrix} = \begin{bmatrix} 4k^2 - 12k + 8 \\ -k^2 + 2k - 1 \end{bmatrix}$$

Put $k = 1$

(adj A) $B = \begin{bmatrix} 4-12+8 \\ -1+2-1 \end{bmatrix} = \begin{bmatrix} 0 \\ 0 \end{bmatrix}$ $\quad$ [not true]

Put $k = 3$

(adj A) $B = \begin{bmatrix} 36-36+8 \\ -9+6-1 \end{bmatrix} = \begin{bmatrix} 8 \\ -4 \end{bmatrix} \neq 0$ $\quad$ [true]

Hence, required value of k is 1.

Alternate Method

Condition for the system of equations has no solution.

$$\frac{a_1}{a_2} = \frac{b_1}{b_2} \neq \frac{c_1}{c_2}$$

$$\therefore \quad \frac{k+1}{k} = \frac{8}{k+3} \neq \frac{4k}{3k-1}$$

Take $$\frac{k+1}{k} = \frac{8}{k+3}$$

$$\Rightarrow k^2 + 4k + 3 = 8k$$

$$\Rightarrow k^2 - 4k + 3 = 0$$

$$\Rightarrow (k-1)(k-3) = 0$$

$$k = 1, 3$$

If $k = 1$, then $\dfrac{8}{1+3} \neq \dfrac{4\cdot1}{2}$ $\quad$ [not true]

If $k = 3$, then $\dfrac{8}{6} \neq \dfrac{4\cdot3}{9-1}$ $\quad$ [true]

$$k = 3$$

Hence, only one value of k exists.

4. (b) If A is matrix of order n, then $|\text{adj } A| = |A|^{n-1}$.

Given, $P = \begin{bmatrix} 1 & \alpha & 3 \\ 1 & 3 & 3 \\ 2 & 4 & 4 \end{bmatrix}$

$\therefore \quad |P| = 1(12-12) - \alpha(4-6) + 3(4-6) = 2\alpha - 6$

$\because \quad P = \text{adj}(A)$ [given]

$\therefore \quad |P| = |\text{adj } A| = |A|^2 = 16$

$\Rightarrow 2\alpha - 6 = 16 \Rightarrow 2\alpha = 22$

$\therefore \quad \alpha = 11$

5. (d) **Given** Matrices are

$$A = \begin{bmatrix} 1 & 0 & 0 \\ 2 & 1 & 0 \\ 3 & 2 & 1 \end{bmatrix}, AU_1 = \begin{bmatrix} 1 \\ 0 \\ 0 \end{bmatrix} \text{ and } AU_2 = \begin{bmatrix} 0 \\ 1 \\ 0 \end{bmatrix}$$

To find Matrix $U_1 + U_2$

Since, both AU_1 and AU_2 are given, hence adding them, we get

$$AU_1 + AU_2 = \begin{bmatrix} 1 \\ 0 \\ 0 \end{bmatrix} + \begin{bmatrix} 0 \\ 1 \\ 0 \end{bmatrix} = \begin{bmatrix} 1+0 \\ 0+1 \\ 0+0 \end{bmatrix}$$

$$\Rightarrow \quad A(U_1 + U_2) = \begin{bmatrix} 1 \\ 1 \\ 0 \end{bmatrix}$$

Since, A is a non-singular matrix i.e. $|A| \neq 0$, hence multiplying both sides by A^{-1}(from RHS), we get

$$A^{-1}A(U_1 + U_2) = A^{-1}\begin{bmatrix} 1 \\ 1 \\ 0 \end{bmatrix}$$

$$\Rightarrow U_1 + U_2 = \begin{bmatrix} 1 & 0 & 0 \\ 2 & 1 & 0 \\ 3 & 2 & 1 \end{bmatrix}^{-1} \times \begin{bmatrix} 1 \\ 1 \\ 0 \end{bmatrix} \quad \ldots\text{(i)}$$

$$\text{Now,} \quad |A| = \begin{vmatrix} 1 & 0 & 0 \\ 2 & 1 & 0 \\ 3 & 2 & 1 \end{vmatrix} = 1 \times \begin{vmatrix} 1 & 0 \\ 2 & 1 \end{vmatrix} - 0 + 0$$

[by expanding the determinant along row 1]

$\Rightarrow \quad |A| = 1$

Now, cofactor matrix of A (i.e. the matrix in which every element is replaced by corresponding cofactor)

$$= \begin{bmatrix} \begin{vmatrix} 1 & 0 \\ 2 & 1 \end{vmatrix} & -\begin{vmatrix} 2 & 0 \\ 3 & 1 \end{vmatrix} & \begin{vmatrix} 2 & 1 \\ 3 & 2 \end{vmatrix} \\ -\begin{vmatrix} 0 & 0 \\ 2 & 1 \end{vmatrix} & \begin{vmatrix} 1 & 0 \\ 3 & 1 \end{vmatrix} & -\begin{vmatrix} 1 & 0 \\ 3 & 2 \end{vmatrix} \\ \begin{vmatrix} 0 & 0 \\ 1 & 0 \end{vmatrix} & -\begin{vmatrix} 1 & 0 \\ 2 & 0 \end{vmatrix} & \begin{vmatrix} 1 & 0 \\ 2 & 1 \end{vmatrix} \end{bmatrix} = \begin{bmatrix} 1 & -2 & 1 \\ 0 & 1 & -2 \\ 0 & 0 & 1 \end{bmatrix}$$

$$\therefore \text{adj}(A) = \begin{bmatrix} 1 & -2 & 1 \\ 0 & 1 & -2 \\ 0 & 0 & 1 \end{bmatrix}^T = \begin{bmatrix} 1 & 0 & 0 \\ -2 & 1 & 0 \\ 1 & -2 & 1 \end{bmatrix}$$

$$\Rightarrow \quad A^{-1} = \frac{\text{adj}(A)}{|A|}$$

$$\Rightarrow \quad A^{-1} = \begin{bmatrix} 1 & 0 & 0 \\ -2 & 1 & 0 \\ 1 & -2 & 1 \end{bmatrix} \quad [\because |A| = 1]$$

From Eq. (i), we get

$$U_1 + U_2 = \begin{bmatrix} 1 & 0 & 0 \\ 2 & 1 & 0 \\ 3 & 2 & 1 \end{bmatrix}^{-1} \times \begin{bmatrix} 1 \\ 1 \\ 0 \end{bmatrix} = \begin{bmatrix} 1 & 0 & 0 \\ -2 & 1 & 0 \\ 1 & -2 & 1 \end{bmatrix} \times \begin{bmatrix} 1 \\ 1 \\ 0 \end{bmatrix}$$

$$\therefore \quad U_1 + U_2 = \begin{bmatrix} 1+0+0 \\ -2+1+0 \\ 1-2+0 \end{bmatrix} = \begin{bmatrix} 1 \\ -1 \\ -1 \end{bmatrix}$$

6. (c) **Given**

(i) Two matrices P and Q are of order 3×3 such that $P \neq Q$.

(ii) $P^3 = Q^3$ and $P^2Q = Q^2P$

To find The value of determinant of $P^2 + Q^2$.

On subtracting the given equations, we get

$$P^3 - P^2Q = Q^3 - Q^2P$$

$$\Rightarrow \quad P^2(P - Q) = Q^2(Q - P)$$

$$\Rightarrow \quad (P - Q)(P^2 + Q^2) = 0$$

Now, since $\quad P \neq Q$ [given]

$\Rightarrow \quad P - Q \neq 0$

$\Rightarrow \quad |P^2 + Q^2| = 0$

$\therefore \quad P^2 + Q^2 = 0$

7. (a) Since, equation has non-zero solution.

$\Rightarrow \quad \Delta = 0$

$$\therefore \quad \begin{vmatrix} 4 & k & 2 \\ k & 4 & 1 \\ 2 & 2 & 1 \end{vmatrix} = 0$$

$\Rightarrow \quad 4(4-2) - k(k-2) + 2(2k-8) = 0$

$\Rightarrow \quad 8 - k^2 + 2k + 4k - 16 = 0$

$\Rightarrow \quad k^2 - 6k + 8 = 0$

$\Rightarrow \quad (k-2)(k-4) = 0$

$\Rightarrow \quad k = 2, 4$

Hence, number of values of k is two.

8. (b) Since, A and B are symmetric matrices.

$\therefore \quad A^T = A$ and $B^T = B$

Now, to check $A(BA)$ is symmetric.

Consider $[A(BA)]^T = (BA)^T \cdot A^T = (A^T B^T) A^T$

$= (AB)A = A(BA)$

So, $\quad [A(BA)]^T = A(BA)$

$\Rightarrow A(BA)$ is symmetric.

Similarly, $(AB)A$ is symmetric.

So, Statement I is correct. Also, Statement II is correct, as if A and B are symmetric.

$\Rightarrow (AB)$ is symmetric, iff $AB = BA$,

i.e. AB commutative.

Hence, both the statements are correct but Statement II is not the correct explanation of Statement I.

9. (b) $x - ky + z = 0$, $kx + 3y - kz = 0$

and $3x + y - z = 0$ has trivial solution.

$$\Rightarrow \quad \begin{vmatrix} 1 & -k & 1 \\ k & 3 & -k \\ 3 & 1 & -1 \end{vmatrix} \neq 0$$

$\Rightarrow \quad 1(-3+k) + k(-k+3k) + 1(k-9) \neq 0$

$\Rightarrow \quad k - 3 + 2k^2 + k - 9 \neq 0$

$\Rightarrow \quad 2k^2 + 2k - 12 \neq 0$

$\Rightarrow \quad k^2 + k - 6 \neq 0$

$\Rightarrow \quad (k+3)(k-2) \neq 0$

$\therefore \quad k \neq 2, -3$

$k \in R - \{2, -3\}$

10. (c) Determinant of skew-symmetric matrix of odd order is zero and of even order is perfect square.

So, Statement I is correct.

Now, $\quad \det(A^T) = \det(A)$

and $\quad \det(-A) = (-1)^n \det(A)$

So, Statement II is incorrect.

11. (a) Here, $\quad H = \begin{bmatrix} \omega & 0 \\ 0 & \omega \end{bmatrix}$

$$\therefore \quad H^2 = \begin{bmatrix} \omega & 0 \\ 0 & \omega \end{bmatrix}\begin{bmatrix} \omega & 0 \\ 0 & \omega \end{bmatrix} = \begin{bmatrix} \omega^2 & 0 \\ 0 & \omega^2 \end{bmatrix}$$

$$\Rightarrow \quad H^K = \begin{bmatrix} \omega^K & 0 \\ 0 & \omega^K \end{bmatrix}$$

Then, $H^{K+1} = \begin{bmatrix} \omega^{K+1} & 0 \\ 0 & \omega^{K+1} \end{bmatrix}$

$\therefore \quad H^{70} = \begin{bmatrix} \omega^{70} & 0 \\ 0 & \omega^{70} \end{bmatrix} = \begin{bmatrix} \omega & 0 \\ 0 & \omega \end{bmatrix}$

$= H$

12. (d) The given system of linear equations can be put in the matrix form as

$$\begin{bmatrix} 1 & 2 & 1 \\ 2 & 3 & 1 \\ 3 & 5 & 2 \end{bmatrix}\begin{bmatrix} x_1 \\ x_2 \\ x_3 \end{bmatrix} = \begin{bmatrix} 3 \\ 3 \\ 1 \end{bmatrix}$$

$$\sim \begin{bmatrix} 1 & 2 & 1 \\ 0 & -1 & -1 \\ 0 & -1 & -1 \end{bmatrix}\begin{bmatrix} x_1 \\ x_2 \\ x_3 \end{bmatrix} = \begin{bmatrix} 3 \\ -3 \\ -8 \end{bmatrix} \quad \begin{bmatrix} \text{by } R_2 \to R_2 - 2R_1, \\ R_3 \to R_3 - 3R_1 \end{bmatrix}$$

$$\sim \begin{bmatrix} 1 & 2 & 1 \\ 0 & 1 & 1 \\ 0 & 0 & 0 \end{bmatrix}\begin{bmatrix} x_1 \\ x_2 \\ x_3 \end{bmatrix} = \begin{bmatrix} 3 \\ 3 \\ 5 \end{bmatrix} \quad [\text{by } R_3 \to R_3 - R_2]$$

Clearly, the given system of equations has no solution.

Alternate Method

Subtracting the addition of first two equations from third equation, we get

$0 = -5$, which is an absurd result.

Hence, the given system of equations has no solution.

13. (d) Consider $\begin{pmatrix} 1 & * & * \\ * & 1 & * \\ * & * & 1 \end{pmatrix}$. By placing 1 in anyone of the 6 * position and 0 elsewhere, we get 6 non-singular matrices.

Similarly, $\begin{pmatrix} * & * & 1 \\ * & 1 & * \\ 1 & * & * \end{pmatrix}$ gives atleast one non-singular matrix.

Hence, we get atleast 7 non-singular matrix.

14. (c) A satisfies $A^2 - \text{tr}(A) \cdot A + (\det A)I = 0$

On comparing with $A^2 - I = 0$, we get

$\text{tr}(A) = 0, |A| = -1$

Therefore, Statement I is true but Statement II is false.

Alternate Method

Let $A = \begin{bmatrix} a & b \\ c & d \end{bmatrix}$; $a, b, c, d \neq 0$

Now, $A^2 = \begin{bmatrix} a & b \\ c & d \end{bmatrix} \cdot \begin{bmatrix} a & b \\ c & d \end{bmatrix}$

$\Rightarrow A^2 = \begin{bmatrix} a^2 + bc & ab + bd \\ ac + cd & bc + d^2 \end{bmatrix}$

$\because A^2 = I$

$\Rightarrow a^2 + bc = 1$

$bc + d^2 = 1$

and $ab + bd = ac + cd = 0$

Also, $c \neq 0$ and $b \neq 0$

$\Rightarrow a + d = 0$, $\text{tr}(A) = a + d = 0$

and $|A| = ad - bc = -a^2 - bc = -1$

Hence, Statement I is correct but Statement II is incorrect.

15. (c) $\begin{vmatrix} a+1 & b+1 & c-1 \\ a-1 & b-1 & c+1 \\ a & -b & c \end{vmatrix}$

$= \begin{vmatrix} a & a+1 & a-1 \\ -b & b+1 & b-1 \\ c & c-1 & c+1 \end{vmatrix} + (-1)^n \begin{vmatrix} a+1 & a-1 & a \\ b+1 & b-1 & -b \\ c-1 & c+1 & c \end{vmatrix} \quad [\because |A| = |A^T|]$

$= \begin{vmatrix} a & a+1 & a-1 \\ -b & b+1 & b-1 \\ c & c-1 & c+1 \end{vmatrix} + (-1)^{n+1} \begin{vmatrix} a+1 & a & a-1 \\ b+1 & -b & b-1 \\ c-1 & c & c+1 \end{vmatrix} \quad [C_2 \leftrightarrow C_3]$

$= [1 + (-1)^{n+2}] \begin{vmatrix} a & a+1 & a-1 \\ -b & b+1 & b-1 \\ c & c-1 & c+1 \end{vmatrix}$

This is equal to zero only, if $n + 2$ is odd i.e. n is an odd integer.

16. (b) $|\text{adj } A| = |A|^{n-1} = |A|^{2-1} = |A|$

$\text{adj (adj } A) = |A|^{n-2} A = |A|^0 A = A$

Hence, both the statements are true but Statement II is not the correct explanation of Statement I.

17. (c) Let $A = \begin{bmatrix} a & b \\ c & d \end{bmatrix}$

$\therefore \begin{bmatrix} a & b \\ c & d \end{bmatrix}^2 = \begin{bmatrix} 1 & 0 \\ 0 & 1 \end{bmatrix} \quad [\because A^2 = I]$

$\Rightarrow \begin{bmatrix} a^2 + bc & ab + bd \\ ac + cd & bc + d^2 \end{bmatrix} = \begin{bmatrix} 1 & 0 \\ 0 & 1 \end{bmatrix}$

$\Rightarrow b(a + d) = 0$

$c(a + d) = 0$

and $a^2 + bc = 1$

$bc + d^2 = 1$

$\Rightarrow a = 1$

$d = -1$

$b = c = 0$

If $A = \begin{bmatrix} 1 & 0 \\ 0 & -1 \end{bmatrix}$, then

$A^2 = \begin{bmatrix} 1 & 0 \\ 0 & -1 \end{bmatrix}\begin{bmatrix} 1 & 0 \\ 0 & -1 \end{bmatrix} = \begin{bmatrix} 1 & 0 \\ 0 & 1 \end{bmatrix} = I \quad A \neq I, A \neq -I$

$\det(A) = -1$ (Statement I is correct)

Statement II, $\text{tr}(A) = 1 - 1 = 0$, Statement II is incorrect.

18. (a) Given equations are

$x - cy - bz = 0$

$cx - y + az = 0$

and $bx + ay - z = 0$

For non-zero solution,

$\begin{vmatrix} 1 & -c & -b \\ c & -1 & a \\ b & a & -1 \end{vmatrix} = 0$

$\Rightarrow 1(1 - a^2) + c(-c - ab) - b(ac + b) = 0$

$\Rightarrow 1 - a^2 - c^2 - abc - abc - b^2 = 0$

$\Rightarrow a^2 + b^2 + c^2 + 2abc = 1$

19. (d) As $\det(A) = \pm 1$, A^{-1} exists.

And $A^{-1} = \dfrac{1}{\det(A)}(\text{adj } A) = \pm (\text{adj } A)$

All entries in adj (A) are integers.

Hence, A^{-1} has integer entries.

20. (b) Given, $D = \begin{vmatrix} 1 & 1 & 1 \\ 1 & 1+x & 1 \\ 1 & 1 & 1+y \end{vmatrix}$

Applying $C_2 \to C_2 - C_1$ and $C_3 \to C_3 - C_1$

$= \begin{vmatrix} 1 & 0 & 0 \\ 1 & x & 0 \\ 1 & 0 & y \end{vmatrix} = xy$

Hence, D is divisible by both x and y.

21. (c) Since, $A = \begin{bmatrix} 5 & 5\alpha & \alpha \\ 0 & \alpha & 5\alpha \\ 0 & 0 & 5 \end{bmatrix}$

$$\therefore \quad A^2 = \begin{bmatrix} 5 & 5\alpha & \alpha \\ 0 & \alpha & 5\alpha \\ 0 & 0 & 5 \end{bmatrix}\begin{bmatrix} 5 & 5\alpha & \alpha \\ 0 & \alpha & 5\alpha \\ 0 & 0 & 5 \end{bmatrix}$$

$$= \begin{bmatrix} 25 & 25\alpha + 5\alpha^2 & 10\alpha + 25\alpha^2 \\ 0 & \alpha^2 & 5\alpha^2 + 25\alpha \\ 0 & 0 & 25 \end{bmatrix}$$

$$\Rightarrow \quad |A^2| = \begin{vmatrix} 25 & 25\alpha + 5\alpha^2 & 10\alpha + 25\alpha^2 \\ 0 & \alpha^2 & 5\alpha^2 + 25\alpha \\ 0 & 0 & 25 \end{vmatrix}$$

$$= 25\begin{vmatrix} \alpha^2 & 25\alpha + 5\alpha^2 \\ 0 & 25 \end{vmatrix} = 625\alpha^2$$

$$\Rightarrow \quad 625\alpha^2 = 25 \qquad \text{[given]}$$

$$\therefore \quad |\alpha| = \frac{1}{5}$$

22. (a) Since, $A^2 - B^2 = (A - B)(A + B)$

$$\therefore \quad A^2 - B^2 = A^2 - B^2 + AB - BA$$

$$\Rightarrow \quad AB = BA$$

23. (c) Given, $A = \begin{bmatrix} 1 & 2 \\ 3 & 4 \end{bmatrix}$ and $B = \begin{bmatrix} a & 0 \\ 0 & b \end{bmatrix}$

Now, $AB = \begin{bmatrix} 1 & 2 \\ 3 & 4 \end{bmatrix}\begin{bmatrix} a & 0 \\ 0 & b \end{bmatrix} = \begin{bmatrix} a & 2b \\ 3a & 4b \end{bmatrix}$

and $BA = \begin{bmatrix} a & 0 \\ 0 & b \end{bmatrix}\begin{bmatrix} 1 & 2 \\ 3 & 4 \end{bmatrix} = \begin{bmatrix} a & 2a \\ 3b & 4b \end{bmatrix}$

If $AB = BA$, then $a = b$.

Hence, $AB = BA$ is possible for infinitely many values of B's.

24. (a) $\because \quad A^2 - A + I = O$

$$\Rightarrow \quad A^{-1}A^2 - A^{-1}A + A^{-1}I = O$$

$$\Rightarrow \quad (A^{-1}A)A - (A^{-1}A) + A^{-1} = O$$

$$\Rightarrow \quad A - I + A^{-1} = O$$

$$\Rightarrow \quad A^{-1} = I - A$$

25. (a) Given,

$$f(x) = \begin{bmatrix} 1 + a^2x & (1 + b^2)x & (1 + c^2)x \\ (1 + a^2)x & 1 + b^2x & (1 + c^2)x \\ (1 + a^2)x & (1 + b^2)x & 1 + c^2x \end{bmatrix}$$

Applying $C_1 \to C_1 + C_2 + C_3$, we get

$$f(x) = \begin{bmatrix} 1 + a^2x + x + b^2x + x + c^2x & (1 + b^2)x & (1 + c^2)x \\ x + a^2x + 1 + b^2x + x + c^2x & 1 + b^2x & (1 + c^2)x \\ x + a^2x + x + b^2x + 1 + c^2x & (1 + b^2)x & 1 + c^2x \end{bmatrix}$$

$$= \begin{bmatrix} 1 + (a^2 + b^2 + c^2 + 2)x & (1 + b^2)x & (1 + c^2)x \\ 1 + (a^2 + b^2 + c^2 + 2)x & 1 + b^2x & (1 + c^2)x \\ 1 + (a^2 + b^2 + c^2 + 2)x & (1 + b^2)x & 1 + c^2x \end{bmatrix}$$

$$= \begin{bmatrix} 1 & (1 + b^2)x & (1 + c^2)x \\ 1 & 1 + b^2x & (1 + c^2)x \\ 1 & (1 + b^2)x & 1 + c^2x \end{bmatrix} \qquad [\because a^2 + b^2 + c^2 = -2, \text{ given}]$$

Applying $R_1 \to R_1 - R_3$, $R_2 \to R_2 - R_3$, we get

$$= \begin{bmatrix} 0 & 0 & x - 1 \\ 0 & 1 - x & x - 1 \\ 1 & (1 + b^2)x & 1 + c^2x \end{bmatrix} = \begin{vmatrix} 0 & x - 1 \\ 1 - x & x - 1 \end{vmatrix} = (x - 1)^2$$

Hence, $f(x)$ is of degree 2.

26. (d) The system of given equations has no solution, if

$$\begin{vmatrix} \alpha & 1 & 1 \\ 1 & \alpha & 1 \\ 1 & 1 & \alpha \end{vmatrix} = 0$$

Applying $C_1 \to C_1 + C_2 + C_3$

$$\begin{vmatrix} \alpha + 2 & 1 & 1 \\ \alpha + 2 & \alpha & 1 \\ \alpha + 2 & 1 & \alpha \end{vmatrix} = 0$$

Applying $R_2 \to R_2 - R_1$, $R_3 \to R_3 - R_1$

$$(\alpha + 2)\begin{vmatrix} 1 & 1 & 1 \\ 0 & \alpha - 1 & 0 \\ 0 & 0 & \alpha - 1 \end{vmatrix} = 0$$

$$\Rightarrow \quad (\alpha + 2)(\alpha - 1)^2 = 0$$

$$\therefore \quad \alpha = 1, -2$$

But $\alpha = 1$ makes given three equations same. So, the system of equations has infinite solutions. So, answer is $\alpha = -2$ for which the system of equations has no solution.

27. (c) Since, $a_1, a_2, \ldots, a_n, \ldots$ are in GP, then

$\log a_n, \log a_{n+1}, \log a_{n+2}, \ldots, \log a_{n+8}, \ldots$ are in an AP.

Given,

$$\Delta = \begin{vmatrix} \log a_n & \log a_{n+1} & \log a_{n+2} \\ \log a_{n+3} & \log a_{n+4} & \log a_{n+5} \\ \log a_{n+6} & \log a_{n+7} & \log a_{n+8} \end{vmatrix}$$

$$\therefore \quad \Delta = \begin{vmatrix} a & a + d & a + 2d \\ a + 3d & a + 4d & a + 5d \\ a + 6d & a + 7d & a + 8d \end{vmatrix} = 0$$

where, a and d are the first term and common difference of an AP.

Applying $C_2 \to C_2 - C_1, C_3 \to C_3 - C_1$

$$\Rightarrow \quad \Delta = \begin{vmatrix} a & d & 2d \\ a + 3d & d & 2d \\ a + 6d & d & 2d \end{vmatrix} = 0 \quad \text{[since, two columns are similar]}$$

28. (d) The given matrix, $A = \begin{bmatrix} 0 & 0 & -1 \\ 0 & -1 & 0 \\ -1 & 0 & 0 \end{bmatrix}$

(a) It is clear that, A is not a zero matrix.

(b) Now, $(-1)\, I = -1\begin{bmatrix} 1 & 0 & 0 \\ 0 & 1 & 0 \\ 0 & 0 & 1 \end{bmatrix}$

$$= \begin{bmatrix} -1 & 0 & 0 \\ 0 & -1 & 0 \\ 0 & 0 & -1 \end{bmatrix} \neq A$$

$$\therefore \quad (-1)\, I \neq A$$

(c) Now, $|A| = 0 + 0 - 1\begin{vmatrix} 0 & -1 \\ -1 & 0 \end{vmatrix}$

$$= -1(-1) = 1$$

Since, $|A| \neq 0$, so A^{-1} exists.

(d) Now, $A^2 = \begin{bmatrix} 0 & 0 & -1 \\ 0 & -1 & 0 \\ -1 & 0 & 0 \end{bmatrix}\begin{bmatrix} 0 & 0 & -1 \\ 0 & -1 & 0 \\ -1 & 0 & 0 \end{bmatrix}$

$$\Rightarrow \quad A^2 = \begin{bmatrix} 1 & 0 & 0 \\ 0 & 1 & 0 \\ 0 & 0 & 1 \end{bmatrix}$$

$$\therefore \quad A^2 = I$$

29. (d) Since, B is the inverse of matrix A, i.e. $B = A^{-1}$.

$$\therefore \quad 10A^{-1} = \begin{bmatrix} 4 & 2 & 2 \\ -5 & 0 & \alpha \\ 1 & -2 & 3 \end{bmatrix}$$

$$\Rightarrow \quad 10A^{-1}A = \begin{bmatrix} 4 & 2 & 2 \\ -5 & 0 & \alpha \\ 1 & -2 & 3 \end{bmatrix} A$$

$$\Rightarrow \quad 10I = \begin{bmatrix} 4 & 2 & 2 \\ -5 & 0 & \alpha \\ 1 & -2 & 3 \end{bmatrix}\begin{bmatrix} 1 & -1 & 1 \\ 2 & 1 & -3 \\ 1 & 1 & 1 \end{bmatrix} \quad [\because A^{-1}A = I]$$

$$\Rightarrow \begin{bmatrix} 10 & 0 & 0 \\ 0 & 10 & 0 \\ 0 & 0 & 10 \end{bmatrix} = \begin{bmatrix} 10 & 0 & 0 \\ -5+\alpha & 5+\alpha & -5+\alpha \\ 0 & 0 & 10 \end{bmatrix}$$

$$\Rightarrow \quad -5 + \alpha = 0 \text{ and } 5 + \alpha = 10 \Rightarrow \alpha = 5$$

30. (c) Since, the system of linear equations has a non-zero solution, then

$$\begin{vmatrix} 1 & 2a & a \\ 1 & 3b & b \\ 1 & 4c & c \end{vmatrix} = 0$$

Applying $R_2 \to R_2 - R_1, R_3 \to R_3 - R_1$, we get

$$\Rightarrow \begin{vmatrix} 1 & 2a & a \\ 0 & 3b-2a & b-a \\ 0 & 4c-2a & c-a \end{vmatrix} = 0$$

$$\Rightarrow (3b - 2a)(c - a) - (4c - 2a)(b - a) = 0$$

$$\Rightarrow \quad 3bc - 3ba - 2ac + 2a^2 = 4bc - 2ab - 4ac + 2a^2$$

$$\Rightarrow \quad 2ac = bc + ab$$

On dividing by abc both sides, we get

$$\frac{2}{b} = \frac{1}{a} + \frac{1}{c}$$

So, a, b and c are in HP.

31. (b) Given,

$$A = \begin{bmatrix} a & b \\ b & a \end{bmatrix}$$

$$A^2 = \begin{bmatrix} \alpha & \beta \\ \beta & \alpha \end{bmatrix}$$

$$\therefore \quad A^2 = AA = \begin{bmatrix} a & b \\ b & a \end{bmatrix}\begin{bmatrix} a & b \\ b & a \end{bmatrix}$$

$$\Rightarrow \quad A^2 = \begin{bmatrix} a^2+b^2 & ab+ba \\ ba+ab & b^2+a^2 \end{bmatrix}$$

$$\Rightarrow \quad \begin{bmatrix} \alpha & \beta \\ \beta & \alpha \end{bmatrix} = \begin{bmatrix} a^2+b^2 & 2ab \\ 2ab & a^2+b^2 \end{bmatrix}$$

$$\Rightarrow \quad \alpha = a^2 + b^2$$

and $\quad \beta = 2ab$

32. (a) Given,

$$\Delta = \begin{vmatrix} 1 & \omega^n & \omega^{2n} \\ \omega^n & \omega^{2n} & 1 \\ \omega^{2n} & 1 & \omega^n \end{vmatrix}$$

$$= 1(\omega^{3n} - 1) - \omega^n(\omega^{2n} - \omega^{2n}) + \omega^{2n}(\omega^n - \omega^{4n})$$

$$= 1(1-1) - 0 + \omega^{2n}(\omega^n - \omega^n) \quad [\because \omega^3 = 1]$$

$$= 0$$

33. (d) Since, l, m and n are the pth, qth and rth terms of a GP whose first term is A and common ratio is R.

$$\therefore \quad l = AR^{p-1}$$

$$\Rightarrow \quad \log l = \log A + (p-1)\log R$$

Similarly, $\quad \log m = \log A + (q-1)\log R$

and $\quad \log n = \log A + (r-1)\log R$

Now, $$\begin{vmatrix} \log l & p & 1 \\ \log m & q & 1 \\ \log n & r & 1 \end{vmatrix} = \begin{vmatrix} \log A + (p-1)\log R & p & 1 \\ \log A + (q-1)\log R & q & 1 \\ \log A + (r-1)\log R & r & 1 \end{vmatrix}$$

Applying $C_1 \to C_1 - [C_3 \log A + (C_2 - C_3)\log R]$,

$$\begin{vmatrix} 0 & p & 1 \\ 0 & q & 1 \\ 0 & r & 1 \end{vmatrix} = 0$$

34. (a) Let $$\Delta = \begin{vmatrix} 1 & 1+i+\omega^2 & \omega^2 \\ 1-i & -1 & \omega^2 - 1 \\ -i & -1+\omega-i & -1 \end{vmatrix}$$

Applying $R_1 \to R_1 + R_3$, we get

$$= \begin{vmatrix} 1-i & -1 & \omega^2-1 \\ 1-i & -1 & \omega^2-1 \\ -i & -1+\omega-i & -1 \end{vmatrix} \quad [\because \omega + \omega^2 = -1]$$

$$= 0 \quad \text{[since, two rows are identical]}$$

JEE Advanced & IIT JEE Archive

1. (c,d) $X^T = -X$, $Y^T = -Y$, $Z^T = Z$

(a) Let $P = Y^3Z^4 - Z^4Y^3$

$$\therefore P^T = (Y^3Z^4 - Z^4Y^3)^T = (Y^3Z^4)^T - (Z^4Y^3)^T$$
$$= (Z^4)^T(Y^3)^T - (Y^3)^T(Z^4)^T$$
$$= -Z^4Y^3 + Y^3Z^4 = P$$

$\Rightarrow$ Hence, it is Symmetric.

(b) Let $\quad P = X^{44} + Y^{44}$

$$\therefore \quad P^T = (X^{44})^T + (Y^{44})^T = X^{44} + Y^{44} = P$$

$\Rightarrow$ Hence, it is Symmetric.

(c) Let $\quad P = X^4Z^3 - Z^3X^4$

$$\therefore \quad P^T = (Z^3)^T(X^4)^T - (X^4)^T(Z^3)^T$$
$$= Z^3X^4 - X^4Z^3 = -P$$

$\Rightarrow$ Hence, it is skew-symmetric.

(d) Let $\quad P = X^{23} + Y^{23}$

$$\Rightarrow P^T = (X^{23})^T + (Y^{23})^T = -(X^{23} + Y^{23}) = -P$$

$\Rightarrow$ Hence, it is skew-symmetric.

2. (b,c)

$$\because \begin{vmatrix} (1+\alpha)^2 & (1+2\alpha)^2 & (1+3\alpha)^2 \\ (2+\alpha)^2 & (2+2\alpha)^2 & (2+3\alpha)^2 \\ (3+\alpha)^2 & (3+2\alpha)^2 & (3+3\alpha)^2 \end{vmatrix} = -648\alpha$$

$$\Rightarrow \begin{vmatrix} 1 & \alpha & \alpha^2 \\ 4 & 2\alpha & \alpha^2 \\ 9 & 3\alpha & \alpha^2 \end{vmatrix}\begin{vmatrix} 1 & 1 & 1 \\ 2 & 4 & 6 \\ 1 & 4 & 9 \end{vmatrix} = -648\alpha$$

$$\Rightarrow \alpha^3\begin{vmatrix} 1 & 1 & 1 \\ 4 & 2 & 1 \\ 9 & 3 & 1 \end{vmatrix}\begin{vmatrix} 1 & 1 & 1 \\ 2 & 4 & 6 \\ 1 & 4 & 9 \end{vmatrix} = -648\alpha$$

$$\Rightarrow \quad -8\alpha^3 = -648\alpha$$
$$\Rightarrow \quad \alpha^3 = 81\alpha$$
$$\Rightarrow \quad \alpha^3 - 81\alpha = 0$$
$$\Rightarrow \quad \alpha = 0, 9, -9$$

3. (c,d) **Plan** A square matrix M is invertible if $\det(M)$ or $|M| \neq 0$.

Let $M = \begin{bmatrix} a & b \\ b & c \end{bmatrix}$

(a) Given, $\begin{bmatrix} a \\ b \end{bmatrix} = \begin{bmatrix} b \\ c \end{bmatrix}$

$\Rightarrow \quad a = b = c = \alpha \quad$ (let)

$\Rightarrow \quad M = \begin{bmatrix} \alpha & \alpha \\ \alpha & \alpha \end{bmatrix} \Rightarrow |M| = 0$

$\Rightarrow M$ is non-invertible.

(b) Given, $[bc] = [ab]$

$\Rightarrow \quad a = b = c = \alpha \quad$ (let)

Again $\quad |M| = 0$

$\Rightarrow M$ is non-invertible.

(c) As given, $M = \begin{bmatrix} a & 0 \\ 0 & c \end{bmatrix} \Rightarrow |M| = ac \neq 0$ [$\because a$ and c are non-zero]

$\Rightarrow M$ is invertible.

(d) $M = \begin{bmatrix} a & b \\ b & c \end{bmatrix} \Rightarrow |M| = ac - b^2 \neq 0$

$\because$ ac is not equal to square of an integer.

$\therefore$ M is invertible.

4. (a,b) **Plan**

(i) If A and B are two non-zero matrices and $AB = BA$, then $(A-B)(A+B) = A^2 - B^2$.

(ii) The determinant of the product of the matrices is equal to product of their individual determinants,

i.e. $\quad |AB| = |A||B|$

Given, $\quad M^2 = N^4$

$\Rightarrow \quad M^2 - N^4 = 0$

$\Rightarrow \quad (M - N^2)\ (M + N^2) = 0 \quad$ [as $MN = NM$]

Also, $\quad M \neq N^2$

$\Rightarrow \quad M + N^2 = 0$

$\Rightarrow \quad \det(M + N^2) = 0$

Also, $\det(M^2 + MN^2) = (\det M)\ (\det(M + N^2))$

$= (\det M)\ (0) = 0$

As $\quad \det(M^2 + MN^2) = 0$

Thus, there exists non-zero matrix U such that

$(M^2 + MN^2)U = 0$

5. (b,c,d) Here, $P = [p_{ij}]_{n\times n}$ with $p_{ij} = w^{i+j}$

$\therefore$ When $n = 1$

$P = [p_{ij}]_{1\times 1} = [w^2] \Rightarrow P^2 = [w^4] \neq 0$

$\therefore$ When $\quad n = 2$

$$P = [p_{ij}]_{2\times 2} = \begin{bmatrix} p_{11} & p_{12} \\ p_{21} & p_{22} \end{bmatrix} = \begin{bmatrix} w^2 & w^3 \\ w^3 & w^4 \end{bmatrix} = \begin{bmatrix} w^2 & 1 \\ 1 & w \end{bmatrix}$$

$$P^2 = \begin{bmatrix} w^2 & 1 \\ 1 & w \end{bmatrix}\begin{bmatrix} w^2 & 1 \\ 1 & w \end{bmatrix} \Rightarrow P^2 = \begin{bmatrix} w^4+1 & w^2+w \\ w^2+w & 1+w^2 \end{bmatrix} \neq 0$$

when $n = 3$

$$P = [p_{ij}]_{3\times 3} = \begin{bmatrix} w^2 & w^3 & w^4 \\ w^3 & w^4 & w^5 \\ w^4 & w^5 & w^6 \end{bmatrix} = \begin{bmatrix} w^2 & 1 & w \\ 1 & w & w^2 \\ w & w^2 & 1 \end{bmatrix}$$

$$P^2 = \begin{bmatrix} w^2 & 1 & w \\ 1 & w & w^2 \\ w & w^2 & 1 \end{bmatrix}\begin{bmatrix} w^2 & 1 & w \\ 1 & w & w^2 \\ w & w^2 & 1 \end{bmatrix} = \begin{bmatrix} 0 & 0 & 0 \\ 0 & 0 & 0 \\ 0 & 0 & 0 \end{bmatrix} = 0$$

$\therefore$ $P^2 = 0$, when n is a multiple of 3.

$P^2 \neq 0$, when n is not a multiple of 3.

$\Rightarrow \quad n = 57$ is not possible.

$\therefore \quad n = 55, 58, 56$ is possible.

6. (c,d) (a) $(N^TMN)^T = N^TM^T(N^T)^T = N^TM^TN$, is symmetric as M is symmetric and skew-symmetric as M is skew-symmetric.

(b) $(MN - NM)^T = (MN)^T - (NM)^T = NM - MN = -(MN - NM)$

$\therefore$ Skew-symmetric, when M and N are symmetric.

(c) $(MN)^T = N^TM^T = NM \neq MN$ $\therefore$ Not correct

(d) $(\text{adj } MN) = (\text{adj } N)\cdot(\text{adj } M)$ $\therefore$ Not correct.

7. (a,d) **Concept Involved** If $|A_{n\times n}| = \Delta$, then $|\text{adj } A| = \Delta^{n-1}$

Here, $\quad \text{adj } P_{3\times 3} = \begin{bmatrix} 1 & 4 & 4 \\ 2 & 1 & 7 \\ 1 & 1 & 3 \end{bmatrix}$

$\Rightarrow \quad |\text{adj } P| = |P|^2$

$$\therefore \quad |\text{adj } P| = \begin{vmatrix} 1 & 4 & 4 \\ 2 & 1 & 7 \\ 1 & 1 & 3 \end{vmatrix} = 1(3-7) - 4(6-7) + 4(2-1)$$

$= -4 + 4 + 4 = 4$

$\Rightarrow \quad |P| = \pm 2$

8. (d) Given, $\quad P^T = 2P + I \quad \dots$(i)

$\therefore \quad (P^T)^T = (2P + I)^T = 2P^T + I$

$\Rightarrow \quad P = 2P^T + I$

$\Rightarrow \quad P = 2(2P + I) + I$

$\Rightarrow \quad P = 4P + 3I \quad$ or $\quad 3P = -3I$

$\Rightarrow \quad PX = -IX$

9. (d) **Concept Involved** It is a simple question on scalar multiplication i.e.

$$\begin{vmatrix} ka_1 & ka_2 & ka_3 \\ b_1 & b_2 & b_3 \\ c_1 & c_2 & c_3 \end{vmatrix} = k\begin{vmatrix} a_1 & a_2 & a_3 \\ b_1 & b_2 & b_3 \\ c_1 & c_2 & c_3 \end{vmatrix}$$

Description of Situation Construction of Matrix,

i.e. If $A = [a_{ij}]_{3\times 3} = \begin{bmatrix} a_{11} & a_{12} & a_{13} \\ a_{21} & a_{22} & a_{23} \\ a_{31} & a_{32} & a_{33} \end{bmatrix}$

Here, $\quad P = [a_{ij}]_{3\times 3} = \begin{bmatrix} a_{11} & a_{12} & a_{13} \\ a_{21} & a_{22} & a_{23} \\ a_{31} & a_{32} & a_{33} \end{bmatrix}$

$Q = [b_{ij}]_{3\times 3} = \begin{bmatrix} b_{11} & b_{12} & b_{13} \\ b_{21} & b_{22} & b_{23} \\ b_{31} & b_{32} & b_{33} \end{bmatrix}$, where, $b_{ij} = 2^{i+j}a_{ij}$

$$\therefore \quad |Q| = \begin{vmatrix} 4a_{11} & 8a_{12} & 16a_{13} \\ 8a_{21} & 16a_{22} & 32a_{23} \\ 16a_{31} & 32a_{32} & 64a_{33} \end{vmatrix} = 4\times 8\times 16\begin{vmatrix} a_{11} & a_{12} & a_{13} \\ 2a_{21} & 2a_{22} & 2a_{23} \\ 4a_{31} & 4a_{32} & 4a_{33} \end{vmatrix}$$

$$= 2^9 \times 2\times 4\begin{vmatrix} a_{11} & a_{12} & a_{13} \\ a_{21} & a_{22} & a_{23} \\ a_{31} & a_{32} & a_{33} \end{vmatrix} = 2^{12}\cdot|P| = 2^{12}\cdot 2 = 2^{13}$$

10. (c) Given, $M^T = -M, N^T = -N$ and $MN = NM \quad \dots$(i)

$\therefore \quad M^2N^2(M^TN)^{-1}(MN^{-1})^T$

$\Rightarrow \quad M^2N^2N^{-1}(M^T)^{-1}(N^{-1})^T\cdot M^T$

$\Rightarrow \quad M^2N(NN^{-1})(-M)^{-1}(N^T)^{-1}(-M)$

$\Rightarrow \quad M^2NI(-M^{-1})(-N)^{-1}(-M) \Rightarrow -M^2NM^{-1}N^{-1}M$

$$\Rightarrow \quad -M\cdot(MN)M^{-1}N^{-1}M \quad \Rightarrow \quad -M(NM)M^{-1}N^{-1}M$$

$$\Rightarrow \quad -MN\,(NM^{-1})N^{-1}M$$

$$\Rightarrow \quad -M(NN^{-1})M$$

$$\Rightarrow \quad -M^2$$

Note *Here, non-singular word should not be used, since there is no non-singular 3×3 skew-symmetric matrix.*

11. (a) $|A| \neq 0$, as non-singular

$$\therefore \quad \begin{vmatrix} 1 & a & b \\ \omega & 1 & c \\ \omega^2 & \omega & 1 \end{vmatrix} \neq 0$$

$$\Rightarrow 1(1-c\omega) - a(\omega - c\omega^2) + b(\omega^2 - \omega^2) \neq 0$$

$$\Rightarrow \quad 1 - c\omega - a\omega + ac\omega^2 \neq 0$$

$$\Rightarrow \quad (1-c\omega)(1-a\omega) \neq 0$$

$$\Rightarrow \quad a \neq \frac{1}{\omega}, c \neq \frac{1}{\omega}$$

$\Rightarrow$ $a = \omega, c = \omega$ and $b \in \{\omega, \omega^2\}$ $\Rightarrow$ 2 solutions

12. (9) Let $\quad M = \begin{bmatrix} a_1 & a_2 & a_3 \\ b_1 & b_2 & b_3 \\ c_1 & c_2 & c_3 \end{bmatrix}$

$$\therefore \quad M\begin{bmatrix}0\\1\\0\end{bmatrix} = \begin{bmatrix}-1\\2\\3\end{bmatrix}, M\begin{bmatrix}1\\-1\\0\end{bmatrix} = \begin{bmatrix}1\\1\\-1\end{bmatrix}, M\cdot\begin{bmatrix}1\\1\\1\end{bmatrix} = \begin{bmatrix}0\\0\\12\end{bmatrix}$$

$$\Rightarrow \quad \begin{bmatrix}a_2\\b_2\\c_2\end{bmatrix} = \begin{bmatrix}-1\\2\\3\end{bmatrix}, \begin{bmatrix}a_1-a_2\\b_1-b_2\\c_1-c_2\end{bmatrix} = \begin{bmatrix}1\\1\\-1\end{bmatrix}, \begin{bmatrix}a_1+a_2+a_3\\b_1+b_2+b_3\\c_1+c_2+c_3\end{bmatrix} = \begin{bmatrix}0\\0\\12\end{bmatrix}$$

$\Rightarrow a_2 = -1, b_2 = 2, c_2 = 3, a_1 - a_2 = 1,$
$b_1 - b_2 = 1, c_1 - c_2 = -1$
$a_1 + a_2 + a_3 = 0, b_1 + b_2 + b_3 = 0,$
$c_1 + c_2 + c_3 = 12$
$\therefore \quad a_1 = 0, b_2 = 2, c_3 = 7$
$\Rightarrow$ Sum of diagonal elements $= 0 + 2 + 7 = 9$

13. (4) $|A| = (2k+1)^3, |B| = 0$

But det (adj A) = det (adj B) = 10^6

$$\Rightarrow \quad (2k+1)^6 = 10^6 \quad \Rightarrow \quad k = \frac{9}{2}$$

$$\Rightarrow \quad [k] = 4$$

14. (d) Given, $A = \begin{bmatrix} a & b \\ c & a \end{bmatrix}$, $a, b, c \in \{0, 1, 2, \ldots, p-1\}$

If A is skew-symmetric matrix, then $a = 0, b = -c$

$$\therefore \quad |A| = -b^2$$

Thus, P divides $|A|$, only when $b = 0$ ...(i)

Again, if A is symmetric matrix, then $b = c$ and $|A| = a^2 - b^2$.

Thus, p divides $|A|$, if either p divides $(a-b)$ or p divides $(a+b)$.

p divides $(a-b)$, only when $a = b$.

i.e. $\quad a = b \in \{0, 1, 2, \ldots, (p-1)\}$.

i.e. $\quad p$ choices ...(ii)

$\therefore \quad p$ divides $(a+b)$

$\Rightarrow p$ choices, including $a = b = 0$ included in (i)

$\therefore$ Total number of choices are $(p + p - 1) = 2p - 1$.

15. (c) Trace of $A = 2a$, will be divisible by p, iff $a = 0$.

$|A| = a^2 - bc$, for $(a^2 - bc)$ to be divisible by p. There are exactly $(p-1)$ ordered pairs (b, c) for any value of a.

$\therefore$ Required number is $(p-1)^2$.

16. (d) The number of matrices for which p does not divide $tr\,(A) = (p-1)p^2$ of these $(p-1)^2$ are such that p divides $|A|$.

The number of matrices for which p divides $tr\,(A)$ and p does not divides $|A|$ are $(p-1)^2$.

$\therefore$ Required number $= (p-1)p^2 - (p-1)^2 + (p-1)^2 = p^3 - p^2$

17. (a) Since, $A\begin{bmatrix}x\\y\\z\end{bmatrix} = \begin{bmatrix}1\\0\\0\end{bmatrix}$ is linear equation in three variables and that could have only unique, no solution or infinitely many solution.

$\therefore$ It is not possible to have two solutions.

Hence, number of matrices of A is zero.

18. (a) The given system of equations can be expressed as

$$\begin{bmatrix} 1 & -2 & 3 \\ 1 & -3 & 4 \\ -1 & 1 & -2 \end{bmatrix}\begin{bmatrix}x\\y\\z\end{bmatrix} = \begin{bmatrix}-1\\1\\k\end{bmatrix}$$

Applying $R_2 \to R_2 - R_1, R_3 \to R_3 + R_1$

$$\sim \begin{bmatrix} 1 & -2 & 3 \\ 0 & -1 & 1 \\ 0 & -1 & 1 \end{bmatrix}\begin{bmatrix}x\\y\\z\end{bmatrix} = \begin{bmatrix}-1\\2\\k-1\end{bmatrix} \sim \begin{bmatrix} 1 & -2 & 3 \\ 0 & -1 & 1 \\ 0 & 0 & 0 \end{bmatrix}\begin{bmatrix}x\\y\\z\end{bmatrix} = \begin{bmatrix}-1\\2\\k-3\end{bmatrix}$$

$$R_3 \to R_3 - R_2$$

When $k \neq 3$, the given system of equations has no solution.

$\Rightarrow$ Statement I is true. Clearly, Statement II is also true as it is rearrangement of rows and columns of $\begin{bmatrix} 1 & -2 & 3 \\ 1 & -3 & 4 \\ -1 & 1 & -2 \end{bmatrix}$.

19. (a) Every square matrix satisfied its characteristic equation,

$$\text{i.e. } |A - \lambda I| = 0 \Rightarrow \begin{vmatrix} 1-\lambda & 0 & 0 \\ 0 & 1-\lambda & 1 \\ 0 & -2 & 4-\lambda \end{vmatrix} = 0$$

$$\Rightarrow \quad (1-\lambda)\{(1-\lambda)(4-\lambda) + 2\} = 0$$

$$\Rightarrow \quad \lambda^3 - 6\lambda^2 + 11\lambda - 6 = 0$$

$$\Rightarrow \quad A^3 - 6A^2 + 11A - 6I = 0 \quad \ldots(i)$$

Given, $6A^{-1} = A^2 + cA + dI$, multiplying both sides by A, we get

$$6I = A^3 + cA^2 + dA$$

$$\Rightarrow \quad A^3 + cA^2 + dA - 6I = O \quad \ldots(ii)$$

On comparing Eqs. (i) and (ii), we get

$$c = -6 \text{ and } d = 11$$

20. (a) Now, $P^TP = \begin{bmatrix} \sqrt{3}/2 & -1/2 \\ 1/2 & \sqrt{3}/2 \end{bmatrix}\begin{bmatrix} \sqrt{3}/2 & 1/2 \\ -1/2 & \sqrt{3}/2 \end{bmatrix}$

$$\Rightarrow \quad P^TP = \begin{bmatrix} 1 & 0 \\ 0 & 1 \end{bmatrix} \Rightarrow P^TP = I \Rightarrow P^T = P^{-1}$$

Since, $\quad Q = PAP^T$

$$\therefore \quad P^TQ^{2005}P \quad \ldots(i)$$

$$= P^T[(PAP^T)(PAP^T)\ldots 2005 \text{ times}]P$$

$$= \underbrace{(P^TP)\,A\,(P^TP)\,A\,(P^TP)\ldots(P^TP)\,A\,(P^TP)}_{2005 \text{ Times}}$$

$$= IA^{2005} = A^{2005} \quad \text{[from Eq. (i)]}$$

$$\therefore \quad A = \begin{bmatrix} 1 & 1 \\ 0 & 1 \end{bmatrix}$$

$$A^2 = \begin{bmatrix} 1 & 1 \\ 0 & 1 \end{bmatrix}\begin{bmatrix} 1 & 1 \\ 0 & 1 \end{bmatrix} = \begin{bmatrix} 1 & 2 \\ 0 & 1 \end{bmatrix} \Rightarrow A^3 = \begin{bmatrix} 1 & 2 \\ 0 & 1 \end{bmatrix}\begin{bmatrix} 1 & 1 \\ 0 & 1 \end{bmatrix} = \begin{bmatrix} 1 & 3 \\ 0 & 1 \end{bmatrix}$$

... ...
... ...

$$A^{2005} = \begin{bmatrix} 1 & 2005 \\ 0 & 1 \end{bmatrix}$$

$$\therefore \quad P^TQ^{2005}P = \begin{bmatrix} 1 & 2005 \\ 0 & 1 \end{bmatrix}$$

Last 3 Years' Questions
JEE Main & Advanced
(2016-2018)

JEE Main

2016

1. If $f(x) + 2f\left(\frac{1}{x}\right) = 3x, x \neq 0$ and $S = \{x \in R : f(x) = f(-x)\}$; then S **[Relations and Functions]**

(a) is an empty set
(b) contains exactly one element
(c) contains exactly two elements
(d) contains more than two elements

2. If $A = \begin{bmatrix} 5a & -b \\ 3 & 2 \end{bmatrix}$ and A adj $A = AA^T$, then $5a + b$ is equal to **[Matrices and Determinants]**

(a) -1 (b) 5 (c) 4 (d) 13

3. The system of linear equations

$$x + \lambda y - z = 0;\ \lambda x - y - z = 0;\ x + y - \lambda z = 0$$

has a non-trivial solution for **[Matrices and Determinants]**

(a) infinitely many values of λ
(b) exactly one value of λ
(c) exactly two values of λ
(d) exactly three values of λ

2017

4. If S is the set of distinct values of b for which the following system of linear equations

$$x + y + z = 1, \Rightarrow x + ay + z = 1 \quad \text{and} \quad ax + by + z = 0$$

has no solution, then S is **[Matrices and Determinants]**

(a) an infinite set
(b) a finite set containing two or more elements
(c) singleton set
(d) a empty set

5. Let ω be a complex number such that $2\omega + 1 = z$, where $z = \sqrt{-3}$. If $\begin{vmatrix} 1 & 1 & 1 \\ 1 & -\omega^2 - 1 & \omega^2 \\ 1 & \omega^2 & \omega^7 \end{vmatrix} = 3k$,

then k is equal to **[Matrices and Determinants]**

(a) $-z$ (b) z (c) -1 (d) 1

6. If $A = \begin{bmatrix} 2 & -3 \\ -4 & 1 \end{bmatrix}$, then adj $(3A^2 + 12A)$ is equal to **[Matrices and Determinants]**

(a) $\begin{bmatrix} 72 & -84 \\ -63 & 51 \end{bmatrix}$
(b) $\begin{bmatrix} 51 & 63 \\ 84 & 72 \end{bmatrix}$
(c) $\begin{bmatrix} 51 & 84 \\ 63 & 72 \end{bmatrix}$
(d) $\begin{bmatrix} 72 & -63 \\ -84 & 51 \end{bmatrix}$

7. The function $f : R \to \left[-\frac{1}{2}, \frac{1}{2}\right]$ defined as $f(x) = \frac{x}{1 + x^2}$ is

[Inverse Trigonometric Functions]

(a) invertible
(b) injective but not surjective
(c) surjective but not injective
(d) neither injective nor surjective

8. Let $a, b, c \in R$. If $f(x) = ax^2 + bx + c$ be such that $a + b + c = 3$ and $f(x+y) = f(x) + f(y) + xy, \forall\, x, y \in R$, then $\sum_{n=1}^{10} f(n)$ is equal to

[Relations and Functions]

(a) 330 (b) 165 (c) 190 (d) 255

2018

9. If $\begin{vmatrix} x-4 & 2x & 2x \\ 2x & x-4 & 2x \\ 2x & 2x & x-4 \end{vmatrix} = (A + Bx)(x - A)^2$, then the ordered pair (A, B) is equal to

[Matrices and Determinants]

(a) $(-4, -5)$ (b) $(-4, 3)$ (c) $(-4, 5)$ (d) $(4, 5)$

10. If the system of linear equations

$$x + ky + 3z = 0$$
$$3x + ky - 2z = 0$$
$$2x + 4y - 3z = 0$$

has a non-zero solution (x, y, z), then $\frac{xz}{y^2}$ is equal to

[Matrices and Determinants]

(a) -10 (b) 10 (c) -30 (d) 30

JEE Advanced

2016

1. Let $P = \begin{bmatrix} 3 & -1 & -2 \\ 2 & 0 & \alpha \\ 3 & -5 & 0 \end{bmatrix}$, where $\alpha \in R$. Suppose $Q = [q_{ij}]$ is a matrix such that $PQ = kI$, where $k \in R, k \neq 0$ and I is the identity matrix of order 3. If $q_{23} = -\frac{k}{8}$ and $\det(Q) = \frac{k^2}{2}$ then

[Matrices and Determinants, Type 2]

(a) $\alpha = 0, k = 8$
(b) $4\alpha - k + 8 = 0$
(c) $\det(P \operatorname{adj}(Q)) = 2^9$
(d) $\det(Q \operatorname{adj}(P)) = 2^{13}$

2. The total number of distincts $x \in R$ for which $\begin{vmatrix} x & x^2 & 1+x^3 \\ 2x & 4x^2 & 1+8x^3 \\ 3x & 9x^2 & 1+27x^3 \end{vmatrix} = 10$ is

[Matrices and Determinants, Type 3]

3. Let $z = \frac{-1+\sqrt{3}i}{2}$, where $i = \sqrt{-1}$, and $r, s \in \{1, 2, 3\}$.

Let $P = \begin{bmatrix} (-z)^r & z^{2s} \\ z^{2s} & z^r \end{bmatrix}$ and I be the identity matrix of order 2. Then, the total number of ordered pairs (r, s) for which $P^2 = -I$ is

[Matrices and Determinants, Type 3]

4. Let $P = \begin{bmatrix} 1 & 0 & 0 \\ 4 & 1 & 0 \\ 16 & 4 & 1 \end{bmatrix}$ and I be the identity matrix of order 3. If $Q = [q_{ij}]$ is a matrix, such that $P^{50} - Q = I$, then $\frac{q_{31} + q_{32}}{q_{21}}$ equals

[Matrices and Determinants, Type 1]

(a) 52 (b) 103 (c) 201 (d) 205

Note ***Type-1*** *Only one correct option,* ***Type-2*** *More then one correct option,* ***Type-3*** *Single digit Answer,* ***Type-4*** *Numerical value based*

2017

5. Which of the following is(are) NOT the square of a 3×3 matrix with real entries?

[Matrices and Determinants, Type 2]

(a) $\begin{bmatrix} 1 & 0 & 0 \\ 0 & 1 & 0 \\ 0 & 0 & -1 \end{bmatrix}$ (b) $\begin{bmatrix} 1 & 0 & 0 \\ 0 & -1 & 0 \\ 0 & 0 & -1 \end{bmatrix}$

(c) $\begin{bmatrix} -1 & 0 & 0 \\ 0 & -1 & 0 \\ 0 & 0 & -1 \end{bmatrix}$ (d) $\begin{bmatrix} 1 & 0 & 0 \\ 0 & 1 & 0 \\ 0 & 0 & 1 \end{bmatrix}$

6. For a real number α, if the system

$$\begin{bmatrix} 1 & \alpha & \alpha^2 \\ \alpha & 1 & \alpha \\ \alpha^2 & \alpha & 1 \end{bmatrix}\begin{bmatrix} x \\ y \\ z \end{bmatrix} = \begin{bmatrix} 1 \\ -1 \\ 1 \end{bmatrix}$$

of linear equations, has infinitely many solutions, then $1+\alpha+\alpha^2 =$

[Matrices and Determinants, Type 3]

7. How many 3×3 matrices M with entries from $\{0, 1, 2\}$ are there, for which the sum of the diagonal entries of $M^T M$ is 5 ? **[Matrices and Determinants, Type 1]**

(a) 198 (b) 162 (c) 126 (d) 135

2018

8. The number of real solutions of the equation $\sin^{-1}\left(\sum_{i=1}^{\infty} x^{i+1} - x\sum_{i=1}^{\infty}\left(\frac{x}{2}\right)^i\right)$ $= \frac{\pi}{2} - \cos^{-1}\left(\sum_{i=1}^{\infty}\left(-\frac{x}{2}\right)^i - \sum_{i=1}^{\infty}(-x)^i\right)$ lying in the interval $\left(-\frac{1}{2}, \frac{1}{2}\right)$ is ....

(Here, the inverse trigonometric functions $\sin^{-1} x$ and $\cos^{-1} x$ assume values in $\left[-\frac{\pi}{2}, \frac{\pi}{2}\right]$ and $[0, \pi]$, respectively.) **[Inverse Trigonometric Functions, Type 4]**

9. For any positive integer n, define $f_n : (0, \infty) \to R$ as $f_n(x) = \sum_{j=1}^{n} \tan^{-1}\left(\frac{1}{1+(x+j)(x+j-1)}\right)$ for all $x \in (0, \infty)$. (Here, the inverse trigonometric function $\tan^{-1} x$ assumes values in $\left(-\frac{\pi}{2}, \frac{\pi}{2}\right)$).

Then, which of the following statement(s) is (are) TRUE?

[Inverse Trigonometric Functions, Type 2]

(a) $\sum_{j=1}^{5} \tan^2(f_j(0)) = 55$

(b) $\sum_{j=1}^{10} (1+f'_j(0)) \sec^2(f_j(0)) = 10$

(c) For any fixed positive integer n, $\lim_{x\to\infty} \tan(f_n(x)) = \frac{1}{n}$

(d) For any fixed positive integer n, $\lim_{x\to\infty} \sec^2(f_n(x)) = 1$

10. Let S be the set of all column matrices $\begin{bmatrix} b_1 \\ b_2 \\ b_3 \end{bmatrix}$ such that $b_1, b_2, b_3 \in R$ and the system of equations (in real variables)

$$-x+2y+5z = b_1$$
$$2x-4y+3z = b_2$$
$$x-2y+2z = b_3$$

has at least one solution. Then, which of the following system(s) (in real variables) has (have) at least one solution for each $\begin{bmatrix} b_1 \\ b_2 \\ b_3 \end{bmatrix} \in S$?

[Matrices and Determinants, Type 2]

(a) $x + 2y + 3z = b_1$, $4y + 5z = b_2$ and $x + 2y + 6z = b_3$
(b) $x + y + 3z = b_1$, $5x + 2y + 6z = b_2$ and $-2x - y - 3z = b_3$
(c) $-x + 2y - 5z = b_1$, $2x - 4y + 10z = b_2$ and $x - 2y + 5z = b_3$
(d) $x + 2y + 5z = b_1$, $2x + 3z = b_2$ and $x + 4y - 5z = b_3$

11. Let P be a matrix of order 3×3 such that all the entries in P are from the set $\{-1, 0, 1\}$. Then, the maximum possible value of the determinant of P is ....

[Matrices and Determinants, Type 4]

12. Let $E_1 = \{x \in R : x \neq 1 \text{ and } \frac{x}{x-1} > 0\}$ and $E_2 = \left\{ x \in E_1 : \sin^{-1}\left(\log_e\left(\frac{x}{x-1}\right)\right) \text{ is a real number} \right\}$

(Here, the inverse trigonometric function $\sin^{-1} x$ assumes values in $\left[-\frac{\pi}{2}, \frac{\pi}{2}\right]$.).Let $f : E_1 \to R$ be the function defined by $f(x) = \log_e\left(\frac{x}{x-1}\right)$ and $g : E_2 \to R$ be the function defined by $g(x) = \sin^{-1}\left(\log_e\left(\frac{x}{x-1}\right)\right)$.

[Matrices and Determinant]

	List-I		List-II
P.	The range of f is	1.	$\left(-\infty, \frac{1}{1-e}\right] \cup \left[\frac{e}{e-1}, \infty\right)$
Q.	The range of g contains	2.	$(0, 1)$
R.	The domain of f contains	3.	$\left[-\frac{1}{2}, \frac{1}{2}\right]$
S.	The domain of g is	4.	$(-\infty, 0) \cup (0, \infty)$
		5.	$\left(-\infty, \frac{e}{e-1}\right]$
		6.	$(-\infty, 0) \cup \left(\frac{1}{2}, \frac{e}{e-1}\right]$

The correct option is
(a) P $\to$ 4; Q $\to$ 2; R $\to$ 1; S $\to$ 1
(b) P $\to$ 3; Q $\to$ 3; R $\to$ 6; S $\to$ 5
(c) P $\to$ 4; Q $\to$ 2; R $\to$ 1; S $\to$ 6
(d) P $\to$ 4; Q $\to$ 3; R $\to$ 6; S $\to$ 5

Answer *with* Explanations

JEE Main

1. *(c)* We have, $f(x) + 2f\left(\frac{1}{x}\right) = 3x,$

$x \neq 0$ ...(i)

On replacing x by $\frac{1}{x}$ in the above Eq. (i),

we get $f\left(\frac{1}{x}\right) + 2f(x) = \frac{3}{x}$

$\Rightarrow \quad 2f(x) + f\left(\frac{1}{x}\right) = \frac{3}{x}$ ...(ii)

On multiplying Eq. (ii) by 2 and subtracting Eq. (i) from Eq. (ii), we get

$$4f(x) + 2f\left(\frac{1}{x}\right) = \frac{6}{x}$$

$$f(x) + 2f\left(\frac{1}{x}\right) = 3x$$

$$- \quad - \quad -$$

$$3f(x) = \frac{6}{x} - 3x$$

$\Rightarrow \quad f(x) = \frac{2}{x} - x$

Now, consider $f(x) = f(-x)$

$\Rightarrow \quad \frac{2}{x} - x = -\frac{2}{x} + x$

$\Rightarrow \quad \frac{4}{x} = 2x$

$\Rightarrow \quad 2x^2 = 4 \Rightarrow x^2 = 2$

$\Rightarrow \quad x = \pm\sqrt{2}$

Hence, S contains exactly two elements.

2. *(b)* Given, $A = \begin{bmatrix} 5a & -b \\ 3 & 2 \end{bmatrix}$ and $A \text{ adj } A = AA^T$

Clearly, $A(\text{adj } A) = |A| I_2$

[$\because$ if A is square matrix of order n, then $A(\text{adj } A) = (\text{adj } A) \cdot A = |A| I_n$]

$$= \begin{vmatrix} 5a & -b \\ 3 & 2 \end{vmatrix} I_2 = (10a + 3b) I_2$$

$$= (10a + 3b)\begin{bmatrix} 1 & 0 \\ 0 & 1 \end{bmatrix}$$

$$= \begin{bmatrix} 10a + 3b & 0 \\ 0 & 10a + 3b \end{bmatrix} \quad \text{...(i)}$$

and $AA^T = \begin{bmatrix} 5a & -b \\ 3 & 2 \end{bmatrix}\begin{bmatrix} 5a & 3 \\ -b & 2 \end{bmatrix}$

$$= \begin{bmatrix} 25a^2 + b^2 & 15a - 2b \\ 15a - 2b & 13 \end{bmatrix} \quad \text{...(ii)}$$

$\because \quad A(\text{adj } A) = AA^T$

$$\therefore \begin{bmatrix} 10a + 3b & 0 \\ 0 & 10a + 3b \end{bmatrix} = \begin{bmatrix} 25a^2 + b^2 & 15a - 2b \\ 15a - 2b & 13 \end{bmatrix}$$

[using Eqs. (i) and (ii)]

$\Rightarrow \quad 15a - 2b = 0 \Rightarrow a = \frac{2b}{15}$ ...(iii)

and $\quad 10a + 3b = 13$ ...(iv)

On substituting the value of 'a' from Eq. (iii) in Eq. (iv), we get

$$10 \cdot \left(\frac{2b}{15}\right) + 3b = 13$$

$\Rightarrow \quad \frac{20b + 45b}{15} = 13$

$\Rightarrow \quad \frac{65b}{15} = 13$

$\Rightarrow \quad b = 3$

Now, substituting the value of b in Eq. (iii), we get $5a = 2$

Hence, $\quad 5a + b = 2 + 3 = 5$

3. *(d)* Given, system of linear equation is

$$x + \lambda y - z = 0; \; \lambda x - y - z = 0; \; x + y - \lambda z = 0$$

Note that, given system will have a non-trivial solution only if determinant of coefficient matrix is zero, i.e.

$$\begin{vmatrix} 1 & \lambda & -1 \\ \lambda & -1 & -1 \\ 1 & 1 & -\lambda \end{vmatrix} = 0$$

$\Rightarrow 1(\lambda + 1) - \lambda(-\lambda^2 + 1) - 1(\lambda + 1) = 0$

$\Rightarrow \quad \lambda + 1 + \lambda^3 - \lambda - \lambda - 1 = 0$

$\Rightarrow \quad \lambda^3 - \lambda = 0$

$\Rightarrow \quad \lambda(\lambda^2 - 1) = 0$

$\Rightarrow \quad \lambda = 0 \text{ or } \lambda = \pm 1$

Hence, given system of linear equation has a non-trivial solution for exactly three values of λ.

4. *(d)* $\because \Delta = \begin{vmatrix} 1 & 1 & 1 \\ 1 & a & 1 \\ a & b & 1 \end{vmatrix} = 1(a - b) - 1(1 - a) + 1(b - a^2) = -(a - 1)^2$

$$\Delta_1 = \begin{vmatrix} 1 & 1 & 1 \\ 1 & a & 1 \\ 0 & b & 1 \end{vmatrix} = 1(a - b) - 1(1) + 1(b)$$

$$= -(a - 1)$$

$$\Delta_2 = \begin{vmatrix} 1 & 1 & 1 \\ 1 & 1 & 1 \\ a & 0 & 1 \end{vmatrix} = 1(1) - 1(1 - a) + 1(0 - a) = 0$$

and $\Delta_3 = \begin{vmatrix} 1 & 1 & 1 \\ 1 & a & 1 \\ a & b & 0 \end{vmatrix} = 1(-b) - 1(-a) + 1(b - a^2)$

$$= -a(a - 1)$$

For $\quad a = 1$

$\Delta = \Delta_1 = \Delta_2 = \Delta_3 = 0$

Δ for $b = 1$ only

$x + y + z = 1,$

$x + y + z = 1$

and $\quad x + y + z = 0$

i.e. no solution ($\because$ RHS is not equal)

Hence, for no solution $b = 1$ only

5. *(a)* Given, $2\omega + 1 = z$

$$\Rightarrow \quad 2\omega + 1 = \sqrt{-3} \qquad [\because z = \sqrt{-3}]$$

$$\Rightarrow \quad \omega = \frac{-1+\sqrt{3}i}{2}$$

Since, ω is cube root of unity.

$$\therefore \quad \omega^2 = \frac{-1-\sqrt{3}i}{2}$$

and $\omega^{3n} = 1$

Now, $\begin{vmatrix} 1 & 1 & 1 \\ 1 & -\omega^2 - 1 & \omega^2 \\ 1 & \omega^2 & \omega^7 \end{vmatrix} = 3k$

$$\Rightarrow \quad \begin{vmatrix} 1 & 1 & 1 \\ 1 & \omega & \omega^2 \\ 1 & \omega^2 & \omega \end{vmatrix} = 3k$$

$[\because 1 + \omega + \omega^2 = 0 \text{ and } \omega^7 = (\omega^3)^2 \cdot \omega = \omega]$

On applying $R_1 \to R_1 + R_2 + R_3$, we get

$$\begin{vmatrix} 3 & 1+\omega+\omega^2 & 1+\omega+\omega^2 \\ 1 & \omega & \omega^2 \\ 1 & \omega^2 & \omega \end{vmatrix} = 3k \Rightarrow \begin{vmatrix} 3 & 0 & 0 \\ 1 & \omega & \omega^2 \\ 1 & \omega^2 & \omega \end{vmatrix} = 3k$$

$$\Rightarrow \quad 3(\omega^2 - \omega^4) = 3k$$

$$\Rightarrow \quad (\omega^2 - \omega) = k$$

$$\therefore \quad k = \left(\frac{-1-\sqrt{3}i}{2}\right) - \left(\frac{-1+\sqrt{3}i}{2}\right)$$

$$= -\sqrt{3}i = -z$$

6. *(b)* We have, $A = \begin{bmatrix} 2 & -3 \\ -4 & 1 \end{bmatrix}$

$$\therefore A^2 = A \cdot A = \begin{bmatrix} 2 & -3 \\ -4 & 1 \end{bmatrix}\begin{bmatrix} 2 & -3 \\ -4 & 1 \end{bmatrix}$$

$$= \begin{bmatrix} 4+12 & -6-3 \\ -8-4 & 12+1 \end{bmatrix}$$

$$= \begin{bmatrix} 16 & -9 \\ -12 & 13 \end{bmatrix}$$

Now, $3A^2 + 12A = 3\begin{bmatrix} 16 & -9 \\ -12 & 13 \end{bmatrix} + 12\begin{bmatrix} 2 & -3 \\ -4 & 1 \end{bmatrix}$

$$= \begin{bmatrix} 48 & -27 \\ -36 & 39 \end{bmatrix} + \begin{bmatrix} 24 & -36 \\ -48 & 12 \end{bmatrix}$$

$$= \begin{bmatrix} 72 & -63 \\ -84 & 51 \end{bmatrix}$$

$$\therefore \text{adj}(3A^2 + 12A) = \begin{bmatrix} 51 & 63 \\ 84 & 72 \end{bmatrix}$$

7. *(c)* We have, $f(x) = \dfrac{x}{1+x^2}$

$$\therefore \quad f\left(\frac{1}{x}\right) = \frac{\frac{1}{x}}{1+\frac{1}{x^2}} = \frac{x}{1+x^2} = f(x)$$

$$\therefore \quad f\left(\frac{1}{2}\right) = f(2) \text{ or } f\left(\frac{1}{3}\right) = f(3) \text{ and so on.}$$

So, $f(x)$ is many-one function.

Again, let $y = f(x)$

$$\Rightarrow \quad y = \frac{x}{1+x^2}$$

$$\Rightarrow \quad y + x^2 y = x$$

$$\Rightarrow \quad yx^2 - x + y = 0$$

As, $x \in R$

$$\therefore \quad (-1)^2 - 4(y)(y) \ge 0$$

$$\Rightarrow \quad 1 - 4y^2 \ge 0$$

$$\Rightarrow \quad y \in \left[\frac{-1}{2}, \frac{1}{2}\right]$$

$$\therefore \quad \text{Range} = \text{Codomain} = \left[\frac{-1}{2}, \frac{1}{2}\right]$$

So, $f(x)$ is surjective.

Hence, $f(x)$ is surjective but not injective.

8. *(a)* We have, $f(x) = ax^2 + bx + c$

Now, $f(x+y) = f(x) + f(y) + xy$

Put $y = 0 \Rightarrow f(x) = f(x) + f(0) + 0$

$$\Rightarrow \quad f(0) = 0 \Rightarrow c = 0$$

Again, put $y = -x$

$$\therefore \quad f(0) = f(x) + f(-x) - x^2$$

$$\Rightarrow \quad 0 = ax^2 + bx + ax^2 - bx - x^2$$

$$\Rightarrow \quad 2ax^2 - x^2 = 0$$

$$\Rightarrow \quad a = \frac{1}{2}$$

Also, $a + b + c = 3$

$$\Rightarrow \quad \frac{1}{2} + b + 0 = 3$$

$$\Rightarrow \quad b = \frac{5}{2}$$

$$\therefore \quad f(x) = \frac{x^2 + 5x}{2}$$

Now, $f(n) = \dfrac{n^2 + 5n}{2} = \dfrac{1}{2}n^2 + \dfrac{5}{2}n$

$$\therefore \quad \sum_{n=1}^{10} f(n) = \frac{1}{2}\sum_{n=1}^{10} n^2 + \frac{5}{2}\sum_{n=1}^{10} n$$

$$= \frac{1}{2} \cdot \frac{10 \times 11 \times 21}{6} + \frac{5}{2} \times \frac{10 \times 11}{2}$$

$$= \frac{385}{2} + \frac{275}{2} = \frac{660}{2} = 330$$

9. *(c)* Given,

$$\begin{vmatrix} x-4 & 2x & 2x \\ 2x & x-4 & 2x \\ 2x & 2x & x-4 \end{vmatrix} = (A + Bx)(x - A)^2$$

$\Rightarrow$ Apply $C_1 \to C_1 + C_2 + C_3$

$$\begin{vmatrix} 5x-4 & 2x & 2x \\ 5x-4 & x-4 & 2x \\ 5x-4 & 2x & x-4 \end{vmatrix} = (A + Bx)(x - A)^2$$

Taking common $(5x - 4)$ from C_1, we get

$$(5x-4)\begin{vmatrix} 1 & 2x & 2x \\ 1 & x-4 & 2x \\ 1 & 2x & x-4 \end{vmatrix} = (A + Bx)(x - A)^2$$

Apply $R_2 \to R_2 - R_1$ and $R_3 \to R_3 - R_1$

$$\therefore\ (5x-4)\begin{vmatrix}1 & 2x & 0\\ 0 & -x-4 & 0\\ 0 & 0 & -x-4\end{vmatrix}$$

$$=(A+Bx)(x-A)^2$$

Expanding along C_1, we get

$(5x-4)(x+4)^2=(A+Bx)(x-A)^2$

Equating, we get

$$A=-4 \text{ and } B=5$$

10. *(b)* We have,

$x+ky+3z=0;\ 3x+ky-2z=0;\ 2x+4y-3z=0$

System of equation has non-zero solution, if

$$\begin{vmatrix}1 & k & 3\\ 3 & k & -2\\ 2 & 4 & -3\end{vmatrix}=0$$

$\Rightarrow (-3k+8)-k(-9+4)+3(12-2k)=0$

$\Rightarrow -3k+8+9k-4k+36-6k=0$

$\Rightarrow$

$-4k+44=0 \Rightarrow k=11$

Let $\quad z=\lambda$, then we get

$$x+11y+3\lambda=0 \quad \text{...(i)}$$

$$3x+11y-2\lambda=0 \quad \text{...(ii)}$$

and $\quad 2x+4y-3\lambda=0 \quad$ ...(iii)

Solving Eqs. (i) and (ii), we get

$$x=\frac{5\lambda}{2},\ y=\frac{-\lambda}{2},\ z=\lambda$$

$$\Rightarrow \quad \frac{xz}{y^2}=\frac{5\lambda^2}{2\times\left(-\frac{\lambda}{2}\right)^2}=10$$

JEE Advanced

1. *(b, c)* Here, $P=\begin{bmatrix}3 & -1 & -2\\ 2 & 0 & \alpha\\ 3 & -5 & 0\end{bmatrix}$

Now, $\quad |P|=3(5\alpha)+1(-3\alpha)-2(-10)$

$$=12\alpha+20 \quad \text{...(i)}$$

$$\therefore \quad \text{adj }(P)=\begin{bmatrix}5\alpha & 2\alpha & -10\\ -10 & 6 & 12\\ -\alpha & -(3\alpha+4) & 2\end{bmatrix}^T$$

$$=\begin{bmatrix}5\alpha & -10 & -\alpha\\ 2\alpha & 6 & -3\alpha-4\\ -10 & 12 & 2\end{bmatrix} \quad \text{...(ii)}$$

As, $\quad PQ=kI$

$\Rightarrow \quad |P||Q|=|kI|$

$\Rightarrow \quad |P||Q|=k^3$

$\Rightarrow \quad |P|\left(\frac{k^2}{2}\right)=k^3 \quad \left[\text{given, } |Q|=\frac{k^2}{2}\right]$

$\Rightarrow \quad |P|=2k \quad$ ...(iii)

$\because \quad PQ=ki$

$\therefore \quad Q=kp^{-1}I$

$$=k\cdot\frac{\text{adj}P}{|P|}=\frac{k(\text{adj }P)}{2k} \quad \text{[from Eq. (iii)]}$$

$$=\frac{\text{adj }P}{2}$$

$$=\frac{1}{2}\begin{bmatrix}5\alpha & -10 & -\alpha\\ 2\alpha & 6 & -3\alpha-4\\ -10 & 12 & 2\end{bmatrix}$$

$\therefore \quad q_{23}=\frac{-3\alpha-4}{2} \quad \left[\text{given, } q_{23}=-\frac{k}{8}\right]$

$\Rightarrow \quad -\frac{(3\alpha+4)}{2}=-\frac{k}{8}$

$\Rightarrow \quad (3\alpha+4)\times 4=k$

$\Rightarrow \quad 12\alpha+16=k \quad$ ...(iv)

From Eq. (iii), $\quad |P|=2k$

$\Rightarrow \quad 12\alpha+20=2k \quad$ [from Eq. (i)] ...(v)

On solving Eqs. (iv) and (v), we get

$$\alpha=-1 \text{ and } k=4 \quad \text{...(vi)}$$

$\therefore \quad 4\alpha-k+8=-4-4+8=0$

$\therefore$ Option (b) is correct.

Now, $|P\text{ adj }(Q)|=|P||\text{adj }Q|$

$$=2k\left(\frac{k^2}{2}\right)^2=\frac{k^5}{2}=\frac{2^{10}}{2}=2^9$$

$\therefore$ Option (c) is correct.

2. (2) Given, $\begin{vmatrix}x & x^2 & 1+x^3\\ 2x & 4x^2 & 1+8x^3\\ 3x & 9x^2 & 1+27x^3\end{vmatrix}=10$

$$\Rightarrow \quad x\cdot x^2\begin{vmatrix}1 & 1 & 1+x^3\\ 2 & 4 & 1+8x^3\\ 3 & 9 & 1+27x^3\end{vmatrix}=10$$

Apply $R_2\to R_2-2R_1$ and $R_3\to R_3-3R_1$, we get

$$x^3\begin{vmatrix}1 & 1 & 1+x^3\\ 0 & 2 & -1+6x^3\\ 0 & 6 & -2+24x^3\end{vmatrix}=10$$

$$\Rightarrow \quad x^3\cdot\begin{vmatrix}2 & 6x^3-1\\ 6 & 24x^3-2\end{vmatrix}=10$$

$\Rightarrow \quad x^3(48x^3-4-36x^3+6)=10$

$\Rightarrow \quad 12x^6+2x^3=10$

$\Rightarrow \quad 6x^6+x^3-5=0$

$\Rightarrow \quad x^3=\frac{5}{6},-1$

$\therefore \quad x=\left(\frac{5}{6}\right)^{1/3},-1$

Hence, the number of real solutions is 2.

3. *(1)* Here, $z = \frac{-1 + i\sqrt{3}}{2} = \omega$

$\because \quad P = \begin{bmatrix} (-\omega)^r & \omega^{2s} \\ \omega^{2s} & \omega^r \end{bmatrix}$

$P^2 = \begin{bmatrix} (-\omega)^r & \omega^{2s} \\ \omega^{2s} & \omega^r \end{bmatrix} \begin{bmatrix} (-\omega)^r & \omega^{2s} \\ \omega^{2s} & \omega^r \end{bmatrix}$

$= \begin{bmatrix} \omega^{2r} + \omega^{4s} & \omega^{r+2s}[(-1)^r + 1] \\ \omega^{r+2s}[(-1)^r + 1] & \omega^{4s} + \omega^{2r} \end{bmatrix}$

Given, $P^2 = -I$

$\therefore \quad \omega^{2r} + \omega^{4s} = -1$

and $\omega^{r+2s}[(-1)^r + 1] = 0$

Since, $r \in \{1, 2, 3\}$ and $(-1)^r + 1 = 0$

$\Rightarrow \quad r = \{1, 3\}$

Also, $\omega^{2r} + \omega^{4s} = -1$

If $r = 1$, then $\omega^2 + \omega^{4s} = -1$

which is only possible, when $s = 1$.

As, $\omega^2 + \omega^4 = -1$

$\therefore \quad r = 1, s = 1$

Again, if $r = 3$, then

$\omega^6 + \omega^{4s} = -1$

$\Rightarrow \quad \omega^{4s} = -2$ [never possible]

$\therefore \quad r \neq 3$

$\Rightarrow (r, s) = (1, 1)$ is the only solution.

Hence, the total number of ordered pairs is 1.

4. *(b)* Here, $P = \begin{bmatrix} 1 & 0 & 0 \\ 4 & 1 & 0 \\ 16 & 4 & 1 \end{bmatrix}$

$\therefore \quad P^2 = \begin{bmatrix} 1 & 0 & 0 \\ 4 & 1 & 0 \\ 16 & 4 & 1 \end{bmatrix} \begin{bmatrix} 1 & 0 & 0 \\ 4 & 1 & 0 \\ 16 & 4 & 1 \end{bmatrix}$

$= \begin{bmatrix} 1 & 0 & 0 \\ 4+4 & 1 & 0 \\ 16+32 & 4+4 & 1 \end{bmatrix}$

$= \begin{bmatrix} 1 & 0 & 0 \\ 4 \times 2 & 1 & 0 \\ 16\,(1+2) & 4 \times 2 & 1 \end{bmatrix}$ ...(i)

and $P^3 = \begin{bmatrix} 1 & 0 & 0 \\ 4 \times 2 & 1 & 0 \\ 16\,(1+2) & 4 \times 2 & 1 \end{bmatrix} \begin{bmatrix} 1 & 0 & 0 \\ 4 & 1 & 0 \\ 16 & 4 & 1 \end{bmatrix}$

$= \begin{bmatrix} 1 & 0 & 0 \\ 4 \times 3 & 1 & 0 \\ 16\,(1+2+3) & 4 \times 3 & 1 \end{bmatrix}$ ...(ii)

From symmetry,

$P^{50} = \begin{bmatrix} 1 & 0 & 0 \\ 4 \times 50 & 1 & 0 \\ 16\,(1+2+3+\ldots+50) & 4 \times 50 & 1 \end{bmatrix}$

$\because \quad P^{50} - Q = I$ [given]

$\therefore \quad \begin{bmatrix} 1 - q_{11} & -q_{12} & -q_{13} \\ 200 - q_{21} & 1 - q_{22} & -q_{23} \\ 16 \times \frac{50}{2}(51) - q_{31} & 200 - q_{32} & 1 - q_{33} \end{bmatrix} = \begin{bmatrix} 1 & 0 & 0 \\ 0 & 1 & 0 \\ 0 & 0 & 1 \end{bmatrix}$

$\Rightarrow \quad 200 - q_{21} = 0,$

$\frac{16 \times 50 \times 51}{2} - q_{31} = 0, \; 200 - q_{32} = 0$

$\therefore \quad q_{21} = 200, q_{32} = 200, q_{31} = 20400$

Thus, $\frac{q_{31} + q_{32}}{q_{21}} = \frac{20400 + 200}{200}$

$= \frac{20600}{200} = 103$

5. *(a,c)* For a matrix to be square of other matrix its determinant should be positive.

(a) and (c) $\rightarrow$ Correct

(b) and (d) $\rightarrow$ Incorrect

6. *(1)* $\begin{vmatrix} 1 & \alpha & \alpha^2 \\ \alpha & 1 & \alpha \\ \alpha^2 & \alpha & 1 \end{vmatrix} = 0$

$\Rightarrow \quad \alpha^4 - 2\alpha^2 + 1 = 0$

$\Rightarrow \quad \alpha^2 = 1$

$\Rightarrow \quad \alpha = \pm 1$

But $\alpha = 1$ not possible [Not satisfying equation]

$\therefore \quad \alpha = -1$

Hence, $1 + \alpha + \alpha^2 = 1$

7. *(a)* Sum of diagonal entries of $M^T M$ is $\sum a_i^2$.

$$\sum_{i=1}^{9} a_i^2 = 5$$

Possibilities

I. 2, 1, 0, 0, 0, 0, 0, 0, 0, which gives $\frac{9!}{7!}$ matrices

II. 1, 1, 1, 1, 1, 0, 0, 0, 0, which gives $\frac{9!}{4! \times 5!}$ matrices

Total matrices $= 9 \times 8 + 9 \times 7 \times 2 = 198$

8. *(2)* We have,

$\sin^{-1}\left(\sum_{i=1}^{\infty} x^{i+1} - x \sum_{i=1}^{\infty} \left(\frac{x}{2}\right)^i\right)$

$= \frac{\pi}{2} - \cos^{-1}\left(\sum_{i=1}^{\infty} \left(\frac{-x}{2}\right)^i - \sum_{i=1}^{\infty} (-x)^i\right)$

$\Rightarrow \quad \sin^{-1}\left[\frac{x^2}{1-x} - \frac{x \cdot \frac{x}{2}}{1 - \frac{x}{2}}\right] = \frac{\pi}{2} - \cos^{-1}\left[\frac{\frac{-x}{2}}{1 + \frac{x}{2}} - \frac{(-x)}{1+x}\right]$

$\left[\because \sum_{i=1}^{\infty} x^{i+1} = x^2 + x^3 + x^4 + \ldots = \frac{x^2}{1-x}\right.$ using sum of infinite terms of GP $\left.\right]$

$$\Rightarrow \sin^{-1}\left[\frac{x^2}{1-x}-\frac{x^2}{2-x}\right]=\frac{\pi}{2}-\cos^{-1}\left[\frac{x}{1+x}-\frac{x}{2+x}\right]$$

$$\Rightarrow \sin^{-1}\left[\frac{x^2}{1-x}-\frac{x^2}{2-x}\right]=\sin^{-1}\left(\frac{x}{1+x}-\frac{x}{2+x}\right)$$

$$\left[\because \sin^{-1}x=\frac{\pi}{2}-\cos^{-1}x\right]$$

$$\Rightarrow \frac{x^2}{1-x}-\frac{x^2}{2-x}=\frac{x}{1+x}-\frac{x}{2+x}$$

$$\Rightarrow x^2\left(\frac{2-x-1+x}{(1-x)(2-x)}\right)=x\frac{(2+x-1-x)}{(1+x)(2+x)}$$

$$\Rightarrow \frac{x}{2-3x+x^2}=\frac{1}{2+3x+x^2} \text{ or } x=0$$

$$\Rightarrow x^3+3x^2+2x=x^2-3x+2$$

$$\Rightarrow x^3+2x^2+5x-2=0 \text{ or } x=0$$

Let $f(x)=x^3+2x^2+5x-2$

$f'(x)=3x^2+4x+5$

$f'(x)>0, \forall\, x\in R$

$\therefore$ x^3+2x^2+5x-2 has only one real roots

Therefore, total number of real solution is 2.

9. *(d)* We have,

$$f_n(x)=\sum_{j=1}^{n}\tan^{-1}\left(\frac{1}{1+(x+j)(x+j-1)}\right) \text{ for all } x\in(0,\infty)$$

$$\Rightarrow f_n(x)=\sum_{j=1}^{n}\tan^{-1}\left(\frac{(x+j)-(x+j-1)}{1+(x+j)(x+j-1)}\right)$$

$$\Rightarrow f_n(x)=\sum_{j=1}^{n}[\tan^{-1}(x+j)-\tan^{-1}(x+j-1)]$$

$$\Rightarrow f_n(x)=(\tan^{-1}(x+1)-\tan^{-1}x)+(\tan^{-1}(x+2)-\tan^{-1}(x+1))+(\tan^{-1}(x+3)-\tan^{-1}(x+2))+\ldots+(\tan^{-1}(x+n)-\tan^{-1}(x+n-1))$$

$$\Rightarrow f_n(x)=\tan^{-1}(x+n)-\tan^{-1}x$$

This statement is false as $x\neq 0$. i.e., $x\in(0,\infty)$

(b) This statement is also false as $0\notin(0,\infty)$

(c) $f_n(x)=\tan^{-1}(x+n)-\tan^{-1}x$

$$\lim_{x\to\infty}\tan(f_n(x))=\lim_{x\to\infty}\tan(\tan^{-1}(x+n)-\tan^{-1}x)$$

$$\Rightarrow \lim_{x\to\infty}\tan(f_n(x))=\lim_{x\to\infty}\tan\left(\tan^{-1}\frac{n}{1+nx+x^2}\right)$$

$$=\lim_{x\to\infty}\frac{n}{1+nx+x^2}=0$$

$\therefore$ (c) statement is false.

(d) $\lim_{x\to\infty}\sec^2(f_n(x))=\lim_{x\to\infty}(1+\tan^2 f_n(x))$

$$=1+\lim_{x\to\infty}\tan^2(f_n(x))$$

$$=1+0=1$$

$\therefore$ (d) statement is true

10. *(a, d)* We have,

$$-x+2y+5z=b_1$$
$$2x-4y+3z=b_2$$
$$x-2y+2z=b_3$$

has at least one solution.

$$\therefore \quad D=\begin{vmatrix}-1&2&5\\2&-4&3\\1&-2&2\end{vmatrix}$$

and $D_1=D_2=D_3=0$

$$\Rightarrow D_1=\begin{vmatrix}b_1&2&5\\b_2&-4&3\\b_3&-2&2\end{vmatrix}$$

$$=-2b_1-14b_2+26b_3=0$$

$$\Rightarrow b_1+7b_2=13b_3 \quad \ldots\text{(i)}$$

(a) $D=\begin{vmatrix}1&2&3\\0&4&5\\1&2&6\end{vmatrix}=1(24-10)+1(10-12)$

$=14-2=12\neq 0$

Here, $D\neq 0\Rightarrow$ unique solution for any b_1, b_2, b_3.

(b) $D=\begin{vmatrix}1&1&3\\5&2&6\\-2&-1&-3\end{vmatrix}$

$=1(-6+6)-1(-15+12)+3(-5+4)=0$

For atleast one solution

$D_1=D_2=D_3=0$

Now, $D_1=\begin{vmatrix}b_1&1&3\\b_2&2&6\\b_3&-1&-3\end{vmatrix}$

$=b_1(-6+6)-b_2(-3+3)+b_3(6-6)$

$=0$

$D_2=\begin{vmatrix}1&b_1&3\\5&b_2&6\\-2&b_3&-3\end{vmatrix}$

$=-b_1(-15+12)+b_2(-3+6)-b_3(6-15)$

$=3b_1+3b_2+9b_3=0$

$\Rightarrow b_1+b_2+3b_3=0$

not satisfies the Eq. (i)

$\therefore$ It has no solution.

(c) $D=\begin{vmatrix}-1&2&-5\\2&-4&10\\1&-2&5\end{vmatrix}$

$=-1(-20+20)-2(10-10)-5(-4+4)=0$

Here, $b_2=-2b_1$ and $b_3=-b_1$ satisfies the Eq. (i)

Planes are parallel.

(d) $D=\begin{vmatrix}1&2&5\\2&0&3\\1&4&-5\end{vmatrix}=1(0-12)-2(-10-3)+5(8-0)=54$

$D\neq 0$

$\therefore$ It has unique solution for any b_1, b_2, b_3.

11. *(4)* Let $\text{Det}(P)=\begin{vmatrix}a_1&b_1&c_1\\a_2&b_2&c_2\\a_3&b_3&c_3\end{vmatrix}$

$=a_1(b_2c_3-b_3c_2)-a_2(b_1c_3-b_3c_1)+a_3(b_1c_2-b_2c_1)$

Now, maximum value of Det $(P)=6$

If $a_1=1, a_2=-1, a_3=1,$

$$b_2c_3 = b_1c_3 = b_1c_2 = 1$$

and $$b_3c_2 = b_3c_1 = b_2c_1 = -1$$

But it is not possible as

$$(b_2c_3)(b_3c_1)(b_1c_2) = -1$$

and $$(b_1c_3)(b_3c_2)(b_2c_1) = 1$$

i.e., $$b_1b_2b_3c_1c_2c_3 = 1 \text{ and } -1$$

Similar contradiction occurs when

$$a_1 = 1, a_2 = 1, a_3 = 1, b_2c_1 = b_3c_1 = b_1c_2 = 1$$

and $$b_3c_2 = b_1c_3 = b_1c_2 = -1$$

Now, for value to be 5 one of the terms must be zero but that will make 2 terms zero which means answer cannot be 5

Now,

$$\begin{vmatrix} 1 & 1 & 1 \\ -1 & 1 & 1 \\ 1 & -1 & 1 \end{vmatrix} = 4$$

Hence, maximum value is 4.

12. (a) We have,

$$E_1 = \left\{x \in R : x \neq 1 \text{ and } \frac{x}{x-1} > 0\right\}$$

$\therefore$ $$E_1 = \frac{x}{x-1} > 0$$

$+ \quad 0 \quad - \quad 1 \quad +$

$$E_1 = x \in (-\infty, 0) \cup (1, \infty)$$

and $$E_2 = \left\{\begin{array}{l} x \in E_1 : \sin^{-1}\left(\log_e\left(\frac{x}{x-1}\right)\right) \\ \text{is a real number} \end{array}\right\}$$

$$E_2 = -1 \leq \log_e \frac{x}{x-1} \leq 1$$

$\Rightarrow$ $$e^{-1} \leq \frac{x}{x-1} \leq e$$

Now, $$\frac{x}{x-1} \geq e^{-1} \Rightarrow \frac{x}{x-1} - \frac{1}{e} \geq 0$$

$\Rightarrow$ $$\frac{ex - x + 1}{e(x-1)} \geq 0 \Rightarrow \frac{x(e-1)+1}{(x-1)e} \geq 0$$

$+ \quad \frac{-1}{e-1} \quad - \quad 1 \quad +$

$\Rightarrow$ $$x \in \left(-\infty, \frac{1}{1-e}\right] \cup (1, \infty)$$

Also, $$\frac{x}{x-1} \leq e$$

$\Rightarrow$ $$\frac{(e-1)x - e}{x-1} \geq 0$$

$+ \quad 1 \quad - \quad \frac{e}{e-1} \quad +$

$\Rightarrow$ $$x \in (-\infty, 1) \cup \left[\frac{e}{e-1}, \infty\right)$$

So, $$E_2 = \left(-\infty, \frac{1}{1-e}\right] \cup \left[\frac{e}{e-1}, \infty\right)$$

$\therefore$ The domain of f and g are

$$\left(-\infty, \frac{1}{1-e}\right] \cup \left[\frac{e}{e-1}, \infty\right)$$

and Range of $\frac{x}{x-1}$ is $R^+ - \{1\}$

$\Rightarrow$ Range of f is $R - \{0\}$ or $(-\infty, 0) \cup (0, \infty)$

Range of g is $\left[-\frac{\pi}{2}, \frac{\pi}{2}\right] - \{0\}$ or $\left[-\frac{\pi}{2}, 0\right) \cup \left(0, \frac{\pi}{2}\right]$

Now, $P \rightarrow 4, Q \rightarrow 2, R \rightarrow 1, S \rightarrow 1$

Hence, option (a) is correct answer.